Waste Gas Treatment for Resource Recovery

Integrated Environmental Technology Series

The *Integrated Environmental Technology Series* addresses key themes and issues in the field of environmental technology from a multidisciplinary and integrated perspective.

An integrated approach is potentially the most viable solution to the major pollution issues that face the globe in the 21st century.

World experts are brought together to contribute to each volume, presenting a comprehensive blend of fundamental principles and applied technologies for each topic. Current practices and the state-of-the-art are reviewed, new developments in analytics, science and biotechnology are presented and, crucially, the theme of each volume is presented in relation to adjacent scientific, social and economic fields to provide solutions from a truly integrated perspective.

The *Integrated Environmental Technology Series* will form an invaluable and definitive resource in this rapidly evolving discipline.

Series Editor

Dr Ir Piet Lens, The University of Wageningen, P.O. Box 8129,
6700 EV Wageningen, The Netherlands (piet.lens@wur.nl).

Published titles

Advanced Biological Treatment Processes for Industrial Wastewaters: *Principles and applications*

Biofilms in Medicine, Industry and Environmental Biotechnology: *Characteristics, analysis and control*

Biofuels for Fuel Cells: *Renewable energy from biomass fermentation*

Decentralised Sanitation and Reuse: *Concepts, systems and implementation*

Environmental Technologies to Treat Sulfur Pollution: *Principles and engineering*

Phosphorus in Environmental Technology: *Principles and applications*

Pond Treatment Technology

Resource Recovery and Reuse in Organic Solid Waste Management

Soil and Sediment Remediation: *Mechanisms, technologies and applications*

Water Recycling and Resource Recovery in Industries: *Analysis, technologies and implementation*

Waste Gas Treatment for Resource Recovery

Edited by **Piet N.L. Lens, Christian Kennes, Pierre Le Cloirec and Marc A. Deshusses**

Publishing
London • Seattle

Published by IWA Publishing, Alliance House, 12 Caxton Street, London SW1H 0QS, UK

Telephone: +44 (0) 20 7654 5500; Fax: +44 (0) 20 7654 5555; Email: publications@iwap.co.uk; Web: www.iwapublishing.com

First published 2006

Typeset by Gray Publishing, Tunbridge Wells, Kent
Index prepared by Indexing Specialists, Hove

British Library Cataloguing in Publication Data
A CIP catalogue record for this book is available from the British Library

Library of Congress Cataloging-in-Publication Data
A catalog record for this book is available from the Library of Congress

ISBN-10: 1 84339 127 9
ISBN-13: 9781843391272

Front cover photographs:
Top, right: Bioreactor treating CS_2 containing waste gas of a cellulose casing production plant (courtesy of Bart Kraakman, Bioway, The Netherlands).

Top, left: CH_4 gas molecule entrapped in the cavity formed by water molecules in a hydrate structure (courtesy of Prof. Kuhs, University of Göttingen, Germany).

Bottom: Fungal biofilm (*Ophiostoma* sp.) treating α-pinene polluted air in a packed bed bioreactor (courtesy of Christian Kennes, University of La Coruña, Spain).

Contents

Preface

Environmental contamination of air, water and soil by (eco)toxic pollutants is a serious threat to the quality of our primary resources for life. Of particular concern is the pollution of air with organic and inorganic volatile pollutants or particulate matter. This is a major contribution to climate change, acid rain, smog formation in industrialized cities, and the appearance of new diseases, among others. There is now overwhelming evidence that human health directly depends on the quality of the air we breath and the general state of our environment. Consequently, a better balance is needed between disease-centred (therapy) and environment-centred (prevention) measures. We can never approach sustainable development, a goal constantly reiterated by almost all governments, if we do not find efficient ways and means of either preventing pollutants entering the environment or eliminating them from contaminated ecosystems.

In recent years, energy and feedstock materials for the chemical industry are in increasing demand. With constraints related to availability and use of oil, the energy and chemical industry is undergoing considerable changes. The need for the use of cheaper and widely available feedstocks, and the development of sustainable and environmentally friendly chemical processes is rapidly growing under both economical and public pressure. Therefore, waste gas treatment has gradually been integrated into process design. Instead of discharging their waste gases into the atmosphere, industries increasingly attempt to become self-sufficient and recover compounds from their own wastestreams or use (upgraded) wastestreams of neighbouring industries as raw material. Sustainable gas treatment concepts are under development, and can lead to the recovery of useful by-products like energy in the form of biogas, hydrogen or electricity, and chemicals in the form of fertilizers (ammonia, phosphates) or raw materials (elemental sulfur, sulfuric acid, etc.), for example. Obviously, adding value to waste gas by upgrading the recovered compounds will only be a reality if it is demonstrated that there is a fundamental basis and a tangible advantage in using these recovered compounds rather than buying raw materials and feedstock.

To understand the treatment alternatives available to control and solve air pollution problems, and to decide on the abatement strategy to be adopted, several aspects need to be considered. This book volume covers these aspects, including socio-economic considerations, legal incentives in European and other recent directives concerning air pollution, characterization of waste gases, as well as technical solutions available to treat contaminated air and to recover valuable compounds from these polluted waste gases.

Promising new uses of different physical, chemical and biological technologies, or a combination of them, are suitable for resource recovery from waste gases. In many cases, it was shown that bioprocesses are able to permanently eliminate pollutants inexpensively as well as to convert them into a recoverable form. Intriguing properties of bacteria, fungi and protozoa have been described, which enable expansion of biofilters and bioscrubbers into sectors still rather unfamiliar with environmental biotechnology, as the chemical industry. Probably the best example for this is the striking commercial success of air and off-gas treatment in bioreactors, a field in which many companies have established as the technology has proven to be cheap and reliable. Significant developments have recently also been made in non-biological processes based, among others, on catalytic activities and mass transfer processes. All these advances in the development of new sustainable routes for product recovery from gases will be strongly promoted by increasing integration of efforts and expertise. This book covers many of these different aspects and includes case studies that demonstrate the technical and economic feasibility.

We wish to thank all the contributors of the book for their enthusiastic support and timely submission of their manuscripts. This book is based on the Euro Summer School, 'Closing water and resources cycles: options for gas treatment', 26 June to 1 July 2005, held in Wageningen, The Netherlands. This Euro Summer School was financially supported by the 'Improving the Human Potential' programme of the EU (grant number HPCF-2002-00405). In addition to many of the oral presentations from the Euro Summer School, a number of invited contributions are included in this volume. We are grateful to Alan Click and Alan Peterson of IWA Publishing for their help and editorial support in realizing this book.

Piet Lens
University of Wageningen, The Netherlands

Christian Kennes
University of La Coruña, Spain

Pierre Le Cloirec
Ecole de Mines de Nantes, France

Marc Deshusses
University of California, USA

Contributors

Douglas Aaron
Georgia Institute of Technology
School of Civil and Environmental Engineering, 200 Bobby Dodd Way
Atlanta, GA 30332-0373, USA

Stephane Abanades
CNRS-PROMES
7 Rue du Four Solaire
66120 Font Romeu Odeillo, France

Francesca Accettola
4 square Boileau
78150 Le Chesnay, France

Werner Ahrer
Profactor Produktionsforschungs GmbH
Im Stadtgut A2
4407 Steyr/Gleink, Austria

Bernd Calaminus
DFIU – Deutsch-Franzoesisches Institut fuer Umweltforschung
Universitaet Karlsruhe (TH)
Hertzstr. 16, 76187 Karlsruhe, Germany

García A.I. Cardona
Grupo Detoxificación de Aire y Efluentes Gaseosos, Unidad de aplicaciones Medioambientales de la Energía Solar, CIEMAT-PSA
Av. Complutense, 22. Ed. 42
28040 Madrid, Spain

Alfred Bernard
Unit of Industrial Toxicology and Occupational Medicine
Faculty of Medicine
Université Catholique de Louvain
Clos Chapelle-aux-Champs 30
1200 Brussels, Belgium

Peter Brimblecombe
School of Environmental Sciences
University of East Anglia
Norwich NR4 7TJ, UK

Claire de Burbure
Unit of Industrial Toxicology and Occupational Medicine
Faculty of Medicine
Université Catholique de Louvain
Clos Chapelle-aux-Champs 30
1200 Brussels, Belgium

Friso de Jong
Joint Implementation Network
Stichting JIN
Meerkoetlaan 30 A
9765 TD Paterswolde, The Netherlands

Anne-Marie Delort
Laboratoire Synthese Et Etudes de Systemes à Interet Biologique (SEESIB)
UMR 6504 CNRS-Universite Blaise Pascal
63177 Aubiere cedex, France

Karel J.A. de Waal
Bioway
Bellstraat 8
6710 BJ Ede, The Netherlands

Marc Deshusses
Department of Chemical and
Environmental Engineering
University of California Riverside
CA 92521, USA

Paul H.M. Feron
TNO Science & Industry
Department of Separation Technology
PO Box 342
7300 AH Apeldoorn, The Netherlands

Gilles Flamant
CNRS-PROMES
7 Rue du Four Solaire
66120 Font Romeu Odeillo, France

Daniel Gauthier
CNRS-PROMES
7 Rue du Four Solaire
66120 Font Romeu Odeillo, France

Jutta Geldermann
Deutsch-Franzosisches Institut fur
Umweltforschung (DFIU/IFARE)
Institut fur Industriebetriebslehre und
Industrielle Produktion (IIP)
Universitat Karlsruhe (TH)
Hertzstr. 16
76187 Karlsruhe, Germany

Armando González-Sánchez
Department of Process Engineering
UAM- Iztapalapa
Av. San Rafael Atlixco No. 186
Col. Vicentina
09340 Mexico D.F., Mexico

Carlos Henriques
Zeolite Group, CEBQ
Chemical Engineering Department
Instituto Superior Técnico
Av. Rovisco Pais
1049-001 Lisboa, Portugal

Phil Hobbs
IGER – North Wyke Research Station
Okehampton EX20 2SB, UK

Albert Jansen
TNO Science & Industry
Department of Separation Technology
PO Box 342
7300 AH Apeldoorn, The Netherlands

Catrinus J. Jepma
University of Groningen
Meerkoetlaan 30
9765 TD Paterswolde, The Netherlands

Yaomin Jin
University of La Coruña
Department of Chemical Engineering
Campus da Zapateira
15071 La Coruña, Spain

Peral José
Univ. Autónoma de Barcelona
Departamento de Química
Unidad de Química Física
08193 Bellaterra, Barcelona, Spain

Christian Kennes
University of La Coruña
Department of Chemical Engineering
Campus da Zapateira
15071 La Coruña, Spain

Bart Kraakman
Bioway, Bellstraat 8
6710 BJ Ede, The Netherlands

Claude Lamy
Laboratoire d'Electrocatalyse
Université de Poitiers-CNRS UMR 6503
40 avenue du Recteur Pineau
86022 Poitiers Cedex, France

Mihalis Lazaridis
Technical University of Crete
Department of Environmental
Engineering
Polytechneioupolis
73100 Chania, Greece

Pierre Le Cloirec
Ecole des Mines de Nantes
GEPEA UMR CNRS 6144
4 rue Alfred Kastler, BP 20722
44307 Nanes cedex 03, France

Laurence Le Coq
Ecole des Mines de Nantes
GEPEA UMR CNRS 6144
4 rue Alfred Kastler, BP 20722
44307 Nantes cedex 03, France

Piet N.L. Lens
University of Wageningen
Section Environmental Technology
PO Box 8129
6700 EV Wageningen, The Netherlands

Marta I. Litter
Unidad de Actividad Química
Centro Atómico Constituyentes
Av. Gral. Paz 1499
San Martín Prov. de Buenos Aires, Argentina

Ann-Lise Norman
Physics and Astronomy
The University of Calgary
Calgary, Alberta T2N 1N4, Canada

Óscar Prado
University of La Coruña
Department of Chemical Engineering
Campus da Zapateira
15071 La Coruña, Spain

F. Ramôa Ribeiro
Zeolite Group, CEBQ
Chemical Engineering Department
Instituto Superior Técnico, Av. Rovisco Pais, 1049-001 Lisboa, Portugal

Sergio Revah
Department of Process and Technology
UAM – Cuajimalpa, c/o UAM-Iztapalapa
Av. San Rafael Atlixco No. 186
Col. Vicentina
09340 Mexico D.F., Mexico

M.F. Ribeiro
Zeolite Group, CEBQ
Chemical Engineering Department
Instituto Superior Técnico
Av. Rovisco Pais
1049-001 Lisboa, Portugal

Michel Roustan
INSA de Toulouse, LIPE
135 Avenue de Rangueil
31077 Toulouse cedex 04, France

Benigno Sánchez
Grupo Detoxificación de Aire y Efluentes Gaseosos
Unidad de aplicaciones Medioambientales de la Energía Solar
CIEMAT-PSA
Av. Complutense, 22. Ed. 42
28040 Madrid, Spain

Rainer Schöftner
Profactor Produktionsforschungs GmbH
Im Stadtgut A2
4407 Steyr/Gleink, Austria

Jan Sipma
University of Wageningen
Section Environmental Technology
PO Box 8129
6700 EV Wageningen, The Netherlands

Sjaak Slanina
University of Wageningen
Meteorology and Air Quality Group
Duivendaal 2
6701 AP Wageningen, The Netherlands

Alfons J.M. Stams
University of Wageningen
Laboratory for Microbiology
Hesseling van Suchtelenweg 4
6703 CT Wageningen, The Netherlands

Albert Subrenat
Ecole des Mines de Nantes
GEPEA UMR CNRS 6144
4 rue Alfred Kastler, BP 20722
44307 Nantes cedex 03, France

Steven Trogisch
Protecma, Energía y medio ambiente, S.L.
Edificio Centros Tecnologicos
Parque Cientifico y Tecnologico
33203 Gijón, Spain

Costas Tsouris
Oak Ridge National Laboratory
P.O. Box 2008
Oak Ridge, TN 37831-6181, USA

Wytze van der Gaast
Joint Implementation Network
Stichting JIN
Meerkoetlaan 30 A
9765 TD Paterswolde, The Netherlands

María C. Veiga
University of La Coruña
Department of Chemical Engineering
Campus da Zapateira
15071 La Coruña, Spain

Vladimir Yakushev
VNIIGAZ
142717, p.Razvilka
Moscow region, Russia

PART I: Socio-economic aspects of gas pollution and gas recovery

Section IA: Gas: pollutant or resource?

Section IB: Pollution control policy

1

Introduction on air pollution: the emission–effect relation

J. Slanina

1.1 INTRODUCTION: WHAT IS AIR POLLUTION?

Atmospheric chemistry in the last decades was an important subject of study, not only because of scientific fascination with a very complex subject, but also because of the increased problems with air pollution during the past 50 years. Air pollution is somewhat difficult to define, as many air pollutants (at low concentrations) are essential nutrients for sustainable development of ecosystems. So air pollution could be defined as:

A state of the atmosphere which leads to exposure of human beings and/or ecosystems to such high levels or loads of specific compounds or mixtures thereof that damage is caused.

With a very few exceptions all compounds which are considered air pollutants have both natural as well as man-made origins (Figure 1.1). And air pollution is not a new phenomenon; it was forbidden in medieval times to burn coal in London while

 Waste Gas Treatment for Resource Recovery edited by Piet N.L. Lens, Christian Kennes, Pierre Le Cloirec and Marc Deshusses. ISBN: 1843391279. Published by IWA Publishing, London, UK.

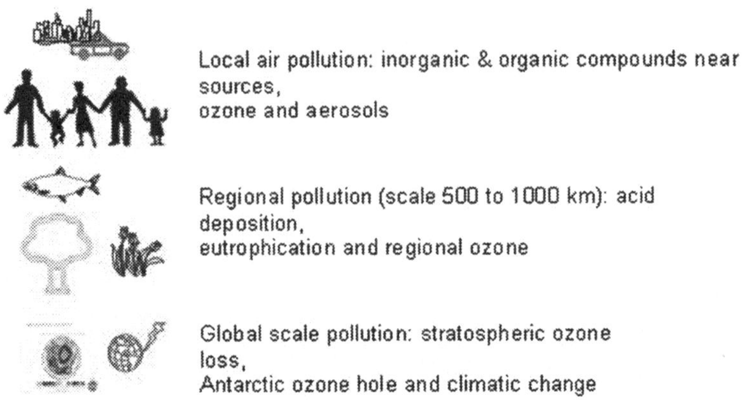

Figure 1.1 Effects of air pollutants for different scales of space.

parliament was in session. Air pollution problems have dramatically increased in intensity as well as scale due to the increase of emissions since the industrial revolution.

All reports on air pollution in the 19th and early 20th century indicate that the problems were local, in or around the industrial centers and the big cities. Even the infamous environmental catastrophes in the area of Liege (the first recorded occurrence of death by air pollution) in the thirties, or in London in the fifties were essentially local phenomena. In the London smog episode stagnant air accumulated such as high sulfur dioxide and sulfuric acid concentrations that probably several thousand inhabitants died. Epidemiological research has quite recently shown that there are effects of aerosols on the respiratory tract (inducing asthma and bronchitis and early demise). This evidence that aerosols increase mortality, have prompted research on the composition and sources of aerosol, in order to find adequate control measures.

It is a feature of the second half of the last century that the effects of air pollution were detected on regional (>500 km), continental and global scale. Acid deposition caused around 1960 the first observed effects on regional to continental scales. Lakes in Scandinavia as well as in North America lost their fish populations as the lakes were acidified by acid deposition to the point that fish eggs would not produce young fish anymore. About 10 years later damage in forest, loss of vitality of trees, was also contributed to acid deposition.

Smog episodes in US cities as Los Angeles were reported in the same period. Reactions of volatile organic compounds (VOC) and nitrogen oxides produced high concentrations of ozone and peroxides, which are harmful for human beings as well as for ecosystems. Around the same period the first high oxidant concentrations (the complex mixture of ozone, peroxides and other products of the reactions of organics and nitrogen oxides is called oxidants) were more and more frequently taking place in Europe during stagnant meteorological conditions (Figure 1.2).

At the same time severe eutrophication (damage and changes in ecosystems due to availability of too large amounts of nutrients) was encountered in Europe and the US. Deposition of ammonium and nitrates was shown to contribute substantially to high

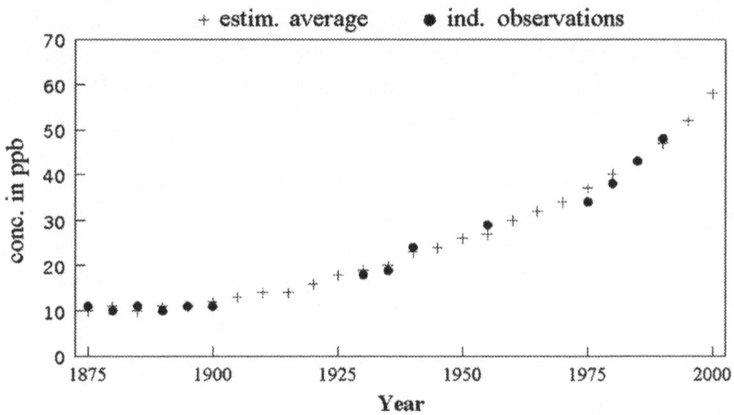

Figure 1.2 Ozone concentrations in free troposphere in Europe since 1870.

nutrient concentrations in soil and groundwater. Nitrate and ammonium are beneficial, even essential, for development of vegetation but in too high concentration they lead the loss of diversity especially in oligotrophic (=adapted to low nutrient availability) ecosystems. This problem was mainly encountered in The Netherlands, Belgium, and parts of Germany, Denmark and Southern Sweden.

The next scale of air pollution is effects on global dimensions, such as the destruction of stratospheric ozone due to emissions of chlorofluorocarbon compounds (CFCs). This issue was given a lot of attention in the period 1985–1995, as it was revealed that the destruction of stratospheric ozone leads to higher UV-light intensities and hence to skin cancer.

From 1990 onward the increase of the concentrations of radiative active substances (compounds which alter the radiative balance of the earth, greenhouse gasses but also aerosols and water in gaseous and liquid form: as clouds) and the connected climatic consequences induced new research on air pollution. Greenhouse gases absorb long-wave infrared radiation emitted from the earth, thereby retaining heat in the atmosphere and increasing the total radiative flux at the surface of the earth. Aerosols and clouds reflect incoming short-wave sunlight and influence the optical properties of clouds toward more reflection of sunlight and hence increasing concentrations of aerosol lead to a decrease of the radiative flux at the surface.

This sequence in time of air pollution problems could give the impression of sudden increases of air pollutant concentration, but that is probably not the case as can be explained in the case of ozone. By careful characterizing old methodologies, Voltz-Thomas and Kley have been able to reconstruct ozone concentrations in the free troposphere (the air not directly influenced by processes taking place at the earth surface).

The ozone concentrations in Europe slowly increased at a rate of 1–2% per year from 10 to over 50 ppb. It is well documented that the effects of ozone start at levels of about 40 ppb (ppb is a mixing ratio of 1 molecule ozone in a billion molecules of air).

So it is not surprising that the effects of ozone were detected in the seventies, as background continental ozone was already 30 ppb, and additional oxidant formation would increase the ozone concentrations locally or regionally. But the increase of the continental background concentrations was already going on for a long period.

In general, effects of pollution in general and of air pollution also are a function of the degree of transgression of the limits over which effects can be expected. The reader can find additional information on the effects of air pollution in Chapter 7 and measures to abate air pollution are treated in Chapter 3 of this book.

1.2 OVERVIEW OF THE EMISSION EFFECT RELATIONS

Pollutants which affect human health and ecosystems in the form in which they are initially emitted are called primary pollutants. Sulfur dioxide is a good example of a primary air pollutant.

Ozone is a good example of so-called secondary air pollutants. Ozone is a product of a large set of atmospheric chemical reactions involving nitrogen oxides (NO_x) and VOC. These precursors (compounds which lead to a certain product, in this case, ozone) are emitted by a wide variety of sources, but in the US and Europe mainly by traffic. These precursors are transported by air movement and will be involved in an intricate and a very large set of atmospheric reactions under the influence of sunlight. Ozone is one of the products.

The next stage is the deposition of primary and secondary air pollutants. Deposition can take place by way of precipitation. Air pollution is first incorporated in clouds or precipitation and transported this way to the earth surface (wet deposition), or air pollutants are deposited directly on vegetation and the earth surface (dry deposition).

The deposition load (time integrated deposition fluxes per area unit; a popular though not very logical way of expressing loads, is in moles per hectare per year) of different compounds describes the total amount of air pollutants, which reaches ecosystems.

Effects caused by acid deposition and eutrophication are directly linked to the deposition loads of pollutants, as will be explained later. Other effects of pollutants (e.g. the effects of oxidants) are probably also tied in with the load of pollutants reaching the relevant parts of the vegetation, but this connection has not yet been demonstrated in all cases. The removal of pollutants by chemical transformations and deposition processes is essential. If for instance sulfur dioxide emitted by natural causes like volcanoes would not be removed this way, a concentration of 1000 ppb or more would be reached in atmosphere in a few months, and ecosystems and the total human population would be endangered.

Once the effects of air pollutants are related to transgression of critical loads or concentrations, it is possible to derive the necessary emission reductions. Models describing the complete process from emissions, via transport, chemical conversions, to deposition and exposure plus the effects are used for this purpose.

1.3 EMISSIONS

Most pollutants are emitted both by natural as well as by anthropogenic sources. Natural sources are not influenced by man or by man-induced activities. Volcanoes are a good example of this kind. A lot of emissions are biogenic, that is produced by living organisms, but these emissions are very often influenced by man's activities.

N_2O is a greenhouse gas, which is for a large part emitted during nitrification and de-nitrification processes (the conversion of ammonium to nitrate and of nitrate to N_2 and to ammonium, respectively) taking place in the soil. But largest N_2O emissions are observed where nitrogen-containing fertilizer is applied in agriculture. The ratio between anthropogenic and natural emissions is very important, as only the anthropogenic part can be influenced for example, by abatement measures. The anthropogenic sources of SO_2 (see Table 1.1), are much larger compared to the natural ones, in fact they exceeded the natural ones as early as 1950. The natural emissions of NO_x are according IPCC smaller than the anthropogenic emissions.

Table 1.1 Overview emissions of sulfur, nitrogen and organic compounds according to IPCC (2005).

	Northern Hemisphere	Southern Hemisphere	Global*	Range	Source
NO_x (as Tg N/year)	32	9	41		(See also Chapter 4)
Fossil fuel (1985)	20	1.1	21		Benkovitz et al. (1996)
Aircraft (1992)	0.54	0.04	0.58	0.4–0.9	Penner et al. (1999b); Daggett et al. (1999)
Biomass burning (ca. 1990)	3.3	3.1	6.4	2–12	Liousse et al. (1996); Atherton (1996)
Soils (ca. 1990)	3.5	2.0	5.5	3–12	Yienger and Levy (1995)
Agricultural soils			2.2	0–4	Yienger and
Natural soils			3.2	3–8	Levy (1995)
Lightning	4.4	2.6	7.0	2–12	Price et al. (1997); Lawrence et al. (1995)
NH_3 (as Tg N/year)	41	13	54	40–70	Bouwman et al. (1997)
Domestic animals (1990)	18	4.1	21.6	10–30	Bouwman et al. (1997)
Agriculture (1990)	12	1.1	12.6	6–18	Bouwman et al. (1997)
Human (1990)	2.3	0.3	2.6	1.3–3.9	Bouwman et al. (1997)
Biomass burning (1990)	3.5	2.2	5.7	3–8	Bouwman et al. (1997)
Fossil fuel and industry (1990)	0.29	0.01	0.3	0.1–0.5	Bouwman et al. (1997)
Natural soils (1990)	1.4	1.1	2.4	1–10	Bouwman et al. (1997)
Wild animals (1990)	0.10	0.02	0.1	0–1	Bouwman et al. (1997)
Oceans	3.6	4.5	8.2	3–16	Bouwman et al. (1997)
SO_2 (as Tg S/year)	76	12	88	67–30	
Fossil fuel and industry (1985)	68	8	76	60–100	Benkovitz et al. (1996)

(Continued)

Table 1.1 (Continued)

	Northern Hemisphere	Southern Hemisphere	Global*	Range	Source
Aircraft (1992)	0.06	0.004	0.06	0.03–1.0	Penner et al. (1998a); Penner et al. (1999b); Fahey et al. (1999)
Biomass burning (ca. 1990)	1.2	1.0	2.2	1–6	Spiro et al. (1992)
Volcances	6.3	3.0	9.3	6–20	Andres and Kasnoc (1998) (incl. H_2S)
DMS or H_2S (as TgS/yr)	11.6	13.4	25.0	12–42	
Oceans	11	13	24	13–26	Kettle and Andreae (2000)
Land biota and soils	171	65	236	0.4–5.6	Bates et al. (1992); Andreae and Jaschke (1992)
Volatile organic emissions (as TgC/yr)	171	65	236	100–560	
Anthropogenic (1985)	104	5	109	60–160	Piccot et al. (1992)
Terpenes (1990)	67	60	127	40–400	Guenther et al. (1995)

*The global figure may not equal the sum of the N. hemisphere and S. hemisphere totals due to rounding.

Table 1.2 Overview of the emissions of greenhouse gases.

	CO_2 (in Pg C/year)	CH_4 (in Tg/year)	N_2O (in Tg N/year)
Anthropogenic			
Fossil fuel	5.5	100	
Biomass burning	1.6	40	
Ruminants		80	
Waste treatment		80	
Rice fields		60	
Total anthropogenic	7.1	360	4.4*
Natural			
Wetlands/continents		120	6
Oceans		10	2
Others		20	
Total natural	(C-cycles, large)	150	8

*Sources are not very well quantified.

NO_x emissions together with the emissions of VOC are responsible for the formation of oxidants like ozone and PAN (peroxide acetyl nitrate). Some of the NO_x sources, especially the natural ones, are quite uncertain, and the contribution of natural sources to ambient VOC is not very well known either. The VOC emissions, especially the natural/biogenic emissions, are quite uncertain, but Table 1.1 makes clear that for the time being the natural VOC emissions are predominant. Near industrialized areas anthropogenic sources are dominant nonetheless. In this table the differences

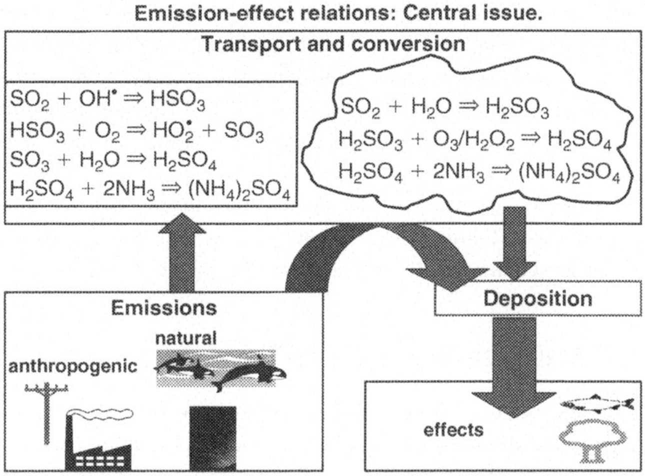

Figure 1.3 Cycle of sulfur dioxide and secondary sulfate aerosol formation.

in emissions between Northern and Southern Hemisphere are highlighted, which makes clear that in general the contribution of the North to emissions is much larger. An extreme case is reactive chlorine (the main compound responsible for destroying stratospheric ozone), where at least 90% of the relevant emissions are anthropogenic. CFCs were after 1945 widely applied until they were banned in the Montreal protocol (in 1998). No natural source of these compounds is known.

A simplified overview of the emissions of greenhouse gasses is given in Table 1.2, natural sources for CO_2 are not mentioned. The exchange between atmosphere and ecosystems consists actually of large C-cycles (e.g. decay of biomass in fall and uptake during leaf formation and growth in spring) and these are in equilibrium.

1.4 FORMATION OF AEROSOLS

Aerosols, defined as solid or liquid particles which have a certain lifetime suspended in air (particle size varying from a few nanometers to several hundred micrometers, for example, dust particles), are either emitted as such (primary aerosol), or formed by way of the transformations of sulfur dioxide, nitrogen oxides and ammonia into respectively sulfates, nitrates and ammonium (secondary aerosol). Many VOC are converted to oxidized organic species with low volatility and then become part of ambient aerosol also.

Ambient aerosol concentrations are partly of primary and partly of secondary nature. Sea salt particles caused by waves and dust swept up by wind are examples of primary aerosol. A large part of the secondary aerosols, however, consists of sulfates and, depending on the specific conditions, nitrate and organic compounds, which are formed by atmospheric conversion with sulfur dioxide, nitrogen oxides and VOC as precursors.

Sulfates (see Figure 1.3) used to represent the largest fraction of the anthropogenic fraction. However, the present composition of aerosol reflects a shift in emissions especially in the developed countries in the past 10 years. Emissions of sulfur oxides have gone down, due to measures to reduce, for example, acid deposition. Emissions of traffic have been stabilized or even increased, not withstanding the introduction of exhaust catalysts in cars, as in, for example, the US and Europe the number of cars and the total driven distances per car has clearly increased. So there is a change in aerosols from sulfate to nitrate.

In many developing countries, in East Asia and South America for instance, the same shift in emissions is taking place right now. One could say that acid deposition problems are being exchanged with exposure to increased oxidant concentrations.

1.5 LIFETIMES OF POLLUTANTS

Conversion in troposphere and stratosphere and removal by deposition processes determine the fate and the lifetime of pollutants. A simplified approach is used here where analogous to a resistance model (parallel resistances), the overall lifetime can be expressed as:

$$\frac{1}{\tau_{\text{result}}} = \frac{1}{\tau_{\text{w}}} + \frac{1}{\tau_{\text{d}}} + \frac{1}{\tau_{\text{c}}}$$

Here $\tau_{\text{(resultant)}}$ denotes the total lifetime, τ_c is lifetime as function of conversion, τ_d lifetime as function of removal by dry deposition and τ_w lifetime as function of removal by wet deposition. An overview on lifetimes of some important pollutants is given in Table 1.3.

Next to estimates of τ_c, τ_w and τ_d also an estimate of the total lifetime, $\tau_{\text{(resultant)}}$ and of the transport distance (dist.) are given. If a compound lives so long that a global distribution is observed, this is indicated by "global". Stratosph. means that the compound is not destroyed by OH-radicals in the troposphere but only by short-wave UV radiation in the stratosphere.

Very small aerosol particles, as generated by combustion engines, live very shortly because these very small particles more or less behave like gas molecules, including

Table 1.3 Lifetimes and transport distances of air pollutants.

Compound	τ_c	τ_w	τ_d	$\tau_{\text{(resultant)}}$	Distance
Aerosol (<0.1 μm)	Min.-hours	—	Hours	Hours	10
Aerosol (>10 μm)	—	Large	Hours	Hours	100
Aerosol (≈1 μm)	—	Week	Week	Week	1000
Sulfur dioxide	Days	Days	Days	Days	1000
Nitrogen oxides	Days	—	Days	Days	1000
Ammonia	Hours	Hours	Hours	Hours	250
VOC reactive	Hours-days	Large	—	Hour-days	200–1000
Ozone	Hours-days	—	Days	Hours	200–1000
Carbon dioxide	—	—	200 years	200 years	Global
Methane	10 years	—	Large	12 years	Global
N_2O	Stratosph.	—	—	200 years	Global
CFCs	Stratosph.	—	—	100–200 years	Global

transport by Brownian motion (diffusion) and coagulate rapidly. The big ones are removed by gravity but the fraction around 1 μm, where most anthropogenic produced sulfates, nitrates and organics are found, live quite long and are generally removed by cloud formation.

For sulfur dioxide, deposition processes and conversion are of equal weight, while nitrogen oxides are removed by conversion. The transport distance of ammonia is very limited due to short deposition and conversion lifetimes.

The reactivity of different classes of organic compounds varies considerably, hence a large difference in lifetimes and transport distance exists. Conversion of ozone is important in near sources, as ozone reacts rapidly, see above, with NO as emitted by traffic. Far away from sources dry deposition is the main loss term.

Carbon dioxide is not so much deposited as well taken up by vegetation, for example in spring, and again released as in fall litter decomposes. The total cycle of carbon tied in with vegetation, ocean uptake and release, etc. is so large that a lifetime of emitted CO_2 of 100 to 200 years is calculated.

Methane is mainly converted to CO and next to CO_2 in reaction with OH-radicals. There is uptake of methane in soils too, but this probably only a small fraction. N_2O and CFCs are not destroyed in the troposphere but are decomposed in the stratosphere.

1.6 AMBIENT CONCENTRATION OF POLLUTANTS

In Figure 1.4 the contributions of different sources are schematically represented. In stagnant conditions, when regional or over-regional transport cannot occur, local sources are responsible. In many cases a mix of different sources is important. To give an example, 5 years ago ozone was attributed to local sources in Hong Kong and Guangzhou. Recent research has clearly shown that the origin of ozone in these locations is often of regional origin. It is clear that this knowledge has large consequences on abatement measures: local measures do not reduce the ozone concentrations if they have a regional origin.

The contributions of different regions and concentrated emissions near very industrialized cities lead to specific concentration patterns (and so to deposition patterns). The different contributions are easily recognized in Figure 1.5 where acid deposition in Europe is given.

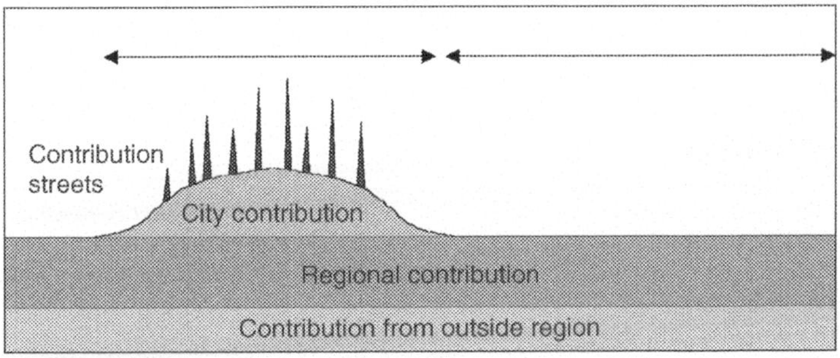

Figure 1.4 The contributions of different origins of air pollution.

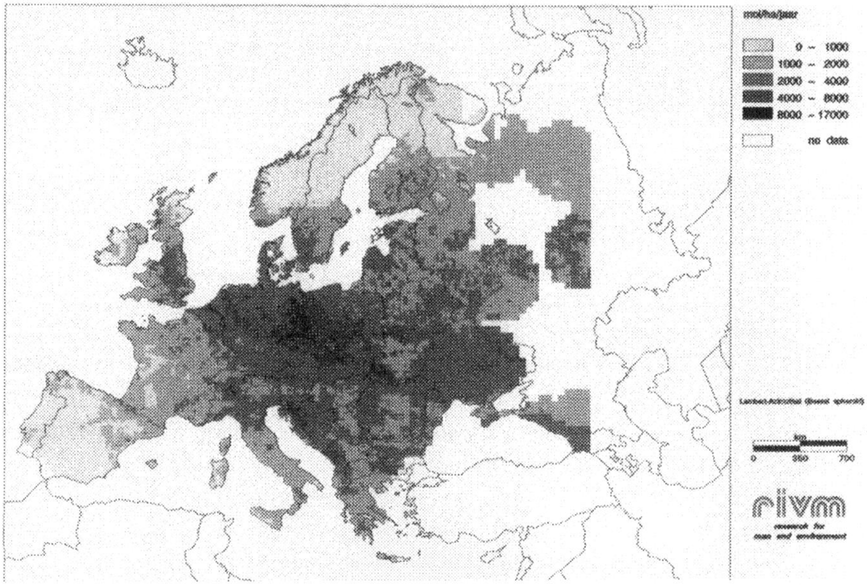

Figure 1.5 Acid deposition in Europe. Yearly acid deposition in Europe (in moles per hectare per year).

The regions affected by larger scale pollution as well as the local polluted areas (very dark) adjacent to large sources are easily recognized.

But the most severe problems are at present encountered in what are called megacities in developing countries where particulate and ozone concentrations are encountered of several hundred microgram per cube meter.

1.7 CONCLUSION

The assessment of the impact of emissions of air pollutants and the formulation of effective abatement measures for air pollutants are dependent on sufficient knowledge on the complete cycle of emissions, transport, transformation deposition and impact.

Even though our knowledge of this complete chain has been expanded greatly for many compounds, we are still confronted by many uncertainties which hinder severe prediction of future situation, see the debates on climatic change, and the necessity of effective abatement policies.

REFERENCES

Finlayson-Pitts, B.J. and Pitts, J.N. (2000) *Chemistry of the Upper and Lower Atmosphere*, 969 pp. Academic Press, New York.

Graedel, T.E. and Crutzen, P.J. (1993) *Atmospheric Change; An Earth System Perspective*, 446 pp. W.H. Freeman, New York.

ICCP (2001) Technical summary WG1 and references therein.

Jacob, D.J. (1999) *Introduction to Atmospheric Chemistry.* Harvard University. Princeton University Press, Princeton, NJ.

2

Transboundary particulate matter pollution in Europe

M. Lazaridis

2.1 INTRODUCTION

2.1.1 Airborne particulate matter

Airborne particulate matter (PM) is a complex mixture of many different chemical species originating from a variety of sources. Composition, morphology, physical and thermodynamic properties of PM vary in different geographical places and have a seasonal variability (Alpert & Hopke, 1981; EPA, 2003; EMEP, 2005; Finlayson-Pitts & Pitts, 1986; IPCC, 2001; Position Paper on Particles, 2004; Seinfeld & Pandis, 1998). Sources of PM can be either primary or secondary in nature. Primary particles can be furthermore divided as anthropogenic or natural depending on their origin. Secondary formed particles in the atmosphere are from both natural and anthropogenic origin as well, and originate from chemical transformations of gaseous precursors such as sulfur dioxide, nitrogen oxides and volatile organic compounds (VOCs). Recent research studies highlight also the importance of biogenic hydrocarbons (such as terpenes) in the formation of organic aerosols (Hoffmann et al., 1997). Aerosol is defined as a suspension of solid and liquid particles in gas but has used commonly to refer to the particulate component only.

Natural sources of primary aerosols in Europe include sea spray, fugitive dust (e.g. soil re-suspension by the wind), long-range transport of Sahara dust, volcanic and biogenic emissions (Coen et al., 2004; Lazaridis et al., 2006; Sotiropoulou et al., 2004). In Europe, anthropogenic sources of aerosols are dominant because of the urbanization of many countries and the large number of vehicle and combustion sources, both industrial and residential (Dulac & Chazette, 2003; Dusek et al., 2004; Eleftheriadis et al., 2006; Position Paper on Particles, 1998, 2004). There is a consistent pattern of geographical variability in Europe with lower concentrations of PM_{10} (particles with aerodynamic diameter larger than 10 μn) in the far north and higher concentrations in the southern countries (Position Paper on Particles, 1998, 2004). This is possible due to natural emissions of unsaturated hydrocarbons (including isoprene) which are highly reactive. The condensable organic species are formed in the gas phase by the reaction of VOCs with the principal atmospheric oxidizing agents, that is ozone, OH radicals and to a small extent NO_3 radicals (Kanakidou et al., 2005).

However, the majority of monitoring data is from urban networks in some parts of Europe and there is no systematic monitoring programme with representative rural sites in most countries. In addition, many research studies have been performed in north-western Europe, where the difference in aerosol concentrations between urban and non-urban areas does not exceed 20% (e.g. Van Der Zee et al., 1998). This can be attributed to high emissions from a densely populated area, the small weather variability between the measurement sites and the importance of long-range transport of air pollutants. Furthermore, a number of studies on wintertime concentrations of PM_{10} and black smoke (BS) in 14 urban and 14 non-urban locations in Europe indicate a relative small difference, on average 22% for PM_{10} and 43% for BS (Hoek et al., 1997). Similar observations on the regional character of PM are reported in United States (EPA, 2003).

2.1.2 PM size distribution

An important characteristic of atmospheric PM is the tremendous variation in size ranging from tens of micrometers to a nanometer size (EPA, 2003). For example, combustion-generated particles (vehicle emissions, power generation) are ranging in size between 0.003 and 1 μm. Pollens and soil dust are composed of particles mainly above 2 μm, whereas fly ash from coal combustion produces particles ranging from 0.1 to 50 μm. Aerosols in the atmosphere undergo changes in their chemical composition and size. This is due to a variety of physical and chemical processes such as nucleation (new particle formation), condensation, evaporation, coagulation, deposition (both wet and dry), activation due to water, and other gaseous species and aqueous phase reactions (Finlayson-Pitts & Pitts, 1986; Fiedler et al., 2005; Seinfeld & Pandis, 1998).

2.1.3 PM chemical composition

The lifetime of PM in the atmosphere is ranging from a few days to a few weeks and these relative long residence times result to differences between the average total mass of $PM_{2.5}$ between urban and non-urban continental aerosols (Heintzenberg, 1989). Table 2.1 shows data on mass concentration and composition of tropospheric aerosols as summarized by Heintzenberg (1989). The data show that the average $PM_{2.5}$ mass in remote areas is about 6 times lower than the urban concentration whereas the non-urban

Table 2.1 Mass and chemical composition of tropospheric PM (Heintzenberg, 1989).

Region	Mass ($\mu g\,m^{-3}$)	Percentage composition					
		Unspecified	C (elements)	C (organics)	NH_4^+	NO_3^-	SO_4^{2-}
Remote (11 areas)	4.8	57	0.3	11	7	3	22
Non-urban Continental (14 areas)	15	19	5	24	11	4	37
Urban (19 areas)	32	18	9	31	8	6	28

Table 2.2 Parameters for model aerosol distributions expressed with the help of three log-normal modes (I, II, III) (Seinfeld & Pandis, 1998). Each log-normal mode is characterized with the number of aerosol (N), their mean diameter (D_p) and standard deviation (σ_g).

Type	Mode I			Mode II			Mode III		
	N (cm^{-3})	D_p (μm)	Log(σ)	N (cm^{-3})	D_p (μm)	Log(σ)	N (cm^{-3})	D_p (μm)	Log(σ)
Urban	9.93E4	0.013	0.245	1.11E3	0.014	0.666	3.64E4	0.05	0.337
Marine	133	0.008	0.657	66.6	0.266	0.210	3.1	0.58	0.396
Rural	6650	0.015	0.225	147	0.054	0.557	1990	0.084	0.266
Remote-continental	3200	0.02	0.161	2900	0.116	0.217	0.3	1.8	0.380
Free troposphere	129	0.007	0.645	59.7	0.250	0.253	63.5	0.52	0.425
Polar	21.7	0.138	0.245	0.186	0.75	0.300	3E-4	8.6	0.291
Desert	726	0.002	0.247	114	0.038	0.770	0.178	21.6	0.438

continental is 2 times lower. These observations show the importance of the regional component of PM, which is highly affected by long-range transport of pollutants.

The chemical complexity of atmospheric aerosols requires a consideration of their composition and sources. In addition, because of the considerable influence of natural sources, with larger size in general than the anthropogenic emissions (Position Paper, 1998, 2004; Van Dingenen et al., 2004), there is a need for measurements of smaller particles such as $PM_{2.5}$. This standardization will allow a better understanding of the anthropogenic influence and further implementation of control strategies. A detailed presentation of the aerosol size distribution for different environmental conditions is given by Seinfeld and Pandis (1998) and is summarized in Table 2.2.

The complexity of the processes controlling the shape and size of airborne particles, the enormous number of different sources and the difficulties of their measurements resulted to critical reactions against proposed national ambient air quality standards in the United States (Abelson, 1998). Even though the complexity of the factors affecting exposures to PM is recognized, there is an extensive research effort in recent years in basic aerosol research, in examining their physico-chemical properties and the relationship between exposure to PM and resulting health effects (EPA,

2003). Recent aerosol studies focus on the detailed number size distribution characterization of PM (Kulmala et al., 2004; Stanier et al., 2004).

2.2 EUROPEAN AEROSOL PATTERNS AND TRENDS

2.2.1 Aerosol monitoring

Measurements of PM are insufficient to have a complete picture of the PM concentrations across Europe. Measurements of PM are concentrated in urban areas and in addition there is no sufficient reporting of the data to international organizations, which could allow a comprehensive geographical overview of PM concentrations in Europe. However, the ambient concentration of PM at a defined spatial and temporal point is a complex functional of emissions, dispersion, physico-chemical processes and removal. The aerosol residence time is an important factor that determines the ambient aerosol concentration. Fine aerosols in the size range of 0.1–1.0 μm, have a residence time of several days and if they are transported in the upper troposphere their residence time increases to several weeks.

There is a vast literature concerning patterns and trends of ambient aerosols and their concentration characteristics. The interested reader can refer to the reviews by Seinfeld and Pandis (1998), Position Paper (1998, 2004), Putaud et al. (2004), Van Dingenen et al. (2004), EMEP (2005) and EPA (2003).

Measurements of PM_{10} and BS wintertime (winter 1993/1994) concentrations in Europe during the PEACE study (Hoek et al., 1997) show small differences between urban and non-urban locations. Similar winter studies for PM_{10} $PM_{2.5}$ and BS in the Netherlands (winters of 1992/1993, 1993/1994 and 1994/1995) show the same trend (Van Der Zee et al., 1998): PM_{10} and BS concentrations were on average 13% and 19% higher in the urban areas than in the non-urban areas. Non-organic secondary aerosol concentrations were on average 8% lower in the urban areas whereas $PM_{2.5}$ concentrations were similar in between urban and non-urban sites. Also higher elemental concentrations in PM_{10} were found in the urban area for all elements except Si. Meteorological conditions seem to play a crucial role in the observations. Easterly winds resulted to smaller differences between urban and non-urban sites. This shows the importance of the regional aerosol component from Eastern Europe. Winds from the sea lead to lower average PM concentrations and larger differences between urban and non-urban locations.

Measurements of PM_{10} have been also performed in Switzerland during 1993 at a dozen urban, rural and alpine sites. The PM_{10} concentrations ranged between 10 $\mu g\,m^{-3}$ (alpine region) and 33 $\mu g\,m^{-3}$ (urban site). An interesting observation is that the rural levels were one-third of the urban levels and homogeneously distributed. In the Alpine sites, the levels were considerably lower. Aircraft measurements over the Swiss plateau during photosmog episodes were also performed and showed that under conditions of strong air pollution and strong photochemical activity the aerosol particles grow with secondary processes (such as condensation) into the diameter range 0.12–1.5 μm. It was also found that in Switzerland the ratio between PM_{10} and total suspended particulates (TSP) ranged between 0.6 and 0.75 with the highest ratios in the urban sites. Seasonally there were higher concentrations of PM during winter and autumn due to cold temperature inversions.

Sampling of aerosols in an European Monitoring and Evaluation Programme (EMEP) station, Iskrba (Slovenia), has occurred in spring 1997 and winter 1998 using a low-pressure Berner impactor. Major components include sulfate, organic and elemental carbon, nitrate and chloride. The highest concentrations of SO_4^{2-} and NH_4^+ are in the accumulation mode, while in the coarser fraction there are high concentrations of Na^+, Mg^{2+}, Ca^{2+} and heavy metals. A diurnal black carbon cycle as in an urban site was not observed and the concentration of PM was higher in spring than in the winter.

Long-range transport of PM also has been studied intensively in southern Scandinavia (Pakkanen et al., 1996). Air masses originating from the British Isles, central and eastern Europe were mainly responsible for the long-range transport of pollution such as sulfur. A long-range transport model has been developed to in the EMEP framework to study the impact of sulfur pollution in Scandinavia and in Europe in general. During long-range transport episodes, 20–50 times higher concentrations of pollutants are observed compared to the background levels at the Birkenes site (Southern Norway). Measurements of the aerosol size distribution using a Berner impactor showed one or two modes in both the fine mode and coarse mode. The use of a stacked filter unit (SFU) and low-pressure impactors (LPIs) in two locations in Norway (Birkenes and Nordmoen) showed similar average values for elemental fine to total aerosol ratios. Furthermore, the elemental ratios for several pollutant elements of the air masses in southern Norway (Pakkanen et al., 1996) and southern Sweden (Swietlicki et al., 1989) are very similar.

Furthermore, biological aerosols may also have a considerable influence in the mass size distribution and studies in an urban/rural-influenced region in Mainz indicated that biological aerosols can constitute close to 30% of the total aerosol mass. These results are valid for an urban/rural-influenced region. Results from measurements on the Atlantic Ocean show a percentage of biological aerosols to total aerosol of only 14% in number and 8% in volume (Matthias-Maser, 1998).

2.2.2 Composition of non-urban aerosol

A chemical mass balance of fine particles (diameter smaller than 1.5 μm) collected at three European sites was reported recently (Zappoli et al., 1998). The sites were located in southern Sweden (S), in a rural area in the Great Hungarian Plain (H) and in the polluted Po Valley, northern Italy (I). The total fine aerosol mass ranged from a value of 38 $\mu g\,m^{-3}$ (I) in the most polluted area to around 6 $\mu g\,m^{-3}$ (S) in the background site. The rural station in Hungary showed a value close to 23 $\mu g\,m^{-3}$ (H). Speciation of the aerosol components with reference to water solubility of the different species is presented in Table 2.3. The results show a very high water solubility (WSOC) and higher fraction of WSOC in the background site compared to the rural and polluted sites (Zappoli et al., 1998).

2.2.3 Urban European aerosol pattern

The concentration of urban aerosols is a mixture of emissions from industry, traffic, various combustion sources, secondary formation, natural sources and from long-range transport. The number size distribution is dominated by fine particles, whereas

Table 2.3 Speciation (in % units) of different aerosol classes (Zappoli et al., 1998).

Site	EC	WISOC	WSOC	OSII	NH_4^+	SO_4^{2-}	NO_3^-	WISIC	ND
S	1	7	36	11	4	20	4	15	–
H	2	13	14	2	10	32	8	n. a.	18
I	3	7	14	14	10	20	16	11	6

EC: elemental carbon; WISOC: water-insoluble organics; WSOC: water-soluble organics; OSII: other soluble inorganic ions; WISIC: water-insoluble inorganics; ND: non-detected; na: not applicable.

Table 2.4 Weekly average mass concentrations of aerosol component at Leeds 1995/1996 ($\mu g\, m^{-3}$) measures (Airborne Particles Expert Group – Interim Report, 1998).

Week beginning	Urban NO_3^-	Rural NO_3^-	Urban SO_4^{2-}	Rural SO_4^{2-}	Urban NH_4^+	Rural NH_4^+	Total Mass	Sum of salts*
15 May 95	1.16		2.32				18	4.7
21 Jun. 95	4.13		3.73				26	10.5
05 Jul. 95	4.53		5.48				54	13.4
26 Jul. 95	7.2		6.07				55	17.7
09 Aug. 95	2.95	2.42	6.68	5.87	1.18	0.83	42	13.0
18 Oct. 95	4.35	2.31	4.34	4.12	0.67	0.36	42	11.6
15 Nov. 95	2.77	1.27	3.96	2.80	1.22	0.94	44	9.0
17 Jan. 96	5.26	4.43	11.18	9.57	4.89	2.35	69	22.2
Average	4.04	–	5.47	–	–	–	44	12.7

* Sum of ammonium sulfate and nitrate: $1.29\ NO_3^- + 1.375\ SO_4^{2-}$. Also $NH_4NO_3 = 1.29 \times NO_3$ and $(NH_4)_2SO_4 = 1.375 \times SO_4^{2-}$.

the mass size distribution has two modes: the accumulation and the coarse modes (Seinfeld & Pandis, 1998).

There are recent attempts for monitoring urban air quality in the European cities with the establishment of the European air quality monitoring and information network of the European Environment Agency (EEA), called EUROAIRNET (EEA, 2002). Until now, over 800 stations in Europe have been selected which monitor various gaseous pollutants, as well as, TSP, PM_{10}, $PM_{2.5}$ and BS. Literature concerning monitoring and levels of pollutants in urban areas in Europe include the reviews by Putaud et al. (2004) and the Position Paper (2004).

In UK, the urban sulfate concentrations are similar to rural levels. The rural to urban ratio for nitrate varies from 0.46 to 0.84 showing therefore an urban source of nitrate measures (Airborne Particles Expert Group – Interim Report, 1998). The mass median diameters of sulfate, nitrate and ammonium measured in Leeds (UK) were 1.1, 1.6 and 1.0 μm, respectively (Table 2.4). Model calculations show that both UK and continental European sources of sulfur dioxide contribute to sulfate pollution episodes in the UK measures (Airborne Particles Expert Group – Interim Report, 1998). An overview of the urban patterns in Europe can be found in the Position Paper (1998, 2004). In the United States there is currently a comprehensive database for PM and various gaseous pollutants through the Internet (www.epa.gov/ttn/airs).

2.3 AIR POLLUTION OBSERVATION: EXISTING NETWORKS

2.3.1 European Monitoring and Evaluation Programme

The Co-operative Programme for Monitoring and Evaluation of the Long-Range Transmission of Air Pollutants in Europe (EMEP) is a regional policy-driven monitoring programme under the Convention on Long-Range Transboundary Air Pollution. The main goal of PM measurements within EMEP is to monitor transboundary long-range transport of aerosols in Europe and its long-term trends. Monitoring of PM_{10} was added to the EMEP work programme in 1998. In 2000, PM_{10} measurements were reported to EMEP from three different countries (Germany, Switzerland and Spain) from a total of 26 stations. In 2002, Austria, the Czech Republic, Denmark, Ireland, the Joint Research Centre (JRC) in Ispra (Italy), The Netherlands, Norway, Portugal, Sweden, Slovakia and Great Britain were measuring PM_{10} in addition to Germany, Spain and Switzerland. PM_{10} monitoring is expected to start soon in Finland, Greece, Slovenia and possibly Kazakhstan.

According to the EU Directive 1999/30/EC, the First Daughter Directive of the Air Quality Framework Directive 1996/62/EC, the Member States (including Norway and other EEA-Member-States) have to start monitoring, among other pollutants, ambient air PM_{10}. The Directive requires that the measurements have to be performed according to the CEN standard EN12341, which means by use of 1 of the 3 reference methods described in the standard or with a method shown to be equivalent to one of the reference methods, or display a consistent relationship to one of the references. For instance, it is required by EN 12341 that the filters are equilibrated, at 20°C (±1) and 50% R.H. (±5), for 48 h. This equilibration should be performed before the filters are weighed previous to the sample collection, and after sampling, before the filter is weighed again with the collected sample.

The EMEP monitoring programme has started to prepare its measurement requirements for mass and chemical analysis of aerosol filters and the first aerosol chemical speciation results have been already performed (EMEP, 2005). In the United States, there is currently a lot of work in progress for the evaluation and field performance of $PM_{2.5}$ samplers. Many national inter-comparison studies have been organized and further studies are in progress.

Figure 2.1 illustrates measured levels of PM_{10} at rural, urban and roadside sites for 1999. Switzerland, Germany, Spain and Italy performed PM measurements in 1999 using gravimetric measurement methods. Italy reports TSP data from only one station located at Ispra, although as of 2001 PM_{10} and $PM_{2.5}$ data will be reported. Spain began reporting PM_{10} data from 10 stations across the country in 1999. In addition, Germany (8 stations) and Switzerland (5 stations) continued to report PM_{10} data following gravimetric methods. Furthermore, daily measurements of PM sulfate and total nitrate are performed within the EMEP framework. Data outside the EMEP framework were obtained from the European Topic Centre (ETC/ACC) under contract from the EEA. It is evident that different measurement methods have been used to quantify aerosol levels at different locations across Europe. Therefore, a comparison of data is not straightforward and only qualitative results can be obtained

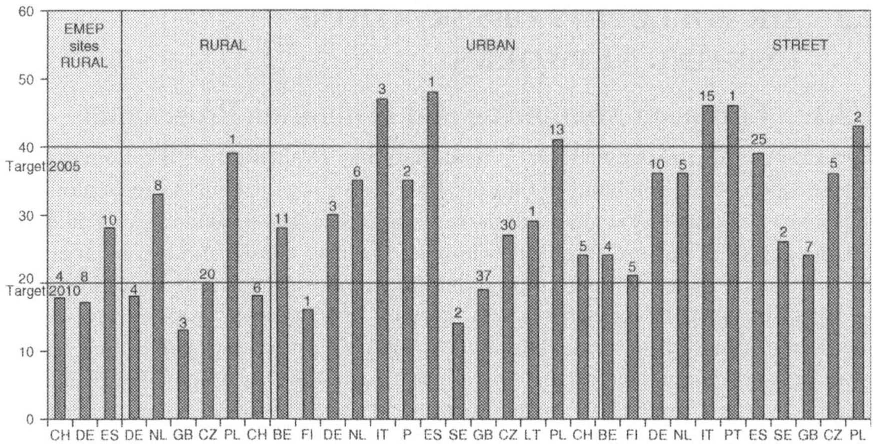

Figure 2.1 Annual average PM_{10} concentrations in different European countries or various site types (in $\mu g\,m^{-3}$) during 1999. The upper number corresponds to the number of stations in each country.

from Figure 2.1. In addition, for some countries the number of urban and street sites is very limited.

More recent data in the EMEP framework are presented in Figure 2.2. Annual averages of PM_{10} concentrations measured at the different EMEP stations during 2003 are presented. Only those stations that have a geographical and time data coverage of more than 50% are shown. This leaves only 7 out of 13 Spanish stations. Annual averages are based on concentrations measured over a 24 h sampling period (EMEP, 2005; Lazaridis et al., 2002).

In addition, Figure 2.3 shows the $PM_{2.5}$ annual average values for the year 2003 as depicted from available data of the EMEP sites. There are currently no PM_1 measurements on a regular basis within EMEP. However, PM_1 is likely to be less affected by local re-suspension than $PM_{2.5}$, and allows us to separate the accumulation and coarse modes typically seen in rural aerosol size distributions. Thus, PM_1 monitoring may have important advantages over $PM_{2.5}$.

Although measurements are not comparable between countries, there are some interesting patterns with relatively high concentrations in the most populated areas of Europe (EMEP, 2005). This is in agreement with data from 1997 and 1998. Concentrations in Scandinavia and northern Europe are lower but it is interesting to point out the higher PM_{10} concentrations in southern countries such as Italy and Spain.

It is expected that resuspended dust may contribute a significant fraction of PM_{10} in these cases due to dry conditions. It is apparent from the above measurement data and previous studies that more precise, harmonized and quality-controlled measurements are needed to obtain a comprehensive picture of geographical and seasonal characteristics of PM in Europe (Position Paper, 2004). The annually averaged PM_{10} concentrations in Figure 2.2 are for most countries still far above the limit value set for 2010, in many cases even twice as high as the limit value (see Table 2.5).

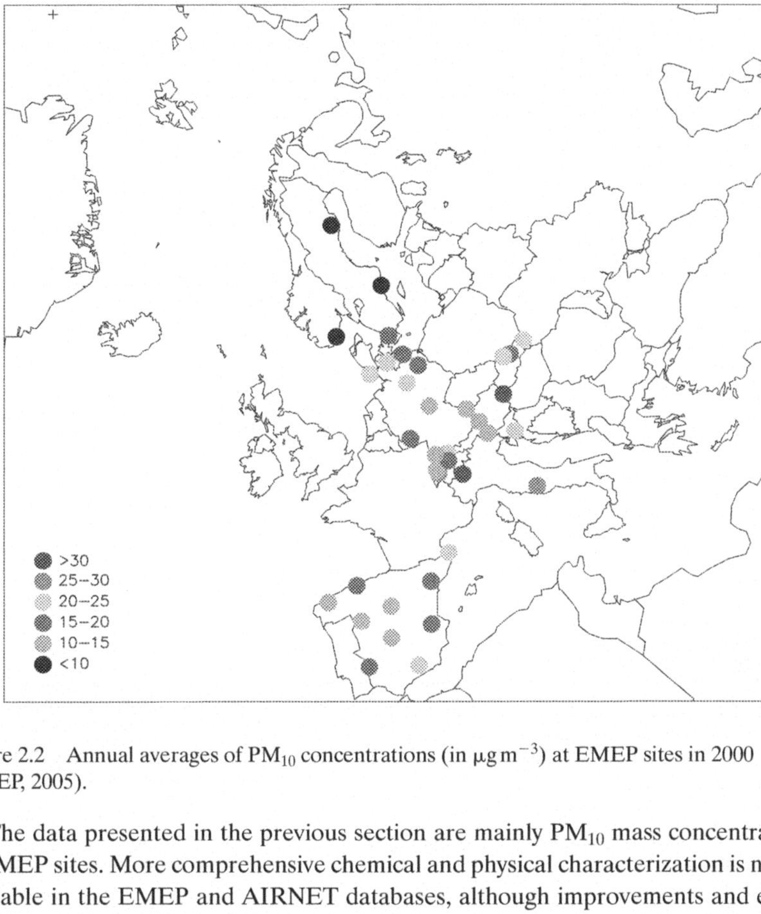

Figure 2.2 Annual averages of PM_{10} concentrations (in $\mu g\,m^{-3}$) at EMEP sites in 2000 (EMEP, 2005).

The data presented in the previous section are mainly PM_{10} mass concentrations at EMEP sites. More comprehensive chemical and physical characterization is not yet available in the EMEP and AIRNET databases, although improvements and extensions have been proposed. More comprehensive data are available from national and scientific studies (Position Paper, 2004; Putaud et al., 2004).

2.3.2 Global Atmosphere Watch

The Global Atmosphere Watch (GAW) programme is a World Meteorological Organization (WMO) sponsored activity and currently supported by about 80 WMO member countries. It is the goal of GAW to develop and maintain long-term measurements of atmospheric constituents in order to detect trends, develop aerosol predictive capabilities and understand processes of aerosol dynamics (WMO, 2001, 2004). With respect to aerosols, the objective of GAW is to support a global network determining the spatio-temporal distribution of aerosol properties related to climate forcing and air quality up to multi-decadal time scales. Guidelines on aerosol chemical measurements at WMO and GAW stations are documented by Baltensperger et al. (2003). The GAW network consists of 22 global stations and some 300 regional stations. According to the recommendations of the Scientific Advisory Group (SAG) for aerosols, regional stations measure aerosol optical depth, as well as the aerosol light scattering and absorption coefficient. If possible these should be complemented by routine mass concentration

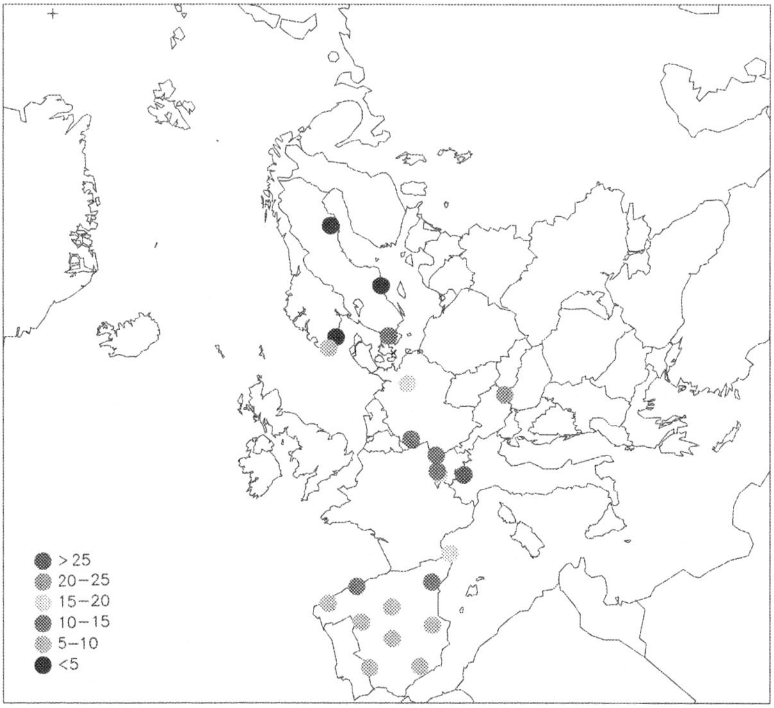

Figure 2.3 Annual 2003 averages of daily $PM_{2.5}$ mass measurements (in $\mu g\, m^{-3}$) (EMEP, 2005).

Table 2.5 EU limit values for PM_{10} and date to be met, for health protection.

Compound	Limit/target value	Value	Remarks	Date to be met
PM_{10}	Annual average	$40\, \mu g\, m^{-3}$		1 Jan. 2005
Stage 1	Daily average	$50\, \mu g\, m^{-3}$	May be exceeded up to 35 days a year	1 Jan. 2005
PM_{10}	Annual average	$20\, \mu g\, m^{-3}$	Indicative	1 Jan. 2010
Stage 2	Daily average	$50\, \mu g\, m^{-3}$	Indicative; may be exceeded up to 7 days a year	1 Jan. 2010

and composition measurements in two aerosol size fractions. At global stations, a larger number of measurements are desirable. These include the regional parameters listed above as well as the light scattering, hemispheric backscattering and absorption coefficients at various wavelengths, aerosol number concentration, cloud condensation nuclei (CCN) concentration at 0.5% supersaturation, and diffuse, global and direct solar radiation. Additional parameters such as the aerosol size distribution, detailed size fractionated chemical composition, dependence of aerosol properties on relative humidity, CCN concentration at various supersaturations and the vertical distribution of aerosol properties should be measured intermittently at global stations.

Within region VI (Europe), there are 4 Global stations (with some of them existing of several substations) and about 150 regional stations contributing to GAW. Many of these regional stations are both GAW and EMEP sites. Information on any station as well as on the parameters measured at each site can be obtained via the GAW station information system (GAWSIS) at http://www.empa.ch/gaw/gawsis/. Data are delivered to and made available by the World Data Centre for Aerosols (WDCA, located in Ispra, Italy http://ies.jrc.cec.eu.int/wdca/). The Institute for Tropospheric Research in Leipzig (Germany) hosts a GAW World Calibration Centre (WCC) for physical aerosol parameters. A host for the chemical parameters still must be located.

The SAG provided initial recommendations on the measurement parameters within the context of all GAW components (WMO, 2001). An aerosol guide is being produced, with detailed recommendations on the suggested aerosol parameters. This activity is done in conjunction with the EC project CREATE (construction, use and delivery of a European aerosol database).

2.3.3 EuroAirnet

The European Air Quality Monitoring Network (EuroAirnet) has been developed in close cooperation between the EEA and the European countries (EEA, 2002). EuroAirnet comprises a selection from already existing stations in the countries. Data from EuroAirnet are reported to AIRBASE (http://etc-acc.eionet.eu.int/databases/airbase.html), a database managed by the European Topic Centre on Air Quality and Climate Change under contract to EEA. In 2000, PM_{10} data from 425 stations in 17 countries were in the database. In 2001, there were already 928 stations from 24 countries reporting PM_{10} to AIRBASE. Of those, 816 stations were located in the 15 EU member countries, 101 stations reported from 7 of the ascension countries and 11 stations were from other countries (Iceland and Bulgaria). In contrast to EMEP, which focuses on regional monitoring, and in contrast to GAW, which mainly has a global focus, EuroAirnet is strongly dominated by monitoring at urban and traffic sites. In 2001, 427 of 928 stations were located at urban sites, 292 were traffic sites, 100 were rural sites and 109 stations were other or non-specified sites. $PM_{2.5}$ data are still very sparse within AIRBASE.

2.4 MODELLING AMBIENT PM TRANSPORT

2.4.1 Air quality modelling

There are currently several three-dimensional (3-D) Eulerian air quality models that include a PM module. The modelling of PM is mostly concentrated in a revised chemistry module and deposition module but recently several models include aerosol dynamic routines. For example, an air quality model developed by the US EPA, the Regional Particulate Model (RPM), has the ability to model detailed aerosol dynamic processes using a modal method to describe the aerosol size distribution (Binkowski & Shankar, 1995).

The description of the pollutant transport in the 3-D air quality models is performed using the atmospheric diffusion equation including a source term and a scavenging term. The atmospheric diffusion equation can be written as (Seinfeld & Pandis, 1998):

$$\frac{\partial c_i}{\partial t} + u_j \frac{\partial c_i}{\partial x_j} = \frac{\partial}{\partial x_j}\left(K_{jj} \frac{\partial c_i}{\partial x_j}\right) + R_i(x,t) + S_i(x,t)$$

where c_i is the mean concentration of species i, x_j the space coordinate at the direction j ($j = 1-3$), K_{jj} is the eddy diffusivity at the direction j, $R_i(x, t)$ denotes the scavenging of species during chemical reactions and deposition, and the $S_i(x, t)$ is the source term (emissions). The numerical evaluation of the atmospheric diffusion equation is usually performed using an operator splitting technique, which employs a separation of the horizontal and vertical transport terms from the chemical reaction and emission terms, so that each term is solved separately.

2.4.2 Model application

Long-range transport and fate of photochemical gaseous air pollutants and PM have been studied extensively in Europe the past decades under the framework of several national and international efforts (Beck et al., 1999; Berdowski et al., 1998; EU, 1996, 1997; EMEP-WMO, 1997). A number of modelling studies for combined aerosol and gaseous species with Eulerian regional and mesoscale models have also been performed during the past several years (Ackermann et al., 1998; Binkowski & Shankar, 1995; Lurmann et al., 1997; Pai et al., 2000). However, there is scarce information concerning consistent modelling studies in mesoscale geographical areas in Europe to reveal the atmospheric composition/variability of ozone and PM (Ackermann et al., 1998). Furthermore, available information on characteristics of photochemical pollutants/fine particles in Southern Europe and their effects on air quality are also limited.

In the major part of modelling studies, the photo-oxidants and PM have been studied separately, even though the fine fraction of the PM is directly controlled by the airborne concentrations of photo-oxidants and other gaseous pollutants (Pandis et al., 1992; Seinfeld & Pandis, 1998; Zhang et al., 2004). Therefore, a combined intensive modelling study and comparison with available experimental data for ozone/fine particles in the Mediterranean area should offer valuable information for characterizing their dynamics and association. Furthermore, the Mediterranean is characterized by a specific natural aerosol load, namely sea spray and North African Desert dust. These natural PM emissions are involved in heterogeneous reactions with anthropogenic gaseous pollutants and may modify the processes leading to gas to particle conversion.

An application of the UAM-AERO model has been made at the Eastern Mediterranean area (Lazaridis et al., 2005). The emphasis was given on 24 h average predictions for particulate mass and chemical composition. The contribution of both anthropogenic and natural sources to the spatial distribution of PM in the area under study is shown in Figure 2.4. Specifically, Figure 2.4 shows the average spatial distribution of fine ($d < 1.25\,\mu m$) and coarse ($1.25\,\mu m < d < 10\,\mu m$) aerosols for 14th July 2000 at surface level at 01:00 a.m. The highest concentration for the coarse aerosol fraction is observed above the Aegean sea and at certain areas above the island of Crete and the Greek mainland. Sources of the coarse fraction are mainly sea salt particles and resuspended dust emissions. High concentrations of fine aerosols

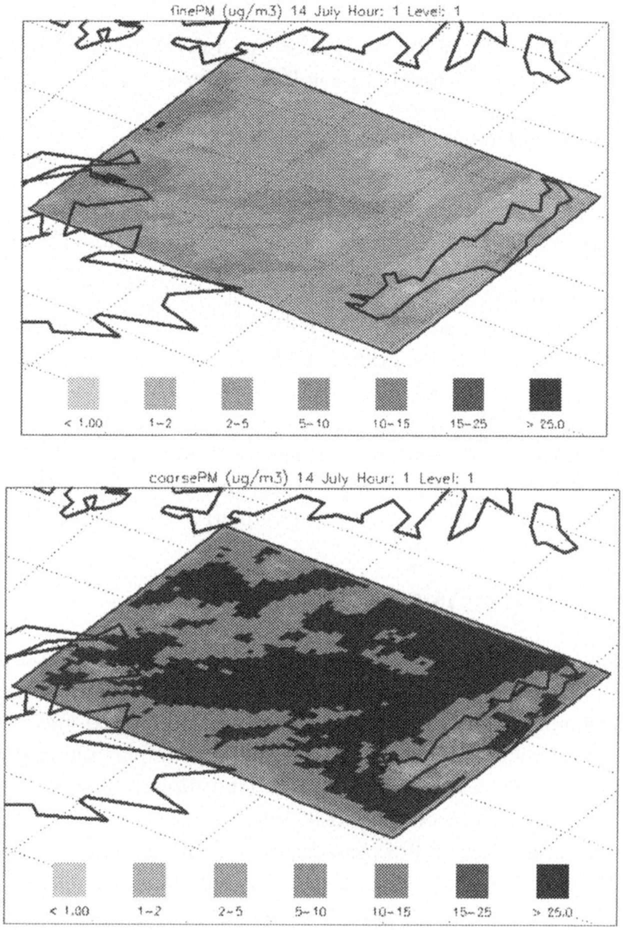

(a)

(b)

Figure 2.4 Spatial distribution of (a) coarse and (b) fine aerosols in the modelling domain on 14th July at 01:00 a.m. (Lazaridis et al., 2005, 2006).

are observed above the greater Athens region and the Greek mainland which are related mainly to anthropogenic sources.

2.4.3 Air quality software

The available 3-D air quality models including modelling of aerosol processes (in addition to EMEP work) are mesoscale models and long-range transport models (Seigneur et al., 1999).

2.4.3.1 Mesoscale models

(1) The California Institute of Technology model (CIT).
(2) The Denver Air Quality Model (DAQM).
(3) The Gas, Aerosol, Transport and Radiation model (GATOR).

(4) The Regional Particulate Model (RPM).
(5) The SARMAP Air Quality Model with aerosols (SAQM-AERO).
(6) The Urban Airshed Model Version IV with aerosols (UAM-AERO).

2.4.3.2 Long-range transport models

(1) The Regulatory Modelling System for Aerosols and Deposition (REMSAD).
(2) The Urban Airshed Model Version IV with linearized chemistry (UAM-LC).
(3) The Visibility and Haze in the Western Atmosphere model (VISHWA).

More recent models include the models 3 from the US EPA and the model CAMx (EPA, 2003). At European level, the EUTREND model (Eerens et al., 1996) has been also used for calculating primary PM_{10}. However, the model has not the possibility to include the formation of secondary aerosols.

Currently, the EMEP model is the main regional regulatory modelling tool in studying photochemical pollutants and acidification in Europe and treats the PM_{10} and $PM_{2.5}$ fractions of the atmospheric sulfate and nitrate aerosols using a 3-D model with detailed physico-chemical modules (Kahnert & Tarrason, 2003).

2.5 CONCLUSIONS

The dynamics of the PM in the atmosphere involve various physical and chemical processes under different time scales. Aerosol dynamics involve a wide range of particle sizes ranging from few nanometers to several hundred micrometers, different composition and chemical reactivity and undergo complex physical transformations (nucleation, condensation, coagulation and deposition processes).

There is an urgent need to provide the policy-making community in Europe with state-of-the-art data on PM characteristics. PM_{10} and $PM_{2.5}$ monitoring data have to be quality assured and the different measurement methodologies to be assessed. The regional component of the ambient PM mass is important part even for urban areas. Therefore the determination of air quality policies for the reduction of PM levels has to take into account the long-range transport of pollutants. Natural emissions such as sea salt, resuspended dust, Saharan dust outbreaks, forest fires and emissions from the vegetation have to be better quantified for determining their importance in the PM mass closure.

Modelling or measuring atmospheric aerosols involves many challenging tasks and is a fast evolving scientific area. Meteorological processes affect substantially the physical and chemical processes of atmospheric aerosols as well as the geographical and temporal variation of their sources. There is a substantial scientific literature examining various aspects of the physico-chemical properties of atmospheric aerosols and an overview is presented in this chapter.

REFERENCES

Abelson, P.H. (1998) Airborne particulate matter. *Science* **281**, 1609.

Ackermann, I.J., Hass, H., Memmesheimer, M., Ebel, A., Binkowski, F.S. and Shankar, U. (1998) Modal aerosol dynamics model for Europe: development and first applications. *Atmos. Environ.* **32**, 2981–2999.

Airborne Particles Expert Group – Interim Report (1998) Source apportionment of airborne particulate matter in United Kingdom.

Alpert, D.J. and Hopke, P.K.A. (1981) Determination of the sources of airborne particles collected during the regional air pollution study. *Atmos. Environ.* **15**, 675–687.

Baltensperger, U. et al. (2003) Aerosol measurement procedures, guidelines and recommendations. WMO/GAW Report No. 153. WMO, Geneva.

Beck, J.P., Krzyzanowski, M., Koffi, B., Hjellbrekke, A.-G., Hootsen, H., Millan, M.M., Tombrou, M. and Simpson, D. (1999) Tropospheric Ozone in the European Union "The Consolidated Report". European Environment Topic Centre on Air Quality.

Berdowski, J.J.M., Mulder, W., Veldt, C., Visschedijk, A.J.H. and Zandveld, P.Y.J. (1998) Particulate emissions (PM_{10}–$PM_{2.5}$–$PM_{0.1}$) in Europe in 1990 and 1993 RIVM.

Binkowski, F.S. and Shankar, U. (1995) The regional particulate model 1. Model description and preliminary results. *J. Geophys. Res.* **100**, 26191–26209.

Coen, M.C., Weingartner, E., Schaub, D., Hueglin, C., Corrigan, C., Henning, S., Schwikowski, M., Baltensperger, U. (2004) Saharan dust events at the Jungfraujoch: detection by wavelength dependence of the single scattering albedo and first climatology analysis. *Atmos. Chem. Phys.* **4**, 2465–2480.

Dulac, F., Chazette, P. (2003) Airborne study of a multi-layer aerosol structure in the eastern Mediterranean observed with the airborne polarized lidar ALEX during a STAAARTE campaign (7 June 1997). *Atmos. Chem. Phys.* **4**, 1817–1831.

Dusek, U., Covert, D.S., Wiedensohler, A., Neususs, C., Weise, D. (2004) Aerosol number to volume ratios in Southwest Portugal *during ACE-2. Tellus B*, **56**, 477–491.

European Environment Agency (2002) *Environmental signals 2002.* Benchmarking the millennium. No. 9 Environmental Assessment Report, 2002. European Environment Agency, Copenhagen.

Eerens et al. (1996) *EUTREND Model*. Bilthoven, RIVM.

Eleftheriadis, K., Colbeck, I., Housiadas, C., Lazaridis, M., Mihalopoulos, N., Mitsakou, C., Smolik, J. and Zdimal, V. (2006) Size distribution, composition and origin of the submicron aerosol in the marine boundary layer during the eastern Mediterranean "SUB-AERO" experiment. *Atmos. Environ.* (In press).

EMEP-WMO (1997) *Workshop on Strategies for Monitoring of Regional Air Pollution in Relation to the Need within EMEP, GAW and Other International Bodies*. EMEP/CCC-Report 10/97. Norwegian Institute for Air Research.

EMEP/CCC (2002) *Measurements of Particulate Matter: Status Report 2002* (ed. M. Kahnert). EMEP/CCC-Report 4/2002. Norwegian Institute for Air Research.

EMEP/CCC-MSC-W (2005) *Transboundary Particulate Matter in Europe: Status Report 2005* (eds. Yttri and Trseth). EMEP Report 4/2005. Norwegian Institute for Air Research and Norwegian Meteorological Institute.

EPA (2003) Air quality criteria for particulate matter. EPA/600/P-99/002aD. US Environmental Protection Agency.

Fiedler, V., Dal Maso, M., Boy, M., Aufmhoff, H., Hoffmann, J., Schuck, T., Birmili, W., Hanke, M., Uecker, J., Arnold, F., Kulmala, M. (2005) The contribution of sulphuric acid to atmospheric particle formation and growth: a comparison between boundary layers in Northern and Central Europe. *Atmos. Chem. Phys*. **5**, 1773–1785.

Finlayson-Pitts, B.J. and Pitts, J.N. (1986) *Atmospheric Chemistry: Fundamentals and Experimental Techniques*. John Wiley & Sons.

Heintzenberg, J. (1989) Fine particles in the global troposphere – a review. *Tellus* **41B**, 149–160.

Hoek, G., Forsberg, B., Borowska, M., Hlawiczka, S., Vaskovi, E., Welinder, H., Branis, M., Benes, I., Kotesovec, F., Hagen, L.E., Cyrys, J., Jantunen, M., Roemer, W. and Brunekreef, B. (1997) Wintertime PM_{10} and black smoke concentrations across Europe: results from the PEACE Study. *Atmos. Environ.* **31**, 3609–3622.

Hoek, G. et al. (2002). The association between mortality and indicators of traffic-related air pollution in a Dutch cohort study. *Lancet* **360**, 1203–1209.

Hoffmann, T., Odum, J.R., Bowman, F., Collins, D., Klockow, D., Flagan, R.C. and Seinfeld, J.H. (1997). Formation of organic aerosols from oxidation of biogenic hydrocarbons. *J. Atmos. Chem.* **26**, 189–222.

IPCC (Intergovernmental Panel on Climate Change) (2001). Third Assessment Report – Climate Change 2001.

Kahnert, M. and Tarrasón, L. (eds.) (2003) *Transboundary Particular Matter in Europe: Status Report 2003*. EMEP Status Report 4/2003. Norwegian Institute of Air Research, Kjeller. http://www.emep.int

Kanakidou, M., Seinfeld, J.H., Pandis, S.N., Barnes, I., Dentener, F.J., Facchini, M.C., Van Dingenen, R., Ervens, B., Nenes, A., Nielsen, C.J., Swietlicki, E., Putaud, J.P., Balkanski, Y., Fuzzi, S., Horth, J., Moortgat, G.K., Winterhalter, R., Myhre, C.E.L., Tsigaridis, K., Vignati, E., Stephanou, G. and Wilson, J. (2005). Organic aerosol and global climate modelling: a review. *Atmos. Chem. Phys.* **5**, 1053–1123.

Kulmala, M., Vehkamaki, H., Petaja, T., Dal Maso, M., Lauri, A., Kerminen, V.M., Birmili, W. and McMurry, P.H. (2004) Formation and growth rates of ultrafine atmospheric particles: a review of observations. *J. Aerosol Sci.* **35**, 143–176.

Lazaridis, M., Semb, A., Larssen, S., Hjellbrekke, A.-G., Hov, Ø., Hanssen, J.E., Schaug, J. and Tørseth, K. (2002) Measurements of particulate matter within the framework of the European Monitoring and Evaluation Programme (EMEP) I. First results. *Sci. Total Environ.* **285**, 209–235.

Lazaridis, M., Spyridaki, A., Solberg, S., Smolik, J., Zdimal, V., Eleftheriadis, K., Aleksandropoulou, V., Hov, O. and Georgopoulos, P.G. (2005) Mesoscale modelling of combined aerosol and photooxidant processes in the eastern Mediterranean. *Atmos. Chem. Phys.* **5**, 927–940.

Lazaridis, M., Eleftheriadis, K., Smolik, J., Colbeck, I., Kallos, G., Drossinos, Y., Zdimal, V., Vecera, Z., Mihalopoulos, N., Mikuska, P., Bryant, C., Housiadas, C., Spyridaki, A., Astitha, M., Havranek, V. (2006) Dynamics of fine particles and photo-oxidants in the Eastern Mediterranean (SUB-AERO) *Atmos. Environ*. (in press).

Lurmann, F.W., Wexler, A.S., Pandis, S.N., Mussara, S., Kumar, N. and Seinfeld, J.H. (1997) Modelling urban and regional aerosols-II. Application to California's South Coast Air Basin. *Atmos. Environ.* **31**, 2695–2715.

Matthias-Maser, S. (1998) Primary biological aerosol particles: their significance, sources, sampling methods and size distribution in the atmosphere. In *Atmospheric Particles* (eds. R.M. Harrison and R.E. van Grieken).

Pai, P., Vijayaraghavan, K. and Seigneur, C. (2000) Particulate matter modeling in the Los Angeles basin using the SAQM-AERO. *J. Air Waste Manag. Assoc.* **50**, 32–42.

Pakkanen, T.A., Hillamo, R.E., Keronen, P., Maenhaut, W., Ducastel, G. and Pacyna, J.M. (1996) Sources and physico-chemical characteristics of the atmospheric aerosol in southern Norway. *Atmos. Environ.* **30**, 1391–1405.

Pandis, S.N., Harley, R.A., Cass,G.R. and Seinfeld, J.H. (1992) Secondary organic aerosol formation and transport. *Atmos. Environ.* **26A**, 2269–2282.

Position Paper on PM (Second) (2004). CAFE Working Group on Particles.

Position Paper on Particles (1998) *Ambient Air Pollution by Particulate Matter*, Technical Working Group on Particles.

Putaud, J.P., Raes, F., Van Dingenen, R., Brüggemann, E., Facchini, M.C., Decesari, S., Fuzzi, S., Gehrig, R., Huglin, C., Laj, P., Lorbeer, G., Maenhaut, W., Mihalopoulos, N., Müller, K., Querol, X., Rodriguez, S., Schneider, J., Spindler, G., ten Brink, H., Tørseth, K., Wiedensohler, A. (2004) A European aerosol phenomenology-2: chemical characteristics of particulate matter at kerbside, urban, rural and background sites in Europe. *Atmos. Environ.* **38**, 2579–2595.

Seigneur, C., Pai, P., Hopke, P. and Grosjean, D. (1999) Modeling atmospheric particulate matter. *Environ. Sci. Technol.* **331**, 80A–86A.

Seinfeld, J.H. and Pandis, S.N. (1998) *Atmospheric Chemistry and Physics, from Air Pollution to Climate Change*. John Wiley and Sons.

Sotiropoulou, R.E.P., Tagaris, E., Pilinis, C., Andronopoulos, S., Sfetsos, A. and Bartzis, J.G. (2004) The BOND project: biogenic aerosols and air quality in Athens and Marseille greater areas. *J. Geophys. Res.* **109**, D05205.

Stanier, C.O., Khlystov, A.Y. and Pandis, S.N. (2004) Ambient aerosol size distributions and number concentrations measured during the Pittsburgh Air Quality Study (PAQS). *Atmos. Environ.* **38**, 3275–3284.

Swietlicki, E., Hansson, H.C., Svantesson, B. and Asking, L. (1989) European source region identification of long range transported aerosol. In *European Source Region Identification of Long Range Transported Ambient Aerosol Based on PIXE Analysis and Related Techniques* (ed. E. Swietlicki), pp. 79–117. Thesis, Department of Nuclear Physics, Lund University, Lund, Sweden (LUTFD2/(TFKF-1015)).

United Kingdom Department of Health (2001) Committee on the Medical Effects of Air Pollutants (COMEAP). *Statement and Report on Long-Term Effects of Particles on Mortality*. Department of Health, UK.

Van der Zee, S.K., Hoek, G., Harssema, H. and Brunekreef, B. (1998) Characterization of particulate air pollution in urban and non-urban areas in the Netherlands. *Atmos. Environ.* **32**, 3717–3729.

Van Dingenen, R., Raes, F., Putaud, J.P., Baltensperger, U., Charron, A., Facchini, M.C., Decesari, S., Fuzzi, S., Gehrig, R., Hansson, H.C., Harrison, R.M., Huglin, C., Jones, A.M., Laj, P., Lorbeer, G., Maenhaut, W., Palmgren, F., Querol, X., Rodriguez, S., Schneider, J., ten Brink, H., Tunved, P., Tørseth, K., Wehner, B., Weingartner, E., Wiedensohler, A. and Wåhlin, P. (2004) A European aerosol phenomenology-1: physical characteristics of particulate matter at kerbside, urban, rural and background sites in Europe. *Atmos. Environ.* **38**, 2561–2577.

WMO (2001) *Global Atmosphere Watch Measurements Guide*. WMO No. 143. World Meteorological Organization, Geneva.

World Meteorological Organisation (2004) *The changing atmosphere. An integrated global atmospheric chemistry observation theme for the IGOS partnership*. Report GAW No. 159 (WMO TD No. 1235). URL: http://www.wmo.ch/web/arep/gaw/gawreports.html

Zappoli, S., Andracchio, A., Fuzzi, S., Facchini, M.C., Gelencser, A., Kiss, G., Krivacsy, Z., Molnar, A., Barcza, T., Meszaros, E., Hansson, H.-C., Rosman, K. and Zebuhr, Y. (1998) Organic components and chemical mass balance of fine aerosol in different areas of Europe. *J. Aerosol Sci.* **29**, S731–S732.

Zhang, Y., Pun, B., Wu, S.-Y., Vijayaraghavan, K. and Seigneur, C. (2004) Application and evaluation of two air quality models for particulate matter for a Southeastern US episode. *J. Air Waste Manage. Assoc.* **54**, 1478–1493.

3

Abatement strategies for air pollution

B. Calaminus

3.1 INTRODUCTION

At a first glance, abatement of air pollutants on the one hand and material recovery on the other hand are not directly linked. Airborne anthropogenic pollutants are undesired chemical components, particles or droplets released into the atmosphere mostly through off-gases (or also re-suspended in the case of particulate matter (PM)) which may lead to a large variety of harmful effects, for example to ecosystems or human beings through acidification, eutrophication, respiratory health impacts, corrosion, etc. Further adverse effects like global warming, increased ultraviolet (UV) radiation, etc. not only result from the "classical" pollutants, but also from greenhouse gases (GHGs like CO_2, CH_4, N_2O,…) and stratospheric ozone depleting substances like chlorofluorocarbons (CFCs). In the context of this chapter, the term "pollutant" comprises also GHG.

Material recovery in combination with abatement of air pollution may occur mainly in three different ways:

1. Direct recovery (e.g. regenerative adsorption or cold traps for VOC (solvent) off-streams).

 Waste Gas Treatment for Resource Recovery edited by Piet N.L. Lens, Christian Kennes, Pierre Le Cloirec and Marc Deshusses. ISBN: 1843391279. Published by IWA Publishing, London, UK.

2. Indirect recovery (e.g. gypsum or H_2SO_4/H_2SO_3 production during flue gas desulfurisation, metal-containing dusts from steel-making processes).
3. Resource efficiency (e.g. substitution of fossil fuel use by more efficient processes, regenerative power supply, etc.).

The essential purpose of abating pollutants from exhausts is to minimise their harmful effects. A secondary objective may be the recovery of usable material side streams which are characteristic for specific abatement techniques and can have a substantial impact on operating costs. Biosphere plant nutrition through uptake of CO_2 and nitrogen components (mainly nitrates and ammonium) is not considered as material recovery in this context.

In this chapter, typical origins of emissions to air are laid out, as well as different aspects of source–receptor relationships and strategies for abating transboundary air pollution on different (political) scales.

3.2 ORIGIN AND FATE OF EMISSIONS

3.2.1 Emission sources and transport to receptors

To a large extent and depending on the type of pollutant the emissions to air originate from chemical processes involving medium- or high-temperature reactions. These source categories are very versatile, reaching for example from small household heating installations and mobile internal combustion engines in cars, ships, etc. to large combustion plants (LCP) as big point sources. Their specific emission patterns strongly depend on the level and performance quality of the technical equipment, on the characteristics of the fuel used as well as on adequate operating practice. For instance, chemical formation of sulfur oxides, diesel soot or polychlorinated dioxins/furans (PCDD/PCDF) are directly linked to the sulfur content in fuels, to the share of aromatic soot precursors in diesel and to the presence of chlorine in incinerated waste, respectively.

"Cold" processes can also lead to the formation of air emissions. This can be microbial conversion of manure to ammonia and methane in cattle farming. Abiotic examples include mechanical abrasion of car brakes and tires setting free PM, which may then be mobilised into ambient air several times by re-suspension along street canyons. Evaporation of volatile substances such as solvents represents another important type of mobilisation mechanisms for pollutants, in other words of emission.

In contrast to typical natural emission sources like sulfur oxides from volcano activity or thermal NO_x from lightning, anthropogenic sources can be influenced or even controlled, especially when they are not accidental. Figure 3.1 structures the main types of emission sources and illustrates the fate of long-range emissions from their source through the transition/transport/deposition phase to the impact at the location of immission.

Following different emission inventories like CORINAIR (European Atmospheric Emission Inventory), combustion of fossil fuels represents the major anthropogenic source of most pollutants. Table 3.1 gives indicative figures for the main source categories.

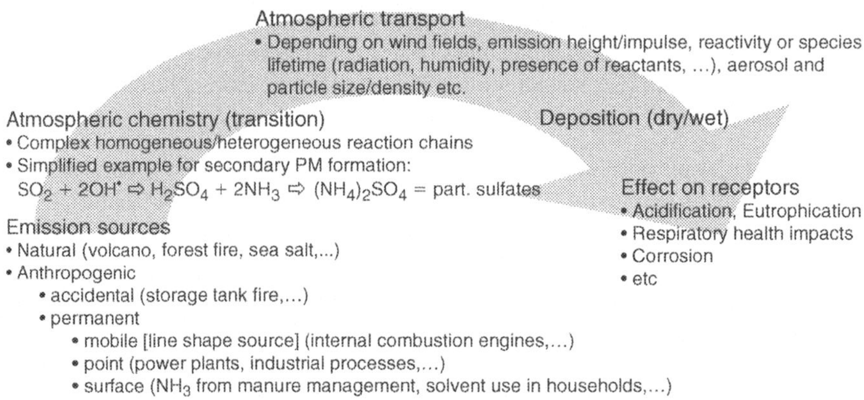

Figure 3.1 Main types of emission sources and emission behaviour.

Table 3.1 Anthropogenic emissions in Europe from the EDGAR 3.2 emission inventory, reference year 1995 (Huntrieser & Schlager, 2004).

Source	CO_2 (Pg/year)	CO (Tg/year)	NO_x (Tg/year)	SO_2 (Tg/year)	NMHC (Tg/year)
Total (fractions of global emissions)	4.7 (18%)	61 (7%)	17 (15%)	27 (19%)	20 (12%)
Fossil fuel	4.6	45	15	20	12
Biofuel	0.01	3	0.1	0	0.4
Industrial processes	0.1	7	1	7	6
Agriculture	0.007	2	0.1	0	0.3
Biomass burning	0.001	3	0.1	0	0.4
Waste handling	0	0.3	0	0	0.6

3.2.2 Transport scales and pollutant interaction

Transfer from source to receptor can be very fast and at small scale, but emissions can also be transboundary, including intercontinental (or hemispheric) up to global. A systemisation of scales is proposed in Figure 3.2, linking types of pollutants with transfer distances, as well as – on the inner graph – with typical residence times for some species. In Stohl (2004) several authors comprehensively analyse the state of knowledge on long-range transport of pollutants putting special emphasis on the comparison of model predictions with scientific pollution observations based on measurements at surface sites, in airborne devices like aircrafts or balloons, or from satellites.

By nature, the "cocktail" of various air pollutants is characterised by complex interactions in the atmosphere mainly induced by solar radiation, humidity and diffusion or convective mixing with other components/radicals, including vertical exchange between the boundary layer (ground level), troposphere and stratosphere (see also Chapter 1). Homogeneous gas–gas reactions take place as well as heterogeneous gas–liquid (droplet) or gas–solid reactions. A general overview of the most important formation pathways for air pollutants is given, for example, by Fontan (2003).

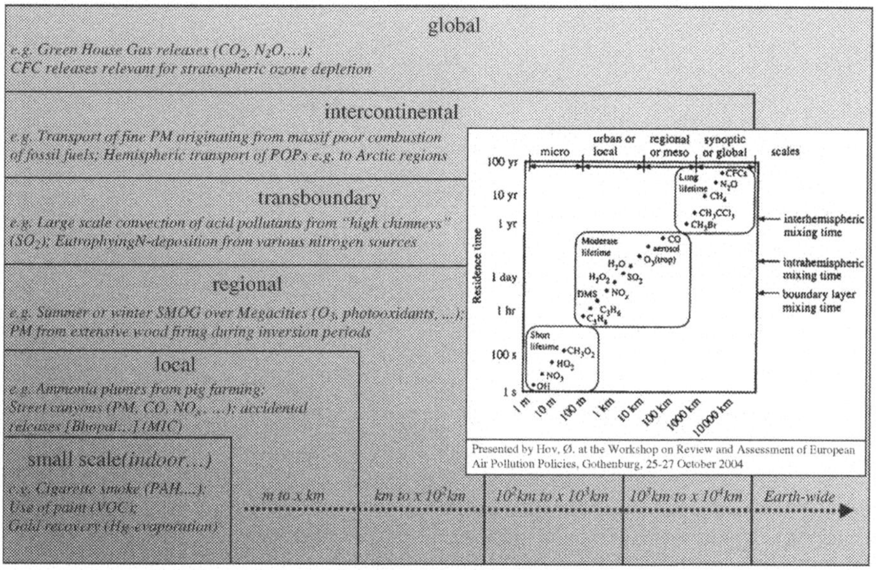

Figure 3.2 Scales of emission transfer and residence times.

Throughout the past three decades of intensified scientific work on air pollution, it became clear that on both the formation and the effect side, systems of interacting polluting components had to be studied and understood. As can be seen in Figure 3.3, nitrogen oxides not only account for the "classical" effects of acidification, eutrophication and population exposure, but also play an important role as a precursor for the formation of tropospheric ozone and secondary PM.

Deep knowledge of precursor interactions and of multiple impacts on pollution receptors requests a so-called multi-pollutants/multi-effects approach, which will be readdressed when dealing with abatement strategies. Intensive reduction of single pollutants may even lead to adverse or trade-off effects when taking into consideration the whole picture of effects. Two examples may better illustrate what is meant when talking about trade-off effects:

1. Secondary particles like sulfates and nitrates certainly are unwanted pollutants, but with regard to global warming, they tend to have an overall cooling effect due to reflection of solar radiation.
2. Huntrieser and Schlager (2004) explain, that the secondary pollutant ozone is photochemically produced by hydroxyl radical oxidation of the precursors CO, CH_4 and non-methane hydrocarbons (NMHCS) in the presence of NO_x. In the vicinity of highly industrialised areas, the O_3 production is mainly controlled by NMHC, whereas NO_x mainly controls the O_3 abundance in remote regions. *Reductions* in ozone precursors may also lead to an ozone *increase* in some very polluted regions since high NO_x levels depress photochemical activity by reducing HO_X (hydroperoxyl radicals) in the formation of HNO_3. As a consequence, there is evidence for a reduction in O_3 peak concentrations in recent years in some areas in Europe (e.g. UK) and an increase in O_3 background abundance.

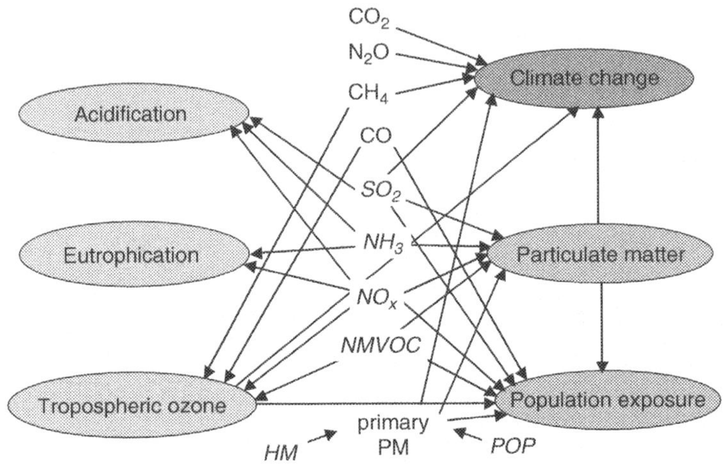

Figure 3.3 Multi-pollutants/multi-effects approach for integrated pollution abatement strategy development. NMVOC: non-methane volatile organic compounds; HM: heavy metals; PM: particulate matter; POP: persistant organic pollutants.

It is obvious that serious emission reduction politics should not be tempted to consider not to abate individual pollutants because some trade-off effect might lead to an adverse effect in another context, or to use PM emissions as a "countermeasure" to global warming. But these inter-linkages should be fully understood in order to be able to develop integrated models allowing to predict the most important effects and to derive multi-pollutant/multi-effect-based abatement strategies.

To reduce negative impacts of pollutant immissions generally only two strategies can be followed: abate emissions at their sources or take measures to decrease effects on receptors. The first is in general terms by far more efficient and intelligent but often characterised by a long response time. The second may become necessary as an intermediate solution to prevent acute irreversible damage, as is practised for example when liming fresh water lakes or forest soils to compensate acidification. Various national and international approaches and strategies to abate emissions are dealt with in the following paragraph.

3.3 ABATEMENT STRATEGIES

3.3.1 High chimneys

Certainly already the early human users of fire places learned quite quickly that proper evacuation of combustion gases from closed rooms is important. More densely populated areas requested even stacks to release toxic gases at heights above ground were sufficient dilution could be assured. High chimney policy was probably the main emission "control" measure during the entire period of industrialisation until the 1970s, when long-range source–receptor relationships became more and more evident. Due to steep industrial growth after World War II, the massive increase in emissions of air

Table 3.2 Protocols to the CLRTAP.

Title	Entered into force on	Number of parties
The 1999 Gothenburg Protocol to abate acidification, eutrophication and ground-level ozone	17 May 2005	20
The 1998 Aarhus Protocol on persistent organic pollutants (POPs)	23 Oct. 2003	25
The 1998 Aarhus Protocol on heavy metals	29 Dec. 2003	27
The 1994 Oslo Protocol on further reduction of sulfur emissions	5 Aug. 1998	26
The 1991 Geneva Protocol concerning the control of emissions of Volatile organic compounds (VOCs) or their transboundary fluxes	29 Sep. 1997	21
The 1988 Sofia Protocol concerning the control of emissions of nitrogen oxides or their transboundary fluxes	14 Feb. 1991	30
The 1985 Helsinki Protocol on the reduction of sulfur emissions or their transboundary fluxes by at least 30%	2 Sep. 1987	22
The 1984 Geneva Protocol on Long-Term Financing of the Cooperative Programme for Monitoring and Evaluation of the Long-Range Transmission of Air Pollutants in Europe (EMEP)	28 Jan. 1988	41

pollutants made acidification become the most pressing environmental problem in many European countries and in North America in the 1980s. Particularly Scandinavian countries in which acidification of lakes through import of sulfur dioxide emissions could be observed were among those who claimed for urgent international action.

3.3.2 International protocols for emission control

Coordinated international abatement strategies were at that time not only asked for between countries, but also and in particular between the political (and ideological and military) blocks of Western and Eastern Europe. So much the higher was the importance of the fact that Parties to the United Nations Economic Commission for Europe (UNECE) after a long political process were successful in signing the 1979 Convention on Long-Range Transboundary Air Pollution (CLRTAP), but it did not allow to achieve short-term improvements. As Jost describes in Economic Commission for Europe (2004), in fact, acid deposition and long-range transfer of pollutants had been the focus of scientific studies and cooperation, for example, by the Air Management Policy Group of the Organisation for Economic Co-operation and Development (OECD); but environmental policy-making in European countries did not consider the problem. Saving Scandinavian fish in Scandinavian freshwaters had been deemed too costly, and frequently high chimneys still had been the choice for solving local air pollution problems.

Undeniable forest damage linked to air pollution not only in Germany but also in Scandinavia, together with increasing "green" public awareness of these linkages made acid deposition become a major policy concern, as a result of which finally "The 1985 Helsinki Protocol on the Reduction of Sulfur Emissions or their Transboundary Fluxes by at least 30%" was born (see also Table 3.2).

The first sulfur protocol was the result of an urgent, pragmatic approach which was more technique-oriented based on reduction efficiencies obtainable by flue gas desulfurisation units on fossil-fired medium- or large-power plants. It did not yet consider the relation between ecological objectives on the receptor side and emission reduction obligations on the source side. Nor were economic consequences for individual countries taken into account, since the same flat-rate emission reductions were applied to all Parties, irrespective of cost-effectiveness.

Abatement strategies in an international context should be based on answers to questions like the following:

- Starting from strong scientific evidence: what are the environmental objectives to be achieved in which time horizon?
- Which single substances have to be considered in a multi-pollutants/multi-effects approach?
- Considering available and emerging technical and non-technical measures, and realistically assuming that overall benefits (including global warming reduction, etc.) exceed costs: what is the most cost-effective way of reducing emissions?
- What instruments or regimes (guiding, supporting and pressing) are adequate to ensure compliance of all Parties to their individual reduction obligations?

3.3.3 Multi-pollutants/multi-effects approach

The 1999 Gothenburg Protocol to Abate Acidification, Eutrophication and Ground-level Ozone (the latter with impacts on vegetation and human health) is until now the protocol for which above-outlined integrated approach was applied to the highest extent. The Convention provided the forum for intense scientific networking and for science-policy exchange as a prerequisite for integrated assessment-based negotiations, which helped and still help to overcome national egoisms.

EU countries are submitted to emission reduction obligations not only resulting from UNECE-Protocols, but also from proper EU legislation. These two supranational frameworks apply to the Parties/Members and request them to "translate" the obligations into national law or programmes, which then may directly concern the various categories like agriculture, transport, industry, power sector, households, etc. in a country. The following graph (Figure 3.4) illustrates the three frameworks UNECE, EU and National Law for the case of Germany irrespectively of the fact that of course not all national laws related to emission reduction are necessarily linked to EU or to UNECE context, but at least they should not be in contrast to.

In general terms, man starts rethinking his behaviour at the moment when diagnostics of harmful consequences of his activites become evident. Since air pollution formation and behaviour involve complex processes, since nearly all human activities are somehow related to use of energy and mobilisation of emissions and finally due to dilution, transport, "quasi-invisibility" and delayed impact response of pollution, the "polluters" rarely face direct consequences. Therefore, it is of primary importance to assure and communicate hard scientific evidence of the phenomena linked to air emissions. Reliable inventories of pollutants, source types and locations are as important as time variant source–receptor relationships including atmospheric chemistry and transport matrices. Deposition mecanisms of pollutants and their synergy or

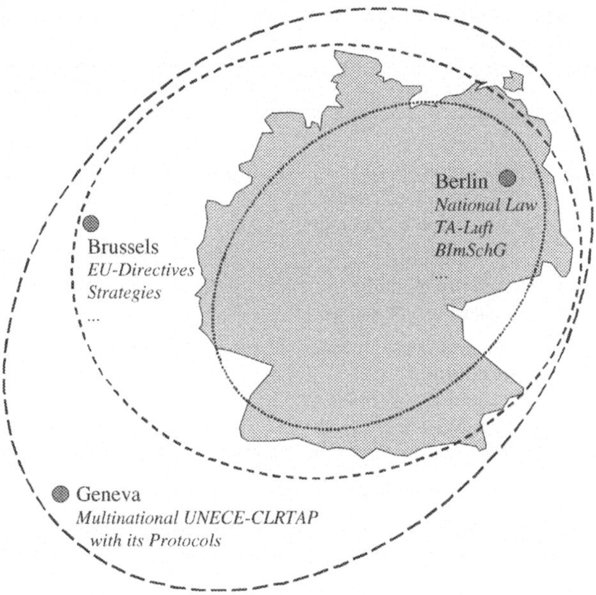

Figure 3.4 Different political frameworks of emission reduction obligations (e.g. Germany).

trade-off effects over the short or long term need to be well understood. In other words: the recognition of pollution-related problems must be consensual, and in most cases it actually is.

Once a problem is identified and countries convene to react in the framework of a convention, a simple but pragmatic approach could be to set a flat-rate emission reduction of *X*% for each Party (see e.g. the first Sulpur Protocol (1985) in Table 3.2). Much more efficient but ambitious is the integrated approach which starts from setting effect-based targets at the receptors (e.g. reduction of loss in life expectancies by *Y* months) and then calculates backward the most overall cost-efficient way to reach the targets. These results obtained through complex integrated assessment modelling (IAM) are then subject to negotiations and finally lead to settings of National Emission Ceilings (NEC), which are mandatory for each individual country. In the context of the NEC Directive 2001/81/EU, the EU-25 is requested to develop national emission reduction programmes which demonstrate how the ceiling for specific pollutants can be met.

3.3.4 Top-down and bottom-up approaches

Based on the findings of a large international scientific community, the European Commission (Clean Air For Europe – CAFE thematic strategy), the broader UNECE (Protocols) or – even more globally – the Parties to the Kyoto-Protocol have developed and adopted programmes and strategies to reduce emissions of pollutants/GHG. In a kind of top-down approach the respective objectives of these frameworks are transposed into national reduction obligations, leading finally to the application of reduction measures at source level in different sectors.

Table 3.3 Classification of technical and non-technical measures as suggested in Sternhufvud and Åström (2005).

Main type	Classification	Example in energy sector
Technological	End of pipe	Selective catalytic reduction
	Fuel quality	Low sulfur coal
	Conversion efficiency	Good practice – flaring
	Conversion technology	Combustion modifications
	Demand	Improved insulation
Miscellaneous	Land use planning	Biofuel plantations
	Time of use planning	Timer ventilation
Behavioural	Demand/consumption reduction	Lower indoor temperature
	Substitution	Change expenditure patterns
	Technology of choice	Low-energy refrigerator
	Technology use	Shorter showers

As a general, preventive request, emission or performance standards for new (and existing) installations are set at various levels: in the annexes to the UNECE-Protocols, in the EU-Directives for LCPs or Waste Incinerators, in the Integrated Pollution Prevention and Control Directive (96/61/EU), which relates to best available techniques (BAT) not exclusively but particularly addicted to "low emissions to air". Internationally harmonised obligations to impose high standards in licensing procedures for new emitting installations are an important instrument and bottom-up approach for preventive emission control.

Inherent high emission performances of processes (primary measures at the source) and classical secondary add-on techniques ("end of pipe") are not the only way of controlling pollutant emissions. Also non-technical abatement measures (NTM) gain importance for future reduction strategies and have been recently overviewed for IAM, suggesting the classification of measures (Sternhufvud & Åström 2005) shown in Table 3.3.

National governments that want (or are obliged through international treaties) to achieve reductions in air emissions have at hand a certain bundle of policy instruments to promote application of technical/non-technical measures. The most restrictive ones are based on command-and-control. They impose, for example, stringent emission limit values and performances for specific activities/plant types and sizes. Further instruments, particularly in the transport sector, can be emission-related taxes or tolls as well as incentives for accelerated change, elevated feed-in tariffs for boosting regenerative and clean power generation, or also governmental subsidies for promising research (e.g. fuel cells…). National authorities normally dispose of a set of severe sanctions to enforce compliance with emission control obligations, whereas in a multi-national (and therefore very political) context like UNECE non-compliance with mandatory obligations does not entail sanctions other than "blaming and shaming". Discussions are going on concerning the development of a system of stronger sanctions for international treaties.

3.3.5 Decision aid by integrated optimising modelling

When a large number of industrial sectors is affected by enforced environmental requirements, a cost-optimal transformation is essential for the economic and environmental efficiency both of companies and regions. Techno-economic dynamic optimisation models have been developed and implemented for various European countries (Zanini et al., 2004). The resulting cost functions for different scenarios are important for the multi-national allocation of emission reductions. Cost-discounting effects and the temporal pathway of the implementation of emission reduction options within a given planning horizon (up to 2020) can be considered. The results can be used to analyse the cost-effectiveness of environmental legislation as well as for emission projections and inventories (Geldermann et al., 2003).

Actually, to support the development of cost-optimal abatement strategies under the Convention and for CAFE, the integrated assessment model used is RAINS (Regional Air Pollution Information and Simulation Model) (Amann et al., 2005). To analyse synergies between air pollution and GHGs, the new tool GAINS has been developed by IIASA/Laxenburg additionally (Klaassen et al., 2005). Alternatively, the multi-pollutant multi-effect modelling of European air pollution control strategies can also be done with the MERLIN model (Multi-Pollutant Multi-Effect Modeling of European AiR Pollution ControL Strategies – an INtegrated Approach) (Reis et al., 2004).

Specifically dedicated to the analysis and optimisation of energy systems which are restricted by emission reduction obligations is the Program Package for Emission Reduction Strategies in Energy Use and Supply (PERSEUS) developed and operated at the Universität Karlsruhe (TH) (Rentz et al., 2003). Dynamic energy and mass flow optimisation models have been designed to assess emission reduction strategies on the national level (Geldermann et al., 2003). They are based on a detailed representation of all relevant emission processes and corresponding emission reduction options, including structural options. They provide the "cost optimal" evolution of the production system (production technologies and abatement options in place) over a given planning horizon (up to 2020) which allows for the achievement of emission reduction ceilings and the fulfilment of the demand of products or services specified exogenously on the sector level. The dynamic energy and mass flow model ARGUS (Allocation module for a computer-aided generation of environmental strategies for emissions) has been developed, too, at the Universität Karlsruhe (TH) for the elaboration of emission reduction strategies for VOC from stationary sources on a national level (Geldermann et al., 2003).

Reliable techno-economic data on emission abatement measures represent an important input to such models. The Expert Group on Techno-economic Issues (EGTEI) has been set up under CLRTAP to provide respective databases as described by Ball et al., (2005).

3.3.6 Flexible instruments

In comparison with stringent emission reduction obligations or emission limit values, a certain set of more flexible instruments has been thought out particularly in the framework of GHG reduction schemes, but not restricted to them. The project-based instruments of joint implementation (JI) and clean development mechanism (CDM) foresee the possibility of investing in cost-effective emission abatement projects in other countries in order to realise emission credits which then are deduced from own reduction obligations. Since the geographical location of emissions of GHG is irrelevant for global warming in contrast to the effects of pollutant emissions, JI and CDM practically do not apply on pollutant abatement activities.

However, emission certificates trading systems nowadays do exist for both GHG and pollutants (e.g. in the US and The Netherlands). Apparently, the European Commission is actually reflecting about the introduction of such a system for classical pollutants on community scale.

3.4 OUTLOOK

Throughout the past two decades emission reduction efforts in Europe have been quite successful for example with regard to sulfur oxides, some heavy metals and POPs. Other pollutants like nitrous oxides, ammonia and PM have now reached the top of the agenda on European level. Based on IAM calculation runs made by IIASA with RAINS in the framework of the CAFE thematic strategy, effect-related problem areas in the UNECE-region have been identified and need to be further addressed in future emission abatement efforts (Figure 3.5).

In addition, flattening of the increase in GHG emissions will remain in the focus on global scale for still a long time. A decrease in overall GHG emissions can under the actual political commitments not be expected for many decades.

Local air quality hot spots in densely populated areas like megacities are now paid major attention to, both scientifically and politically. Due to transboundary transfer of long-living pollutants or their precursors, it is necessary to consider emission

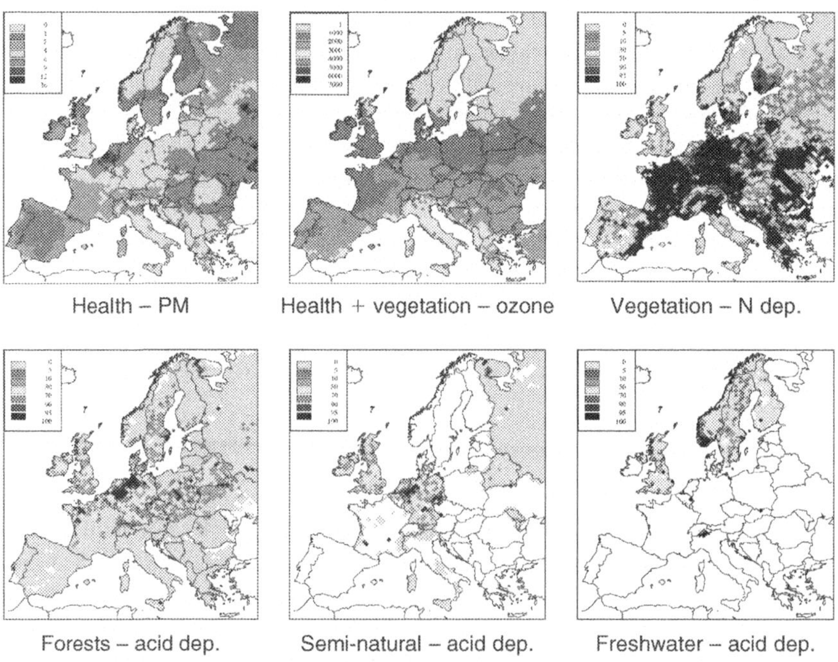

Figure 3.5 Remaining immission effect problem areas in 2020 based on IAM calculation runs made by IIASA with RAINS (lighter colour: no risk; dep.: deposition).

release trends in upwind countries or continents when local immission limits are to be set. Ozone background levels are a typical example for such a link. But also episodes of high-PM background concentrations may counterbalance the effectiveness of imposed urban quality measures.

The development and application of clean processes, high energy and resource efficiencies (BAT, emerging technologies, etc.) and of stringent emission standards should be seen as preventive and therefore – at least on the long run – the most cost-efficient approach. Europe should extend its leading role in emission abatement technologies and practises for the sake of environmental protection and economic success. Harmonisation of high-abatement technology and product standards should allow to maintain competitiveness and prevent from market distortion. When considering the full picture of direct and indirect costs induced by the release of air emissions, cost–benefit analyses clearly show that preventing and controlling air emission pays off (see also Cofala et al., 2004).

Specific emission reductions achieved per unit of activity are often balanced out by increase of total activity. Particularly with regard to the great economies in transition like China and India ways must be found to decouple economic growth and standard of living from emissions and unsustainable use of resources. Capacity building through scientific and technological exchange are in this context as important as significant improvements in public awareness.

Considering the complex multitude of different emission sources, the transboundary aspects and the large number of imaginable counter measures mostly involving costs, there certainly is not one single solution for the problem, but an optimised package of measures to prevent the environment from harmful effects of emissions is needed.

REFERENCES

Amann, M., Bertok, I., Cabala, R., Cofala, J., Heyes, C., Gyarfas, F., Klimont, Z., Schöpp, W. and Wagner, F. (2005) *A Final Set of Scenarios for the Clean Air For Europe (CAFE) Programme*. CAFE Report No. 6. International Institute for Applied Systems Analysis (IIASA). http://www.iiasa.ac.at/rains/CAFE_files/CAFE-D3.pdf

Ball, M., Calaminus, B., Kerdoncuff, P. and Rentz, O. (2005) *Techno-Economic Databases in the Context of Integrated Assessment Modelling*, pp. 1189–1193. Environmental Modelling and Software 20, Elsevier.

Cofala, J., Amann, M., Gyarfas, F., Schöpp, W., Boudri, J.C., Hordijk, L., Kroeze, C., Junfeng, L., Lin, D., Panwar, T.S. and Gupta, S. (2004) Cost-Effective Control of SO_2 Emissions in Asia. *J Environ. Manag.* **72**, 149–161.

Economic Commission for Europe (2004) *Clearing the Air – 25 Years of the Convention on Long-Range Transboundary Air Pollution*. Economic Commission for Europe, United Nations, New York and Geneva.

Fontan, J. (2003) *Les Pollutions de l'Air – Les Connaître pour les Combattre*, Vuibert.

Geldermann, J., Avci, N., Wenzel, S. and Rentz, O. (2003) *Decision support for the national implementation of emission reduction measures by the dynamic mass flow optimisation model ARGUS*. In *Operations Research Proceedings 2002* (eds. U. Leopold-Wildburger, F. Rendl and G. Wäscher), pp. 296–301. Springer Verlag, Heidelberg.

Huntrieser, H. and Schlager, H. (2004) Air pollution form and import to Europe: experimental evidence. In *Intercontinental Transport of Air Pollution* (ed. A. Stohl), *The Handbook of Environmental Chemistry* (Vol. 4: Air Pollution, Part G), pp. 69–98. Springer-Verlag, Berlin, Heidelberg.

Klaassen, G., Berglund, C. and Wagner, F. (2005) *The GAINS Model for Greenhouse Gases – Version 1.0: Carbon Dioxide* (CO_2). IIASA Interim Report IR-05-53, Laxenburg.

Reis, S., Nitter, S. and Friedrich, R. (2004) Application of genetic algorithms for the optimisation of multi-pollutant multi-effect problems. In *iEMSs International Congress: "Complexity and Integrated Resources Management"* (eds. C. Pahl-Wostl, S. Schmidt and T. Jakeman). International Environmental Modelling and Software Society, Osnabrueck.

Rentz, O., Fleury, A. and Fichtner, W. (2003) *Etude de la réduction des émissions de* NO_x *et* SO_2 *du parc thermique à flamme d'EDF*. Final Report on behalf of Electricité de France (EDF), Karlsruhe.

Sternhufvud, C. and Åström, S. (2005) *Workshop on the importance of non-technical measures for reduction in emissions of air pollutants and how to consider them in Integrated Assessment Modelling*. IVL Report B1664.

Stohl, A. (ed.) (2004) *Intercontinental Transport of Air Pollution. The Handbook of Environmental Chemistry* (Vol. 4: Air Pollution, Part G). Springer-Verlag, Berlin, Heidelberg.

Zanini, G., Monforti-Ferrario, F., Ornelli, P., Pignatelli, T., Vialetto, G., Brusacasa, G., Calori, G., Finardi, S., Radice, P. and Silibello, C. (2004) *The MINNI Project. Proceedings of the 9th International Conference on Harmonisation within Atmospheric Dispersion Modelling for Regulatory Purposes*, Garmisch-Partenkirchen, pp. 243–247.

4

Dealing with waste gas in enhancing sustainable development: the role of the Kyoto mechanisms

C.P. Jepma, W.P. van der Gaast and F. de Jong

4.1 INTRODUCTION

Sustainable development is a key concern to the global community, as it would enable countries to achieve economic growth with a lower burden on the environment, both locally/regionally and globally. The importance of this policy issue for both industrialised and developing countries has, among others, been underscored by the EU Sustainable Development Strategy 2005 (to be adopted by the EU Council in June 2006, http://europa.eu.int/comm/sustainable/sds2005-2010/index_en.htm) and at the World Summit on Sustainable Development in Johannesburg (South Africa, 2002 http://www.johannesburgsummit.org/). Sustainable development also holds a prominent place in the UN Framework Convention on Climate Change (UNFCCC, adopted in 1992 in Rio de Janeiro, Brazil) and its Kyoto Protocol (adopted in 1997 in Kyoto, Japan), which both state that industrialised countries must facilitate a transfer of sustainable technologies to developing countries.

In the context of the Kyoto Protocol, a specific mechanism to enhance sustainable technology transfers to developing countries is the clean development mechanism (CDM). Under the CDM, industrialised countries, which have quantified emission reduction or limitation commitments under the Kyoto Protocol, can achieve part of their commitments through investments in projects that reduce greenhouse gas (GHG) emissions in developing countries. An important condition for the CDM is that projects contribute to sustainable development in the countries where they are implemented (the so-called host countries). Typical examples of CDM activities are: installing a wind power plant which replaces coal-fired power production capacity; a co-generation plant using the residues from rice production ("rice husk") for power and heat production, thereby replacing fossil-fuel-based capacity; the capture of methane from a landfill site for the purpose of flaring, gas use or power production, etc. Such activities not only reduce GHG emissions, they also contribute to improving the local environment in the host countries as replacing fossil fuels or capturing methane also reduces local air pollution and improves health conditions for the population.

The CDM became operational on an interim basis in 2000 (pending the entry-into-force of the Kyoto Protocol). In 2001, at the *Seventh Conference of the Parties to the UNFCCC* (held in Marrakech, Morocco) modalities and procedures were adopted for the CDM, such as accounting rules for the calculation of GHG emission reductions achieved through projects and the monitoring of project performance (UNFCCC, 2001). This "rulebook" has now been officially adopted by the first meeting of the Parties to the Kyoto Protocol, which was held in Montreal (Canada, 2005). During its interim regime, the CDM made a rather slow start, but as per January 2006, the global pipeline of projects has grown to almost 600 activities (see Section 4.3; JIN, 2005a).

As per January 2006, around 10% of the CDM pipeline consists of projects aiming at capturing waste gas which would otherwise be released into the atmosphere. This chapter discusses how such projects contribute to sustainable development in the host countries. Section 4.2 briefly introduces the concept of sustainable development and how it has been applied in policy making, in particular under the Kyoto Protocol. Section 4.3 gives an over-view of the present CDM pipeline, including the share of waste gas management projects. Section 4.4 illustrates the concept of sustainable development through waste gas management projects by exploring a number of example projects.

4.2 DEFINING SUSTAINABLE DEVELOPMENT

Several industrialised countries are exploring ways to stimulate research and development of sustainable energy technologies and facilitate the implementation of these technologies in practice. In the context of the EU Integrated Pollution Prevention and Control Directive, standards have been set for technologies (see also http://eippcb.jrc.es/) and also individual Member States have initiated study programmes on energy transition management. For example, the Netherlands Government has established a programme on Transition towards a Sustainable Energy Future, which analyses the state of play of sustainable energy technologies and the medium- to long-term implementation perspectives of these technologies in the Netherlands. An important aspect of the Dutch analysis is the exchange of thoughts and experience among stakeholders to ensure that the analysis clearly reflects the opinions, concerns, preferences and knowledge of the persons and organisations who have a responsibility in this area of sustainable energy technology and who are affected by the technology decisions (SenterNovem, 2005). The European Commission also actively supports research and policy making on global sustainable development, especially in developing countries. In its Sixth Framework Programme, the European Commission has allocated considerable funding to research projects that aim at exploring paths towards a decoupling of economic growth and pressure on environmental conditions on a global scale. One specific example of such an activity is the ENTTRANS project on the potential of transferring and implementing sustainable energy technologies (including waste gas treatment) through the CDM. The activity objective is to explore for a number of technologies: first, what capacity, institutional and social acceptability requirements need to be met before they can achieve their full market potential; second, to what extent have these requirements been met in developing countries; third, what particular support actions would be needed to enable the countries to meet these requirements and fourth, how could the technology and financial transfer potential of the CDM contribute to sustainable energy technology diffusion to developing countries.

All these activities might suggest that a proper definition of sustainable development exists or that the CDM contains a clear structure for judging whether a project is sustainable or not. Reality, however, shows that such a clear definition in many cases does not exist. In fact, when defining the CDM contribution to sustainable development, it was decided (at the climate negotiations of November 2000 in the Hague, The Netherlands) that such a definition should be left to the host countries themselves. This decision was taken despite the concern that competition among developing countries to host CDM projects could lead to a "race to the bottom" in terms of an increasing focus on low-cost emission reduction credits and a decreasing focus on sustainable development (JIN, 2000).

Not surprisingly, the application of sustainable development definitions by CDM host countries largely differs. In fact, by far most attention is paid to the "economics" of the projects in terms of how many emission reduction credits projects achieve at what price and transaction costs (JIN, 2005b). Although some CDM host countries have formulated development criteria which projects must fulfil, these are often of a general nature and have hardly served as a bottleneck for project implementation. Several investor parties (governments and multilateral organisations) take the development context of a project into consideration mainly as "due diligence"; only the Federal Government of Belgium has explicitly formulated in its CDM tender guidelines that project proposals will be judged on their sustainable development contribution. In practice, this probably implies that sustainable development is often considered a quality rather than a key criterion and projects that score high on this quality could just receive a higher price for the emission reductions offered (JIN, 2005c).

Table 4.1 Sustainable development criteria used by the Government of South Africa.

Ecological	• Impact on local environmental quality • Change in usage of natural resources • Impacts on biodiversity and ecosystems
Economic	• Economic impacts • Appropriate technology transfer
Social	• Alignment with national provincial and local development priorities • Social equity and poverty alleviation
General project acceptability	• Alignment with policies and plans of government • Reasonability/fairness of distribution of benefits

Source: South African Designated National Authority for the CDM (2004).

Table 4.1 shows, as an example, the list of sustainable development criteria which the Government of South Africa uses in judging CDM project proposals. The generic overview of criteria in Table 4.1 can be specified by adding weights to the criteria depending on the project type and the region in which the project is located. For waste gas capture projects, such a weighting assessment has been carried out based on a literature review by the US Environmental Protection Agency (EPA, 2000) and by a research for the UK Government (Begg et al., 2002). This assessment has resulted in the following key sustainable development criteria for this project type:

- *Safety concerns*: reduction of explosion hazard at landfills and coal beds/coal mines.
- *Financial returns*:
  - selling energy from the methane capture process to grid-connected and off-grid customers.
  - reduction of compliance costs with government regulations, for example, EU landfill directive.
- *Employment*: job creation.
- *Local air/environment quality*: reduction of smog formation and local air pollution.
- *Reduction of GHG emissions*: methane emission reduction by flaring or energy substitution.

The case studies described in Section 4.4, which have been derived from activities carried out under the Kyoto Protocol, will be assessed on their contribution to sustainable development with a view to the criteria listed above.

4.3 WASTE GAS RECOVERY UNDER THE CDM

Under the CDM, an elaborate pipeline of projects is being developed. As of 17 January 2006, 589 projects are included of which 71 projects have been registered, 63 have requested registration and the remaining projects in the pipeline are undergoing validation of the project design documents (Fenhann, 2006). Registered projects have successfully passed through the CDM project procedures from inception to final

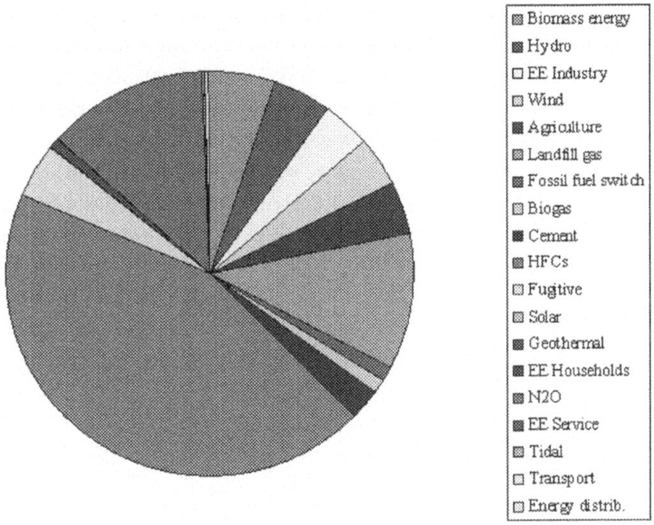

Figure 4.1 Sectoral share of CDM activities: annual emission reductions (*Source*: UNEP Risø 2006).

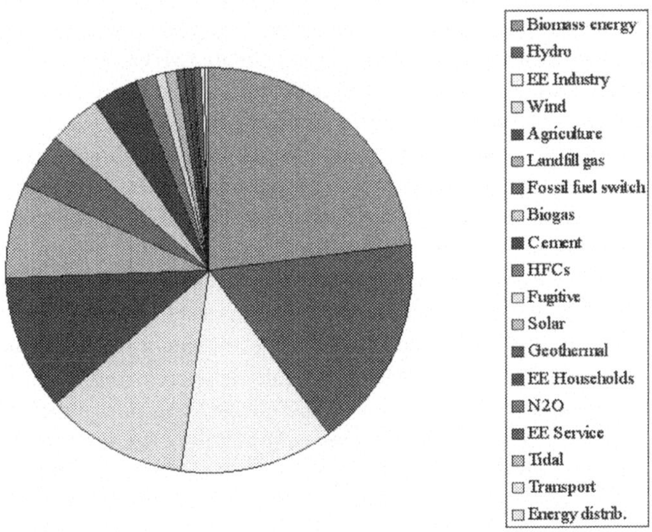

Figure 4.2 Sectoral share of CDM activities: project share (*Source*: UNEP Risø 2006).

registration with the UNFCCC CDM Executive Board (CDM EB). Methodologies for these project activities to calculate the GHG emission reductions have been approved. Figures 4.1 and 4.2 show the share of the different project categories under development in volumetric terms (annually) and in terms of the number of proposed projects per category, respectively.

Clearly, highly potent GHGs such as HFC-23 (a by-product from the production of the refrigerant HCFC-22). Its decomposition implies a substantial GHG emission reduction given its global warming potential (GWP) of 11,700 and nitrous oxide (N_2O; GWP: 310) dominate the total number of emission reductions in the current project pipeline (these are often referred to as "low hanging fruit" as little sustainable development benefits are to be expected). A more evenly distribution among project categories can be seen in Figure 4.2. Except for coal mining activities, all "gas recovery type of projects" are represented in the project pipeline and elaborated on below.

4.3.1 Biogas

Biogas is a gaseous mixture generated during the digestion process of organic matter such as human and animal excrements, rice straw or leaves under anaerobic conditions in the absence of oxygen. In nature, biogas is generated at the bottom of stagnated ponds, lakes, swamps or in the digestive system of animals. It is composed of 60% of methane (CH_4; GWP: 21) and for the remaining share of carbon dioxide (CO_2) and some other gases.

Biogas activities make up 1% of the total number of emission reduction credits that are likely to be generated from the 71 registered projects (Fenhann, 2006). The same percentage applies to the project pipeline. In number, different figures apply. Biogas type projects make up 6% and 4% of registered projects and proposed projects, respectively. In total, 24 biogas projects are being proposed with a total yearly emission reduction capacity of 1.269 Mt CO_2-eq. (or 10.3 Mt CO_2-eq. by 2012). Out of these 24 projects, four have been registered of which only one project falls under the large-scale category. Overall, the majority of these projects are developed in the category of small-scale activities, to which simplified GHG accounting procedures apply. Nevertheless, on average, over 50 kt CO_2-eq. is planned to be reduced per project per year. Figure 4.3 shows the number of projects that are either under validation or have already been registered per host country.

Biogas projects contribute significantly to sustainable development of the communities using them. In most host countries (and least developed countries in particular), especially fuel wood and charcoal and to a lesser extent kerosene and LPG are used for cooking purposes or electricity generation, whereas this is detrimental for in-house living conditions. In fact, poor indoor air quality is one of the major causes of acute respiratory infections among women and children. Besides, costs of such off-grid electricity generation are typically higher than if such communities would have a connection to the grid (e.g. in Cambodia, the average tariff for electricity is about \$0.14/kWh in Phnom Penh, and \$0.30–0.92/kWh in rural areas). Using a simple technology such as a biodigester (see Box 4.1) allows rural communities to benefit from manure management and use the resulting methane for cooking purposes and lighting, thus reducing expenditures on fuel for such purposes. Moreover, the use of biofertiliser (slurry) contributes to a high yield of crops and vegetables, potentially allowing extra income generation (the internal rate of return of a biogas plant can be 39%). Other benefits include a reduced workload for its users, in particular for women in rural areas. It has been calculated that the use of biogas can save a typical family up to 3 h a day, mainly due to the reduction in the time needed for collecting firewood.

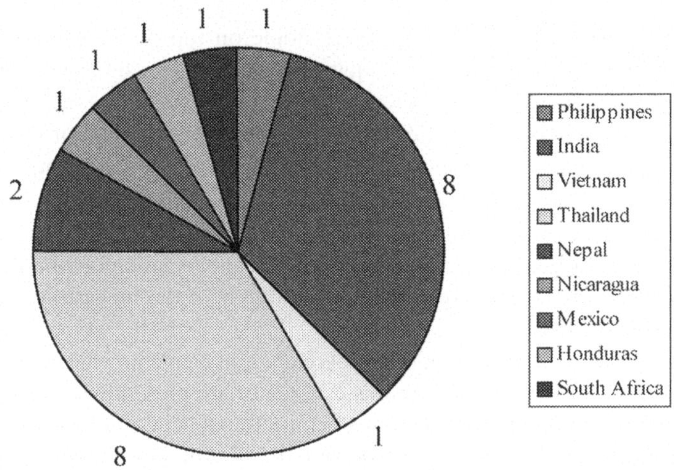

Figure 4.3 Geographical distribution of proposed biogas projects under the CDM (*Source*: UNEP Risø 2006).

Box 4.1 Formation of biogas in a fixed dome biogas plant

In the initial stage of the operation cycle, the surface of the slurry in the digester and the surface of the slurry having contact with the atmosphere at the inlet and outlet are equal and at the "zero level" (the biogas pressure in the digester is equal to 0 ($P = 0$)). The generated gas accumulating in the upper part of the digestion tank will push the slurry up into the compensation tank and the inlet pipe. Thus, the surface of the slurry in the digester lowers while the surface of the slurry in the compensation tank is rising. The height difference generates the gas pressure. The more gas is generated, the higher the pressure.

When the slurry in the compensation tank reaches the highest level, part of the digested slurry is automatically discharged through the overflow (the gas pressure reaches the maximum value ($P = P_{max}$)). When the gas is used, the slurry from the compensation tank flows back into the digester. The level of the slurry in the compensation tank lowers down; while the level of the slurry in the digester rises. As the difference between the two levels decreases, the gas pressure gradually decreases.

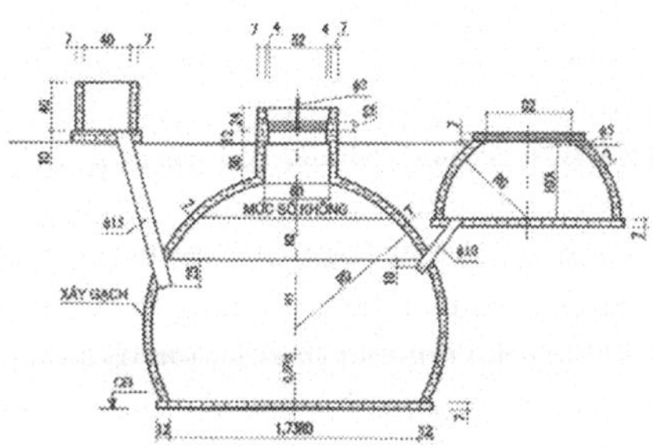

This plant's (KT.2 plant) semi-spherical dome on top of the conical bottom makes it better resistant against the upward forces typical connected with high groundwater levels. The plant is therefore particularly suitable for the Mekong delta in South Vietnam. The KT.2 plant is a further development of the TG-BP model designed by a Thai–German project in Thailand (www.biogas.org.vn).

Finally, the use of biogas significantly improves the sanitary situation of the farmyard and the environment, as manure is deposited directly in the biogas digester rather than dumped in the open.

The two registered biodigester CDM programmes in Nepal provide good examples in this regard and lessons learnt from this experience are presently being explored in Asian countries such as Lao People's Democratic Republic, Cambodia and Vietnam (e.g. in the context of the Asia Pro-eco project "EU – Asia institutional co-operation and multinational dialogues on enabling the meaningful participation of Cambodia, Lao People's Democratic Republic and Vietnam in the CDM": url: www.hwwi.org). However, a decision by the CDM EB has complicated the application of this project category for small-scale development as the reference to non-renewable biomass has been deleted in the indicative list of simplified baseline and monitoring methodologies for selected small-scale CDM project activities. Deletion of the reference means that no longer use can be made of a standardised baseline calculation (a baseline scenario needs to be constructed in order to show how one expects emission paths would have developed without implementation of the project). Given the fact that least developed countries often lack reliable data (for baseline calculations) the further development of this project activity is thus at stake (JIN, 2005d).

4.3.2 Landfill gas

Since the 1970s, technologies have been developed for the recovery of LFG emissions from landfill sites (Boyle, 2004). In developing a site, each area is covered with a layer of impervious material after it is filled, and the gas is collected by an array of interconnected perforated pipes at depths of up to 20 m in the refuse. The system operates under low pressure, which also implies that there will be no physical leakage of methane from the extraction system (although leakage at the generator is possible). Digestion in a landfill takes place over years and the (rather volatile, unpredictable) release of gases (oxygen, nitrogen, CO_2, hydrogen and methane) differs (in number and composition) over time (JIN, 2005e). Generally speaking, the site would be at maximum production after 8–10 years. The LFG extraction needs to be managed because the low pressure attracts oxygen into the landfill, which could possibly lead to explosions. An essential characteristic of LFG extraction is that both the volume of gas extracted and its methane content can vary and cannot be predicted precisely. If generators for electricity production are installed, the generating capacity has to match the expected amount of methane, but because of variations in production, the remaining gas has to be flared.

LFG account for 9% in the total number of emission reductions from registered projects. Its (volumetric) share in the project pipeline stands at 11%. In terms of number of projects, LFG activities make up 17% and 8% of registered and proposed projects,

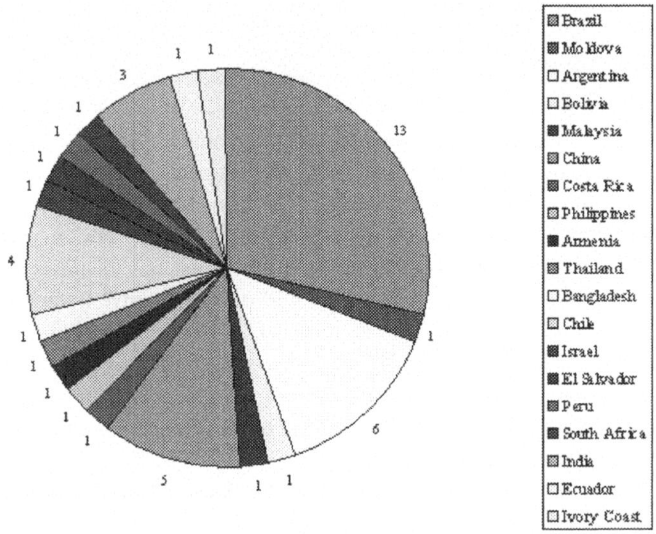

Figure 4.4 Geographical distribution of proposed LFG projects under the CDM (*Source*: UNEP Risø 2006).

respectively. In total, 45 projects have been proposed which in total aim at reducing some 12.9 Mt CO_2-eq. on a yearly basis (or 99 Mt CO_2-eq. by 2012 cumulatively). Twelve projects have been registered (of which three are small-scale activities), which on average reduce around 300 kt CO_2-eq. per project per year. Figure 4.4 reflects the geographical distribution of the proposed activities in this project category.

Given the fact that the maximum yield of methane from landfill sites is reached after about 8–10 years, its further development under the CDM highly hinges on a "post-Kyoto" regime. As the current Kyoto Protocol regime will end by the end of 2012, the highest productivity rate at the site (and thus the highest number of emission reductions) will be achieved outside the "Kyoto" regime. Out of the present pipeline (as per January 2006), about half of the projects are planned to start in 2006 and only three in 2007.

4.3.3 Fugitive emissions

Five fugitive gas capture projects are included in the present CDM project pipeline of which one project in Vietnam (Rang Dong Oil Field Associated Gas Recovery and Utilisation Project) has been registered recently (see Box 4.2). On average, over 1 Mt CO_2-eq. is reduced per project per year and a total of about 35 Mt CO_2-eq. is projected to be reduced during the period up to 2012. Other host countries include Brazil, Bangladesh and Nigeria.

Recovery of gases that would otherwise be flared is only one category for development under this type of emission reduction projects. For example, in the activities implemented jointly (AIJ) phase, which was a pilot programme under the UNFCCC aimed at experimenting with project-based emissions trading, one project (the RUSAGAS Project) was developed to seal the leaking valves at two compressor

Box 4.2 Rang dong oil field gas recovery and utilisation project (Petrovietnam 2004)

This project recovers and utilises gases produced in oil fields near the south-eastern coast of Vietnam. Prior to 2002, this by-product gas was disposed at the platform via a combustion process (flaring). During this process, waste gas was burned on site to produce a waste stream consisting largely of CO_2 but also containing uncombusted remnants of the original waste stream. The project activity itself includes construction of a gas pipeline and compressor facilities to recover and transport the by-product gas, which would otherwise have been flared. The gas recovered is processed into dry gas (mostly CH_4), as well as LPG (butane and propane), and condensate (hydrocarbon molecules containing five or more carbons). The dry gas is supplied to nearby power plants and will be sent to a local fertiliser plant, whereas LPG and condensate is consumed domestically as home cooking fuel and is used to produce gasoline.

This project contributes to the sustainable development of Vietnam in several ways. The project provides an additional source of clean-burning natural gas to support the development needs of Vietnam and contributes to the reduction of import dependency of petroleum products through provision of gas to the power plants at a significantly reduced price – about half the rate obtained through other gas fields. The elimination of flaring offshore and the substitution of gas for existing fuel in domestic electricity production result in reduced atmospheric pollution. Finally, the net volume of CO_2 emissions eliminated is approximately 6.77 Mt over the (10-year) crediting period.

stations, thereby reducing methane emissions in the natural gas transmission and distribution systems in the Russian Federation.

4.3.4 Coal mine/bed methane

Methane is stored within coal as a result of a process whereby plant material is converted to coal (JIN, 2005e). During coal mining activities, the stored methane is released and, without countermeasures, contributes substantially to the build up of GHG emissions. Consideration of this type of projects in countries such as India, China and Vietnam is – apart from the emission reduction component – particularly interesting as high concentrations of methane in mines are potentially dangerous. In China alone, explosions as a consequence of high methane content resulted in the death of over 2000 miners in 2003. Besides, a valuable source of energy is wasted, if not captured.

Traditionally, for safety reasons, the methane is released into the atmosphere through mine ventilation and degasification systems. In the case of China (as the world's largest producer of coal responsible for about one-third of worldwide methane emissions from mining activities), regulatory standards do not require coal mines to flare or use any of the vented gas. However, in 2003, China's coal mines produced and vented about 12 billion m^3 of coal-mine methane (CMM) into the atmosphere. If only 10% of the country's CMM resources could be used economically, this amount would provide a new GHG reducing energy supply equivalent to 100 years of China's current natural gas production rate (Carbon Finance, 2005; JIN, 2005e).

Apart from the methane emission reductions, other environmental concerns should be considered as well. In the process of reducing the pressure in the coal (through which methane is released), water has to be removed from the coal bed, which may be saline with detrimental effects for the surroundings of the coal bed. Besides, the production rate of methane from coal beds may be highly variable because of the heterogeneous nature of the coal beds.

Although a number of submitted CDM methodologies have been consolidated in this project category and now have a general applicability (coal-bed methane and CMM capture and use for power (electrical or motive), and heat and/or destruction by flaring; see http://cdm.unfccc.int/methodologies/PAmethodologies/approved.html), no project has been submitted for validation as of yet. However, commercial production of coal-bed methane has been undertaken in both the USA and China and projects have also been developed in Central and Eastern Europe.

4.4 KYOTO-BASED CASE STUDIES OF WASTE GAS CAPTURE PROJECTS

This section shows some examples of projects that have been set up in the context of the Kyoto Protocol, both via the CDM and joint implementation (JI). Both CDM and JI are project-based emissions trading mechanisms, with CDM enabling project co-operation between industrialised and developing countries and JI facilitating project co-operation among industrialised countries. The projects described in this section are: the Liepaja Solid Waste Management Project in Latvia, the Salvador da Bahia Landfill Gas Project in Brazil and the Jincheng Coal-Mine Methane Power Generation Project in China. These projects are illustrative since they have either completed the CDM and JI project cycles as envisaged under the Kyoto Protocol (including a comprehensive stakeholders' assessment with a view to exploring local aspects of the projects, such as reduction of air pollution, hazardous waste landfilling and economic aspects, such as financial returns and job creation) or are in an advanced stage of development.

4.4.1 Solid waste management in Latvia

Latvia became an EU Member State in May 2004, which, among others, implied that it must comply with the environmental standards included in the EU *Acquis Communautaire*, which is the set of legal documents based on EU laws and directives, which together set the standards for a broad range of topics within the EU, among others for energy and environment. All Member States must incorporate the *Acquis* standards in their domestic legislation. For Candidate Member States, the *Acquis* forms the basis for the pre-accession negotiations. In the area of waste, plans were subsequently developed to concentrate municipal solid waste on safe, well-controlled regional sites. With a contribution from the World Bank's Prototype Carbon Fund (PCF), the Liepaja Solid Waste Management Project could go one step further and install a state-of-the-art system with collection and use of generated methane. In the 1990s, the Government of Latvia had already started a programme to upgrade or close all existing disposal sites, many of which posed a risk to local groundwater resources. To support this programme institutionally, a *National Solid Waste Management Strategy* was prepared with the help of the Danish government. The strategy recommended about 10 regional waste management solutions in Latvia.

With 65,000 tonnes per year, the Liepaja region (on the Baltic Sea coast) generates about 10% of the country's annual municipal waste. It has 29 disposal sites, all of which are inappropriately located due to their geology and high water table. None of the sites has an effective natural barrier or artificial lining to protect the groundwater against percolating leachate. A large site at Skede and a medium size site at Grobina account together for 94% of the total disposed waste volume in the region. The purpose of the *Liepaja Solid Waste Management Project* is to implement a self-sustaining waste management system in that region. In addition to protecting the quality of groundwater, the project aims to use the collected methane for electricity and heat generating purposes.

In 2001, the PCF signed an agreement with Latvia about the financing of the project and the purchase of resulting GHG emission reductions. It contains requirements regarding monitoring, reporting and independent third-party verification of emission reductions and other environmental benefits achieved by the project. Finally, based on a risk assessment study, the agreement also addresses the equitable sharing of risks and benefits. The project design consists of the remediation of one existing landfill and the closing down of the other sites. In order to select the site for remediation, a profitability analysis was made that compared a number of different project design options at two sites, Skede and Grobina. This analysis indicated that the sanitary landfill at Skede (without landfill gas collection) was the most cost-effective alternative.

However, the Environmental Impact Assessment found social and environmental reasons at the Skede site that excluded it as a possible option. For instance, Skede borders at a national park which serves as the habitat for several endangered and rare species, and is adjacent to a summer colony whose inhabitants at the later stage of project preparation started to raise objections. The same option at the Grobina site was the next-best alternative and therefore proposed as regional landfill site for the Liepaja region. At the Grobina site a solid waste management and disposal centre are implemented and ten energy cells installed. These energy cells are filled with waste over a period of 5 years and the recovered gas as a result of anaerobic degradation of the waste is used to generate power, which will be delivered to the power grid (PCF, 2000).

The total project cost was estimated at €13 million, of which the PCF contributes around €4 million in return for the GHG credits (estimated at 34 ktonnes per year, which would at the current market price for Kyoto Protocol-based emission reductions of approximately €5 per tonne CO_2-eq. have a value of €170,000 per year). Apart from the GHG reduction, other benefits of the project include: facilitating the separation of recyclable material; reducing environmental problems for the neighbours of the 28 disposal sites to be closed; demonstrating how an obsolete disposal site can be turned into a sanitory one; and arresting the ongoing contamination of groundwater.

4.4.2 Salvador da Bahia landfill, Brazil

The Salvador da Bahia landfill is located about 20 km northeast of the city of Salvador in the Bahia State of Brazil. It has a total capacity of 18 million m^3, and receives about

850,000 tonnes a year of municipal solid waste. In 1999, the city launched a tender for operating the landfill over the next 20 years at a maximum price of R$ (Real) 16.69 per tonne (US$ 5.60 per tonne). The Brazilian company VEGA won the bid with a price of R$ 15.86 (US$ 5.30) per tonne – a price which includes landfill design, licensing, construction, operation and aftercare (20 years following closure).

In Brazil, where, according to the World Bank and national studies, more than 90% of the landfills operate without LFG capture and flaring, demonstration of a world standard technology (bottom lining, drainage, water management) represents a significant benefit (performance is top-20% of its category). The environmental license for the landfill states that there should be waste gas capture, though it does not give a percentage capture rate. VEGA's winning bid, on the other hand, contains LFG capture and destruction rates of between 19% and 24% over the lifetime of the landfill. A conservative estimate of 20% recovery of methane gas for passive systems (where the gas comes out as a result of internally built-up pressure) was considered best practice.

Under the CDM project the participants will improve the flare-up rate (i.e. the destruction rate by thermal oxidation) to 80% of the LFG. At some point the project is likely to also include LFG-to-energy equipment in order to produce electricity. However, the technical risk on gas quality and regular quantity availability is still experienced with LFG capture and use in tropical countries, and theoretical models for LFG estimation have not yet been calibrated for use in tropical circumstances (heat, humidity, etc.). Moreover, since the timing of the latter component is also uncertain, it has been excluded from the calculations of the project's GHG emission reductions.

The project's total cost is R$ 45 million for the period 2003–2019, including both investment and operational cost. As for the LFG-to-energy component, VEGA has calculated that the market price for this electricity should be around R$ 0.150 per kWh, while the current market price is between R$ 0.045 and 0.080. Therefore, energy from the LFG would not be competitive by itself so that under business-as-usual circumstances, without the CDM contribution, implementation of the project would have been unlikely.

The operational lifetime of the project is 17 years. The GHG emission reduction achieved by the project activity is the difference between the amount of methane actually destroyed and the amount required to be flared under the terms of the contract corrected for the actual quantity of waste received and the actual methane content of the LFG. The annual tonnage of waste considered in the concession contract varies from 790,000 in 2000 to 1,150,000 in 2019.

The methane generation rate depends on the organic fraction of the waste in this particular landfill (60%) – yielding a value of 180 m^3 per tonne waste – and the decay rate depends on the temperature and humidity of the landfill (Brazilian conditions are quite favourable to biodegradation kinetics). The methane content of the LFG is to be based on the measured content rather than an assumption. While the assumption was 50%, measurements since the start of active gas capture indicate concentrations of between 40% and 50%.

The monitoring plan is based on direct measurement of the amount of LFG captured and destroyed at the flare platform. The monitoring plan provides for continuous measurement of the quantity of waste actually received at the landfill (at the weigh-bridge) and the quantity of methane actually flared. The latter will be determined by monitoring the:

- amount of LFG collected, using a continuous flow meter and monitoring of temperature and pressure
- percentage of LFG that is methane, using a continuous methane analyser
- flare working hours, using a run time meter.

The continuous methane analyser is important because the methane content of captured LFG can vary by more than 20% during a single day due to gas capture network conditions (dilution with air at wellheads, leakage from pipes under the influence of the settling landfill, which can produce cracks, etc.). This methodology is commonly used on landfills with gas-to-energy equipment where it is necessary to have strict control of the fuel for the energy plant. The measuring equipment is sensitive, so a strong quality assurance procedure for the calibration of this equipment is needed.

4.4.3 Coal-mine methane power generation project in China

China is the world's largest coal producer and produces 65% of its energy needs by burning coal. Its coal-bed methane deposits account for an estimated one-third of the world's total, amounting to 30–35 trillion m^3. After the USA, China is the second country to have conducted large-scale field exploration of coal-bed methane.

The first coal-mine methane project in the portfolio of the World Bank's PCF under the CDM (as this project is not yet submitted to a designated operational entity under the CDM for validation, it is not included in the current pipeline of projects) aims to mitigate any negative consequences of coal mining activities by using released methane as a fuel source for a 120 MW on-site power production plant. As such, it replaces imported grid electricity and intends to export the surplus electricity to the power grid, thereby displacing fossil fuel energy sources (the share of coal for electricity generation at the Shanxi Power Grid from which electricity is purchased is about 94%). The project is located near the city of Jincheng in the Shanxi Province of the People's Republic of China and is expected to reduce emissions with about 2 Mt CO_2-eq. per year or 20 Mt CO_2-eq. over its 10-year crediting period. Tangible benefits for the local population, socio-economically dependent on coal, include improved (safety) conditions at the mine and job creation. The Shanxi province, located in northwestern China, is the centre of the country's coal mining industry. The Sihe site alone employs 27,000 people in Jincheng City. An estimated 60 new jobs for re-trained miners and additional specialist staff at the power project are projected to be created (PCF, 2005). China's National Climate Change Coordination Committee has placed coal methane projects in its top four prioritised categories in developing projects under the CDM.

In sum, the following advantages apply:

- Ensuring (much needed) safety in coal production
- Reducing GHGs and environmental pollution
- Recovering energy resource in coal
- Job creation.

As methane has a GWP of 21, methane emission reduction from coal mines is a priority. CMM is not directly suitable for injection into the local gas grid. Upgrading the gas to grid quality is technically possible but not (yet) economically feasible (even with carbon crediting). The gas can be sold to industrial users or to electricity generators, with the latter being the larger market. Industrial users use the gas to produce process heat. For most of the sites this option is not feasible because the industry has to be located very close to the extraction site. Electricity generators (or the project developer capturing the gas himself) can build a small power station (varying from 1 to 10 MW capacity, with the typical size between 1 and 5 MW) adjacent to the gas capture site. Unlike LFG production, the production of CMM is fairly stable and can match the amount needed for electricity production. If the electricity generator is down for technical reasons, the coal mine can act as a buffer, so that there is no need to flare the methane.

4.4.4 Sustainable development through waste gas reuse

In Section 4.2 it was suggested that, based on a literature assessment, the following sustainable development criteria particularly apply to investments in waste gas management projects:

- Safety concerns
- Financial return
- Employment
- Local air/environment quality
- Reduction of GHG emissions.

The description of the CDM and JI projects in Section 4.4 has illustrated that for those projects these criteria seem to have indeed been fulfilled. In particular, the aspects of safety and local environmental quality improvement have turned out to be important reasons to initiate the waste gas management projects. The case of Liepaija shows that the protection of local groundwater conditions is an important envisaged benefit from the project. In the Salvador da Bahia project a large benefit can be expected from the improvement of the local air quality: because the former landfill will be covered under the project, not only GHGs will be captured, but also locally specific pollutants, such as volatile organic compounds and sulfur dioxides. The local population could also benefit from waste gas management projects, especially CMM projects, as these projects improve safety conditions. The Chinese example of CMM capture shows how a coal mine can become self- reliant by using the methane released in the coal mining process and even export the surplus to the grid. Most importantly, safety will improve substantially with the implementation of the project and additional employment is generated. Furthermore, the living conditions in the surroundings of the mines are improved.

The benefit of waste management projects in terms of producing energy to be used as a substitute for heat and power produced by fossil fuels seems less obvious. CMM for instance is not directly suitable for injection into the local gas grid. Upgrading the gas to grid quality is technically possible but not (yet) economically feasible (even with carbon crediting). For LFG capture the technical risk on gas quality and regular quantity availability still remains to be experienced, especially when used in tropical countries, and theoretical models for LFG estimation have not yet been calibrated for use in tropical circumstances (see the Salvador da Bahia project).

4.5 CONCLUSIONS

The role of the Kyoto Protocol's flexibility mechanisms, JI and CDM, could be important in further exploring the sustainable development contribution of waste gas management projects. In several developing countries, there often seems little interest or awareness among key stakeholders of the potential of improving living conditions in the surroundings of coal mines and landfills. Although the potential for waste gas management has been widely recognised, it is likely that the financial uncertainties related to waste-management investments have hampered gas capture activities, unless they were mandatory by law.

In this context, the value of methane emission reductions under a JI or CDM project, and possibly the emission reductions related to replacing other fossil fuels when producing energy from captured methane, could make waste gas management projects financially feasible. Moreover, as the Liepaija case study has shown, this additional funding from JI and CDM emission reductions could also stimulate project developers to invest in projects that go beyond the mandatory standards set by the country or, for example, the EU. From this perspective, waste gas management projects could be considered a clear example of how the Kyoto Protocol flexibility mechanisms contribute to both GHG emission reduction and sustainable development in the host countries, in particular in areas where the methane emissions or leakage used to take place.

REFERENCES

Begg, K.G., Jackson, T., van der Horst, D., Jepma, C.J., van der Gaast, W.P., Bandsma, J., ten Hoopen, M. and Sorrell, S. (2002) *Guidance for UK Emissions Trading Projects Advice to Policy Makers, Phase 2, Technical Document.* A report for the Department of Trade and Industry (DTI), October 2002, University of Surrey, UK, University of Sussex, UK, and Foundation Joint Implementation Network, The Netherlands.

Boyle, G. (ed.) (2004) *Renewable Energy: Power for a Sustainable Future*, second edition, p. 120. Oxford University Press, Oxford.

Carbon Finance (2005) *PDD China Jincheng Coal Mine Methane Power Generation Project*, p. 3. http://carbonfinance.org/Router.cfm?Page=DocLib&CatalogID=5950

EPA (2000) *Climate Change Technologies: Landfill Methane Recovery*, US Environmental Protection Agency, Washington, DC, USA, http://yosemite.epa.gov/oar/globalwarming.nsf/UniqueKeyLookup/SHSU5BUR8R/$File/landfillmethanerecovery.pdf

Fenhann, J. (2006) *CDM Pipeline*, UNEP Risø Centre, 17 January 2006, Copenhagen.

JIN (2000) *Joint Implementation Quarterly*, 6, Special issue prepared for SB-sessions under the UNFCCC of September 2000, Lyon, France, url: http://www.jiqweb.org

JIN (2005a) *Joint Implementation Quarterly* 11(4), 4–5; url: http://jiq.wiwo.nl/jiq4-05.pdf

JIN (2005b) *Joint Implementation Quarterly* 11(4), 2–3; url: http://jiq.wiwo.nl/jiq4-05.pdf

JIN (2005c) *Joint Implementation Quarterly* 11(2), 3; 11(2), 12; url: http://jiq.wiwo.nl/jiq2-05.pdf; http://jiq.wiwo.nl/jiq3-05.pdf

JIN (2005d) *Joint Implementation Quarterly* 11(4), 13; url: http://jiq.wiwo.nl/jiq4-05.pdf

JIN (2005e) Opportunities for Coal-bed/Mining Methane Emission Reduction Projects under the CDM in Vietnam, Paper prepared for *Workshop on Making Projects CDM Viable: Technical Issues for Project Developers*, 21–22 September 2005, Hanoi, Vietnam.

PCF (2000) *Baseline Study for the Greenhouse Gas Component of the Liepaja Regional Solid Waste Management Project*, Prototype Carbon Fund, Washington, DC, USA, revision 15 May 2000, http://carbonfinance.org/Router.cfm?Page=Projport&ProjID=9613

PCF (2005) *China Jincheng CMM – PDD Package*, Prototype Carbon Fund, Washington, DC, USA, http://carbonfinance.org/Router.cfm?Page=PCF&FID=9707&ItemID=9707&ft=Projects&ProjID=9603

Petrovietnam (2004) *Rang Dong Oil Field Associated Gas Recovery and Utilization Project*, project design document, http://cdm.unfccc.int/UserManagement/FileStorage/IDIP6AY1WGT S0532GO19QSS2FXTGFP

SenterNovem (2005) *Energy Transition, Comfort, Continuity and Profit Through Sustainability*, The Hague, The Netherlands, url.http://www.senternovem.nl/energietransitie

South African Designated National Authority for the CDM (2004) *Sustainable Development criteria for the CDM*, see http://cdm.unfccc.int/DNA/view.html?CID=197

UNFCCC (2001) The Marrakech Accords and The Marrakech Declaration, *7th Session of the Conference of the Parties* held in Marrakech, Morocco, November 2001, url: http://unfccc.int/cop7/accords_draft.pdf

US Geological Survey (1997) *Energy Resource Surveys Program*, USGS Fact Sheet FS-019-97, url: http://energy.usgs.gov/factsheets/Coalbed/coalmeth.html

5

Implementation of the European Solvent Directive (1999/13/EC)

J. Geldermann

5.1 INTRODUCTION

Photochemical air pollution represents a major challenge to maintaining clean air in most industrialized and densely populated regions. Photooxidants such as ozone are not directly emitted into the atmosphere, but formed under the influence of sun irradiation out of nitrogen oxides (NO_x) and volatile organic compounds (VOC). VOC are emitted from a wide number of sources, including transportation, solvent use and vegetation. Solvent use includes activities such as coating, surface cleaning, dry cleaning, pharmaceuticals production, etc. (Ebel et al., 1997). These activities are also carried out in small enterprises, and their contribution to the total emissions is considerable.

The European Union (EU) Solvent Directive (Council Directive 1999/13/EC on the limitation of emissions of VOC due to the use of organic solvents) is one of the measures adopted by the EU to improve ambient air quality in Europe. It aims to reduce VOC emissions arising from the use of solvents. Articles 7 and 8.5 of the EU Solvent Directive indicates that the Commission shall ensure an exchange of information between Member States and the activities concerned on the use of organic substances and their potential substitutes. The European Commission proposed for the first time Internet technology for the dissemination of the Solvent Directive (1999/13/EC), in order to display available information for the numerous affected sectors. Moreover, the possibility of an information exchange between the Commission, the Member States and approximately 400,000 companies concerned by the Directive was created.

In this chapter, the requirements for the content of an information exchange network and its implementation are described in Section 5.2. For the sector of vehicle refinishing, decision support for choosing an appropriate strategy to comply with the EU Solvent Directive is presented in Section 5.3. The use of mass and energy flow models for the generation of practical guidelines is described out in Section 5.4.

5.2 INFORMATION EXCHANGE PLATFORM AS A PART OF THE IMPLEMENTATION OF THE EU SOLVENT DIRECTIVE

The prerequisite for informed decision-making is obviously the availability of relevant information. Especially in the European Environmental Policy, the accessibility of information about the legal requirements and responsibilities on European, national and regional level, but also about technical possibilities on how to meet the requirements, is a challenge. While the administration in the European Member States can handle documents in the main European languages, the affected companies (and mainly the small and medium enterprises (SME)) encounter difficulties when using multilingual information required for their decision-making on how to comply with the Directive's requirements. Another problem aggravated by the multilingualism is the constant update of the guiding documents. Here, content management systems can offer suitable support, as first experiences with the implementation of an Internet information exchange platform show (Rentz et al., 2001b).

5.2.1 Development of the internet-based information exchange platform

The European Commission has to ensure an exchange of information between Member States and affected companies regarding the EU Solvent Directive. This directive lays down emission limit values for 20 different categories of installations using solvents. The sectors are coating use (vehicle refinishing, coil coating, other coating, wood coating, leather coating, coating of winding wire and vehicle coating), printing and adhesive use (heatset printing, publication gravure printing, other printing, wood/plastic lamination, adhesive coating) and others (surface cleaning, dry cleaning, wood impregnation, coating manufacture, rubber conversion, vegetable oil extraction, pharmaceutical manufacture). The limit values are determined depending on the production processes and the installation capacity. In general, companies must fulfill the requirements set by the EU Solvent Directive from 2001 (new installations) and 2005 (existing installations). Furthermore, each Member State was required to incorporate the requirements of the EU Solvent Directive into national laws by April 2001.

In practice, more than 400,000 enterprises in the EU are affected by the EU Solvent Directive. In contrast with large companies, SME are generally not well informed about the measures which could aid them in reducing their costs and emissions, primarily because of their limited personal base and cash flow. Consequently, the diffusion of environmentally sound products and technologies is often insufficient. Therefore, the high flexibility of the EU Solvent Directive, due to the enormous range of possible reduction strategies and different national requirements, necessitates information about the process of compliance with the Directive for the encompassed sectors. Consequently, the technical requirements for an information exchange have been defined in close co-operation with industry and the EU. It has been implemented (http://www.voc-infoex.uni-karlsruhe.de, cf. Rentz et al., 2001b) in order to:

- provide generic guidance applicable to all sectors encompassed by the Directive in various languages (cf. Figure 5.1)
- ensure that the flexibility allowed by the Directive is clearly indicated

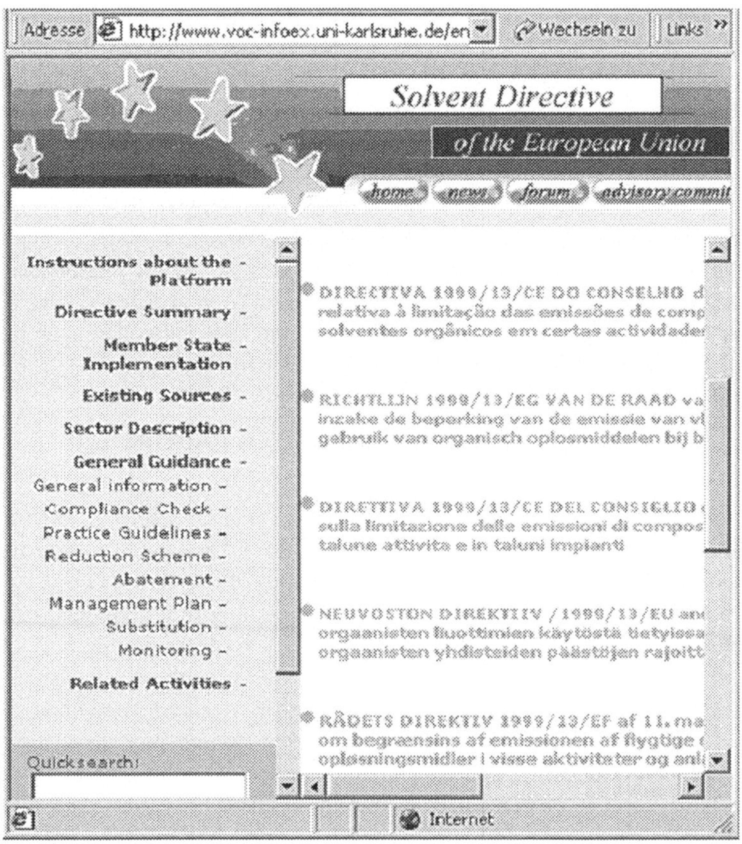

Figure 5.1 Screenshot of the information exchange platform on the European Solvent Directive (1999/13/EC).

- establish the Internet exchange forum on more technologically advanced but easier to use web sites, which are accessible to all users regardless of their Internet capabilities
- allow a decentralized structure, where documents already developed can be published directly by the user
- provide a flexible content and maintenance structure.

In order to allow permanent modifications to the web pages, the information exchange network has been realized with a content management system. Content management systems have been developed in particular as publishing systems for the purpose of maintaining and storing the contents of larger projects and are now also in use for knowledge management. The selected content management system "RedDot Solutions" offers easy design, user authorizations and production of templates (http://www.reddotsolutions.com). The decentralized structure of the information exchange network is realized by the database "MySQL" (http://www.mysql.com/) and applying PHP-scripting language (http://www.php.net). Figure 5.2 shows the implementation of the platform. The documents provided by the users can be published on a defined date.

A content manager has several tasks for ensuring the reliability of the information exchange platform. Drawing the line between technical information and unfair advertisement can be a delicate issue, but generally, it can be observed that suppliers provide helpful data about their products and services. The content manager also moderates discussion forums by varying the theme boards and verifying the comments, and he/she is the contact person for any questions.

5.2.2 Conclusions from the implementation of the information exchange platform

The provision of information about the requirements of the Solvent Directive, the dissemination of existing guidelines, tools and case studies based on mass and energy

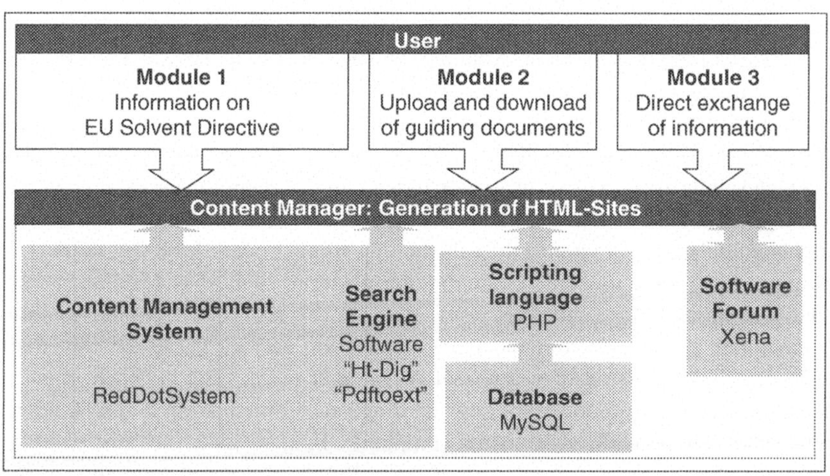

Figure 5.2 Overview of the modules of the information exchange network.

flow management and the possibility of discussion on alternative reduction strategies is made available on thousands of facilities for an enormous range of possible reduction strategies and thus enables informed decisions on the best way of complying with the provisions of the Solvent Directive. This provides a very simple way of exploring problem formulation for the different stakeholders. It became apparent that two different levels of access are needed:

- *Public area*: All users can download and upload public documents and can join a moderated discussion forum for questions, comments and statements.
- *Restricted area*: This area offers a secure space for example for an advisory committee of the European Commission (as the responsible departments of the national environmental agencies in the Member States). Documents (e.g. agendas, notes, etc.) can be exchanged between selected members of the advisory committee.

Some open questions for practical reasons remain, however, for the future. The information exchange depends crucially on keeping the documents and guidelines provided by the platform up-to-date. Also, cross-references to other Internet portals (e.g. http://www.cleaner-production.de) should be included with appropriate links. These would be tasks of the content manager, who also acts as moderator of the discussion forum. It proved important, that he/she ensures the reliability of the information exchange platform, not only during the project phase, but especially afterwards. The content manager is responsible for the information provided on the web sites, and must handle open questions, delete outdated comments and avoid unfair advertisement or other misuse of the Internet platform.

5.3 DECISION SUPPORT THROUGH MASS AND ENERGY FLOW MANAGEMENT IN THE SECTOR OF VEHICLE REFINISHING

In the past few decades, major advances in environmental protection within the painting industry have been made. Large-scale painting of automobiles has been at the forefront of this development. After a period with focus on downstream exhaust air treatment technologies, to reduce the emission of pollutants to the environment, the focus now turns to process integrated measures. Apart from improving the coating efficiency with new application methods, the reduction of the solvent content in paints has a high significance. As a result, water-based paints, except for the clear coat, have become the standard in automobile coating in Germany (Rentz et al., 2003a). Because of the manual labor character of the vehicle refinishing process, however, there is less potential for technical and environmental optimization.

Thus, in spite of this technological progress, approximately 50% of industrial solvent emissions still come from the paint application sector. These emissions contribute significantly to the development of summer smog. The advances made in reducing emissions for plants requiring licensing have unfortunately had no influence on the environmental efforts of smaller companies. In many cases the same paints are still being used as in previous years. Solvent reduced painting systems, such as high solid paints, water-based coating and powder coating have not been able to achieve acceptance, neither have innovative application technologies. The principal arguments against a conversion to these ecologically more favorable alternatives were related to cost and quality.

For this reason, the EU Solvent Directive was brought to life, aiming to significantly reduce industrial solvent emissions. Companies are given the possibility to apply management approaches that appreciably reduce their individual solvent emissions. Up until this point, however, instruments enabling smaller companies to determine their solvent emissions and to simultaneously develop process improvement potentials, while keeping costs in mind, have been missing. Especially in the trade sector, a definite inhibition to change is prevalent, largely due to the belief that environmental protection automatically means higher costs.

Using the mass and energy flow management approach, cost structures and environmental benefits can be made transparent to the entrepreneur. The computer-based mass and energy flow model IMPROVE can be used as a detailed business consultancy tool (Blümel, 2001; Rentz et al., 2001a). Based upon this, practical guidelines are then developed for easy orientation and activity planning. They can be used by companies to help them to fulfill environmental legislation and to display the benefits that can be achieved by various emission reduction measures.

5.3.1 VOC emission reduction measures in the vehicle refinishing sector

Among various other paint application sectors, body shops for vehicle refinishing are also affected by the EU Solvent Directive. Vehicle refinishing is defined as all repair, conservation or decoration outside of manufacturing installations and covers all repainting of vehicles. Original coating of vehicles with refinishing-type materials is also covered, where either the activities are conducted outside of the original manufacturing line, or the vehicle is a trailer, or solvent use is less than 15 tons per year. In Germany, about 12,500 vehicle refinishing body shops exist, of which 65% have less than three employees (Blümel, 2001). The situation in the roughly 90,000 body shops in Europe is comparable (Geldermann, 2006). Compliance with the requirements set by the environmental legislation causes significant difficulties especially for these micro enterprises. Although proven products and techniques for solvent reduction in vehicle refinishing exist, their market diffusion is still low. Due to problems with quality encountered in the early 1990's, strong reservations against the use of water-based coatings arose in the sector. Up to this point, the share of water-based coating systems in Germany is still low, only 2% of all coatings used in vehicle refinishing were water based in 1999. For example, in the Netherlands, a country with similar climatic and economic conditions to Germany, a larger diffusion of water-based coating materials has already been achieved. Under the existing Environmental Dangerous Substances law in the Netherlands, VOC content limits for a number of products have been negotiated between the paint suppliers' association (VVVF) and the users' association (FOCWA) (Ritchie et al., 2000). Thus, water-based coatings quickly reached a market share of about 70%.

In recent years, however, the successful use of new products for the coating of new cars (Rentz et al., 2003a) and the ongoing research of the paint producers helped to reduce these reservations. Another obstacle remaining is the notion that solvent-reduced coating materials lead to high investments and increased material costs. Therefore, the emission reduction targets set by the EU Solvent Directive for the sector of vehicle refinishing appear difficult to achieve.

The most significant reduction of VOC emissions in vehicle refinishing can be accomplished by using water-based coating systems. However, they are not free of organic solvents, and the generally available water-borne base coats must often be used in conjunction with solvent-based primers and/or isocyanate/solvent clear coats, since water-borne clear coats are still under research. Also, the solvents used for cleaning processes, such as solvent hand cleaners, should be reduced.

It should be noted, however, that the potential for technical and environmental optimization is limited because of the manual labor character of the vehicle refinishing process. In comparison, the car manufacturers with their large production sites for serial coating have more technical options for VOC emission reduction. An example is the use of high solid coatings in combination with thermal incineration (Rentz et al., 2003a), which is neither technically nor economically feasible for vehicle refinishing.

Consequently, the consequences of product substitution and process modifications due to the introduction of emission reduction techniques need not be quantified. Therefore, a practical tool for mass and energy flow management should investigate the overall changes in emissions and costs for individual body shops, based on an integrated analysis of all processes from surface cleaning to final finishing and polishing, comprising all materials with VOC content (such as gunwash, pre-cleaner, stopper and body filler, wash primer, pre-coat, primer, primer surfacer, surfacer, filler, topcoat, basecoat, clear coat and several other special coatings).

5.3.2 Development of the process model

In order to quantify the changes in paint consumption and operational costs by use of new materials and techniques for vehicle refinishing, the model IMPROVE has been developed (Blümel, 2001). A large part of the basic data (e.g. consumption figures related to area, components and vehicles, structure of orders, working times and capacity data of the technologies, such as thermal and electrical capacity) was collected in two reference firms in the region Baden-Wuerttemberg in Southern Germany. For an easy customization, a model firm has been defined in close co-operation with branch experts from the paint and equipment manufacturers and from professional associations. This model firm is representative for the majority (about 60–70%) of vehicle refinishing firms in Europe regarding the working structure, spraying processes and corresponding technological equipment.

A basic flow sheet (Figure 5.3) helps to identify the data structures and to investigate the cost structures. Basically, the typical scope of work is classified by the four repair levels, depending on the need for refinishing:

(1) New panel (filler, basecoat and clear coat must be applied).
(2) Color coat (only base- and clear-coat application is necessary).
(3) Body filler <50% (the complete coating must be built up for a surface <50% of the whole part; stopper may be necessary).

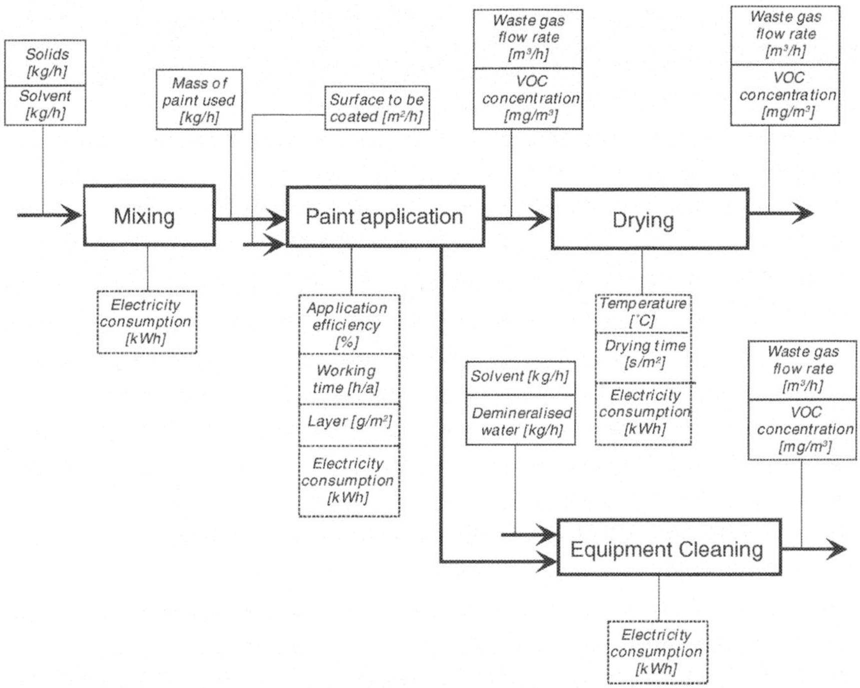

Figure 5.3 Basic flow sheet of vehicle refinishing with cost relevant parameters (Rentz et al., 1999).

(4) Body filler >50% (the complete coating must be built up for a surface >50% of the whole part, stopper may be necessary).

Altogether, 15 different spraying processes, each with a maximum of 55 stages, from receiving the car to handing it over to the customer and a large variety of products and technologies for each stage, were identified and modeled in IMPROVE.

5.3.3 Development of the software model IMPROVE

The model IMPROVE uses the commercial LCA (life cycle assessment) software UMBERTO, which is based on the so-called material flow networks (cf. (Möller & Rolf, 1995); www.umberto.de). The formal basis comes from Petri networks, a special type of network from theoretical informatics whose strict systematics not only allows the set up of complex systems but also a combined material and inventory calculation (Girault & Valk, 2003; Reinecke, 1997; Rosenstengel & Wienand, 1991).

Petri networks can model material flow networks with closed loops, like compressed air in vehicle refinishing, which is used at several process steps. In an LCA, this may lead to problems in the calculation routine according to where information about these contrary material flows are first available. If in a process a product is required which is produced downstream, the system cannot be calculated sequentially against the direction of the material flow since the information for which product quantity this production branch is to be calculated is lacking. (http://www.umberto.de/english/wissen).

Also, other LCA software tools (such as GaBi, Audit and SimaPro) would allow the modeling of production processes in the sector of vehicle refinishing, but the graphical representation and the additional Sankey diagrams are seen as an advantage of the UMBERTO software tool. They display energy flows and help to easily identify where the flows origin and where could be the place for changes with maximum impact in the investigated processes.

It should also be noted that simulation tools for production planning would also fulfill the requirements for the analysis of the considered processes. However, our experience and opinion is that flow sheeting tools such as AspenPlus, Diva and SpeedUp, which are initially designed for chemical engineering, would normally be considered too expensive and over designed for use by SMEs, which often lack the resources to fully utilize the potentials of such tools. Basically, UMBERTO shows the same advantages as standard software for object oriented, graphic and integrated modeling, which improves user acceptance and functionality (such as the process modeling software "EmPlant", www.emplant.de).

5.3.4 Techno-economic evaluation of the material flow network

Based on a graphical model and its mathematical specification, the mass and energy balances can be calculated. Moreover, a complete life cycle analysis can be performed by UMBERTO, by multiplying the input and output flows by impact assessment factors, which are listed in the LCA databases (Guinée et al., 2002).

5.3.4.1 *Material flows*

It is well known that the main environmental concern with vehicle refinishing is the reduction of VOC emissions. Therefore, the transformation of the VOC emissions into the impact category "photochemical oxidant formation potential" (POCP) would not reveal further insights for process improvement. Especially in light of the fact that VOC are treated as a sum parameter and then multiplied by a single linear parameter, which cannot take the different background exposures into account. The latter are for example simultaneous NO_x emission and solar irradiation, which influence different degrees of photooxidant formation (Owen, 1996).

POCP is one of the relative reactivity scales used to measure the potential of a VOC to participate in atmospheric reactions leading to episodes of photochemical ozone. In Europe, the European Commission along with Member State experts and industry, have examined the reactivity concept (as embodied by POCP) as an option for ozone control. It was concluded that although the concept has merits, the state of the science was insufficiently developed to make legislative use of the concept. For these reasons, the Solvent Emissions Directive uses an overall mass reduction of VOCs as a means of achieving VOC emission reductions (ESIG, 2002).

5.3.4.2 *Energy flows*

It should be noted that energy consumption was only calculated based on its demand for the production of compressed air. The analysis focused on the material consumption in the process of vehicle refinishing. Good housekeeping and energy saving measures may include options like building tune-ups (such as calibrating thermostats or lighting upgrades), load reductions (such as insulating roofs and windows, or reducing drafts), fan system and heating/cooling system upgrades, purchase of energy efficient electrical products, shutting off lights when not in use and the use of skylights as natural lighting. Such measures have not been investigated because they do not vary significantly when changing the coating materials.

An overview on energy efficiency issues can be found in Environment Agency (2001) and its "Energy Efficiency Best Practice Programme" (EEBPP, http://www.energy-efficiency.gov.uk). Although this document is oriented towards industrial sectors, general recommendations also apply for SME.

5.3.4.3 *Economic analysis*

For a comprehensive economic and ecological analysis of the mass flows, IMPROVE offers specifically-designed Excel Spreadsheets. This way individual consumption quantities are listed according to special product groups, such as the quantity of waste which has to be particularly monitored. Furthermore, these quantities can be multiplied with specific material prices, which are based on price lists from manufacturers, supplier invoices and billing rates. The wage costs are calculated as a product of the calculated working times and average wage cost rates.

For IMPROVE, the normal prices in the branch were determined for approximately 150 materials, taking into consideration the usual purchasing quantities, size of packaging and unit. One of the project partners, the paint producer Spies Hecker, supplied these prices. Winder and Turner (1992) report about approximately 300 chemical products which were in use among 46 body shops in Australia.

5.3.5 Model customization

The model of the spraying processes in IMPROVE is centrally controlled by approximately 200 parameters. They include aspects such as material consumption, working time, performance characteristics, machine running times and paint mixture ratios. Nevertheless, for modeling a specific body shop only a few parameters need to be adjusted, involving around 20–60 parameters depending on the differences in used techniques and products. In this way, the effort for data collection is significantly reduced. Practical experiences were gained in cooperation with trade associations as well as paint and equipment suppliers, which applied the model to various body shops. Thus with IMPROVE, annual emission and cost reductions for individual body shops can be ascertained in absolute terms (consumption and monetary values), while considering company specific factors. This is, to our knowledge, a first in Germany and even in Europe.

5.4 FURTHER USE OF MASS AND ENERGY FLOW MODELS

Based on case studies data generated by IMPROVE, easy to use diagrams have been developed for an approximate judgment of the benefits or consequences of certain pollution control options for a specific firm. Figure 5.4 illustrates the economic consequences of an investment for a body shop. The amortization (payback period) for automatic spray gun cleaning devices is determined depending on the number of painted parts per year. Such information was not previously available in open literature because there were no such investigations for micro-enterprises in the vehicle refinishing sector. Moreover, the credibility of IMPROVE's results should be appreciated higher than marketing information provided by paint and equipment suppliers, as a survey on participants of the accompanying workshops for the conducted research projects reveals.

5.4.1 Development of a practical guideline for the introduction and use of water-based coatings in small- and medium-sized car refinishing body shops

In order to increase the share of water-based paints in the vehicle refinishing sector and thus reach the emission limit values set by the EU Solvent Directive, the Ministry for Environment and Traffic Baden-Wuerttemberg (Stuttgart, Germany) initiated a project on the development of a practical guideline for an improved use of water-based coating systems and other solvent-reduced coating and auxiliary products. The objective was a direct support of the vehicle refinishing companies.

Therefore, in cooperation with trade associations, nine body shops were chosen with regard to their representation of the sector, taking into account issues such as size, product use from different producers, equipment and organization. The body shops were modeled by IMPROVE and monitored during their conversion from conventional to water-borne coating materials. It became obvious that the total costs for material consumption per order decreased, because of the higher opacity of the water-borne materials. This result was convincing because the statements from the paint suppliers and the first indications from the observations were confirmed by the

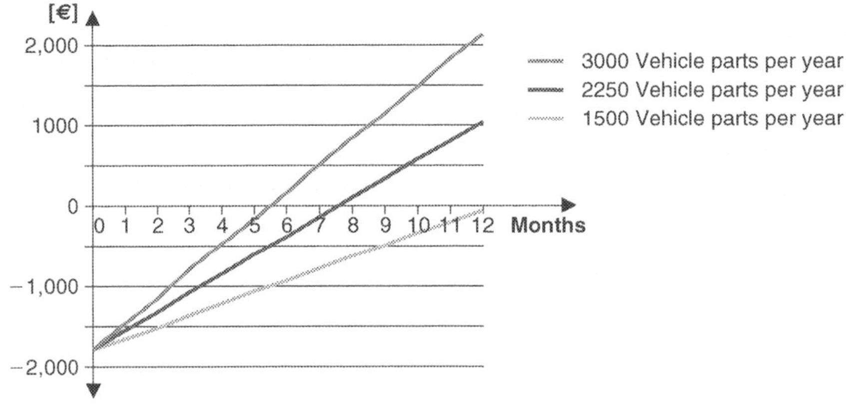

Figure 5.4 Diagram for simple evaluations: determination of the amortization for automatic cleaning devices for spray guns.

IMPROVE flow model. It should be noted that the model's results are to a larger extent accepted as factual information, rather than information offered by paint and equipment suppliers.

Additional measurements were performed in order to control the layer thickness and thus the resulting drying time (Müller, 2000). Without appropriate instruction of the workers, the average layer thickness was more than 80–300% above the target interval of 10–25 μm layer thickness. As a consequence, the energy demand for the drying process increases considerably.

Moreover, interviews and tests after the introduction of water-based coating systems in the nine body shops permitted the evaluation of pros and cons as well as other difficulties caused by the recently introduced system. Based on these broad experiences, the elaborated guideline identifies critical points, suggests solutions for organizational issues, details comprehensible check-lists and supplies contact data for responsible individuals in case of difficulties.

According to our experience, this dissemination of the results of mass and energy flow models in more general terms is an important step towards a more environmentally friendly production especially in SME. General guidelines for environmental protection are broadly available, but it is essential that the advice given are geared to the specific needs of the respective industrial branch. It is of foremost importance that the technical requirements are described in sufficient detail, as well.

5.4.2 Solvent management plan according to the EU Solvent Directive (1999/13/EC)

According to the transposition of the EU Solvent Directive into German law, every vehicle refinishing body shop must establish a solvent management plan (SMP), unless water-based coating systems are utilized. The intent has been to realize defined emission reduction targets and initiate VOC reduction schemes in Europe. The compliance of individual firms is surveyed by an authorized person each year. Each vehicle refinishing shop that establishes a SMP must calculate the input and output categories defined in the EU Solvent Directive (Table 5.1).

Table 5.1 Input and output categories according to Annex III of the EU Solvent Directive.

Input	Output
Quantity of organic solvents used as input into the process/activity (including cleaning solvents) (I1)	Emissions in waste gases (O1)
Input of recovered or reused organic solvent input into the process/activity (I2)	Organic solvents lost in water (O2)
	Organic solvents remaining in products (O3)
	Uncaptured emissions of organic solvents to air (O4)
	Organic solvents/organic compounds lost due to chemical or physical reactions, for example by incineration or waste water treatments (O5)
	Organic solvents contained in collected waste (O6)
	Organic solvents for sale (O7)
	Storage of recovered organic solvents (O8)
	Organic solvents released in other ways (O9)

Practical aids have been developed in order to support vehicle refinishing body shops in establishing solvent balances (Rentz et al., 2001a). Based on the data contained in the IMPROVE model, product information from paint producers has been analyzed and revised. The calculation program developed for drawing up the solvent balance is based on the common software package MS-Excel. In addition, a hard copy version is being offered to firms for manual use. Furthermore, if a body shop is already modeled by IMPROVE, the calculated mass flows can directly be taken as input data for the establishment of the SMP.

5.4.3 Linking IMPROVE with time management

Since paint application in vehicle refinishing is a highly manual process, human resource allocation is essential (Geldermann et al., 2002; Rentz et al., 2003b; Schollenberger et al., 2003). Thus, beside material and energy consumption figures, process time is the most important parameter for the description of the refinishing process. Because of the influences of process modifications on the workflow and process duration, time and motion studies have been carried out on behalf of the German Association of the Automotive Industry (VDA), Working Group for Coating, with the participation of Audi, BMW, Ford, Opel, Porsche and VW. IMPROVE was used for the calculation of transfer prices between body shops and the car manufacturer, especially for operations within warranty. For the calculation of transfer prices and the value of the repair work itself, only vehicle-related process steps must be considered. Maintenance and other operations necessary for the functioning of the installations are already considered in the calculation of the labor costs as an hourly rate (Hennemann, 1994).

Since the most important aspect was the depiction of the variety of working conditions and methods, 28 body shops have been analyzed in several European countries. Time and motion studies based on REFA-methodology (REFA, 1992) have been carried out for 1 week in each body shop. In this way volume fluctuations and biorhythm aspects were covered. Time measurements were carried out with a computer program written and used by fully qualified REFA technicians.

The IMPROVE process model has been extended to a hierarchical network, offering parallel consideration of mass and energy analysis as well as time and motion studies. Additionally, the results of this model offer the possibility of creating environmental figures, with time as the reference variable.

5.5 CONCLUSION

Because of the high flexibility of the EU Solvent Directive and the enormous range of possible reduction strategies, an information exchange between industries, national member state representatives and the EU is needed. In particular SME are overstrained by the generation of cost-effective emission reduction strategies. The provision of information about the requirements of the Solvent Directive, the dissemination of existing guidelines, tools and case studies based on mass and energy flow management and the possibility of discussion about the alternative reduction strategies supports the implementation of the Solvent Directive in a cost-effective way. The content maintenance of the Internet-based network can be realized by a content management system and a decentralized dynamic network structure.

Research on mass and energy flow management demonstrates that these models cannot only support larger companies in their environmental management, but also SMEs. In addition to the ecological and economic aspects, organizational issues can be investigated with a model like IMPROVE. By graphically illustrating the operation and technological environment, as well as supplying a complete view of shop internal material flows, a dissemination of economic and emission-related consequences of using water-based paint and solvent free application devices is feasible. Thereby possible environmental and economic benefits can be identified, even for micro-enterprises. Conventional facility site assessments are untypical for these kinds of companies, therefore making an approach geared to the needs of this sector crucial.

On the basis of the IMPROVE model and its results, we have been able to develop practical guidelines of how to approach these challenges in industry, which have been requested by 1500 interested parties so far. The material is also used in trade schools. Certainly, practical guidelines could have been developed without a software model. However, they are more robust and trustworthy when developed on the basis of analytical models such as the one we have used. The calculation program for the solvent balance (the SMP) is also available for consulting purposes. In this way, a permanent update of the data necessary for determining the input and output categories in the solvent balance, as required by the EU Solvent Directive and their national transpositions, can be ensured. Moreover, a calculation model for drawing up the solvent balance – as requested by the EU Solvent Directive – uses, in simplified terms, the data stored in IMPROVE.

ACKNOWLEDGMENTS

Parts of this paper have been published in Geldermann & Rentz (2004a) and Geldermann & Rentz (2004b). The underlying research projects were conducted by the French–German Institute for Environmental Research (DFIU/IFARE) at the University of Karlsruhe (TH).

REFERENCES

Blümel, F. (2001) *Stoffstrommanagement in der handwerklichen Autoreparaturlackierung – Zur Planung und Steuerung betrieblicher Stoff- und Energieströme in kleinen und mittleren Unternehmen der Autoreparaturlackierung*. Fortschrittsberichte VDI, Reihe 16, Nr. 129, VDI Verlag, Düsseldorf.

Ebel, A., Friedrich, R. and Rohde, H. (1997) *Tropospheric Modelling and Emission Estimation*. Springer, Heidelberg.

Environment Agency (2001) *IPPC H2: Horizontal Guidance Note on Energy Efficiency*. Environment Agency, Bristol.

ESIG (2002) *Questions and Answers Relative to the EU Solvents Emissions Directive*. European Solvents Industry Group (ESIG), Brussels.

Geldermann, J. (2006) *Mehrzielentscheidungen in der industriellen Produktion*. Universitätsverlag Karlsruhe, Karlsruhe.

Geldermann, J. and Rentz, O. (2004a) Environmental decisions and electronic democracy. *J. Multi-criteria Anal.* **12**(2–3), 77–92.

Geldermann, J. and Rentz, O. (2004b) Decision support through mass and energy flow management in the sector of surface treatment. *J. Ind. Ecol.* **8**(4), 173–187.

Geldermann, J., Schollenberger, H. and Rentz, O. (2002) Linking mass and energy flow analysis with time management. In *Part 2: Proceedings of the 16th International Conference of Informatics for Environmental Protection* (eds. W. Pillmann and K. Tochtermann), Environmental Communication in the Information Society, IGU/ISEP, Wien, pp. 424–432.

Girault, C. and Valk, R. (2003) *Petri Nets for Systems Engineering: A Guide to Modeling, Verification, and Applications*. Springer, Heidelberg.

Guinée, J.B., Gorrée, M., Heijungs, R., Huppes, G., Kleijn, R., de Koning, A., van Oers, L., Wegener Sleeswijk, A., Suh, S., Udo de Haes, H.A., de Bruijn, H., Van Duin, R. and Huijbregts, M.A.J. (2002) *Handbook on Life Cycle Assessment – Operational Guide to the ISO Standards*. Kluwer Academic Publishers, Dordrecht.

Hennemann, A. (1994) *Betriebswirtschaftslehre für Fahrzeuglackierbetriebe*. Audin, München.

Möller, A. and Rolf, A. (1995) Methodische Ansätze zur Erstellung von Stoffstromanalyse. In *Stoffstromanalysen in Ökobilanzen und Öko-Audits* (eds. M. Schmidt and A. Schorb). Springer, Berlin.

Müller, J. (2000) *Vorgehensweise bei Materialschichtdickenmessungen*. Institut für Fahrzeuglackierung, Frankfurt.

Owen, J.W. (1996) LCA impact assessment categories – technical feasibility and accuracy. *Int. J. Life Cycle Assess.* **I**(1), 151–159.

REFA – Verband für Arbeitszeitstudien und Betriebsorganisation e.V. (1992) *Methodenlehre des Arbeitsstudiums; Teil 2: Datenermittlung*. **7** Carl Hanser Verlag, München.

Reinecke, M. (1997) Ein Eindruck von Theorie und Anwendbarkeit der Petri-Netze. In *Stoffstromanalyse und -bewertung* (eds. M. Matthies). Institut für Umweltsystemforschung der Universität Osnabrück.

Rentz, O., Nunge, S., Laforsch, M. and Holtmann, T. (1999) Technical background document for the actualisation and assessment of UN/ECE protocols related to the abatement of the transboundary transport of volatile organic compounds from stationary sources. *Report of the Task Force on the Assessment of Abatement Options/Techniques for Volatile Organic Compounds from Stationary Sources*, on behalf of the German Federal Ministry for the Environment, Nature Conservation and Nuclear Safety and the Federal Environmental Agency, Karlsruhe.

Rentz, O., Avci, N. and Geldermann, J. (2001a) Lösemittelbilanz und Reduzierungsplan für kleine und mittlere Unternehmen der KFZ-Reparaturlackierung. Solvent Management Plan according to the EU-Solvent Directive (1999/13/EC) in small and medium-sized car refinishing body shops, Ministerium für Umwelt und Verkehr Baden-Württemberg, Stuttgart.

Rentz, O., Avci, N. and Geldermann, J. (2001b) Study for the implementation and development of an Internet based information exchange between Member States, the European Commission and the activities concerned on the use of organic substances and their potential substitutes according to article 7 Directive 1999/13/EC, on behalf of the DG ENV, http://www.voc-infoex.uni-karlsruhe.de.

Rentz, O., Peters, N.-H., Nunge, S. and Geldermann, J. (2003a) *Bericht über Beste Verfügbare Techniken (BVT) in den Bereichen der Lack- und Klebstoffanwendung in Deutschland/Reports on the Best Available Techniques (BAT) in the Sectors of Paint and Adhesive Application in Germany (Deutsch/English)*. VDI-Verlag, Düsseldorf.

Rentz, O., Schollenberger, H. and Geldermann, J. (2003b) *Arbeitszeitstudie für die Autoreparaturlackierung in Herstellerwerkstätten – Projekt im Auftrag des VDA unter Beteiligung der Marken Audi, BMW, Ford, Opel, Porsche und VW*. Institut für Industriebetriebslehre und Industrielle Produktion, Karlsruhe.

Ritchie, A., Ollerenshaw, B., Hanny, R., Reynolds, F. and Conlin, H. (2000) *Reducing VOC Emissions from the Vehicle Refinishing Sector*. European Commission, Brussels.

Rosenstengel, B. and Wienand, U. (1991) *Petri-Netze – Eine anwendungsorientierte Einführung*. **4** Vieweg, Braunschweig.

Schollenberger, H., Geldermann, J. and Rentz, O. (2003) Analysis of time and motion studies in the sector of vehicle refinishing. *Proceedings of the 17th International Conference on Production Research*, Blacksburg..

Winder, C. and Turner, P.J. (1992) Solvent exposure and related work practices amongst apprentice spray painters in automotive body repair workshops. *Ann. Occup. Hyg.* **36**, 385–394.

6

Integration of biological techniques for air pollution control into sustainable development

M.A. Deshusses

6.1 INTRODUCTION

6.1.1 Biological techniques for air pollution control

Over the past three decades, there has been a rapid maturing and expansion of the use of biological techniques for the control of air pollution. This is illustrated by the evolution from the simple and very crude soil biofilters treating odours described by Carlson and Leiser (1966) to the more advanced processes being researched or already deployed in the field today. Examples of the latter include biotrickling filters using mutant organisms to degrade trichloroethylene (TCE) vapours (Sun & Wood, 1997), high performance biotrickling filters treating H_2S at contact times below 2 s (Gabriel & Deshusses, 2003), innovative foam bioreactor with improved mass transfer characteristics (Kan & Deshusses, 2003), or combined advanced oxidation-bioreactor systems (van Groenestijn, 2001).

 Waste Gas Treatment for Resource Recovery edited by Piet N.L. Lens, Christian Kennes, Pierre Le Cloirec and Marc Deshusses. ISBN: 1843391279. Published by IWA Publishing, London, UK.

The success of biological techniques for air pollution control can be attributed to several reasons:

- Biotreatment of gases works well, it is often very cost effective even at low concentrations of pollutants.
- Biological treatment does not require high temperatures, it does not generate secondary pollutants such as NO_x, particulate matter, spent activated carbon, or additional CO_2.
- Biotreatment of gases is energy efficient and does not require additional fossil fuel.
- Biotreatment of gases is environmentally friendly and is well perceived by the general public.

6.1.2 Sustainable development

While treatment performance and cost are simple to measure, the "environmental friendliness" and greater environmental impact of a given technology is more difficult to quantify, because it requires a comparative assessment of the system's construction and life, and its greater impact on greenhouse gas reduction/emission, energy use, waste generation, etc. Such evaluation is generally done using life-cycle assessments (LCA) (Graedel, 1998) and is motivated by the increasing concerns for sustainable development. Sustainable development refers to "development that meets the needs of the present without compromising the ability of future generations to meet their own needs" (World Commission on Environment and Development, 1987). Ensuring environmental sustainability has been identified as one of the Eight Millennium Development Goals by the United Nations (see http://www.un.org/millenniumgoals). The Goal calls for integration of the principles of sustainable development into country policies and programmes and reversing the loss of environmental resources. Sustainable development is a global economic, social, and environmental concept, yet it is also a controversial topic as the assessment of sustainability requires to assign sometimes subjective impact values to indicators such as resource depletion, global warming, etc. More importantly, the implementation of sustainable developments will drastically change the way we live.

6.1.3 Technology selection for sustainable development

Little has been done in the area of LCA of biological techniques for air pollution control. A preconceived idea is that waste treatment is by nature a non-sustainable activity, because it deals with wastes, and wastes should be avoided in the first place, or at least minimized to achieve sustainability. This is not necessarily true as it misses the point that integrated design, rather than single unit operations should be considered. In other terms, it is the global impact of an industry or processing plant that should be considered and not the individual unit operations that make that industry, as the optimum of the sum is unlikely to be the sum of the optimum of each separate unit operation.

For example, consider the hypothetical case shown in Figure 6.1 which could be a simple processing plant, but the example applies widely. As pollutant emissions are reduced through implementation of any measure such as pollution control, process change, product substitution, operation and maintenance, etc., the environmental impact decreases. At point A, it reaches a minimum. It is important to realize that in reality, the graph in Figure 6.1 contains numerous lines (see inset) corresponding to the

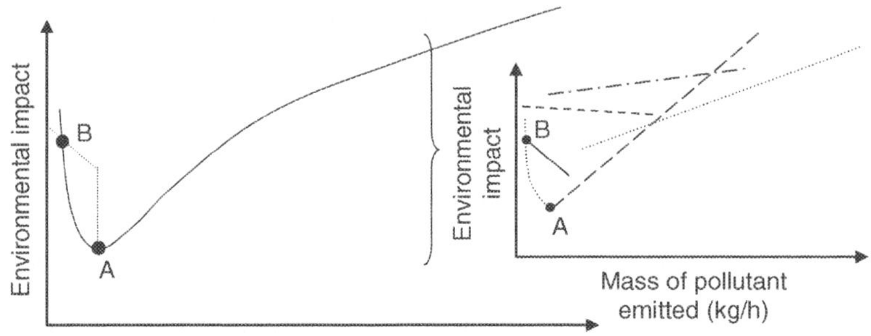

Figure 6.1 Hypothetical example of how global environmental impact (used here as a metric for sustainability) could be affected by pollutant emission. The inset shows the different scenarios corresponding to different design or operation options leading to different environmental impacts. From these scenarios, a compounded line corresponding to the minimum impact can be derived.

many different options available in terms of plant design, operating conditions, maintenance practice, pollution control technology, etc. The compounded minimum line should be considered. As the mass of pollutant emitted further decreases beyond point A, the environmental impact may increase, for example as a result of increased energy use, emission of greenhouse gases, or non-sustainable process changes. Obviously, such increase in the environmental footprint is undesirable. As indicated by the dashed line in Figure 6.1, the compounded line is not necessarily monotonous, because decreasing the overall pollutant emissions may reach the limits of a given technology and require a completely different technological solution. Possible examples include a change of raw products in a processing plant, or switching from biological treatment to incineration for air pollution control.

The determination of rational criteria that allows to select the most environmentally friendly pollution control technique has been remarkably absent in setting pollution control policies, probably because the concepts of sustainability are not well-known and factors affecting it are not well understood. It is a general feeling of those involved in biofiltration that in many cases, air pollution control legislation is forcing tighter controls at the expenses of sustainability (imposing point B versus point A in Figure 6.1). Ideally, some kind of LCA should be conducted as part of the development of environmental policies, however it is not yet done. Obviously, major changes in the thinking of policy makers and in the environmental permitting process are needed before the application of such assessment becomes a routine. In the next section, the impact of the selection of a volatile organic compound (VOC) control method on the environmental emissions is evaluated. Next, recent advances in H_2S control using high performance biotrickling filters are discussed. Specific aspects related to sustainability are emphasized.

6.2 VOC CONTROL, THE CASE FOR AVOIDING UNNECESSARY CO_2 EMISSIONS

A wide range of VOCs can be very effectively controlled using biofilters and biotrickling filters. These include alcohols, ketones, aromatics, esters, ethers, some chlorinated

Table 6.1 Parameters for the hypothetical control of a 50,000 $m^3 h^{-1}$ stream contaminated with 1 $g m^{-3}$ VOC.

Parameters	Value	Unit
Air flow rate	50,000	$m^3 h^{-1}$
VOC inlet concentration	1.0	$g m^{-3}$
Control efficiency required	>97%	–
Mass of VOC controlled	438	tons per year
Mass of CO_2 emitted if using thermal or catalytic oxidation[a]	40,000–55,000	tons per year
Mass of NO_x emitted if using thermal or catalytic oxidation[b]	53	tons per year

[a]Additional CO_2, from fossil fuel combustion.
[b]Assuming 100 ppm_v NO in the exhaust.

solvents, etc. (Cox and Deshusses, 2001; Devinny et al., 1999). Yet, in many cases, thermal or catalytic oxidizers are being installed to control VOC emissions, either by choice of the plant operator or because of local environmental regulations. Let's consider a hypothetical case where the exhaust air flow rate is 50,000 $m^3 h^{-1}$ contaminated with a VOC concentration of 1 $g m^{-3}$, that is, a medium to high concentration. For this example, toluene is taken as contaminant, although the outcome of the discussion does not greatly depend on the nature of the contaminant. Assuming that the plant operates 24 h day^{-1} and 365 days $year^{-1}$, this represents an uncontrolled emission rate of 50 kg h^{-1}, or 438 tons $year^{-1}$.

If thermal oxidation is selected for the control of the exhaust, the calorific value of the air undergoing treatment will need to be increased, usually by adding natural gas or another fossil fuel. Since the concentration of toluene in the exhaust is about 40 times lower than the lower explosive limit (LEL) of toluene (1.1% by volume), a significant amount of additional fuel will be required. In fact, thermal or catalytic oxidation in such case is really about burning fuel rather than the pollutant. As shown in Table 6.1, the additional fossil fuel needed for incineration will result in significant CO_2 and NO_x emissions. Surprisingly, the greater impact of CO_2 and NO_x emissions is seldom considered in the selection of the control technology, although emissions are significant. Other factors such as the manufacture and replacement/disposal of noble metals used for catalytic oxidizers would also need to be included in a more detailed evaluation.

A complete LCA of the VOC control case is outside the scope of this chapter. However, the numbers of Table 6.1 can be placed in perspective using the average damage cost per ton of pollutant emitted reported by Eyre et al. (1997) and Spadaro and Rabl (2002). The average damage cost is a lumped cost that includes environmental damages, and costs to the society associated with morbidity and mortality. These were valued as follows:

- VOCs: \$1000 per ton.
- NO_x: \$18,000 per ton.
- CO_2: \$33 per ton.

If thermal or catalytic oxidation was selected for controlling this air stream, a very removal (>99.5%) of the VOC is expected. However, the resulting damage cost of the CO_2 and NO produced by thermal/catalytic oxidation would be about \$2,500,000 per year, that is, \$1.5 M for CO_2 and \$1 M for NO.

In contrast, if a biological control method was used, and if it achieved only 97% removal of the VOC, about 13 tons of VOC would be released untreated, per year. There would be no NO_x or additional CO_2 produced. As mentioned above, the average damage cost of VOCs is about \$1000 per ton. Thus, the total damage cost of the untreated VOC would be about \$13,000 per year. This is markedly lower compared to conventional treatment. Obviously, the numbers above are subject to some uncertainties (Rabl et al., 2005), and different approaches exist for the evaluation of the global environmental impact. However, the magnitude of the difference between conventional treatment and biological treatment suggests that the topic should be further investigated, and that the results should be considered in setting pollution control policies.

6.3 HIGH-RATE BIOTRICKLING FILTERS FOR H_2S CONTROL

Odour control at wastewater treatment plants has traditionally been conducted using chemical scrubbing. However, chemical scrubbers have several significant drawbacks, including high operating costs, generation of halomethanes (WERF, 1999) and the need for hazardous chemicals on-site. In the past decade, these drawbacks have motivated engineers to look into alternatives such as biological treatment. For H_2S control, biotrickling filters are the most promising alternatives.

To compete with chemical scrubbers, biotrickling filters need to be inexpensive, reliable, and able to treat the same pollutant loads as chemical scrubbers, and with the same efficiency. This may be achieved by careful packing material selection and reactor design, and optimum reactor operation and monitoring (Gabriel & Deshusses, 2003; Gabriel et al., 2004). Over the past years, it has been demonstrated that existing chemical scrubbers could be converted to biotrickling filters, thereby eliminating the need for chemicals, simplifying process control, and improving workers' safety (Gabriel & Deshusses, 2003, 2004). Highly effective biotreatment of H_2S was shown in retrofitted chemical scrubbers operated at gas retention times as short as 1.6 s. Selected results are shown in Figure 6.2, and biotrickling filter characteristics are listed in Table 6.2.

As for the VOC case discussed before, significant differences are expected to exist in the global impact of controlling H_2S when using either chemical or biological scrubbers. However, in this case the assessment does not involve CO_2 or NO_x as these are not emitted from either chemical or biological scrubbers. For this example, the most appropriate measure for assessing the environmental impact is the use of chemicals in the chemical scrubbers, as reported in Table 6.3. The chemical consumption is significant (270–820 m^3 of chemicals per year), and obviously there is an environmental toll for the production and delivery of these chemicals. Chemical scrubbing will also result in secondary pollution. Here, halomethane emissions are estimated at about 700 kg $year^{-1}$. Other factors that have not been considered but that require further definition include, but are not limited to, acid washes that are periodically performed on the chemical

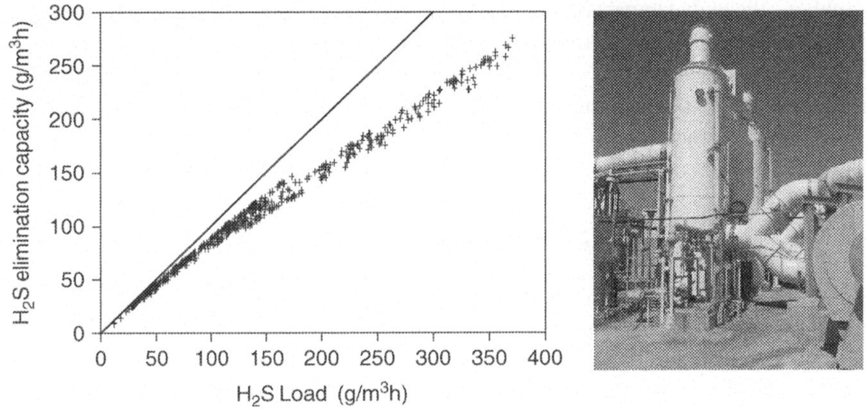

Figure 6.2 Left: H_2S elimination capacity (EC) versus load data for biotrickling filter I during a 6-day period. The diagonal line represents the EC for 100% removal. Right: picture of the system.

Table 6.2 Characteristics of two high-rate biotrickling filters operated at the Orange County Sanitation District (USA) treating sewer gases with high H_2S concentrations. Three other chemical scrubbers were converted, but treat organic odours and low concentrations of H_2S.

Parameters	Biotrickling filter #10	Biotrickling filter #I	Unit
Air flow rate	17,000	17,000	m^3h^{-1}
H_2S inlet concentration	5–40	40–100	ppm
Empty bed residence time	1.6–2.2	1.8–2.0	s
Typical H_2S removal	>95%	75–99%	–
Maximum H_2S elimination capacity	100–110	>260	$g\,m^{-3}h^{-1}$
Packing material	4 cm cubes, open pore polyurethane foam (EDT, Eckental-Eckenhaid, Germany), random dump		
pH of operation	1.8–2.2	1.6–2.5	–

Table 6.3 Cursory evaluation of the environmental factors of the two high-rate biotrickling filters operated at the Orange County Sanitation District.

Parameters	Scrubber #10		Scrubber #I		Unit
Air flow rate	17,000		17,000		m^3h^{-1}
Average H_2S inlet concentration for calculations	20		60		ppm
Scrubber operated	as btf	as chem. scrubber	as btf	as chem. scrubber	
Caustic (25%) consumption[a]	0	33	0	100	m^3year^{-1}
Hypochlorite (12.5%) consumption[a]	0	240	0	720	m^3year^{-1}
Halomethane emission assuming 1 ppm emission (all as chloroform)[b]	0	700	0	700	$kg\,year^{-1}$

[a]Calculated value assuming complete conversion of H_2S to sulfate and neglecting consumption of caustic by CO_2 absorption.
[b]From WERF (1999) and Gabriel and Deshusses (2003).

scrubbers, the global impact of manufacturing and shipping the chemicals to the sewage treatment plant, the liability of having harsh and dangerous chemicals on-site, improvements in workers health and safety, the space occupied by chemical storage. Here again, a complete evaluation is required to fully understand the complex ramifications of the choice of the method for odour control.

6.4 CONCLUDING REMARKS

It is only during the past decade that one has started to seriously discuss the sustainability of selected aspects of our society. A lot of focus has been placed on the effect of pollution, highlighting the fact that pollution control is much needed. As a result, there has been very significant improvements in environmental protection. However, there has been a somewhat narrow minded opinion that more control is always better. This is not necessarily the case, especially in the light of the very large differences between the global impact of biological and chemical or physico-chemical pollutant control methods. This chapter made a case for taking better-informed decisions for the selection of pollution control equipment and for the establishment of environmental policies. This will be a first step in minimizing the impact of pollution and pollution control. In the future, even further efforts will be needed to integrate such reasoning during the design phase of industrial plants and processes, so that the sustainability of the entire process can be optimized, rather than the sustainability of individual units.

REFERENCES

Carlson, D.A. and Leiser, C.P. (1966) Soil beds for the control of sewage odors. *J. Water Pollut. Contr. Fed.* **38**(5), 829–840.

Cox, H.H.J. and Deshusses, M.A. (2001) Biotrickling filters. In *Bioreactors for Waste Gas Treatment* (eds. C. Kennes and M.C. Veiga), pp. 99–131. Kluwer Academic Publisher, The Netherlands.

Devinny, J.S., Deshusses, M.A. and Webster, T.S. (1999) *Biofiltration for Air Pollution Control*, 300 pp. Lewis Publishers, Boca Raton, FL.

Eyre, N., Downing, T., Hoekstra, R., Rennings, K. and Tol, R. (1997) Global warming damages. *Final Report of the ExternE Global Warming Sub-Task, DGXII*. European Commission, Brussels, Belgium.

Gabriel, D. and Deshusses, M.A. (2003) Retrofitting existing chemical scrubbers to biotrickling filters for H_2S emission control. *Proc. Natl. Acad. Sci. USA* **100**, 6308–6312.

Gabriel, D. and Deshusses, M.A. (2004) Technical and economical analysis of the conversion of a full-scale scrubber to a biotrickling filter for odor control. *Wat. Sci. Technol.* **50**, 309–318.

Gabriel, D., Cox, H.H.J. and Deshusses, M.A. (2004). Conversion of full-scale wet scrubbers to biotrickling filters for H_2S control at POTWs. *J. Environ. Eng. ASCE* **130**, 1110–1117.

Graedel, T.E. (1998) *Streamlined Life-Cycle Assessment*, 310 pp. Prentice Hall.

Kan, E. and Deshusses, M.A. (2003) Development of foamed emulsion bioreactor for air pollution control. *Biotechnol. Bioeng.* **84**, 240–244.

Rabl, A., Spadaro, J.V. and van der Zwaan, B. (2005) Uncertainty of air pollution cost estimates: To what extent does it matter? *Environ. Sci. Technol.* **39**(2), 399–408.

Spadaro, J.V. and Rabl, A. (2002) Air pollution damage estimates: the cost per kilogram of pollutant. *Int. J. Risk Assess. Manag.* **3**(1), 75–98.

Sun, A.K. and Wood, T.K. (1997) Trichloroethylene mineralization in a fixed-film bioreactor using a pure culture expressing constitutively toluene ortho-monooxygenase. *Biotechnol. Bioeng.* **55**, 674–685.

van Groenestijn, J.W. (2001) Combined advanced oxidation and biodegradation. In *Bioreactors for Waste Gas Treatment* (eds. C. Kennes and M.C. Veiga), pp. 179–200. Kluwer Academic Publisher, The Netherlands.

Water Environment Research Foundation (WERF) (1999) A critical review of odor control equipment for toxic air emissions reduction. *Water Environment Research Foundation Project 91-VOC-2, WERF Report #D53010*. Alexandria, VA.

World Commission on Environment and Development (1987) *Our Common Future*. Oxford University Press, Oxford. (This document is known as the Brundtland Report.)

PART II: Characterization of waste gases

7

Characterization of pollutants in gases

P. Brimblecombe

7.1 INTRODUCTION

The production of air pollution has often been reduced to a consideration of the products of combustion. This simplification served well as a description of the air in crowded coal burning cities of the late 19th century. These were characterized with fogs generated from sulfur dioxide and airborne soot (Figure 7.1). The figure remains satisfactory as an image of air pollution that arose when solid fuels were burnt in stationary sources. The mid-20th century saw the widespread adoption of liquid fuels in mobile sources with the popularity of the private automobile. The liquid fuel was volatile so evaporated from the engine block or during refuelling, thus adding to organic compounds in exhaust gases. These low-molecular-weight organic compounds acted in the atmosphere as precursors in a series of photochemical reactions that led to the production of pollutants such as ozone and other oxidants. This type of air pollution came to be called *Los Angeles smog* after the city where it was first characterized (Haagen-Smit, 1964).

History tells us that combustion is only one way of polluting the air. The siege of Hermopolis in Egypt came to an end in 734 BC when the stench in the city grew so bad that the citizens surrendered. In early 19th Europe the alkali industry producing

Figure 7.1 Classic air pollution of the early 20th century as seen at Lancaster Place, Birmingham, UK.

caustic soda discharged large quantities of the by-product hydrogen chloride in the air with devastating effects to the surrounding vegetation. The development of air pollution in the city of Auckland (New Zealand) occupying a narrow isthmus between two shallow harbours was dominated by the processes involving reduction reactions on the mud flats that began in the mid-1800s. These became especially serious in the 1950s through sewage discharge to the Manukau Harbour.

Such incidents remind us of the importance of non-combustion sources of pollution. Waste gases or gases arising from industrial discharges can yield large amounts of air pollutants. These can have tragic consequences. A serious incident with emissions from oil and gas processing occurred in Poza Rica (Mexico) in 1950. Large amounts of hydrogen sulfide were discharged, and some 22 people died and many others became ill. There have been subsequent incidents at chemical plants where there were catastrophic emissions from of phenols and dioxin at Seveso (Italy) in 1976 and more fatally methyl isocyanate at Bhopal (India) in 1984.

7.2 CHARACTERIZING ATMOSPHERIC POLLUTANTS

There are a wide range of potential waste emissions. Our discussion thus needs to embrace a wider group of pollutants than are normally considered in accounts of air pollution which may often be restricted to combustion sources.

7.2.1 Composition of the atmosphere

The bulk composition of the atmosphere tends to be rather constant in terms of the major components oxygen and nitrogen and other gases that are relatively inert

Table 7.1 Composition of the atmosphere showing long lived and variable components. Many of the variable gases are pollutants and can be orders of magnitude more concentrated in urban areas.

Gas	Formula	Concentration
Long-lived gases		*% (dry air)*
Nitrogen	N_2	78.08
Oxygen	O_2	20.98
Argon	Ar	0.93
Carbon dioxide	CO_2	0.035
Neon	Ne	0.0018
Methane	CH_4	0.00017
Krypton	Kr	0.00011
Helium	He	0.0005
Hydrogen	H_2	0.00006
Variable gases		*ppb*
Nitrous oxide	N_2O	310
Ozone	O_3	30
Ammonia	NH_3	1
Ethane	C_2H_6	1
Methyl chloride	CH_3Cl	0.6
Carbonyl sulfide	OCS	0.5
Nitrogen dioxide	NO_2	0.3
Sulfur dioxide	SO_2	0.01–0.1
Nitric oxide	NO	0.1
Hydrogen sulfide	H_2S	0.05

(see Table 7.1). Nevertheless carbon dioxide shows diurnal, seasonal and long-term trends, but can be treated as almost constant because compared to many trace gases the range of variation is low.

Water is another gas that is somewhat neglected as the composition of the atmosphere is usually described in terms of the composition of dry air. It is also perhaps that water seems more a part of meteorology than of atmospheric chemistry. The concentration of water varies with temperature and relative humidity. In the tropics under humid conditions water vapour can easily constitute some 5% of the volume. Water occurs in all three phases in the atmosphere because in addition to the vapour, liquid water (cloud, fog droplets) and solids (ice particles, hail) are also present.

In addition to the dominant gases featured in the upper part of Table 7.1, there are trace gases that are a good deal more variable. The chemistry of these gases, which are often reactive, takes place on the background matrix of less variable gases. The concentrations of the variable gases are typically much lower so the units in which their concentration is expressed have to be considered.

7.2.2 Concentration units

The most convenient and physically appropriate unit of concentration for gases is the mole fraction X_a:

$$X_a = N_a/(N_a + N_b + \cdots)$$

where N_a is the number of moles of the gas whose concentration is being expressed and $N_a + N_b + \cdots$ the number of moles of all the rest. The mole fraction is useful because it relates the number of molecules of the gas under consideration to the total number of molecules present. This is rather more practical than it first seems because the most common expressions of concentration such as percentage (c_a [%]), parts per million (c_a [ppm]) or parts per billion (c_a [ppb]) are usually all expressed by volume. Dalton's law of partial pressure tells us that these units are all linked to mole fraction and can be directly converted into pressure (p_a) such that:

$$X_a = c_a\,[\%]/100 = c_a\,[\text{ppm}]/10^6 = c_a\,[\text{ppb}]/10^9 = f(p_a)$$

If we are at the surface of the earth, where the pressure might be expressed in a non-SI unit: atmosphere, $X_a = p_a$, such that the mole fraction of a gas is conveniently equal to its pressure in atmospheres.

Sometimes regulatory tradition or disciplinary practice (especially in terms of describing ecological load) has used mass as a unit for measuring atmospheric concentrations. Thus, the mg, μg and ng per cubic metre have frequently been adopted, but the unit has a number of drawbacks. The concentrations vary with temperature, but more importantly care is always needed to be clear how the mass has been expressed. The gas NO_2 might be expressed as μg(N) m^{-3} or as μg(NO_2) m^{-3} or indeed as the weight of other nitrogen compounds, such as HNO_3 or N_2O_5.

The conversion formulae for changing from units μg m^{-3} to ppb is:

$$1\,\text{ppb} = 0.001 \times M/A\ \mu\text{g(pollutant)}\ \text{m}^{-3}$$

where M is the (g) molecular weight of the pollutant and A is the volume of a mole (i.e. 0.0245 m^3 at 293 K) which can be calculated by the ideal gas law.

Organic gases provide similar problems in that they are sometimes expressed in terms of mass of carbon. As inconvenient as the mass units are, particle concentrations are almost always expressed in mass units, because without knowing the complete composition of the particles it is very difficult to use the mole fraction.

7.2.3 Sources of trace gases

The gases at low concentrations in the atmosphere are usually short lived and have concentrations dependent on where they are measured. Close to their sources they are at high concentration, but much lower at a distance, although some (e.g. carbonyl sulfide or methyl chloride) are sufficiently long lived to become well mixed in the atmosphere. Many trace gases have a natural origin. The oceans have a high concentration of chloride and sulfate, and thus contribute gases such as carbonyl sulfide and methyl chloride to the atmosphere. Sea salt particles driven into the atmosphere by wind or bubble bursting release hydrogen chloride after they become acidified. The oceans are far less rich in nitrogen, so we often look to the land as a predominant natural source of nitrogen compounds. Nitrous oxide (N_2O) arises from soil respiration. Ammonia comes from the degradation of proteinaceous material or urea. Wetlands prove an important source of methane.

Oxides of carbon, sulfur and nitrogen (CO, CO_2, SO_2, NO and NO_2) are produced during combustion of hydrocarbon-based fuels. The oxides of carbon obviously come from the fuel. Sulfur and nitrogen oxides partly arise through the oxidation of traces of these elements in the fuel, but chain reactions during the combustion process oxidize nitrogen in the air to its oxide (known as the Zeldovich cycle). Trace chlorine from the fuel (common in coal) is released as hydrogen chloride and boron as its oxide or boric acid. Metallic elements within the fuels are typically found as oxide particles in the exhaust gases, although mercury in particular is present in its elemental form (Streets et al., 2005).

Combustion products are frequently oxides, but in waste gas streams from the petroleum industry, for example, can contain more reduced gases (e.g. sulfur and nitrogen as their hydrides, hydrogen sulfide and ammonia). Sulfur can also be emitted in its elemental form, as S_2 and even larger polysulfur molecules and as carbonyl sulfide (Brimblecombe, 2003).

Volatile organic compounds are common in waste gases. The more volatile ones have simply evaporated from solvents are part of the production process, such as the hydrocarbons from the petroleum industry. Aromatic and aliphatic hydrocarbons, both saturated and unsaturated are frequent in waste gases from the treatment and refining of crude oil (Speight, 2005). They are often emitted with other compounds such as mercaptans (R-SH) and carbonyl sulfide. Although more oxidized organic compounds are typically less volatile, they can still be found in the gas phase. Most common are the aldehydes, ketones, alcohols and acids with small molecules such as formaldehyde being commonly found (Speight, 2005).

Organohalogen compounds, though perhaps not quite so dominant in petroleum wastes, can be prevalent in other waste gas streams. In particular, they have been widely used as solvents and refrigerants. Thus chloroform, carbon tetrachloride and a range of chlorofluorocarbons are released in substantial quantities, although increasingly regulated, for example by the Montreal Protocol (Derwent et al., 1998). Water treatment frequently involves the use of chlorine, which leads to the release of organochlorine and some chlorobromides (e.g. trichloromethane, dichlorobromomethane, dibromochloromethane and tribromomethane).

7.2.4 Particles

Fine particles found in exhaust gases consist of unburnt fuel (soots), non-combustible components (ash) along with a range of semi- and non-volatile compounds produced during combustion (Lipsky & Robinson, 2006). Soot is largely carbon but other compounds are found adsorbed on the surface. Ash consists of non-combustible material from within the fuel. In the case of coal, there is a large amount of silicon-containing materials that yields glassy particles. More volatile elements such as thallium, antimony arsenic and mercury can concentrate on particles as the exhaust gas cools. Metallic impurities in fuels are usually found as metal oxides in the exhaust gases. In crude oils vanadium and nickel are often at enhanced concentrations so their oxides are prevalent enough to be good tracers of heavy oil combustion. As exhaust gases cool and undergo chemical reaction downstream from the sources further non-volatile compounds become associated with particles. Perhaps most important is sulfuric acid which is found as sulfates (often acidic) on fine particulate material in the atmosphere.

The organic compounds in fuel can be pyrolized or condensed during incomplete combustion and result in particulate organic matter (POM). Much interest in POM has focused around the polycyclic aromatic hydrocarbons (PAH) that are characteristic of a wide range of combustion processes, with the carcinogen benzo(a)pyrene perhaps the most characteristic of these (Viras et al., 1990). Chlorine, sulfur and nitrogen can add to the aromatic rings with the potential to enhance their toxic effect (Kameda et al., 2004). Best known of these are the 2-nitrofluorene and 1-nitropyrene often found in diesel emissions. Chlorine reacts, particularly in the combustion of wastes where phenols are present, to give a range of chlorinated dibenzo compounds. The most notorious of these is *dioxin* or more specifically, 2,3,7,8-tetrachlorodibenzo-p-dioxin (Mocarelli, 2001).

Other processes can generate particles, of which dusts in mining and metallurgical activities are particularly evident. Handling fertilizers and pesticides can give rise to fugitive dust emissions that are especially hazardous (Badr & Harion, 2005). Even more complex and biochemically significant organic compounds are found on aerosol particles from aeration tanks of wastewater treatment plants (e.g. sterols such as cholesterol, coprostanol, campesterol, β-sitosterol, stigmasterol and stigmastanol) and anionic surfactants (Radke, 2005).

7.2.5 Atmospheric chemistry

One very important air pollutant, ozone has not been discussed yet. This is because it is not emitted in substantial quantities from any particular source, but is produced in the atmosphere. It is a secondary pollutant. Primary pollutants act as precursors in their production. Ozone and many other secondary compounds are formed in photochemical smog. The process is driven by sunlight ($h\nu$):

$$NO_2 + h\nu \rightarrow NO + O$$

$$O + O_2 \rightarrow O_3$$

The second reaction involves another molecule (e.g. nitrogen in the air) in the atmosphere to take away the excess energy. This photochemical production of ozone requires a method of producing NO_2 and comes about through the oxidative potential of peroxy radicals in the atmosphere that are produced from volatile organic compounds. The process is initiated by the hydroxyl radical, which is ubiquitous in the atmosphere. Taking ethane as an example this occurs as follows:

$$CH_3CH_3 + {}^{\cdot}OH \rightarrow CH_3CH_2^{\cdot} + H_2O$$

$$CH_3CH_2^{\cdot} + O_2 \rightarrow CH_3CH_2O_2^{\cdot}$$

$$CH_3CH_2O_2^{\cdot} + NO \rightarrow CH_3CH_2O^{\cdot} + NO_2$$

$$CH_3CH_2O^{\cdot} + O_2 \rightarrow CH_3CHO + HO_2^{\cdot}$$

$$HO_2^{\cdot} + NO \rightarrow {}^{\cdot}OH + NO_2$$

Ozone is produced from the NO_2 as shown earlier. In addition to ozone, a range of oxidized products are produced largely as the result of an atmosphere that is essentially oxidizing. The chemistry is driven by radicals and the equations above show the way in which abstraction of a hydrogen atom from organic molecules leads to the formation of oxidized products, such as the acetaldehyde in the schema given above. Organic acids and ketones are also common products. As a general rule, the oxidized products are more soluble and less volatile that the original material, making the oxidation process a mechanism for removing gases from the atmosphere when they dissolve in rain.

Hydroxyl radicals can also take part in addition reactions. For example, the reaction between nitrogen dioxide and the hydroxyl radical leads to the formation of HNO_3, an important contributor to acid rain:

$$NO_{2(g)} + {}^{\cdot}OH_{(g)} \rightarrow HNO_{3(g)}$$

The hydroxyl radical can also add to aromatic compounds. This reaction proceeds through a number of steps, but as an example the oxidation of toluene in the atmosphere can lead to a cresol (among a number of products):

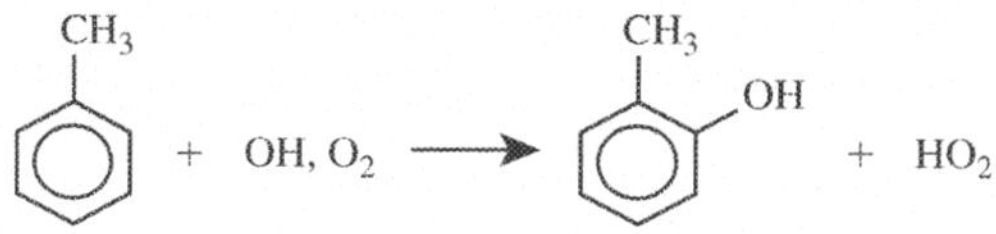

Photochemical oxidation of organic compounds, particularly terpenes (often from coniferous trees) and aromatic molecules (from petrol) in the atmosphere leads to the formation of secondary organic aerosols (SOA) (Baltensperger et al., 2005; Russell et al., 2002). The surfaces of soot particles in the atmosphere seem to be oxidized to yellow-brown coloured materials primarily of macromolecular aromatic polyacids, which are found widely in atmospheric aerosols. These resemble humic acids in soils and are often called "humic-like substances" or HULIS (Samburova et al., 2005) which often have a high surface activity (Latif & Brimblecombe, 2004).

7.2.6 Removal of atmospheric gases from the atmosphere

The reactions described in the previous section represent a form of removal of gas, but it is in reality simply a transformation from one chemical form to another. Ultimately removal requires that the material is deposited at the earth's surface. The solubility and volatility of materials are important characteristics in their removal from the atmosphere. Some gases can deposit directly on the surface of the earth, particularly in the ocean if they are soluble.

Alternatively they are removed after dissolution in rainwater. Soluble gases dissolve in cloud or rain droplets, then are washed out of the atmosphere. The gases have to be very soluble for this to happen. Gas solubility in liquid water in the atmosphere is treated as an equilibrium process, quantified in terms of the Henry's law constant (K_H). The dissolution of a gas such as carbonyl sulfide (OCS) can be written:

$$OCS_{(g)} \leftrightarrow OCS_{(aq)}$$

such that equilibrium is:

$$K_H = cOCS_{(aq)}/pOCS_{(g)}$$

where p is the pressure and c is the concentration.

Taking the atmosphere as the unit of pressure and mol L^{-1} as the unit of concentration, the Henry's law constant will have the units mol L^{-1} atm^{-1}. Although in some disciplines, Henry's law constant is expressed as the gaseous concentration divided by the aqueous concentration, that is $1/K_H$. One should thus take care in using these constants. In the case of the definition used here, the larger the values of this constant, the more soluble the gas. Table 7.2 shows that a gas-like hydrogen peroxide is

Table 7.2 Some Henry's law constants at 15°C.

Gas	K_H (mol L^{-1} atm^{-1})
Hydrogen peroxide	2×10^5
Ammonia	90*
Sulfur dioxide	5.4*
Formaldehyde	1.7*
Carbonyl sulfide	0.035
Ozone	0.02
Oxygen	0.0015
Carbon monoxide	0.001

*Undergoes subsequent hydrolysis equilibria that enhance the amount of dissolution.

very soluble, oxygen very much less. In a storm cloud, there is about a gram of liquid water (i.e. 0.001 L) in every cubic metre of air, which means that a Henry's law constant of 4×10^4 mol L^{-1} atm^{-1} is required to ensure equi-partitioning of the gas between the gaseous and liquid phases. Thus, in Table 7.2 only hydrogen peroxide appears soluble enough to partition effectively into cloud water. However, gases that undergo hydrolysis will also partition effectively into cloud droplets. These processes, and their associated hydrolysis, can be typified by the dissolution of formaldehyde and formic acid in the following reactions:

$$HCHO_{(g)} \leftrightarrow HCHO_{(aq)}$$

$$H_2O + HCHO_{(aq)} \leftrightarrow H_2C(OH)_{2(aq)}$$

$$HCOOH_{(g)} \leftrightarrow HCOOH_{(aq)}$$

$$HCOOH_{(aq)} \leftrightarrow H^+_{(aq)} + HCOO^-_{(aq)}$$

Low-volatility materials such as larger PAH compounds and many inorganic compounds are deposited on particles where they are removed. This process can usefully be represented in an equation:

$$\mathrm{Log}(K_p) = \log(F/A) = m \log p^0 + b$$

where K_p is an equilibrium coefficient for a compound distributed between particle (F, ng m^{-3}) and the gas phases (A, ng m^{-3}), p^0 (mmHg) is sub-cooled vapour pressure of a semi-volatile compound and the slope m typically takes the value of -1. The intercept b includes factors such as particle surface area and enthalpies of desorption. Higher temperatures favour desorption of the semi-volatile compound to the gas phase.

7.3 MONITORING AND ANALYSIS

As with all analysis, determination of pollutants depends on the demands to be made of the results. Methods can be simple for indicative determinations through to those demanded to meet the requirements of legislation (Harrison, 1996; Speight, 2005).

They may be imaginative approaches that are uncertain and as yet at the limits of analytical methodology or follow very stringently laid out standard methods. Some may need to be made rapidly with great sensitivity and the results delivered in real time.

7.3.1 Sampling

Traditionally air pollutants needed to be sampled before they could be measured. This typically meant pumping large quantities of air through filters which could collect the particles for subsequent analysis or gases were absorbed into liquids or onto solids or occasionally condensed directly. This effectively concentrates the pollutant. The sampling train needs to be carefully chosen and constructed of glass, polytetrafluoroethylene (PTFE) or stainless steel because some gases such as SO_2 strongly absorb onto surfaces.

Grab samples are sometimes collected by evacuating a pressure vessel (often of stainless steel) and opening it so that the air is driven in and a virtually instantaneous sample is collected. Syringes, bags and even bottles have been used. John Jeffries an American doctor collected air samples from a balloon for the famous scientist Henry Cavendish in 1784 simply by carrying bottles filled with water aloft and then emptying them at the appropriate height.

7.3.2 Analytical methods

For more than a century a range of very simple methods have been available to analyse trace gases. As an example air-containing SO_2 can be drawn through a hydrogen peroxide solution in a bubbler. The SO_2 is oxidized to sulfuric acid which can then be determined by titration with standard alkali. Classically there have been a wide range of bubbler techniques, which are inexpensive, but subject to interferences and have rather low time resolution (Harrison, 1996).

More recently real-time systems that continuously monitor pollutants have become standard for monitoring networks with commercial instruments purchased from respected manufacturers. The analyses are usually available digitally and can be transmitted to a computer, which may be separated from the sampling site by long distances. The high time resolution means that the structure of air pollution events can be analysed in the kind of detail often required to understand its origins. However, this is an expensive approach and requiring a large capital outlay and skilled technical resource to maintain and calibrate the instruments. Some of the key methods are listed below.

7.3.2.1 Optical methods

There are an increasing number of optical methods (Figure 7.2) that involve absorption of light over long path-lengths (as long as several kilometres). These techniques include: Fourier transform infrared absorption spectroscopy, differential absorption lidar and differential optical absorption spectroscopy (Colls, 1997). A range of commercial instruments are available, but care needs to be given to the sensitivity and path-lengths and any problems with calibration.

Figure 7.2 Measuring air quality continuously on the runway edge in Geneva using differential optical absorption spectroscopy (DOAS). By permission of the Environmental and Legal Affairs division, responsible for promoting the Environmental Management at Geneva International Airport (AIG).

7.3.2.2 Indicative methods

These are potentially simple and sometimes inexpensive methods, although they are not necessarily appropriate for regulatory requirements. They can give an estimate of the extent of any problem and a general indication of average pollution concentrations.

The simplest and most direct may be the electrochemical cells and electronic methods. The latter rely on solid-state sensors to detect the pollutants. The sensors are often heated metal oxide surfaces, perhaps tin or tungsten (Williams, 1999). The gas being measured is absorbed on the surface and the electrical current is measured as it is oxidized. Carbon monoxide can be detected in this way (Ramgir et al., 2006), so the technology has been utilized to detect carbon monoxide from gas leaks indoors. There are also sensors for the nitrogen oxides (Kamei et al., 2004) and a range of other gases such as ozone (Bendahan et al., 2004). However, except at relatively high concentrations there are still problems, as the sensors suffer from drift and interferences. Nevertheless these methods are so convenient that the technology is expected to mature over the coming years.

There are direct indicating devices such as *Draeger* tubes, which became well known in the 1950s when they began to be used to detect alcohol in human breath. These direct indicating devices draw a small quantity of air through a tube packed with an analyte sensitive substance, such that a colour change gives a rough indication of pollutant concentration. Typically they are not sensitive enough to monitor ambient concentrations, but are useful in work places or industrial environments. However, a wide range of colorimetric agents are available, so they can give an immediate indication of the concentration of many pollutant gases (Radojevic & Baskin, 1999).

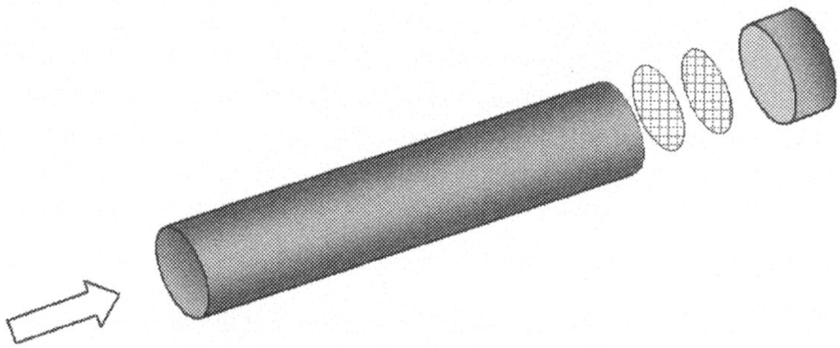

Figure 7.3 An exploded view of a diffusion tube showing from left to right the tube, the two gauzes that support the sorbent and the cap. During transport a further cap seals the open end of the tube, so gases cannot diffuse in (arrow).

Passive samplers, such as diffusion tubes, are somewhat more appropriate to environmental monitoring (Shooter et al., 1997). These are fairly sensitive, but need to integrate measurements over long periods of time (perhaps a week to several months). They do not involve any pumps so they can be placed in remote locations because the pollutants arrive at the adsorbent through diffusion. The low cost per tube also means that they can be set out at a large number of sampling points.

The most commonly used passive samplers are diffusion tubes that measure nitrogen dioxide. The tubes are 71 mm long with an internal diameter of 11 mm (Figure 7.3). One end of the tube is left open to the atmosphere, facing down to prevent rain or dust getting in. The other closed end of the tube has two stainless steel gauzes which act as a support for an absorbent to trap the pollutant. In the case of nitrogen dioxide, triethylamine is used as the absorbent and it converts the nitrogen dioxide to nitrate. This is trapped in the steel gauze to be analysed later by spectrophotometry or ion exchange chromatography. The range of diffusion tubes available will no doubt increase but they have been used to monitor: nitrogen dioxide, sulfur dioxide, hydrogen, carbonyl sulfide, ozone and simple organic acids (Namieśnik et al., 2005).

7.3.2.3 Sulfur dioxide

The air is passed into a hydrogen flame where sulfur atoms radiatively recombine emitting radiation in the 300–425 nm region. The photons can be counted and the number of sulfur atoms detected, but the method detects all sulfur atoms, so is subject to interference from H_2S, for example. An alternative method involves the measurement of fluorescence of SO_2 excited by radiation of 214 nm. Two-minute resolution is possible and concentrations as low as 0.1 ppb are detectable (Harrison, 1996).

7.3.2.4 Nitrogen oxides and ammonia

These are readily detected through a chemiluminescent reaction between nitric oxide (NO) and ozone (Harrison, 1996), which emits photons over the range 600–3000 nm, but the maximum is close to 1200 nm. The time resolution is about a second with a sensitivity of 0.1 ppb. Typically the air supply to the instrument switches back and forward between a direct supply and a stream that passes over a heated stainless steel or molybdenum catalyst. The catalyst converts NO_2 to NO, thus the instrument is able to measure the sum of the concentrations of these cases (i.e. NO_x). At high catalyst temperatures ammonia can also be converted to NO providing a very versatile method for determining the concentration of key nitrogen compounds near their sources.

7.3.2.5 Hydrocarbons

These can be determined by a flame ionization analyser, but this gives a measurement of the total amount of hydrocarbon (Harrison, 1996). Measurements of individual hydrocarbons are possible using gas chromatography (GC). Close to source where concentrations are high this can be done directly, but it typically requires pre-concentration. The GC can be coupled with mass spectrometry, such that the analyses can be conducted in an automated way and the output from the GC interpreted by a computer. However, this is clearly a complex undertaking to maintain and calibrate.

7.3.2.6 Ozone

Ozone is reliably measured through its absorption of the 254 nm line from a low-pressure mercury vapour lamp. The typical time resolution is 30 s and the method is sensitive to around 1 ppb ozone.

7.3.2.7 Carbon monoxide

Carbon monoxide is at ppm concentrations in urban atmosphere and non-dispersive infrared absorbance allows detection in the range 1–50 ppm (Harrison, 1996). There are potential interferences with carbon dioxide and water.

7.3.2.8 Particulate material

The mass of particulate matter in the atmosphere can be monitored continuously using a range of commercially available instruments. Simple ones can monitor the changing colour of filter papers as the air is drawn through (Figure 7.4). These can be effective, but are dependent on the colour of the particles being constant. One popular way to determine the concentration of particles in the air is to deposit them on the tip of the vibrating element of a microbalance. Its frequency can be measured and related to the deposited mass. With the appropriate inlet design the size of particles deposited can be selected. Time resolution is typically an hour and the method can have a sensitivity of about 4 $\mu g\,m^{-3}$. There are problems with hygroscopic particles, which may be heavier in the air than they will be once dried and deposited on the microbalance.

These techniques determine the mass rather than the composition (Harrison, 1996). This demands sampling a large volume of air collecting the particles on filters and using a range of techniques to analyse the inorganic or organic materials as required. These can be technically rather difficult when the analytes are at low concentration. Some analyses (e.g. for the presence of pesticides or polychlorinated dibenzo compounds such as dioxins) can be problematic outside experienced laboratories.

Figure 7.4 An aethelometer which allows near real-time measurements of black carbon aerosol particles that are typical of any combustion sources. The particles are caught on the moving paper tape and the darkness measured optically (photograph: Winter 2005/2006 Crew at Summit Station, Greenland).

7.4 EFFECTS

The primary concern about air pollutants has always been their impact on human health. This focus is understandable, but it has the potential to narrow the view of the effects of air pollutants, making it worthwhile to consider other effects.

7.4.1 Health

In the early 20th century toxicologists studied human response to gaseous pollutants either through deliberate exposure or after exposures that occurred as a result of industrial accidents (Brimblecombe, 1999). Such exposures were usually limited to a few individuals, but gave an understanding of the way in which carbon monoxide, lead and other obvious poisons acted. It was quickly recognized that carbon monoxide binds with haemoglobin in the blood and that this biochemical mechanism gave ways of linking pollution and the length of exposure to effects that were expressed in humans. This has allowed standards for exposure to carbon monoxide to be laid out very clearly and it was possible to relate them to the concentrations of carboxyhaemoglobin in the blood (Maynard & Waller, 1999).

However, the toxicology of most pollutants is not as clear as this. Pollution episodes have provided epidemiological data on the exposure of large number of people allowing detailed statistical analysis. This was particularly important after the smog of 1952 in London, where a range of analyses suggested that there were between 4000 and 12,000 excess deaths. The causes of these deaths are still somewhat of a mystery with different studies assigning them to smoke particles, sulfur dioxide and sulfuric acid

(Speizer, 1999). These early studies initiated a belief that sulfur dioxide acted in synergism with smoke particles to cause health effects. However, the nature of this synergism came to be questioned in more recent times, especially at lower pollution concentrations and a growing importance has been attributed to the fine particles in the air (Pope & Dockery, 1999).

Air pollution has become more complex since the days where it could be reduced to sulfur dioxide and soot. In modern photochemical smogs dominated by the presence of ozone and the nitrogen oxides, there is grave concern over potential for causing respiratory problems and a range of other health outcomes. In the public mind this has become increasingly worrisome because of a rising incidence of asthma in contemporary society (Peden, 2005). The relationship between air pollution and asthma is complex, but ozone impairs lung function, while nitrogen oxides, at high concentrations may affect asthmatics. Oxygen-containing compounds, such as aldehydes, cause eye, nose and throat irritation as well as headaches during periods of smog. Eye irritation has been associated with a group of nitrogen-containing organic compounds, the peroxyacylnitrates. The best known of these nitrogen-containing eye irritants is peroxyacetylnitrate, often called PAN (Taylor, 1969). There are also aromatic equivalents to PAN and peroxybenzoyl nitrate formed from toluene is a particularly potent eye irritant (WHO, 1987).

Since the 1990s, there has been an increasing understanding that the health effects of urban air pollution seem to be driven by fine particles in the atmosphere (Pope & Dockery, 1999). These fine particles are less than 10 μm (PM10), such that they can penetrate into the respiratory system and more often less than 2.5 μm (PM2.5) in diameter, which allows them to find their way into the alveoli of the lung. Improved statistical techniques have allowed a more effective analysis of the relationship between fine particle concentrations in urban air and health outcomes for large populations (Pope & Dockery, 1999). The effects may be most noticeable in people with pre-existing chronic lung and heart disease. Increasing particulate concentrations in air are linked with increases in hospital admissions for respiratory disease, reported symptoms and inhaler use among asthmatic patients. Evidence suggests that if PM10 daily average concentrations increase by 10 $\mu g\,m^{-3}$, mortality increases by 1%.

The underlying reasons why fine particles affect health are really not well understood. Particle size is very important and the PM2.5 particles are especially dangerous because they can penetrate deep into the lung. It may also be that their surface area is large, so there is a potential for oxidative compounds to be available on the surface. Metal oxides at the surface of the particle, such as those of iron and vanadium seem capable of inducing enhanced pulmonary inflammation.

So far, health outcomes that occur very soon after exposure to the pollutant have been discussed. Some air pollutants have health effects that accumulate over decades. In particular those that are carcinogens, such as benzene 1,3-butadiene and benzo(a)pyrene. Benzene is found in air along with other aromatic compounds that are components of gasoline. Although increases in the number of leukaemia case is expected from exposure to benzene in the urban atmosphere, it is complicated by the importance of other sources of benzene to humans such as tobacco smoke. Benzo(a)pyrene is just one of the PAH that are found on urban particles as a result of combustion processes (Viras et al., 1990). Many of these are potential carcinogens.

7.4.2 Material effects

Air pollution damage to many important historic buildings has accumulated where they have been exposed to the ravages of urban environments for centuries (Brimblecombe, 2000). In general, the main damaging agent is sulfur dioxide which oxidizes to sulfuric acid on many materials. The acid rapidly attacks building materials particularly carbonate-based building stone. This sulfation process is usually accompanied by the deposition of soot which leads to the formation of thick black crusts on the more sheltered parts of the building (Sabbioni, 1995). The decline of sulfur dioxide concentrations in urban areas over the last half century and an increase in petroleum-based fuels in cities in recent years has meant that contemporary crusts (Figure 7.5) are richer in organic material. It is likely that the blackening soot particles come from diesel sources in many cities (Bonazza et al., 2006).

Modern photochemical atmospheres can also affect the architecture of modern cities (Brimblecombe, 2000) which are constructed from increasingly diverse materials with polymers, non-ferrous metals, concrete, glass, becoming as widely used as the stone, bricks, mortar, paints of the past. Although iron corrodes rapidly in atmospheres polluted with sulfur dioxide, other metals notably copper and aluminium can be more sensitive to ozone.

The pollutants also penetrate indoors and threaten historic collections in museums and galleries (Brimblecombe, 1990). Particular worries derive from hydrogen sulfide which tarnishes silver and copper and ozone and nitrogen oxides which affect sensitive dyes. In the past, paper and leather were badly affected by high sulfur dioxide concentrations indoors, which was a special problem for libraries with a collection of bound books (Brimblecombe, 1997).

Figure 7.5 The Cathedral of St John in Norwich, England, showing extensive blacking over the last 100 years, increasingly the result of particles from vehicles (photograph: Carlota M. Grossi).

7.4.3 Ecological and agricultural effects

There are a wide range of effects of air pollution on animals and plants (Wellburn, 1987). Farm animals can suffer fluorosis after ingesting grass where high concentrations of fluorine have been deposited (Cronin et al., 2000). This can arise from brick-making, aluminium smelting and in some cases high concentrations of fluorine are emitted in coal combustion.

Pollution effects on vegetation are perhaps more widely known than the effects on animals (Treshow & Anderson, 1989). Natural vegetation has been extensively damaged around areas where large quantities of sulfur dioxide were emitted from burning coal (Gunn et al., 1995). Vegetation, especially lichens are particularly sensitive to sulfur dioxide (Treshow & Anderson, 1989). Since the mid-20th century ozone has been a significant cause of damage or reduction in yield in the case of crops (Ashmore, 2005).

7.4.4 Acid rain and climate change

Many air pollutants also have longer range effects that can occur hundreds of kilometres away from their sources. Sulfur, nitrogen and perhaps more locally chlorine form strong acids (H_2SO_4, HNO_3, HCl) in the atmosphere (Brimblecombe, 1996). These lower the pH of rain and throughout the 1980s the acid rain problem was extremely serious in Europe and North America (Brimblecombe, 2002). The problem has faded somewhat, with the reduction in sulfur emissions, although nitrogen oxides continue to rise (Galloway, 2001). In recent years, acid rain has become more serious in the developing world with the Asia Pacific Rim experiencing increasing concentrations of acid rain precursors as a result of rapid urbanization and industrialization (Brimblecombe, 2002).

Additionally gases and particles can absorb radiation and alter the energy balance of the earth. This is popularly known as the carbon dioxide greenhouse effect (Pearce, 2004), which is set to raise the temperature of the earth over the next century. It is not so widely appreciated that methane, ozone and the halocarbons are also greenhouse gases. The chlorofluorocarbons and the hydrofluorocarbons (HFCs) are widely used and best known as refrigerants are particularly significant greenhouse gases. However, industrial releases can also be important and often these compounds are not covered by the various international protocols that govern greenhouse gases. The manufacture of HFC-23 releases the non-regulated HFC-22 as a by-product. Radiatively active tetrafluoromethane and hexafluoroethane arise in the aluminium production, while more recently hexafluoroethane has been released from plasma etching of silicon (Chang & Yu, 2001).

7.4.5 Odour and other issues

Finally, there is a range of effects that are more difficult to quantify. The impact of odour on the quality of our lives is sufficiently serious to be regulated. Odour is of frequent concern and in rural areas complaint to regulatory authorities can frequently arise from farming practices. Even in cities complaints derive from seemingly innocuous activities such as coffee roasting, chocolate manufacture and food preparation. Domestic and commercial wastes are a further source, so regulations limit potential odour problems from solid waste disposal. The level of public concern over such nuisance has increasingly challenged planning applications.

Figure 7.6 Small electronic nose instrument developed from research at Warwick University at the Sensors Research Laboratory led by Prof. Julian Gardner.

Odour is clearly a problem with some industrial activities and can be problematic during sewage treatment, from industries using solvents, the petroleum industry and activities such as brick-making (Sironi et al., 2005). Many odours are associated with sulfur compounds which can be present as organosulfides in waste gases (Bordado & Gomes, 2002). Mercaptans arising from brick-making give a smell rather like burning rubber. Although petroleum hydrocarbons have very little odour, sulfur compounds and some unsaturated compounds in the waste gases contribute to the smell.

Quantitative assessment of odour is time-consuming, so initial surveys tend to be qualitative and can begin by assembling odour diaries and complaint histories (Van Den Hazel & Waegemaekers, 1991). These are useful in establishing the source of the odour and can be combined with a knowledge of the level of industrial activity or the meteorological situation when the odour was prevalent. More quantitative approaches can use dilution olfactometry and exposure to panels of people (Mannebeck & Mannebeck, 2001). The panel is exposed to carefully diluted odorous air in a sniffing port, until the panel detects the presence of the odour. The approach allows calibration by determining the odour thresholds of individuals by assessing their sensitivity to a gas such as *n*-butanol vapour (Mannebeck & Mannebeck, 2001).

There are also ENoses or electronic noses (Figure 7.6), which often use a number of different polymer films that conduct electricity when materials are deposited on their surfaces (Gardner & Bartlett, 1999). Each film (or element of the sensor array) reacts with the various olfactants in a different way, so the spectrum of responses from the different films allows the odour to be "fingerprinted" (e.g. Wen et al., 2002).

Visibility has generally attracted less concern than odour, although it is an important part of the US Clean Air Act. Here, there has been an attempt to prevent significant detriment to air quality in remote areas. This has been seen as a way of protecting the grand vistas that can be appreciated in the western US.

7.5 CONCLUSIONS

Air pollutants represent a complex mixture with many components spread between different phases and at low concentration. While this presents a range of challenges in understanding, analysis and policy development air pollution is ever changing. The last century has seen dramatic changes in emissions to the environment. The variety of compounds released to the atmosphere in waste streams is probably greater than ever. While it can be difficult to say in what way air pollution may have become better or worse, it is clear that it is changing.

REFERENCES

Ashmore, M.R. (2005) Assessing the future global impacts of ozone on vegetation. *Plant, Cell Environ.* **28**, 949–964.

Badr, T. and Harion, J.L. (2005) Numerical modelling of flow over stockpiles: implications on dust emissions. *Atmos. Environ.* **39**, 5576–5584.

Baltensperger, U., Kalberer, M., Dommen, J., Paulsen, D., Alfarra, M.R., Coe, H., Fisseha, R., Gascho, A., Gysel, M., Nyeki, S., Sax, M., Steinbacher, M., Prevot, A.S.H., Sjögren, S., Weingartner, E. and Zenobi, R. (2005) Secondary organic aerosols from anthropogenic and biogenic precursors. *Faraday Discuss.* **130**, 265–278.

Bendahan, M., Boulmani, R., Seguin, J.L. and Aguir, K. (2004) Characterization of ozone sensors based on WO_3 reactively sputtered films: influence of O_2 concentration in the sputtering gas, and working temperature. *Sensor. Actuator. B: Chem.* **100**, 320–324.

Bonazza, A., Brimblecombe, P., Grossi, C.M. and Sabbioni, C. (2006) Carbon in black layers at the Tower of London, *Environ. Sci. Technol.* (submitted).

Bordado, J.C.M. and Gomes, J.F.P. (2002) Atmospheric emissions of kraft pulp mills. *Chem. Eng. Process.* **41**, 667–671.

Brimblecombe, P. (1990) The composition of museum atmospheres. *Atmos. Environ.* **24B**, 1–8.

Brimblecombe, P. (1996) *Air Composition and Chemistry*. Cambridge University Press.

Brimblecombe, P. (1997) Pollution studies. In *Environment Leather Project – Deterioration and Conservation of Vegetable Tanned Leather.* The Royal Danish Academy of Fine Arts, School of Conservation, Esplanden 34, DK-1263, København, Danmark, K25–K30.

Brimblecombe, P. (1999) Air pollution and health history. In *Air Pollution and Health* (eds. S.T. Holdgate, J.M. Samet, H.S. Koren and R.L. Maynard), pp. 5–18. Academic Press, San Diego, CA.

Brimblecombe, P. (2000) Air pollution and architecture, past, present and future. *J. Archit. Conserv.* **6**, 30–46.

Brimblecombe, P. (2002) Acid drops. *New Sci.* **174**, A1–A4.

Brimblecombe, P. (2003) The global sulphur cycle. In *Treatise on Geochemistry Volume 8 Biogeochemistry* (ed. W.H. Schlesinger), pp. 645–682. Elsevier, Amsterdam.

Chang, M.B. and Yu, S.J. (2001) An atmospheric-pressure plasma process for C_2F_6 removal. *Environ. Sci. Technol.* **35**, 1587–1592.

Colls, J. (1997) *Air Pollution.* E and F.N. Spon, London.

Cronin, S.J., Manoharan, V., Hedley, M.J. and Loganathan, P. (2000) Fluoride: a review of its fate, bioavailability, and risks of fluorosis in grazed-pasture systems in New Zealand. *New Zeal. J. Agricul. Res.* **43**, 295–321.

Derwent, R.G., Simmonds, P.G., O'Doherty, S. and Ryall, D.B. (1998) The impact of the Montreal Protocol on halocarbon concentrations in northern hemisphere baseline and European air masses at Mace Head, Ireland over a ten year period from 1987–1996. *Atmos. Environ.* **32**, 3689–3702.

Galloway, J.N. (2001) Acidification of the world: natural and anthropogenic. *Water, Air, Soil Pollut.* **130**, 17–24.

Gardner, J.W. and Bartlett, P.N. (1999) *Electronic Noses, Principles and Applications*. Oxford University Press.

Gunn, J., Keller, W., Negusanti, J., Potvin, R., Beckett, P. and Winterhalder, K. (1995) Ecosystem recovery after emission reductions: Sudbury, Canada. *Water, Air, Soil Pollut.* **85**, 1783–1788.

Haagen-Smit, A.J. (1964) The control of air pollution. *Sci. Am.* January, 24–31.

Harrison, R.M. (1996) *Pollution Causes, Effects and Control.* Royal Society of Chemistry, London.

Kameda, T., Sanukida, S., Inazu, K., Hisamatsu, Y., Maeda, Y., Takenaka, N. and Bandow, H. (2005) Association of the mutagenicity of airborne particles with the direct emission from combustion processes investigated in Osaka, Japan. *Atmos. Environ.* **38**, 6937–6945.

Kamei, K.-I., Mie, M., Yanagida, Y., Aizawa, M. and Kobatake, E. (2004) Construction and use of an electrochemical NO sensor in a cell-based assessing system. *Sensor. Actuator. B: Chem.* **99**, 106–112.

Latif, M.T. and Brimblecombe, P. (2004) Surfactants in atmospheric aerosols. *Environ. Sci. Technol.* **38**, 6501–6506.

Lipsky, E.M. and Robinson, A.L. (2006) Effects of dilution on fine particle mass and partitioning of semivolatile organics in diesel exhaust and wood smoke. *Environ. Sci. Technol.* **40**, 155–162.

Mannebeck, D. and Mannebeck, H. (2001) Interlaboratory comparison of dynamic olfactometry in Central Europe 2000. *Water Sci. Technol.* **44**, 27–32.

Maynard, R.L. and Waller, R. (1999) Carbon monoxide. In *Air Pollution and Health* (eds. S.T. Holdgate, J.M. Samet, H.S. Koren and R.L. Maynard), pp. 749–796. Academic Press, San Diego, CA.

Mocarelli, P. (2001) Seveso: a teaching story. *Chemosphere* **43**, 391–402.

Namieśnik, J., Zabiegala, B., Kot-Wasik, A., Partyka, M. and Wasik, A. (2005) Passive sampling and/or extraction techniques in environmental analysis: a review. *Anal. Bioanal. Chem.* **381**, 279–301.

Radke, M. (2005) Sterols and anionic surfactants in urban aerosol: emissions from wastewater treatment plants in relation to background concentrations. *Environ. Sci. Technol.* **39**, 4391–4397.

Radojevic, M. and Bashkin, V.N. (1999) *Practical Environmental Analysis*. Royal Society of Chemistry, London.

Ramgir, N.S., Hwang, Y.K., Jhung, S.H., Mulla, I.S. and Chang, J.-S. (2006) Effect of Pt concentration on the physicochemical properties and CO sensing activity of mesostructured SnO_2. *Sensor. Actuator. B: Chem.* **114**, 275–282.

Russell, L.M., Maria, S.F. and Myneni, S.C.B. (2002) Mapping organic coatings on atmospheric particles. *Geophys. Res. Lett.* **29**, 26–31.

Pearce, F. (2004) Harbingers of doom? *New Sci.* **183**, 44–47.

Peden, D.B. (2005) The epidemiology and genetics of asthma risk associated with air pollution. *J. Allergy Clin. Immunol.* **115**, 213–219.

Pope, C.A. and Dockery, D.W. (1999) Epidemiology of particle effects. In *Air Pollution and Health* (eds. S.T. Holdgate, J.M. Samet, H.S. Koren and R.L. Maynard), pp. 673–706. Academic Press, San Diego, CA.

Sabbioni, C. (1995) Contribution of atmospheric deposition to the formation of damage layers. *Sci. Total Environ.* **167**, 49–55.

Samburova, V., Zenobi, R. and Kalberer, M. (2005) Characterization of high molecular weight compounds in urban atmospheric particles. *Atmos. Chem. Phys.* **5**, 2163–2170.

Sepizer, F.E. (1999) Acid sulphate aerosols and health. In *Air Pollution and Health* (eds. S.T. Holdgate, J.M. Samet, H.S. Koren and R.L. Maynard), pp. 603–618. Academic Press, San Diego, CA.

Shooter, D., Brimblecombe, P., Shooter, J., Lowe, D., Day, P.J. and Du, S. (1997) Some characteristics and applications of nitrogen dioxide passive samplers. *Environ. Technol.* **18**, 243–254.

Sironi, S., Capelli, L., Céntola, P., Del Rosso II, R. and Grande, M. (2005) Odour emission factors for assessment and prediction of Italian MSW landfills odour impact. *Atmos. Environ.* **39**, 5387–5394.

Speight, J.G. (2005) *Environmental Analysis and Technology for the Refining Industry.* Wiley, Hoboken, NJ.

Streets, D.G., Hao, J., Wu, Y., Jiang, J., Chan, M., Tian, H. and Feng, X. (2005) Anthropogenic mercury emissions in China. *Atmos. Environ.* **39**, 7789–7806.

Taylor, O.C. (1969) Importance of peroxyacetylnitrate (PAN) as a phytotoxic air pollutant. *J. Air Pollut. Contr. Assoc.* **19**, 347–351.

Treshow, M. and Anderson, F.K. (1989) *Plant Stress from Air Pollution.* John Wiley and Sons, Chichester.

Van Den Hazel, P.J. and Waegemaekers, C.H.F.M. (1991) Odour annoyance in a residential area near a papermill. *Public Health Rev.* **19**, 251–262.

Viras, L.G., Athanasiou, K. and Siskos, P.A. (1990) Determination of mutagenic activity of airborne particulates and of the benzo[a]pyrene concentrations in Athens atmosphere. *Atmos. Environ. B: Urban Atmos.* **24**, 267–274.

Wellburn, A. (1987) *Air pollution and acid rain the biological impact.* Longman Scientific and Technical, London.

Wen, L., Zhang, L., Zhou, F., Lu, Y. and Yang, P. (2002) Quantitative determination of Freon gas using an electronic nose. *Analyst* **127**, 786–791.

WHO (1987) Toluene. In *Air Quality Guidelines for Europe*, pp. 137–147. World Health Organization, Geneva.

Williams, D.E. (1999) Semiconducting oxides as gas-sensitive resistors. *Sensor. Actuator. B: Chem.* **57**, 1–16.

8

Isotope characterization of gaseous pollutants

A.-L. Norman

8.1 INTRODUCTION

8.1.1 Identification of sources of emissions

Characterizing atmospheric pollutants and using this information to develop tools that permit identification of specific industrial or regional emissions from local to global scales should not be overlooked in view of the increasing importance of emissions trading and industrial regulation. Economies and industries are motivated to closing resource cycles as technological advances in removing pollutants improve, and public pressure to implement more stringent standards to protect air and water quality increase. Technology that can identify emission sources and quantify their impacts on air quality and environmental degradation play a role in informing the public of pollutant impacts. This chapter reviews how isotopes of carbon, sulfur, and oxygen can be used to identify, and in some cases, distinguish pollutant emissions and their impacts on the surrounding environment.

Compounds emitted during combustion or released as fugitive emissions from industrial facilities can, in specific instances, be tracked. The method described here relies on distinct differences in natural isotope abundance ratios, or patterns in ratios, between pollutant sources and/or between pollutants and background sources for the same compounds. For example, a common question surrounding climate change and the impact of anthropogenic CO_2 emissions on rising global atmospheric CO_2 levels is "How it is possible to distinguish natural and anthropogenic sources of CO_2?" The answer lies in differing isotope abundances within distinct carbon and oxygen reservoirs contributing to the product, gaseous CO_2. Mass balance equations, combining flux and isotope information, are used to constrain different source contributions to rising atmospheric CO_2 levels. The concepts and assumptions behind isotope mass balance calculations, and how they can be used for source apportionment, are reviewed.

8.1.2 Isotopes

Isotopes of an element contain different number of neutrons in the nucleus. The mass differences that result are more pronounced in light than heavy nuclides. The consequence is that small differences in mass at subatomic scales produce significant and measurable effects at macroscopic scales despite the fact that the isotopes have identical chemical properties.

8.1.2.1 Carbon

Carbon has three isotopes: ^{12}C, ^{13}C, and ^{14}C. The first two are stable and the third is radioactive with a half-life of approximately 5500 years. Although radioactive isotopes are useful tools in apportionment (e.g. Francey et al., 1995; Zondervan & Meijer, 1996), the use of stable isotopes will be the focus of this review. The natural abundance of stable carbon isotopes in terrestrial materials is typically 98.9% and 1.1% for ^{12}C and ^{13}C, respectively. Carbon materials are converted to CO or CO_2 and are measured using an isotope ratio mass spectrometer (IRMS). Slight variations at the parts per thousand level in ratios of $^{13}C/^{12}C$ (or R^{13}) are related to an international standard material. For carbon, the standard material is a mineral with a specific isotope ratio designated by the International Atomic Agency in Vienna, Pee Dee Belamnite or VPDB. The isotope composition, or delta value (δ) in a sample material, X, is compared to the standard, ST, using the relationship:

$$\delta^{13}C = \{(R^{13}_{X}/R^{13}_{ST}) - 1\} \times 10^3$$

Delta values are described in permil (‰) or parts per thousand difference in abundance. Materials containing more ^{13}C than the standard have positive $\delta^{13}C$ values, whereas those with less will be negative.

8.1.2.2 Sulfur

Sulfur has five isotopes: ^{32}S, ^{33}S, ^{34}S, ^{35}S, and ^{36}S. Four are stable (^{32}S, ^{33}S, ^{34}S, ^{36}S) while ^{35}S is radioactive with a half-life of approximately 83 days. Natural abundances for the isotopes of sulfur for ^{32}S to ^{36}S are 95.02%, 0.75%, 4.21%, and 0.017%, respectively. Variations are described relative to the two most abundant isotopes $^{34}S/^{32}S$ (or R^{34}). Typically sulfur compounds are converted to SO_2 or SF_6 and a gas source IRMS is used to relate variations in isotope abundance to the international standard Vienna Cañon Diablo Meteorite (VCDT). The $\delta^{34}S$ for the sample, X, is expressed relative to the standard using a similar relationship to that for carbon:

$$\delta^{34}S = \{(R^{34}_{X}/R^{34}_{ST}) - 1\} \times 10^{3}\ (‰)$$

8.1.2.3 Oxygen

Oxygen has three stable isotopes: ^{16}O, ^{17}O, and ^{18}O. The natural abundances for ^{16}O, ^{17}O and ^{18}O are 99.76%, 0.038%, and 0.204%, respectively. Similar to what is done for carbon and sulfur, oxygen isotope variations are described relative to the two most abundant isotopes; in this case $^{18}O/^{16}O$ (or R^{18}). Oxygen isotope ratios are typically measured as CO_2 or CO using a gas source IRMS. Variations in oxygen isotope abundance are compared to the international standard Vienna Standard Mean Ocean Water (VSMOW). The R^{18} for a sample is compared to SMOW and reported as $\delta^{8}O$ in permil (‰).

$$\delta^{18}O = \{(R^{18}_{X}/R^{18}_{ST}) - 1\} \times 10^{3}\ (‰)$$

8.2 ISOTOPE ABUNDANCE VARIATIONS

8.2.1 Natural carbon isotope abundance variations

Isotope ratios are altered, or fractionated, when one isotope is favored during biological, chemical, or physical processes. Carbon isotopes are fractionated during photosynthesis, for example, where ^{12}C in atmospheric CO_2 diffusing through the stomata is preferentially fixed and incorporated into organic material. Atmospheric CO_2 that is well mixed globally has a carbon isotope composition of ~−8‰. The fractionation factor for carbon during photosynthesis is −20‰ and this holds for the majority of plants, those that use the C-3 (Calvin-Benson) photosynthetic pathway, such as cereals as well as most trees and grasses. Carbon isotope fractionation is less (~−8‰ to −12‰) for succulent plants that have adapted the C-4 (Crusselacean acid metabolism) photosynthetic pathway in response to arid conditions. Therefore, around the globe, bulk organic matter has an isotope composition of ~−25‰ to −27‰, but contributions from C-4 plants may shift this toward −16‰ to −20‰ in present-day organic matter in desert regions. Combustion of organic matter releases CO_2, with a similarly light isotope composition, back to the atmosphere.

8.2.2 Natural oxygen isotope abundance variations

As described above, two sources of carbon can be distinguished based on their distinct isotope composition; atmospheric CO_2 and that from combustion of organic matter. It is sometimes useful to consider the oxygen isotope composition as well as the carbon in atmospheric compounds as it provides additional information. Molecular oxygen in the atmosphere has an isotope composition of +24‰ with respect to SMOW, whereas water at a mid-latitude location may have an oxygen isotope composition of −10‰ with respect to the same standard. Oxygen isotopes in precipitation vary with latitude, altitude, and with distance inland from a coastal site (continental effect) and are a function of the relative humidity, the isotope composition of the source water, and the temperature at which evaporation took place (Dansgaard, 1964; Peng et al., 2005). CO_2 that is equilibrated with water is shifted by approximately 40‰, so CO_2 in equilibrium with water in this example would be expected to have an isotope composition of +30‰. Fractionation occurs during diffusion and exchange in stomata and soil, so the isotope composition from plant and soil respiration can be expected to range between ~+38‰ and +43‰. CO_2 formed during combustion typically incorporates O_2 and retains the isotope signature of atmospheric oxygen, so that it is distinct relative to CO_2 that is respired from plants, or that formed from decay of organic matter. Once in the atmosphere, the probability for exchange of carbon and oxygen atoms in CO_2 with other atmospheric compounds is low (Ciais et al., 1997) and the CO_2 retains its isotope composition.

8.2.3 Natural sulfur isotope abundance variations

Sulfur isotope abundances vary considerably both spatially and with respect to geological time. This is mainly due to the very large sulfur isotope fractionations associated with bacterial sulfate reduction. ^{32}S in sulfate is preferentially reduced to H_2S to meet the energy needs of one of the oldest forms of life on the planet, bacteria, in a process similar to photosynthesis but taking place in anoxic rather than in oxic surroundings. Over geological time, sulfides have generally become enriched in ^{32}S, and the opposite has occurred for many sulfates. For example, seawater sulfate has a $\delta^{34}S$ value of +21‰ while pyrite (FeS) found in glacial tills can be quite negative (Krouse et al., 1984). Undisturbed soils from the plains in southern Alberta, for example, have $\delta^{34}S$ values near −18‰ (Krouse et al., 1991). Plants growing in the soil are isotopically similar to the medium in which they grow, so transport of sulfur to the atmosphere, through a forest fire for example, will result in $\delta^{34}S$ values for SO_2 near −18‰ as well (Krouse et al., 1991). In contrast, gypsum ($CaSO_4.2H_2O$) which is often present in association with petroleum deposits, has $\delta^{34}S$ values ranging between +10‰ and +35‰ (Pilot, 1991). Thermochemical sulfate reduction at depth produces H_2S with the same isotope composition as the sulfate from which it formed (Kiyosu & Krouse, 1990). Again, the occurrence of distinct isotope signatures in a region can be exploited to infer the amount of atmospheric sulfur from each source (Bradley, 2005; Norman et al., 2004a).

8.3 ISOTOPE MASS BALANCE

8.3.1 Two source mixing

If compounds produced during combustion such as CO_2 or SO_2 downwind of a combustion event were sampled for isotope analysis, the isotope composition would be intermediate to the two end members. For CO_2 the end members would be atmospheric CO_2 at −8‰ and CO_2 from combustion, for example, at −27‰. Similarly, the isotope composition of sulfur in SO_2 would fall between the isotope composition of background and combustion SO_2. An example of this type of mixing is illustrated in Figure 8.1. At low concentrations the isotope composition of SO_2 is very similar to

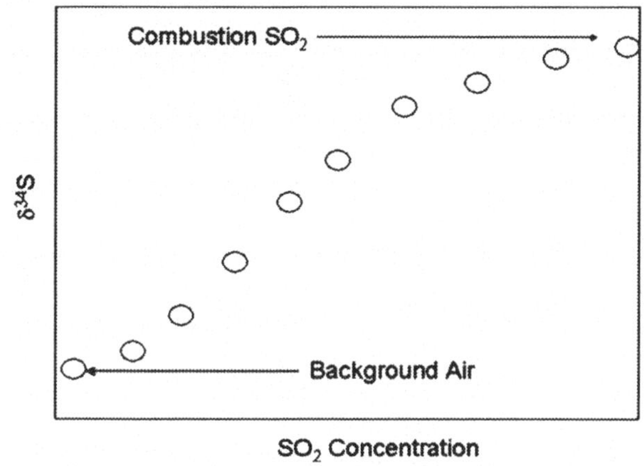

Figure 8.1 An example of two source mixing for the isotope composition of SO_2 is shown. The sulfur isotope composition of the SO_2 in well-mixed air changes with the amount of combustion SO_2 present. In this example, background air has a sulfur isotope composition that is lighter than SO_2 from a combustion source. As the amount of SO_2 from combustion increases, the isotope composition of the mixed SO_2 increases to a limit which is defined by the $\delta^{34}S$ for combustion SO_2.

that of background air. Additional SO_2 from combustion changes the isotope composition of the ambient SO_2 mixture with concentration and approaches a limiting $\delta^{34}S$ value for combustion SO_2 at high concentration. Note, however, that the $\delta^{34}S$ values for the end members may differ with time and location: unlike carbon isotopes they are not globally consistent. In using sulfur isotopes for mass balance it is necessary to ascertain beforehand that the isotope composition of the combustion material and background SO_2 are measurably different, and distinct, on the spatial and temporal scale of the study.

Mathematically, the mixing of two sources with distinct isotope compositions can be expressed in the form of two equations:

$$\delta_T = \delta_A \times f_A + \delta_C \times f_C$$
$$1 = f_A + f_C$$

where the subscripts A and C stand for air and combustion, respectively, and f is the fractional contribution from each source.

8.3.2 Multiple source mixing

Two source mixing is a relatively simple phenomenon that is typically restricted to clean air masses impacted by a single pollutant source. More commonly, multiple emission sources influence the concentration of a compound in air masses. Mathematical expressions for multiple source mixing are:

$$\delta_T = \sum \delta_i \times f_i$$
$$1 = \sum f_i$$

where the sum is over a number of distinct isotope sources (i). Note that with only two equations it is possible to solve for only two variables; the other components must be measured independently or derived from a proxy.

Isotope apportionment techniques are valid once it has been demonstrated that sources are isotopically distinct and that the distinctions are retained over spatial and temporal scales appropriate to the study. It is also important to know whether there is isotope selectivity during chemical or physical transformations, such as oxidation, that may impact the isotope composition of the compound as it is transported through the atmosphere. These effects are well known for CO_2. However, it is often necessary to demonstrate whether or not this is true when working with other atmospheric compounds (e.g. SO_2), particularly if they are short lived.

Reaction products from oxidation or condensation of gaseous emissions can also be tracked using isotope apportionment so long as there is little isotope fractionation during the transformation processes. Sulfur dioxide, a typical combustion emission, is oxidized to sulfate within a few days. Once sulfate has formed it attracts water vapor and forms a fine aerosol. Little isotope selectivity was observed in studies where SO_2 and aerosol sulfate were collected simultaneously (Norman et al., 2004a, b) and demonstrates that fractionation effects on oxidation were small.

Figure 8.2 shows an example where multiple source mixing for aerosol sulfate in coastal region was resolved (Norman et al., 2004b). Three isotopically distinct sulfate sources were apparent and form the points of a triangle bordered by the lines in Figure 8.2. Aerosol sodium concentrations and SO_4/Na ratios for bulk seawater (0.252 by mass) were used to define sulfate from sea spray. The triangular space in which the data fall in Figure 8.2 are bounded by two mixing lines. The upper line is where data would lie on a plot of $\delta^{34}S$ versus percent sea salt when sea salt mixed with sulfate with an isotope composition of approximately +6‰. The lower line represents mixing between sea salt and sulfate with a $\delta^{34}S$ value of −2‰. Measurements of emissions

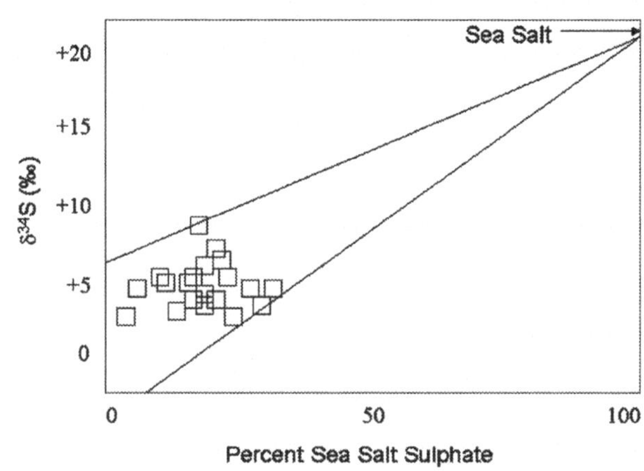

Figure 8.2 The sulfur isotope composition of aerosol sulfate is shown relative to the percent sea salt sulfate present in the sample. The data are bounded by two mixing lines; the upper line represents sea salt mixing with sulfate having a $\delta^{34}S$ value of +6‰, while the lower line demonstrates mixing between sea salt and an emission source with a $\delta^{34}S$ value of −2‰.

downwind from a refinery known to be a major contributor to atmospheric sulfur in the region were found to have a $\delta^{34}S$ value of −1.7‰. The source of the more positive $\delta^{34}S$ value at +6‰ represents well-mixed air in which vehicle exhaust and biogenic gases from the ocean contributed to aerosol sulfate. Apportionment of these distinct sulfur isotope sources was possible using sodium as a constraint for sea salt.

8.3.3 Source apportionment using C and O isotopes

If both $\delta^{13}C$ and $\delta^{18}O$ in atmospheric CO_2 are measured, multiple sources can be distinguished. Consider the situation described before, CO_2 from combustion mixed with CO_2 in background air, but add a third source, from plant and soil respiration and decay. Carbon isotopes can now be used to distinguish combustion (C) plus plant and soil respiration (PS), from background CO_2 (A) in air. Oxygen isotope measurements identify the varying contributions from combustion and plant and soil respiration. Mathematically this is written:

$$\delta^{13}C_T = \delta^{13}C_{C+PS} \times (f_C + f_{PS}) + \delta^{13}C_A \times f_A$$

$$\delta^{18}O_T = \delta^{18}O_C \times f_C + \delta^{18}O_{PS} \times f_{PS} + \delta^{18}O_A \times f_A$$

$$1 = f_C + f_{PS} + f_A$$

$\delta^{13}C_T$ and $\delta^{18}O_T$ are measured quantities, and $\delta^{13}C_A$, $\delta^{13}C_{C+PS}$, $\delta^{18}O_C$, and $\delta^{18}O_A$ are known. $\delta^{18}O_{PS}$ can be determined for the region where the measurements are made. This leaves three equations and three unknowns, f_C, f_{PS}, and f_A, and is a solvable matrix that can be used to find either the proportional contribution to atmospheric CO_2 from each source or, if the total concentration of CO_2 in air is known, the concentration attributable to each of the three sources.

An example of CO_2 mixing and using both carbon and oxygen isotopes to resolve a similar problem is shown in Figure 8.3 (after Huang et al., 2001). Diurnal differences in

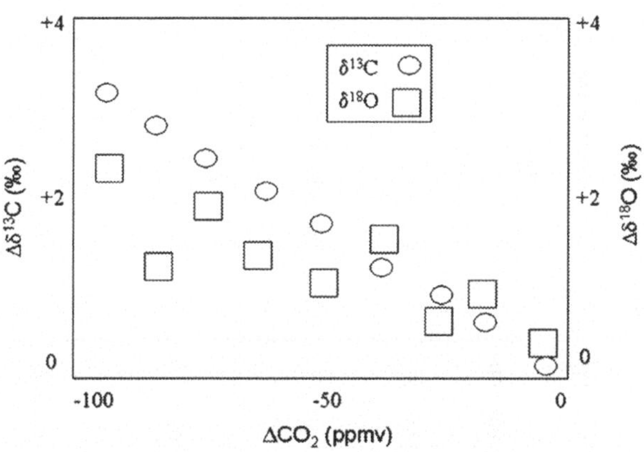

Figure 8.3 Differences in the diurnal carbon and oxygen isotope composition for CO_2 were absent when there was no change in CO_2 mixing ratios. However, as CO_2 was removed from the atmosphere by photosynthesis, changes in the diurnal carbon and oxygen isotope compositions reflected varying effects on plant photosynthesis and respiration.

the carbon and oxygen isotope composition were compared to differences in CO_2 concentration. It is apparent there is no difference in the carbon or oxygen isotope composition of CO_2 during the day and night when little plant respiration or photosynthesis occurs ($\Delta CO_2 = 0\,ppm$). However, as photosynthesis draws CO_2 mixing ratios down during the day, and ΔCO_2 values become negative, the carbon and oxygen isotope compositions vary and reflect contributions from photosynthesis and respiration.

8.3.4 Compound-specific carbon isotope analysis

In the previous section, it was shown that multiple isotopes can be used to identify gases emitted to the atmosphere from different sources. Another method, particularly useful for tracking organic pollutants, is to compare the isotope composition of a large range of compounds in air to those emitted by a particular source (Rudolph et al., 2002). This technique is called compound-specific isotope analysis (CSIA) and it has proved to be a powerful tool in tracking oil spills (Dempster et al., 1997) and is currently a possible tool for tracking gaseous pollutants (Comis, 2003) and their oxidation products. Organic gases released at high temperatures condense and coagulate to form fine aerosols. The isotope composition of the original organic material that was combusted is retained by the aerosols in patterns that can be used to identify their sources (Norman et al., 1999; Okuda et al., 2002). Figure 8.4 illustrates compound-specific isotope measurements for organic aerosols collected downwind of a combustion source. If the isotope pattern and specific organic content of multiple emission sources are known and significantly distinct, then it is possible to resolve the contribution of each source in well-mixed air.

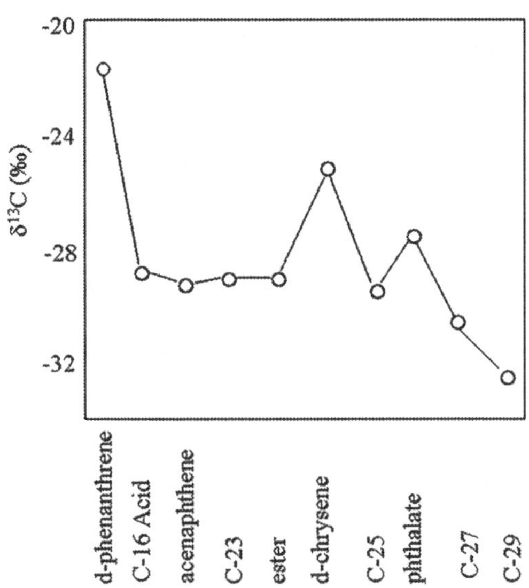

Figure 8.4 Carbon isotope variations in specific organic compounds separated from an aerosol downwind of a C-3 combustion source.

8.4 CONCLUSION

An overview of how C, S, and O can be used in isotope apportionment studies of atmospheric gases and aerosols has been described. In using isotopes in atmospheric studies, two factors must be considered. First, it is important to recognize local, regional, or global emissions of an element and understand its transformation and deposition processes. This is essential to evaluate how different forms of an element contribute to the local pollutant load. Second, it is important to characterize regional emissions that are likely to have a significant impact on concentrations of a compound in the atmosphere to determine whether, and when, emissions sources have distinct isotope compositions. Isotope apportionment techniques comparing CO_2, a long-lived compound important at the global scale, and SO_2, a relatively short-lived compound that is important at regional scales, were described. Methods that can be used to track and quantify emissions that produce fine aerosols are increasingly important to regulatory bodies whose mission is to reduce the potential health effects associated with atmospheric pollutants. High concentrations of submicrometer aerosols in cities have been shown to be associated with increased hospital admissions (Burnett et al., 1998). Some examples of tracking pollutants using inorganic and organic components of aerosols were also reviewed.

REFERENCES

Bradley, C.E. (2005) A multiple isotope approach to tracing natural gas processing emissions. M.Sc. Thesis, University of Calgary, 157 pp.

Burnett, R.T., Cakmak, S. and Brook, J.R. (1998) The effect of the urban ambient air pollution mix on daily mortality rates in 11 Canadian cities. *Can. J. Public Health* **89**(3), 152–156.

Ciais, P., Denning, S., Tans, P.P., Berry, J.A., Randall, D.A., Collatz, G.J., Sellers, P.J., White, J.W.C., Trolier, M., Meijer, H.E.J., Francey, R.J., Monfrey, P. and Heimann, M. (1997) A three-dimensional synthesis study of $\delta^{18}O$ in atmospheric CO_2. 1. Surface fluxes. *J. Geophys. Res.* **102**(D5), 5857–5872.

Comis, T.L. (2003) Chemical and isotopic characterisation of organic natural gas plant emissions in Alberta. M.Sc. Thesis, University of Calgary, 113 pp.

Dansgaard, W. (1964) Stable isotopes in precipitation. *Tellus* **16**, 436–468.

Dempster, H., Sherwood Lollar, B. and Feenstra, S. (1997) Tracing organic contaminants in groundwater – a new methodology using compound specific isotope analysis. *Environ. Sci. Technol.* **31**, 3193–3197.

Francey, R.J., Tans, P.P., Allison, C.E., Enting, I.G., White, J.C. and Trolier, M. (1995) Changes in oceanic and terrestrial carbon uptake since 1982. *Nature* **373**, 326–330.

Huang, L., Norman, A.L., Ernst, D., Chivulescu, A., Shashkov, A., Chan, D. and Higuchi, K. (2001) Isotopic compositions of ($\delta^{13}C$ and $\delta^{18}O$) CO_2 in air and their diurnal changes ($\Delta\delta^{13}C$ and $\Delta\delta^{18}O$) at a boreal forest site, Fraserdale, Canada: implications of processes in a terrestrial ecosystem. Poster presentation, *Proceedings of the 6th International Carbon Dioxide Conference*, Sendai, Japan, 1–5 October.

Kiyosu, Y. and Krouse, H.R. (1990) The role of organic acid in the abiogenic reduction of sulfate and the sulfur isotope effect. *Geochem. J.* **24**, 21–27.

Krouse, H.R., Legge, A.H. and Brown, H.M. (1984) Sulphur gas emissions in the boreal forest: the West Whitecourt case study V. Stable sulphur isotopes. *Water Air Soil Pollut.* **22**, 321–347.

Krouse, H.R., Stewart, B. and Grinenko, V.A. (1991) Pedosphere and biosphere (Chapter 7). In *SCOPE 43 Stable Isotopes: Natural and Anthropogenic Sulphur in the Environment* (eds. H.R. Krouse and V. Grinenko), pp. 267–306. John Wiley & Sons, Toronto, Ont.

Norman, A.L., Hopper, J.F., Blanchard, P., Ernst, D., Brice, K., Alexandrou, N. and Klouda, G. (1999) The stable carbon isotope composition of atmospheric PAHs. *Atmos. Environ.* **33**, 2807–2814.

Norman, A.L., Krouse, H.R. and MacLeod, J. (2004a) Apportionment of pollutant S in an urban airshed: Calgary, Canada, a case study. In *Air Pollution Modeling and Its Application XVI* (eds. C. Borrego and S. Incecik), pp. 107–125.

Norman, A.L., Belzer, W. and Barrie, L. (2004b) Insights into the biogenic contribution to total sulphate in aerosol and precipitation in the Fraser Valley afforded by isotopes of sulphur and oxygen. *J. Geophys. Res.* **109**, DOI: 10.1029/2002JD003072.

Okuda, T., Kumata, H., Naraoka, H. and Takada, H. (2002) Origin of atmospheric polycyclic aromatic hydrocarbons (PAHs) in Chinese cities solved by compound-specific stable carbon isotope analyses. *Org. Geochem.* **33**, 1737–1745.

Peng, H., Mayer, B., Norman, A.L. and Krouse, H.R. (2005) Modeling of hydrogen and oxygen isotope ratios for local precipitation. *Tellus* **57B**, 257–282.

Pilot, J. (1991) Sulphur and oxygen composition of marine evaporite and barite strata. In *SCOPE 43 Stable Isotopes: Natural and Anthropogenic Sulphur in the Environment* (eds. H.R. Krouse and V. Grinenko), pp. 68–95. John Wiley & Sons, Toronto, Ont.

Rudolph, J., Czuba, E., Norman, A.L., Huang, L. and Ernst, D. (2002) Stable carbon isotope composition of nonmethane hydrocarbons in emissions from transportation related sources and atmospheric observations in an urban atmosphere. *Atmos. Environ.* **36**, 1173–1181.

Zondervan, A. and Meijer, H.A.J. (1996) Isotopic characterisation of CO_2 sources during regional pollution events using isotopic and radiocarbon analysis. *Tellus* **48B**, 601–612.

9

Use of NMR to study *in situ* bioconversion of gaseous formaldehyde

A.-M. Delort

9.1 INTRODUCTION

NMR (nuclear magnetic resonance) spectroscopy is a common tool for chemists to analyse the structure of purified organic compounds in solution. However, NMR can be applied *in situ*, that is without any purification of the sample, to analyse complex biological mixtures including whole cells, tissues, cellular extracts, culture media, biofluids but also cells on solid supports and entrapped in gels. In this context, the fields of applications to which NMR has been extended include: medicine and pharmacology, nutrition, biotechnology and more recently environmental sciences. Concerning the environment, most studies are related to the biotransformation of organic pollutants by microorganisms (Delort & Combourieu, 2000, 2001; Grivet et al., 2003), However, only a few papers report gaseous compound bioconversion by *in situ* NMR. These gaseous compounds are almost exclusively CO_2, methane (CH_4), methanol (CH_2OH) and formaldehyde (CH_2O). These studies largely focussed on microbial metabolism using isotopically enriched compounds (^{13}C and deuterium). CO_2 transformation is related to the metabolism of anaerobic and autotrophic microorganisms, that of formaldehyde, methane and methanol to methylotroph and methanotroph microorganisms, but also to the metabolism of *Escherichia*, *Rhodobacter, Staphylococcus* and *Pseudomonas* strains.

This chapter is organised as follows: first, a short overview of the studies dealing with gas bioconversion is presented with the aim of giving general information about NMR specificities (nuclei properties, experimental approaches ...). Second, the case of formaldehyde metabolism is described in detail. Finally, a few perspectives and conclusions are given.

9.2 *IN SITU* NMR STUDY OF BIOCONVERSION OF GASEOUS COMPOUNDS

9.2.1 Generalities

Information about *in situ* NMR spectroscopy is reported here for non-specialists. NMR spectroscopy is based on specific properties of nuclei when placed in a magnetic field. Only some isotopes can give a signal: for instance ^{13}C is observable but not ^{12}C and ^{14}C; on the contrary ^{1}H, Deuterium and ^{3}H are all observable. However, these isotopes have specific properties and cannot be used in a similar way for studying microbial metabolism, and more particularly that of gaseous compounds. For instance, these isotopes differ in their natural abundance but also by their intrinsic properties such as their magnetogyric ratio. These two parameters determine the relative sensitivity of these nuclei (see Table 9.1).

The reader should remember that the interest of NMR spectra is that they are real fingerprints of the organic structures under study. Three main parameters can easily be measured on an NMR spectrum (i) *The chemical shift* which corresponds to the resonance frequency of a specific atom in a molecule. It is characteristic of the electronic environment in this structure (presence of aromatic group, double bonds, oxygen, nitrogen, chloride, ... etc.). This parameter named δ is expressed in ppm (part per million). It determines the position of the NMR signal in the spectrum. (ii) *The coupling constant*, named J is expressed in hertz. This parameter reflects the vicinity of two nuclei in a molecule and it corresponds to the scalar interaction via the electrons in a covalent bond between these two nuclei. This interaction is usually very low after 3 bonds. It can be homonuclear ($J_{H\text{-}H}$, $J_{C\text{-}C}$, ...) or heteronuclear ($J_{H\text{-}C}$, $J_{D\text{-}C}$, $J_{H\text{-}D}$, ...). This parameter determines the shape of the signal in the spectrum (singlet,

Table 9.1 NMR properties of various nuclei and gaseous compounds studied using these nuclei.

Nuclei	Magnetogyric ratio (10^7 rad $T^{-1}.s^{-1}$)	Natural abundance (%)	Relative sensitivity	Gaseous compounds
^{1}H	26.75	99.98	100	CH_2OH
^{13}C	6.73	1.11	1.59	CH_2OH, CO_2, CH_2O, CH_4
Deuterium	4.71	1.50	0.96	CH_2O

doublet, ... multiplet). (iii) *The integral* of the signal which is directly proportional to the number of nuclei which are resonating at a specific frequency. It can thus provide quantitative information.

Apart from this major interest of NMR as a tool for structure determination and quantitative measurements, NMR can be used to trace metabolic pathways using isotopic markers. The advantages in using NMR markers compared to radioactive ones are: (i) it is easier to manipulate stable isotopes for safety. (ii) NMR spectroscopy is non-invasive, meaning that there is no destruction of the sample. Thus, the sample can be used further for complementary experiments, also NMR experiments can be carried out *in vivo* on living cells. (iii) NMR combines isotopic enrichment with structural information. In other words, the NMR analysis of isotopomers identifies which molecule is enriched, the percentage of enrichment and also the position of the enriched nuclei in the molecule. All this information is available in a single spectrum from a sample containing a mixture of compounds. It is absolutely impossible to get such information with other analytical techniques combined with radioactive markers. NMR presents also some drawbacks: (i) an NMR spectrometer and NMR specialists are required to make the analysis and (ii) the NMR sensitivity (in the range of mM) remains very low compared to radioactive isotope detection.

9.2.2 ^{1}H NMR studies

The natural abundance of ^{1}H is almost 100% and its sensitivity is the best one (see Table 9.1), also it is a component of many organic molecules, therefore ^{1}H NMR sounds, *a priori*, like one of the most convenient techniques to study biotransformation of pollutants (Delort & Combourieu, 2000; Grivet et al., 2003). The main advantage is that no enriched molecules are needed. However, because of the narrow range of ^{1}H chemical shifts (about 15 ppm), ^{1}H NMR spectra are complex and many resonances can overlap, including those of the intrinsic microbial metabolites. In addition, the major problem is the presence of the large amount of H_2O molecules (about 55 M!) while metabolites under study are in the range of mM or less. This causes two types of problems: one is a technical problem concerning the saturation of the receptor and the second one is also due to the presence of a huge signal of water masking signals close to its resonance. These drawbacks are partially minimised by modern instruments and multidimensional NMR; but all these difficulties surely explain why almost no studies have been carried out in the case of gaseous compounds. Only methanol metabolism by bacteria isolated from cloud water has been studied recently in our group (Amato et al., 2005). Spectra were recorded directly on incubation media after centrifugation of bacteria. For instance, in the case of a *Pseudomonas* strain, the conversion of methanol into formate was shown. Due to the high sensitivity of ^{1}H NMR, methanol could also be detected in natural samples such as cloud water (unpublished results) and fog samples (Suzuki et al., 1998). Formaldehyde which has been extensively studied by ^{13}C and deuterium NMR cannot be directly measured by ^{1}H NMR in water samples as its resonance is lower than that of water (around 4.8 ppm). The only way to detect it would be to use ^{1}H-^{13}C NMR but in that case the detection limit would be much lower. It should also be noted that CO_2 cannot be studied using ^{1}H NMR because of its lack of protons.

9.2.3 ^{13}C NMR studies

The information content of carbon NMR spectra is very high because of the large chemical shift dispersion (about 250 ppm) and narrow lines that characterise this nucleus. This avoids most of the possible overlapping encountered for ^{1}H NMR. Further, the fact that almost all molecules of living cells have a carbon skeleton means that ^{13}C spectra will reflect most of the chemical rearrangements that may take place between substrate and final product. The main problem is the low sensitivity of the ^{13}C nucleus which prevents most of the studies at natural abundance (Table 9.1), although improvements of NMR spectrometers and NMR sequences (indirect ^{13}C-^{1}H NMR experiments) can partly overcome this problem. In most cases, it is thus necessary to synthesize specifically ^{13}C-enriched xenobiotics which is a limiting factor: these compounds may need to be synthesized for that specific purpose and are expensive. However, ^{13}C NMR remains one of the most interesting approaches to monitor biodegradative metabolism. The analysis of the different isotopomers gives very accurate information about metabolic pathways and metabolic fluxes. Also, NMR can easily be complemented by mass spectral analyses.

In the case of gaseous compounds, only commercially available ^{13}C-enriched molecules have been studied, including $^{13}CO_2$, $^{13}CH_4$, $^{13}CH_2OH$ and $^{13}CH_2O$.

The group of Sanders was pioneer in this field by studying formaldehyde metabolism in *Escherichia coli* (Commodari & Hunter, 1989; Hunter et al., 1984, 1985; Mason et al., 1986) and by *Pseudomonas putida* and *Staphilococcus aureus* (Mason & Sanders, 1989). These studies will be detailed in Section 9.3. Barber and Donohue (1998) continued the experiments on formaldehyde metabolism. They focussed their study on the role of glutathione-dependent formaldehyde dehydrogenase in *Rhodobacter sphaeroides*. A differential metabolism was evidenced when cells were incubated under aerobic or photosynthetic conditions.

The bioconversion of methanol has been investigated by different groups using ^{13}C *in vivo* NMR in *Methylosinus trichosporium* (Cornish et al., 1984) and in a type II methanotroph close to this species (Costa et al., 2001), *Pseudomonas* strains (Narbad et al., 1989), *Hansenula polymorpha* (Jones & Bellion, 1991), *Bacillus methanolicus* (Pluschkell & Flickinger, 2002) and *Methylobacterium extorquens* AM1 (Christoserdova et al., 2004). The methanol degradative pathway generates formaldehyde (Figure 9.1), so that the various approaches and results are very complementary and sometimes close to those concerning the degradation of formaldehyde. This feature will be exemplified in Section 9.3 using the study of Pluschkell and Flickinger (2002). Also in some cases, ^{13}C-enriched methane was used directly as a substrate for methanotrophs (Costa et al., 2000; Vecherskaya et al., 2001).

$^{13}CO_2$ has mainly been studied in metabolic pathways of anaerobic microorganisms. $^{13}CO_2$ incorporation in final volatile fatty acids (acetate, succinate, butyrate, propionate, valerate) has been investigated in bacteria from gut tracks such as pig large-intestinal microflora (De Graeve et al., 1990) and the rumen bacterium *Fibrobacter succinogenes* (Matheron et al., 1997). The unknown autotrophic pathways of CO_2 fixation in the sulfur-reducing Archeabacterium *Thermoproteus neutrophilus* and the phototrophic Eubacterium *Choroflexus aurantiacus* have been studied in detail by ^{13}C NMR (Schäfer et al., 1989; Strauss et al., 1992). However, in that case CO_2 was

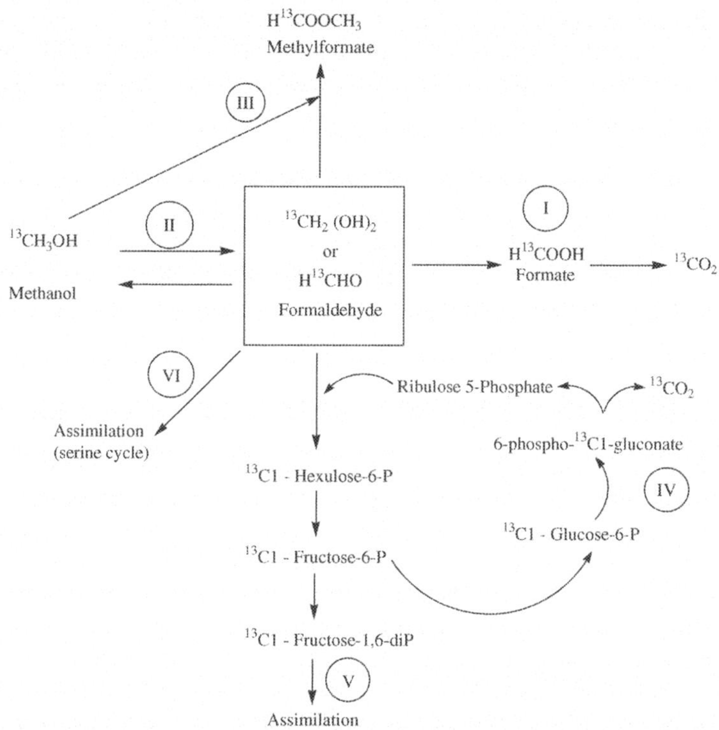

Figure 9.1 Main metabolic routes for the bioconversion of ^{13}C-formaldehyde.

unlabelled and its metabolism was investigated by analysing the ^{13}C isotopic pattern of purified cell constituents (amino acids, acetyl-CoA, pyruvate, oxaloacetate, and 2-oxoglutarate) issued from cells grown in the presence of CO_2 and [1, 4-^{13}C]succinate. [1, 4-^{13}C]succinate was used as a tracer, since this compound was incorporated in small amounts into all cell compounds without affecting the organisms' ability to synthesize all cell constituents from CO_2. This indirect approach proved to be extremely powerful.

9.2.4 Deuterium NMR studies

Similarly to ^{13}C, deuterium (D) is an interesting isotopic tracer allowing identification of metabolic pathways, especially when proton transfers are concerned. In particular, one can take advantage of $J_{H\text{-}D}$ couplings which can distinguish proton and deuterium bared by the same carbon atom (geminal position) and of the isotopic shift due to the quadrupolar effects of deuterium on ^{13}C chemical shifts. In addition, when D-enriched compounds are used, no interference with H_2O or intrinsic metabolite signals is encountered. Also the synthesis of D-enriched xenobiotics is sometimes easier and deuterium usually relaxes rapidly, making it relatively easy to get quantitative data. For these reasons deuterium can be more advantageous than ^{13}C. However, this nucleus presents also some drawbacks: the detection sensitivity and the chemical shift range of deuterium are lower than those of ^{13}C and strong isotopic effects can be observed in enzymatic reactions leading sometimes to toxic effects. Finally, another potential disadvantage of deuterium is that it is usually used as the source of field/frequency lock for a spectrometer. As a consequence, work is very often performed with the spectrometer unlocked which may induce field drift and thus frequency shifts for long-term experiments.

In the case of gaseous compounds, only formaldehyde has been studied. Again, the Sanders group was a pioneer in this area. They studied *in vivo* formaldehyde metabolism by *E. coli*, *P. putida* and *S. aureus* (Hunter et al., 1984; Mason & Sanders, 1989). Full details will be given in Section 9.3. D-formaldehyde NMR resonance is close to that of HDO (deuterium-enriched water). Therefore, when concentrations were very low, D-depleted water was used to avoid overlapping of these resonances.

9.3 NMR STUDY OF FORMALDEHYDE BIOCONVERSION

Formaldehyde is present in the environment (atmosphere, water and soil) both as a result of natural processes (decomposition of plant residues, photochemical reactions in the atmosphere, …) and from man-made sources (industrial production and emissions). It is considered as an environmental contaminant when detected at higher levels than background concentrations. Although this compound is toxic for many microbial species, some of them are able to transform it using various metabolic routes (Figure 9.1). For more details about the enzymes involved in these routes, the reader is referred to KEGG (Kyoto Encyclopaedia of Genes and Genomes) data available on the web (http://www.genome.jp/dbget-bin/show_pathway?rn00680+R01146). Routes I, II, III and IV lead to dissimilation of formaldehyde into final products such as methanol, methylformate or CO_2. Of course, the formation of CO_2 is the most frequent way of dissimilation. Routes I, II and III are "linear dissimilation" while route IV is a "cyclic dissimilation". Routes V and VI, on the contrary, are assimilation pathways of formaldehyde by the microorganisms.

Selected examples of deuterium and ^{13}C NMR studies will be described to show how this technique gave detailed information about these routes.

9.3.1 Identification of formaldehyde metabolites via routes I and II

Figure 9.2 shows *in vivo* ^{13}C NMR spectra of non-aerated *E. coli* suspensions recorded after 20 min and 3 h of incubation with ^{13}C-formaldehyde (Hunter et al., 1984). At time 20 min (lower spectrum), a major signal is recorded at $\delta = 83.2$ ppm due to the formation of formaldehyde hydrate $CH_2(OH)_2$ which is the usual form of formaldehyde in water. This signal is decreasing with time and new signals are visible in the upper spectrum corresponding to methanol (CH_3OH, $\delta = 50.1$ ppm) and formate (HCO_2^-, $\delta = 172.2$ ppm). These results are consistent with routes I and II. Complementary experiments have shown that when cells are aerated in the NMR tube, the formate signal disappeared and signals corresponding to carbon dioxide (CO_2, $\delta = 125.8$ ppm) and bicarbonate (HCO_3^-, $\delta = 161.2$ ppm) appeared (data not shown).

These experiments indicated the presence of a reductive pathway from formaldehyde to methanol and an oxidative pathway to bicarbonate and carbon dioxide via formate, this last reaction is classically catalysed by a formate dehydrogenase. Additional signals (XCH_2Y) were also detected in Figure 9.2 and were identified later (Hunter

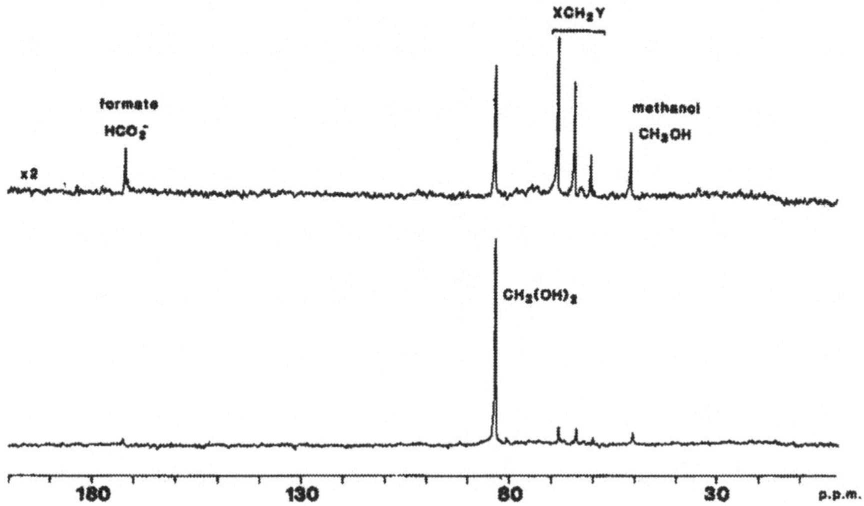

Figure 9.2 ^{13}C NMR spectra of non-aerated *E. coli* suspension (10^{12} cells mL^{-1}) initially containing 10 mM ^{13}C-formaldehyde collected after 20 min (lower spectrum) and 3 h (upper spectrum) (Hunter et al., 1984).

et al., 1985) as [^{13}C1]-1,2-propanediol ($\delta = 68$ ppm), [^{13}C1]-Glycerol ($\delta = 64$ ppm) and [^{13}C1]-1,3-propanediol ($\delta = 60$ ppm). A very interesting analysis of the various isotopomers (^{12}C/^{13}C) showed that these products had a double origin: exogenous formaldehyde and an endogenous bacterial source. Further studies have given more detailed insight in routes I and II in various bacterial strains (see Sections 9.3.2 and 9.3.3).

9.3.2 Synthesis of formate by GSH-formaldehyde dehydrogenase activity (route I)

Mason et al. (1986) have shown by *in vivo* ^{13}C NMR the transient formation of S-(hydroxymethyl)glutathione ($GSCH_2OH$) from glutathione and ^{13}C-formaldehyde (see Scheme I), showing unambiguously that glutathione is the cofactor of the formaldehyde dehydrogenase involved in this pathway in *E. coli*. Barber and Donohue

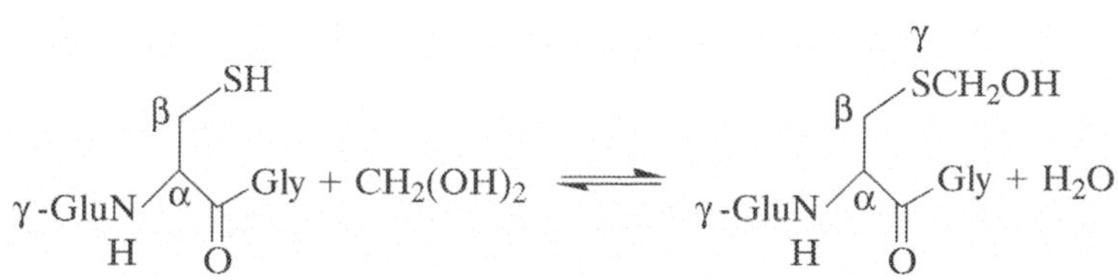

(Scheme I)

(1998) also detected the signal of $GSCH_2OH$ in whole cell ^{13}C NMR spectra of *Rhodobacter sphaeroides* incubated with ^{13}C-formaldehyde.

Figure 9.3 shows the ^{13}Cγ resonance of $GSCH_2OH$ at $\delta = 66.6$ ppm. This signal was present in an *E. coli* bacterial suspension (*in vivo* spectrum, Figure 9.3a) and in the

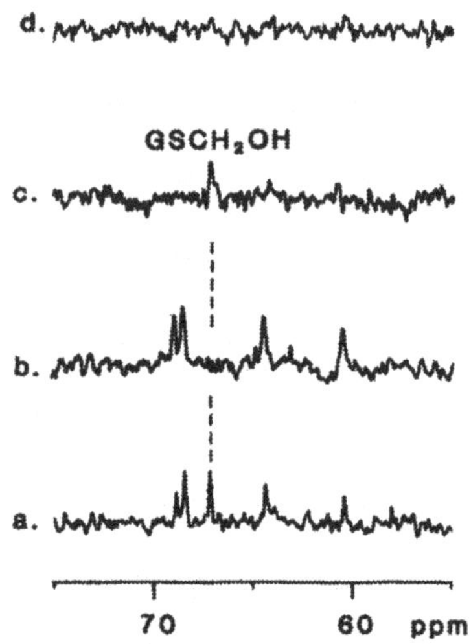

Figure 9.3 (a) ^{13}C NMR spectra of an *E. coli* suspension first incubated with 10 mM ^{13}C-formaldehyde during 1 h and further supplemented with 10 mM ^{13}C-formaldehyde. Spectra were collected after 4 min. The appearance of $GSCH_2OH$ is clearly seen (b) the supernatant from (a) obtained by centrifugation at 4°C. (c) The resuspended bacterial pellet from (a) after centrifugation. (d) Suspension of (c) allowed to warm up to 37°C, the peak due to $GSCH_2OH$ disappeared (Mason et al., 1986).

bacterial pellet kept at 4°C after centrifugation (Figure 9.3c) but was absent in the corresponding supernatant (Figure 9.3b). These results clearly showed the intracellular localization of this adduct. When the cell pellet heated to 37°C, the peak due to $GSCH_2OH$ disappeared, indicating biochemical removal (Figure 9.3d). In contrast, the observation of $GSCH_2OH$ upon addition of formaldehyde to heat-killed cells indicated that the adduct formation is a chemical process. Thus, two processes are involved: the formation of the adduct is chemical, quick and spontaneous, while the consumption of the adduct is slow and biochemical.

9.3.3 Formation of methanol and formate by NADH-oxidoreductase and dismutase ("Cannizzarase") activities (routes I and II)

Two types of enzymatic processes can transform formaldehyde into formate + methanol (see Scheme II):

Oxido-reduction reactions:

$$CD_2(OH)_2 + NAD^+ \rightarrow DCO_2H + NADD$$
$$CD_2(OH)_2 + NADH \rightarrow CD_2HOH + NAD^+$$

Dismutation:

$$2CD_2(OH)_2 \rightarrow DCO_2H + CD_3OH + H_2O$$

(Scheme II)

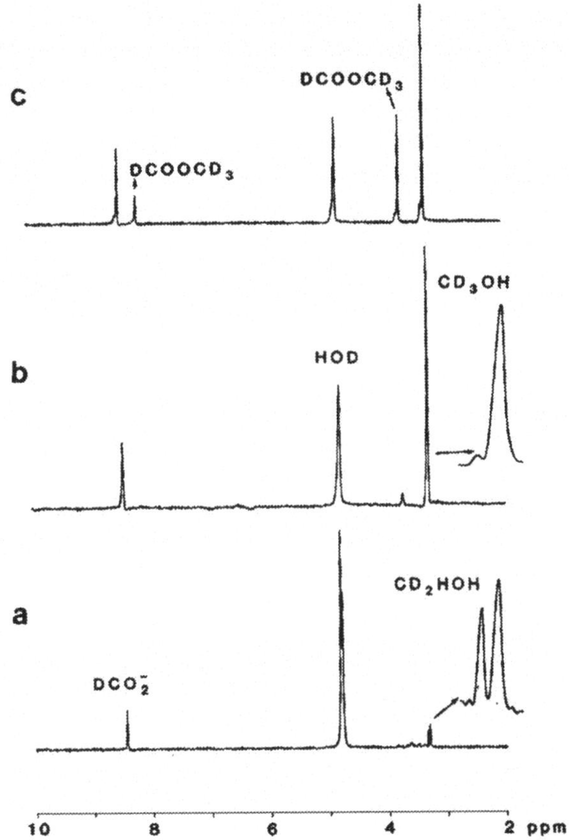

Figure 9.4 Proton-decoupled deuterium NMR spectra of suspensions of (a) *E. coli*, (b) *P. putida* (c) *S. aureus* acquired during anaerobic metabolism of 10 mM D-formaldehyde (Mason & Sanders, 1989).

The first one involves the independent reaction of oxidizing and reducing enzymes, each using NADH or a similar cofactor. In that case, if D-formaldehyde ($CD_2(OH)_2$) is used, one deuterium atom is recovered in the oxidised formate (DCO_2H) and one deuterium atom is transferred on the NAD cofactor (NADD). In parallel, D-formaldehyde can be reduced to methanol (CD_2HOH) by transfer of a proton from NADH. This logic is valid provided that the pool of NADH is large enough and not seriously perturbed by the NADD produced during the previous oxidation/reduction.

The second process corresponds to the activity of a dismutase (sometimes called "Cannizzarase"). A single reaction is involved resulting in the transfer of one hydrogen atom from one formaldehyde molecule to another. Starting from two D-formaldehyde ($CD_2(OH)_2$) molecules, one molecule of DCO_2H (formate) and one of CD_3OH (methanol) are generated.

The basic analysis of the deuterium NMR signal of methanol allows one to distinguish a formaldehyde dismutase activity from classical oxido-reduction systems. The signal of CD_2HOH is a doublet ($J_{H\text{-}D} = 1.7\,Hz$) by virtue of the geminal H-D coupling. The signal of CD_3OH is a simple singlet as no H-D coupling occurs. Mason and Sanders (1989) have clearly shown using this approach that an oxido-reduction system was operating in *E. coli* (Figure 9.4a, doublet) while a dismutase was active in *P. putida* and *S. aureus* (Figure 9.4 b and c, singlet). Note the simplicity of this technique to assess *in vivo* enzymatic activities. Classical biochemical approaches would have been much longer and difficult.

The same authors have studied dismutase properties in more detail. For instance they have performed a cross-dismutation reaction scheme for measurement of kH/kD

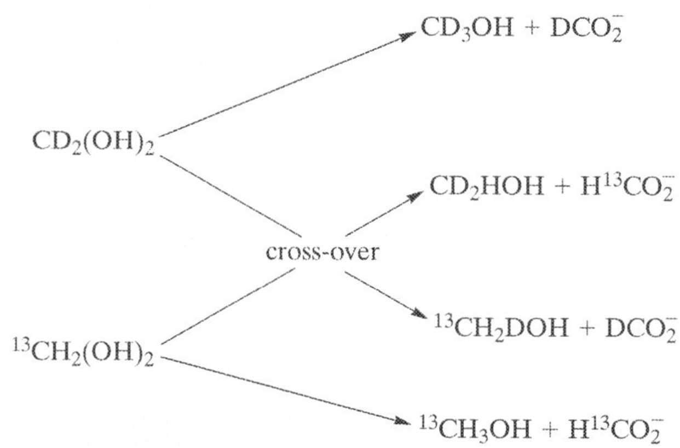

(Scheme III)

hydride transfer to assess the isotopic effect of deuterium in the dismutation reaction. A simple mixed-isotope experiment using ^{13}C- and D-formaldehyde was carried out (see Scheme III).

The ^{13}C and deuterium NMR spectra obtained with *S. aureus* (Figure 9.5) clearly demonstrate the occurrence of this cross-dismutation as all the isotopomers of methanol and carboxylic acid are detected. More precisely in D-NMR spectra (Figure 9.5b) the singlet of CD_3OH, the doublet of CD_2HOH due to $J_{H\text{-}D}$ coupling and finally the doublet of multiplets of $^{13}CH_2DOH$ due to both $J_{H\text{-}D}$ and $J_{H\text{-}C}$ couplings are detected. The $^{13}CH_3OH$ isotopomer cannot be detected by D-NMR as it has no deuterium atom. In ^{13}C NMR spectra (Figure 9.5a) CD_3OH and CD_2HOH isotopomers cannot be detected because they bear a ^{12}C atom. $^{13}CH_3OH$ gives a singlet and $^{13}CH_2DOH$ a doublet (due to $J_{D\text{-}C} = 22\,Hz$). Because ^{13}C NMR spectra are ^1H decoupled, there is no influence of $J_{H\text{-}C}$ on ^{13}C signal patterns. One can also note that the ^{13}C resonance of $^{13}CH_2DOH$ is shifted downfield (0.04 ppm) due to the presence of deuterium. The measurement of the relative concentrations of these methanol isotopomers allowed an estimate for kH/kD of about four for hydride transfer, showing a significant kinetic isotopic effect. Other experiments have been performed for assessing the substrate specificity of dismutase and the stereochemistry of dismutation (not shown).

9.3.4 Formation of methylformate (route III)

The formation of methylformate ($HCOOHCH_3$), which is not a common pathway for formaldehyde dissimilation, was demonstrated in *S. aureus* by Mason and Sanders

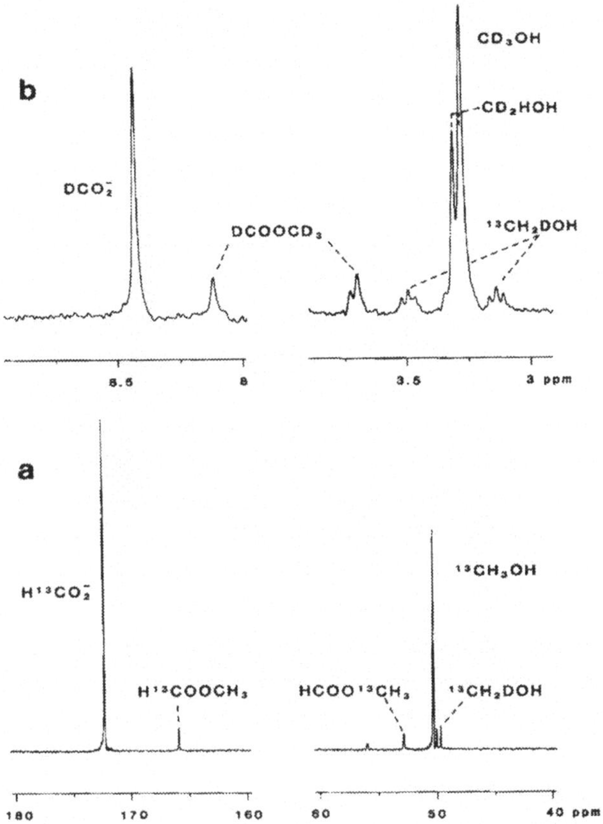

Figure 9.5 (a) ^{13}C and (b) deuterium NMR spectra demonstrating cross-dismutation of a 1:2 mixture of D- and ^{13}C-formaldehyde (total 30 mM) by *S. aureus* (Mason & Sanders, 1989).

(1989). The resonances of this compound were detected *in vivo* on ^{13}C NMR spectra at $\delta = 165.9$ and 52.8 ppm and on D-NMR spectra at $\delta = 8.1$ and 3.8 ppm (Figures 9.4c and 9.5). The various labelling experiments are consistent with a mechanism in which methylformate would result from the enzymatic oxidation of the methanol hemiacetal of formaldehyde (Scheme IV).

$$CH_2O + CH_3OH \longrightarrow CH_3OCH_2OH \xrightarrow{[O]} CH_3OCHO$$

(Scheme IV)

In addition, the analysis of isotope distribution within the methyl group of methylformate shows that methanol is itself the product of dismutation rather than an independent reduction. The same type of reaction has been shown in methylotrophic yeasts (Sakai et al., 1995).

9.3.5 Dissimilation of formaldehyde via the RuMP cycle (route IV)

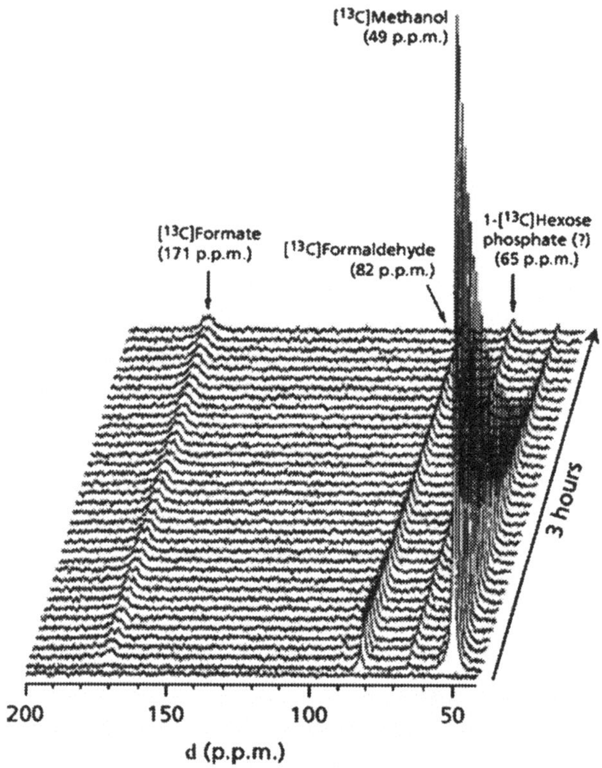

Figure 9.6 ^{13}C NMR analysis of the degradation kinetics of ^{13}C-methanol by *B. methanolicus* suspension (Pluschkell & Flickinger, 2002).

The last possible route (route IV, Figure 9.1) for dissimilation of formaldehyde involves the fixation of formaldehyde on ribulose-5 phosphate, giving rise to hexulose-6-phosphate. This last compound is isomerized in fructose-6-phosphate, which in turn gives glucose-6-phosphate, which is further transformed into 6-phospho-gluconate. This last 6-carbon molecule gives back ribulose-5-phosphate and CO_2. This cycle is named "RuMP cycle" (**Ribu**lose **M**ono**P**hosphate). The operation of such a cycle has been shown in *B. methanolicus* using *in vivo* ^{13}C NMR (Pluschkell & Flickinger, 2002). Figure 9.6 shows ^{13}C NMR analysis of the degradation kinetics of ^{13}C-methanol by *B. methanolicus* whole cells. The integral of ^{13}C-methanol signal ($\delta = 49$ ppm) is decreasing with time while three new signals are appearing in the spectrum at $\delta = 82$, 171 and 65 ppm corresponding, respectively, to ^{13}C-formaldehyde, ^{13}C-formate and $^{13}C_1$-hexose-phosphate. The formation of formaldehyde results from the oxidation of methanol via route II (Figure 9.1). Formaldehyde can be further oxidised into formate via route I (Figure 9.1) but it can also enter the RuMP cycle (route IV) as shown by the detection of $^{13}C_1$-hexose-phosphate.

9.3.6 Transformation of formaldehyde on solid supports

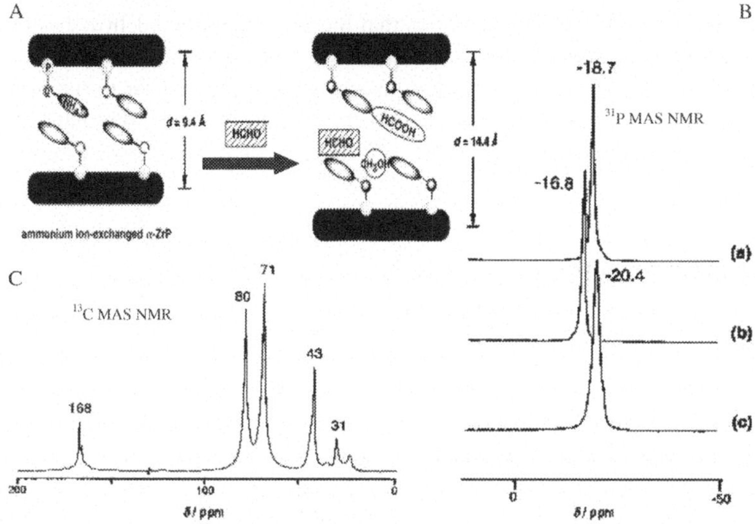

Figure 9.7 (A) Schematic model of intercalation and transformation of formaldehyde in ammonium-ion-exchanged α-zirconium phosphate. (B) ^{31}P MAS NMR spectra of ammonium-ion-exchanged α-zirconium phosphate for the adsorption of formaldehyde: (a) α-zirconium phosphate, (b) ammonium-ion-exchanged α-zirconium phosphate, (c) after adsorption. (C) ^{13}C MAS NMR spectra of ammonium-ion-exchanged α-zirconium phosphate after the adsorption of ^{13}C-enriched formaldehyde (Hayashi et al., 2005).

All the above experiments describe transformations of formaldehyde by microorganisms suspended in liquid media. However, one could imagine that it would be interesting to look at transformations taking place in more complex situations, when microorganisms are attached on solid supports, for instance in biofilters or in packed-bed bioreactors. In that case, classical liquid NMR techniques or *in vivo* NMR cannot be used as the signals are too large due to magnetic susceptibility, huge dipolar couplings and chemical shift anisotropy. Alternatively, special solid-state NMR techniques can be used. Recently Hayashi et al. (2005) have studied the adsorption and transformation of formaldehyde by ammonium-ion-exchanged α-zirconium phosphate (Figure 9.7).

Figure 9.7A shows the structure of this material and also the fate of formaldehyde in this support. First formaldehyde is intercalated within the layers, as shown by the increase of the inter-layered distance (from d = 9.4 to 14.4 Å) measured by XRD and by the variations of ^{31}P chemical shifts measured by ^{31}P MAS NMR (magic angle spinning). Figure 9.7B(a) shows the ^{31}P MAS NMR spectrum of non-exchanged δ-zirconium phosphate, the phosphate ion resonates at $\delta = -18.7$ ppm; after exchange by ammonium ions, this resonance is shifted to $\delta = -16.8$ ppm (Figure 9.7B(b)). Finally the intercalation of formaldehyde in the layers leads to an extra shift of the ^{31}P NMR resonance to $\delta = -20.4$ ppm (Figure 9.7B(c)). Second, the ^{13}C MAS NMR spectrum presented in Figure 9.7C clearly shows that ^{13}C-formaldehyde is transformed *in situ* into ^{13}C-formate (δ = 168 ppm) and ^{13}C-methanol (43 ppm) by a "Cannizzaro reaction". Note that the presence of two distinct ^{13}C NMR resonances of formaldehyde ($\delta = 71$ and 80 ppm) shows that the absorbed formaldehyde has two different environments. Although this experiment was not performed in the presence of microorganisms, the abiotic reaction that takes place is rather similar to the biological "Cannizzarase" or dismutase reaction described in Section 9.3.3 and thus illustrates the potential application of solid-state NMR to investigate bioconversion of gaseous compounds on solid supports.

9.4 CONCLUSION

Although the literature concerning the bioconversion of gaseous compounds by *in situ* NMR studies is rather scarce, the case study of formaldehyde is a very good example of the power of this technique to assess metabolic routes in detail. The analysis of deuterium and ^{13}C isotopomers gives information about the final and intermediate products but also on enzymatic mechanisms. These investigations can be carried out easily, directly *in vivo*, which is a great advantage compared to classical biological approaches. Future development could concern more complex environments, particularly solid media. In addition to CP-MAS NMR, the new HR-MAS NMR (high resolution-MAS) technique could also be used. This later approach allows studying heterogeneous hydrated compounds, including polymers, tissues, hydrated soils or solid matrices.

REFERENCES

Amato, P., Menager, M., Sancelme, M., Laj, P., Mailhot, G. and Delort, A.M. (2005) Microbial population in cloud water at the Puy de Dôme. Implications for the chemistry of clouds. *Atmos. Environ.* **39**, 4143–4153.

Barber, R.D. and Donohue, T.J. (1998) Function of a glutathione-dependent formaldehyde dehydrogenase in *Rhodobacter sphaeroides* formaldehyde oxidation and assimilation. *Biochem.* **37**(2), 530–537.

Christoserva, L., Laukel, M., Portais, J.C., Vorholt, J.A. and Lidstrom, M.E. (2004) Multiple formate dehydrogenase enzymes in the facultative methylotroph *Methylobacterium extorquens* AM1 are indispensable for growth on methanol. *J. Bacteriol.* **186**, 22–28.

Commodari, F. and Hunter, B.K. (1989) Production and interconversion of 1,1-propanediol and hydroxyacetone by *Escherichia coli. Biochem. Cell Biol.* **67**, 468–475.

Cornish, A., Nicholls, K.M., Scott, D., Hunter, B.K., Aston, W.J., Higgins, I.J. and Sanders, J.K. (1984) *in vivo* ^{13}C NMR investigation of methanol oxidation by the obligate methanotroph *Methylosinus trichosporium* OB3b. *J. Gen. Microbiol.* **130**, 2565–2575.

Costa, C., Dijkema, C., Friedrich, M., Garcia-Encina, P., Fernadeza-Polanco, F. and Stams, A.J. (2000) Denitrification with methane as electron donor in oxygen-limited bioreactors. *Appl. Microbiol. Biotechnol.* **53**, 754–762.

Costa, C., Vescherskaya, M., Dijkema, C. and Stams, A.J. (2001) The effect of oxygen on methanol oxidation by an obligate methanotrophic bacterium studied by *in vivo* ^{13}C nuclear magnetic resonance spectroscopy. *J. Ind. Microbiol. Biotechnol.* **26**(1–2), 9–14.

De Graeve, K.G., Grivet, J.P., Durand, M., Beaumentin, P. and Demeyer, D. (1990) NMR study of $^{13}CO_2$ incorporation into short-chain fatty acids by pig large-intestine flora. *Can. J. Microbiol.* **36**(8), 579–582.

Delort, A.-M. and Combourieu, B. (2000) Microbial degradation of xenobiotics. In *NMR in Microbiology: Theory and Applications* (eds. J.N. Barbotin and J.C. Portais), Chapter 19, pp 411–430. Horizon Scientific, UK.

Delort, A.M. and Combourieu, B. (2001) In situ ^{1}H NMR study of the biodegradation of xenobiotics: application to heterocyclic compounds. *J. Ind. Microbiol. Biotechnol.* **26**(1–2), 2–8.

Delort, A.-M., Combourieu, B. and Haroune, N. (2004) Nuclear magnetic resonance studies of interactions between organic pollutants and soil components: a review. *Environ. Chem. Lett.* **1**, 209–213.

Grivet, J.P., Delort, A.M. and Portais, J.C. (2003) NMR and microbiology: from physiology to metabolomics. *Biochimie* **85**(9), 823–840.

Hayashi, A., Fujimoto, Y., Ogawa, Y., Nakayama, H. and Tsuhako, M. (2005) Adsorption of gaseous formaldehyde and carboxylic acids by ammonium-ion-exchanged α-zirconium phosphate. *J. Coll. Interf. Sci.* **283**, 53–63.

Hunter, B.K., Nicholls, K.M. and Sanders, J.K. (1984) Formaldehyde metabolism by *Escherichia coli. in vivo* carbon, deuterium, and two-dimensional NMR observations of multiple detoxifying pathways. *Biochem.* **23**(3), 508–514.

Hunter, B.K., Nicholls, K.M. and Sanders, J.K. (1985) Formaldehyde metabolism by *Escherichia coli*. Carbon and solvent deuterium incorporation into glycerol, 1,2-propanediol, and 1,3-propanediol. *Biochem.* **24**, 4148–4155.

Jones, J.G. and Bellion, E. (1991) Methanol oxidation and assimilation in *Hansenula polymorpha.* An analysis by ^{13}C n.m.r. *in vivo. Biochem. J.* **208**, 475–481.

KEGG: http://www.genome.jp/dbget-bin/show_pathway?rn00680+R01146

Mason, R.P. and Sanders, J.K. (1989) *in vivo* enzymology: a deuterium NMR study of formaldehyde dismutase in *Pseudomonas putida* F61a and *Staphylococcus aureus*. *Biochem.* **28**(5), 2160–2168.

Mason, R.P., Sanders, J.K., Crawford, A. and Hunter, B.K. (1986) Formaldehyde metabolism by *Escherichia coli.* Detection by *in vivo* ^{13}C NMR spectroscopy of S-(hydroxymethyl)glutathione as a transient intracellular intermediate. *Biochem.* **25**(16), 4504–4507.

Matheron, C., Delort, A.-M., Gaudet, G. and Forano, E. (1997) Re-investigation of glucose metabolism in *Fibrobacter succinogenes* using NMR and enzymatic assays. Evidence of pentose phosphate phosphoketolase and pyruvate formate lyase activity. *Biochim. Biophys. Acta* **1335**, 50–60.

Narbad, A., Hewlins, M.J.E. and Callely, A.G. (1989) ^{13}C-NMR studies of acetate and methanol metabolism in methylotrophic *Pseudomonas* strains. *J. Gen. Microbiol.* **135**(6), 1469–1477.

Pluschkell, S.B. and Flinckinger, M.C. (2002) Dissimilation of [(13)C]methanol by continuous cultures of *Bacillus methanolicus* MGA3 at 50 degrees C studied by (13)C NMR and isotope-ratio mass spectrometry. *Microbiol.* **148**, 3223–3233.

Sakai, Y., Murdanoto, A.P., Sembiring, L., Tani, Y. and Kato, N. (1995) A novel formaldehyde oxidation pathway in methylotrophic yeasts: methylformate as a possible intermediate. *FEMS Microbiol. Lett.* **127**, 229–234.

Schäfer, S., Götz, M., Eisenreich, W., Bacher, A. and Fuchs, G. (1989) ^{13}C-NMR study of autotrophic fixation in *Thermoproteus neutrophilus*. *Eur. J. Biochem.* **184**(1), 151–156.

Strauss, G., Eisenreich, W., Bacher, A. and Fuchs, G. (1992) ^{13}C-NMR study of autotrophic fixation pathways in the sulfur-reducing Arcaebacterium *Thermoproteus neutrophilus* and in the phototrophic Eubacterium *Choloflexus aurantiacus*. *Eur. J. Biochem.* **205**(2), 853–866.

Suzuki, Y., Imai, M., Kawakami, Y., Masuda, K. and Akasaka, K. (1998) Identification of low-molecular weight organic compounds in contaminated fog water using proton nuclear magnetic resonance spectroscopy. *Bull. Environ. Contam. Toxicol.* **60**, 355–362.

Vescherskaya, M., Dijkema, C. and Stams, A.J. (2001) Intracellular PHB-conversion in a type II methanotroph studied by *in vivo* ^{13}C nuclear magnetic resonance spectroscopy. *J. Ind. Microbiol. Biotechnol.* **29**, 15–21.

10

Heavy metal release in waste incineration processes

D. Gauthier, S. Abanades and G. Flamant

10.1 INTRODUCTION

10.1.1 Fate of heavy metals during waste incineration

Waste incineration is increasingly implemented because of two main points: it reduces both the volume and the mass of the waste while simultaneously generating power. It is used to eliminate municipal solid waste (MSW), medical wastes (MW) and sewage sludge (SS) as well as industrial solid wastes (ISW). In order to reduce the overall environmental impact of these waste treatment plants, operators have to deal with the fate of heavy metals (HMs) in the plant itself but also in the ultimate residues. In fact, operators of coal combustors have to deal with a similar problem, in spite of lower metal concentrations. Metals contained in wastes are introduced into combustion chambers in many physical and chemical forms. Nevertheless, during incineration, HMs are only transformed and concentrated in the various final residues: bottom ash, filter ash and residues from the air pollution control (APC) devices, as well as in gaseous emissions, which classically all represent potential sources of pollution. The HM partitioning among the various residues depends on the waste composition, the

 Waste Gas Treatment for Resource Recovery edited by Piet N.L. Lens, Christian Kennes, Pierre Le Cloirec and Marc Deshusses. ISBN: 1843391279. Published by IWA Publishing, London, UK.

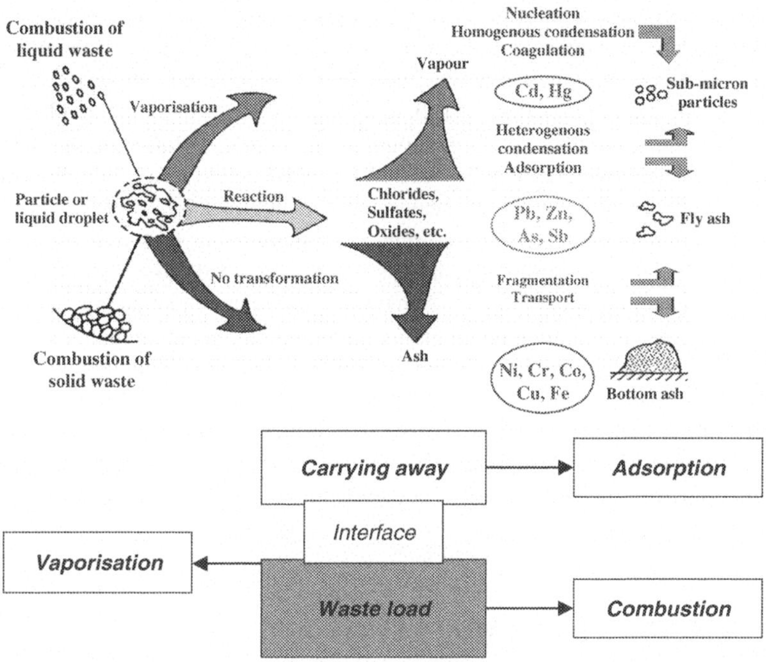

Figure 10.1 Transformations of HMs in an incinerator (scheme adapted after Barton et al., 1990).

physico-chemical properties of the metals and/or their compounds, and the incinerator operating conditions (Barton et al., 1990; Brunner & Mönch, 1986; Chandler et al., 1995). Three main processes affect the fate of metals in an incinerator (Figure 10.1):

(1) The *combustion* of the load, creating the temperature increase inducing the HM transformation and their partial emission.
(2) The *vaporisation* of HMs, leading to their concentration increase in the combustion gases, and then to the formation of submicron particles.
(3) The *adsorption* of HMs on entrained particles.

So, the problem plant operators have to solve is: should the process conditions avoid the vaporisation of toxic metals, so that their concentration will increase in the bottom ash? Or, on the contrary, should the process conditions favour the metal vaporisation and its release in exhaust gases, coupled to the best performing APC system and stabilising materials? This is an important environmental and economical issue since the release of toxic metals is strictly controlled, and the regulation will most probably become even stricter in the future.

10.1.2 Aim of the study

In order to permit the incineration industry to choose among these two possibilities, a complete thermodynamic and experimental study of the behaviour of HMs in waste incinerators and coal combustors was first developed, and the ultimate residues were characterised physico-chemically. This study showed that more stable oxides can be formed that stabilise within them the toxic elements, and it pointed out the influence of operating conditions and gas composition. More recently, a method to determine the kinetics of HM vaporisation and to study the effect of operating conditions on these kinetics was developed. It involves only the measured evolution of the HM concentration in exhaust gases, and a model developed at reactor scale. This method was validated in a lab-scale fluid bed incinerator with metal-spiked mineral matrices, then implemented with spiked organo-mineral matrices and more realistic artificial wastes (RAW). Three representative metals were studied: Cd, Pb and Zn. The metal vaporisation dynamics were obtained from the on-line analysis, and kinetic laws will be determined from these vaporisation rate profiles. Later, these kinetic laws will be introduced in a general model simulating a waste incinerator, developed elsewhere (Menard et al., 2002).

Finally, it is worth noticing that in order to complement this study, a co-operative research programme on the stabilisation of vaporised HMs into ultimate residues by their adsorption on orthophosphate is currently under development (Gauthier et al., 2005).

10.2 THERMODYNAMIC STUDY

The objective of this theoretical approach is to know whether the HM speciation foreseen from thermodynamics can explain the partitioning of HM which is observed experimentally during municipal solid waste incineration (MSWI). The partitioning of several HMs was predicted in a well-defined system whose composition corresponds to that of MSW (Abanades et al., 2002a). This analysis was based on the speciation of HM among various chemical compounds under various incinerator conditions.

Several limitations exist when applying thermodynamics in such a system: assumption of a closed system; physical processes and mixing phenomena are neglected. It is known that equilibrium is never reached in real conditions because of limitations due to transfer phenomena. In spite of these limitations, the equilibrium approach was and is still a useful tool. It can determine the possible chemical reactions and compounds involving HM, and predict their fate in waste incineration and coal combustion (Durlak et al., 1997; Frandsen et al., 1994; Owens et al., 1995; Wey et al., 1996; Wu & Biswas, 1993; Yan et al., 1999).

10.2.1 Equilibrium calculation method

When burning, a solid waste constitutes a multicomponent and multiphase system. The principle of minimising the total Gibbs free energy of the system was used to calculate the thermodynamic equilibrium in order to identify the dominant species of each element. The method of element chemical potentials combined with atom population constraints was operated on GEMINI software (Cheynet, 1988a), involving its extensive attached database COACH (Cheynet, 1988b).

10.2.2 System definition and calculation conditions

The system was defined for representing the burning waste in such a combustion atmosphere that its composition is that measured in an industrial incinerator. The MSW composition (35% moisture, 24.7% ashes, 40.3% carbon and volatiles) was chosen as analysed by ADEME (1998), in which the major elements C, H, O, N, S and Cl constitute the bulk of the burning matrix (Table 10.1). The mineral fraction and minor and trace elements were also considered (Tables 10.2 and 10.3). This complete set of elements represents the very complex system simulating MSW combustion. Equilibrium calculations were performed in the range of 300–2000 K, and the total pressure was 1 atm. Twenty-nine elements were considered in the equilibrium calculations; a total of 1992 species were derived from these elements in the database. In fact, the air excess ratio varied from 0 to 2 to simulate both reducing and oxidising conditions in the furnace, at mean operating temperature $T = 1200$ K, and the influence of the chlorine content was studied.

10.2.3 Results of the theoretical study

The results of the thermodynamic study presented in this section are focused on the three metals considered in the rest of the study (Cd, Pb and Zn).

Table 10.1 Chemical composition of the organic fraction of waste (ADEME, 1998).

Element	Weight (%)
Carbon (C)	35
Hydrogen (H)	5
Oxygen (O)	19.25
Nitrogen (N)	0.76
Chlorine (Cl)	1.7
Sulfur (S)	0.29

Table 10.2 Typical elemental analysis of the mineral part in waste (Pfrang-Stotz & Schneider, 1995).

Compound	SiO_2	Fe_2O_3	CaO	Al_2O_3	MgO	Na_2O	K_2O	TiO_2	MnO	P_2O_5
Content (weight %)	49.2	12.7	15.1	8.8	2.6	4.2	1.04	0.88	0.17	1.04

Table 10.3 Trace element concentration (μg/g dry matter) in MSW (ADEME, 1998).

Element	F	As	B	Cd	Co	Cr	Cu
Concentration	58	5	14	4	113	183	1048
Element	Mn	Hg	Mo	Ni	Pb	Se	Zn
Concentration	412	3	1	48	795	0.02	1000

10.2.3.1 *Influence of temperature on the HM speciation*

Figures 10.2–10.4 show the equilibrium distributions of the HMs in the temperature range 300–2000 K, for a typical air excess ratio $\lambda = 1.6$. In these figures, g refers to the gaseous state, s is the crystalline state and c is the condensed state (liquid or solid). In the temperature range encountered in MSWI, Pb and Cd have a similar thermodynamic behaviour. Their gaseous chlorides are the major species between 600–700 and 1400 K; at higher temperatures, PbO(g) and Cd(g) are the dominant species. In any case, these metals are typically in the vapour phase when T is higher than 600–700 K, and they do not interact significantly with the mineral matter. When the temperature decreases, Pb and Cd exist as chlorides in the condensed phase. This result confirms that both Pb and

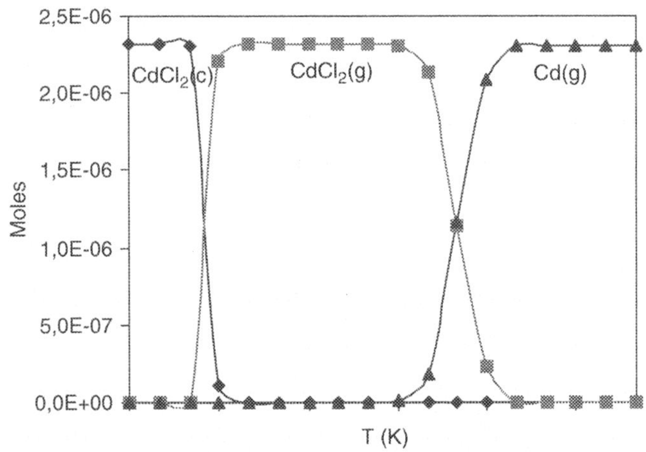

Figure 10.2 Equilibrium distribution of Cd versus temperature.

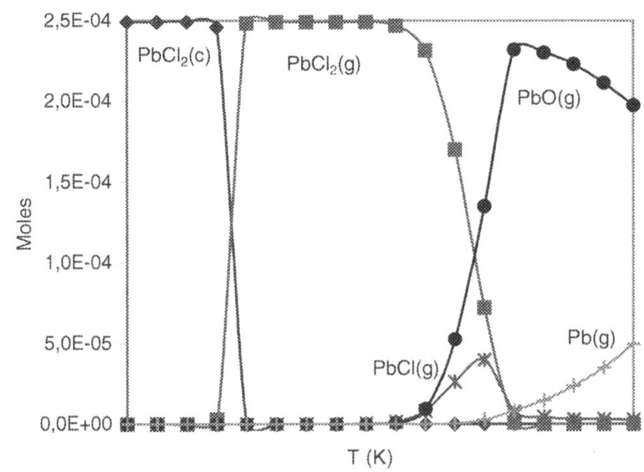

Figure 10.3 Equilibrium distribution of Pb versus temperature.

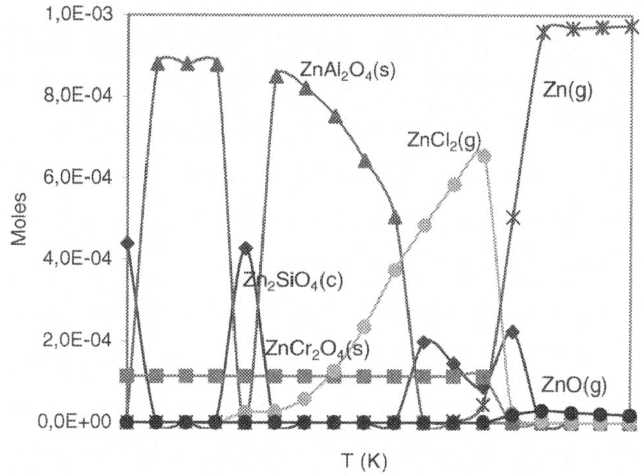

Figure 10.4 Equilibrium distribution of Zn versus temperature.

Cd are easily volatilised, and they are condensed at lower temperature as chlorides. Since metallic chlorides leach easily, Pb and Cd are even more problematic in the ash. Zinc is alternatively found as its silicate Zn_2SiO_4(c) or aluminate $ZnAl_2O_4$(s), depending on temperature. It is trapped as stable species (binary oxides) within the MSW mineral fraction. Therefore, unlike Cd and Pb, Zn is not expected to leach. Elemental zinc and to a lesser extent ZnO(g) are the dominant species above 1500 K. The temperature range for stable chloride formation is not as wide as for Pb and Cd, but $ZnCl_2$(g) is partially formed between 600 and 1500 K. Consequently, Zn volatilisation is partially expected during incineration. A noteworthy interaction occurs between Zn and Cr to form $ZnCr_2O_4$(s). Verhulst et al. (1996) also indicated the possible formation of stable spinels ($ZnO.Fe_2O_3$).

In conclusion, according to the equilibrium analysis, the major part of Zn, which is linked with the mineral elements (Al, Si), should be either in the bottom ash or in fly ash, but Zn may also be slightly released as gaseous compounds.

10.2.3.2 Effect of the oxygen content on the HM speciation

The influence of the total air flow rate (characterised by the air excess number or excess air ratio, λ) was studied at 1200 K under both reducing and oxidising conditions (varying λ between 0 and 2). The results concerning Cd, Pb and Zn are displayed in Figures 10.5–10.7, respectively.

For all three metals, calculations foresee the formation of volatile chlorides under oxidising conditions only. Elemental gaseous forms of these three metals are the major species under reducing conditions. This means these metals can be released whatever the gaseous conditions. Zn exhibits a peculiar behaviour, and solid Zn aluminate is the dominant species under oxidising conditions. Verhulst et al. (1996) found that an excess of Cl or reducing conditions are necessary to completely volatilise Zn, which is generally present as a stable oxide in larger quantities.

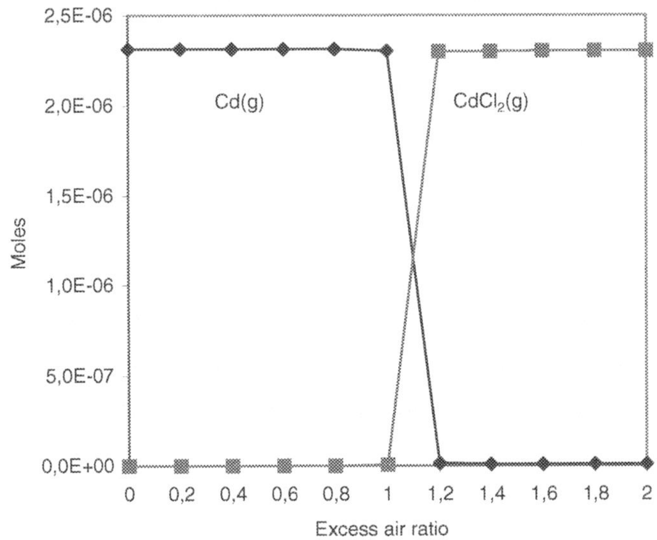

Figure 10.5 Effect of oxygen content on Cd speciation.

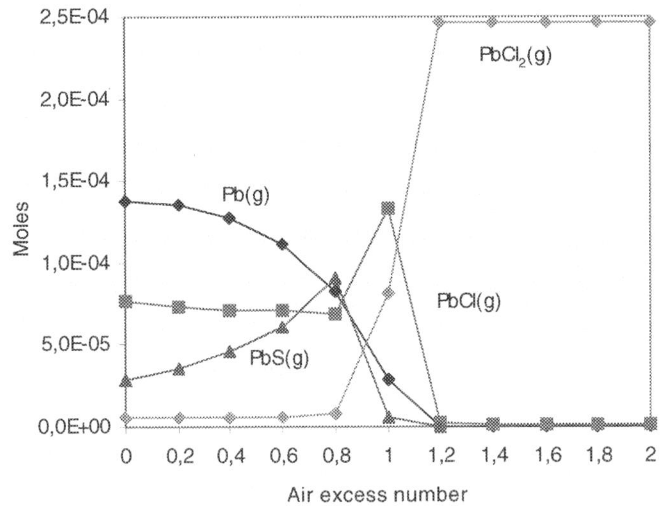

Figure 10.6 Effect of oxygen content on Pb speciation.

10.2.3.3 Influence of the chlorine content on the HM speciation

The effect of the Cl_2 content on the equilibrium distribution of Cd, Pb and Zn was studied at 1200 K, because the interactions between chlorine and these metals determine their extent of release during MSWI.

Chlorine significantly affects Pb and Cd volatilisation since there is a threshold value from which the chloride is the dominant species. Zn interacts mainly with Al and Si, whatever the chlorine content: $ZnAl_2O_4$(s) and $ZnCr_2O_4$(s) are the major stable

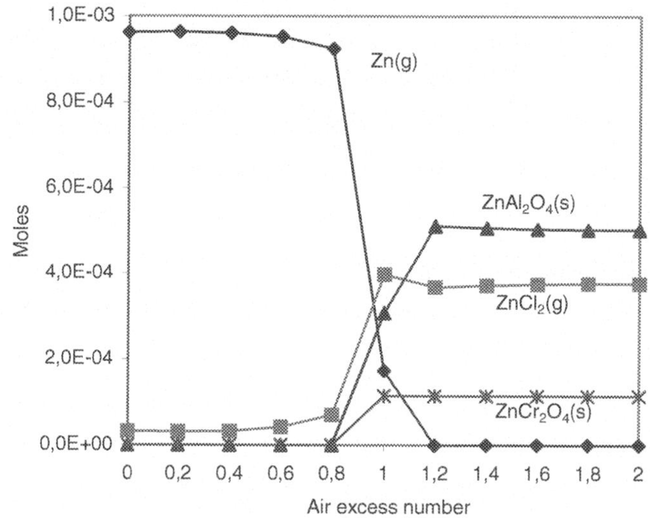

Figure 10.7 Effect of oxygen content on Zn speciation.

species. However, when the chlorine content becomes high enough, Zn chloride is formed, which means it may be released slightly. These results are in agreement with Morf et al.'s results (2000).

10.3 CHARACTERISATION OF ULTIMATE RESIDUES

In this study, the residues coming from a mass-burning incinerator (capacity 28 t/h) were characterised to draw the general tendencies for HM chemical speciation in solid residues from MSWI. They were bottom ashes, boiler ashes and electrostatic precipitator dusts (ESPD). First, the HM chemical trapping and their respective mobility from ashes were investigated with chemical sequential extraction series. Then the ash morphology, HM location, distribution and chemical environment were examined with X-ray microanalysis. Representative samples of 20–30 g were obtained by sieving, coning and quartering samples of residues taken in the industrial incinerator. The analysis procedures were fully detailed and the results discussed by Gagnepain (1998).

10.3.1 Characterisation techniques

The chemical sequential extraction series, based on the methods of Kirby and Rimstidt (1993) and Tessier et al. (1979), were performed on 1 g of dried residue. At each step, a specific reagent was used to extract the various phases in which metals are trapped. Then, the extracted phases were separated from the solid and analysed by inductively coupled plasma optical emission spectroscopy (ICP-OES) to determine the HM content. This method gives an instantaneous view of the HM trapped in the different phases (organic or inorganic). It simulates the various environmental conditions MSWI ashes may undergo, thus showing the trends of HM mobility; and their speciation can be deduced from it.

Polished sections of ash particles embedded in epoxy resin were examined with an electron microprobe "CAMECA" Microbeam, operating in wavelength-dispersive scanning (WDS) mode. The ash morphology was visually observed with the scanning electron microscope (SEM) technique. The HM distribution was mapped in residues and possible associations between the major elements and the HM were shown. This study was focused on Pb and Zn that are representative and concentrated enough to allow their detection via this technique.

10.3.2 HM speciation in residue

As shown in Table 10.4, the combination of the results from the two methods showed that the most probable chemical forms of lead are the pure metal, its oxide, its sulfate, its carbonate and complex silicates in all ashes. Lead chloride was also found in the economiser ash and filter ash. For Zn, possible components are the pure metal, its oxide, its chloride, its sulfate, its carbonate, zinc chromate, and also aluminate and silicate compounds.

Comparing the results of the characterisation (Table 10.4) and of thermodynamic calculations (Figures 10.2–10.7) indicates the following:

1. Experimental data agree well with predictions for Zn. The residues' characterisation showed the presence of $ZnCl_2$ (formed by volatilisation) and of stable binary oxides such as $ZnAl_2O_4$, Zn_2SiO_4 and $ZnCr_2O_4$, as predicted.
2. Carbonates and sulfates are possible speciation forms of Pb and Zn (sequential extraction result), although thermodynamic calculations predicted them at low temperature but never at incineration temperatures.
3. Thermodynamics do not predict correctly the behaviour of Pb. The existence of Pb-enriched nodules in the residues and its frequent association with sulfur (S) may be explained by the formation of Pb(g) and PbS(g), and their condensation as droplets. But the formation of lead silicate in some regions with specific

Table 10.4 Possible Pb and Zn species deduced from the chemical extraction series and XRM.

Identified phase following the extraction step	Corresponding compounds	Possible Pb compounds	Possible Zn compounds
Water soluble	Chlorides, sulfates, nitrates	$PbCl_2$, $Pb(NO_3)_2$	$ZnCl_2$, $ZnSO_4$, K_2ZnCl_4
Ca exchangeable	Adsorbed HM	–	–
Acid soluble	Carbonates, hydroxides	$PbCO_3$	$ZnCO_3$
Organically bound	Organo-metallics, sulfides	–	–
Crystalline and amorphous iron oxides	Sulfates, pure metals, oxides	$PbSO_4$*, PbO*, Pb*	$ZnSO_4$*, ZnO*, Zn*, $ZnCr_2O_4$
Residual fraction	Silicates, alumino-silicates, oxides, pure metal	PbO*, Pb*, Pb silicates Pb_3SiO_5*, $PbSiO_3$*	ZnO*, Zn*, $ZnAl_2O_4$*

*Compounds found by both sequential extraction and XRM.

composition segregation (chlorine depletion for instance) is possible when mass transfer is low, thus explaining the Pb stabilisation in the bottom ash.

10.4 KINETICS OF HM VAPORISATION

Kinetic data on the release of metals during the incineration of MSW are scarce, not to say non-existent. A global inverse method for identifying the kinetics of vaporisation of a metal from model wastes incinerated in a lab-scale fluidised bed is under development. This method involves on-line measurements of the metal in the exhaust gas and modelling of the high-temperature fluidised bed. In the first step, modelling was developed at steady state and dynamics of metals' vaporisation were obtained (Abanades et al., 2003a). The current development involves a transient model, which accounts for the unsteadiness of the metal's concentration in the outlet.

Batch experiments dealt with Cd, Pb and Zn, as before. In order to simplify the approach, model wastes were burned in a lab-scale fluid bed. They were made from both mineral and organic matrices, spiked with metallic salts by impregnation (Abanades, 2001). An on-line gas analysis system, involving ICP-OES, was linked to the lab-scale incinerator. The gas to be analysed was injected into an argon plasma for the spectroscopic detection of the elements. This system allows continuous measurement of the transient concentration of each metal in the off-gas. When organic matrices or real wastes are burned, these measured concentrations of each metal are the only experimental information available, since the burning solids cannot be sampled. The kinetics of vaporisation of a metal from the solid waste (at the particle level) can then be obtained by applying the inverse model.

10.4.1 Experimental study

10.4.1.1 Lab-scale reactor

The lab-scale reactor is a stainless steel fluidised bed (internal diameter 0.105 m, height 0.65 m), as described in detail by Abanades et al. (2003a). The reactor is heated by two Superthal™ electric half-shells, and the fluidising gas is heated before entering the reactor through two high-power electric resistances in series (Figure 10.8). It is either air or a gas mixture simulating the gaseous conditions in an incinerator. The gas mixture composition (vol%) is 4.8% O_2, 70.8% N_2, 8.8% CO_2, 400 mg/Nm^3 SO_2, 15.6% H_2O and 1000–2000 mg/Nm^3 HCl. Temperatures are measured by K-thermocouples and recorded on a PC. The 0.15 m high bed is composed of silica sand (mean diameter: 700 μm; total mass of sand: 1.62 kg). A given mass of metal-spiked particles is injected in the bed when steady state is reached. During experiments with non-burning reactive particles (alumina matrix), solid samples can be withdrawn from the bed at given times for chemical analysis. The metal concentration in each solid sample is measured by ICP-OES, after microwave-assisted digestion in acid, thus giving the temporal evolution of metal concentration $q(t)$ in the model waste.

10.4.1.2 On-line gas analysis system

An elemental gas analysis system using ICP-OES was developed to measure on-line the time course of the metal concentration in the exhaust gases. The ICP spectrometer

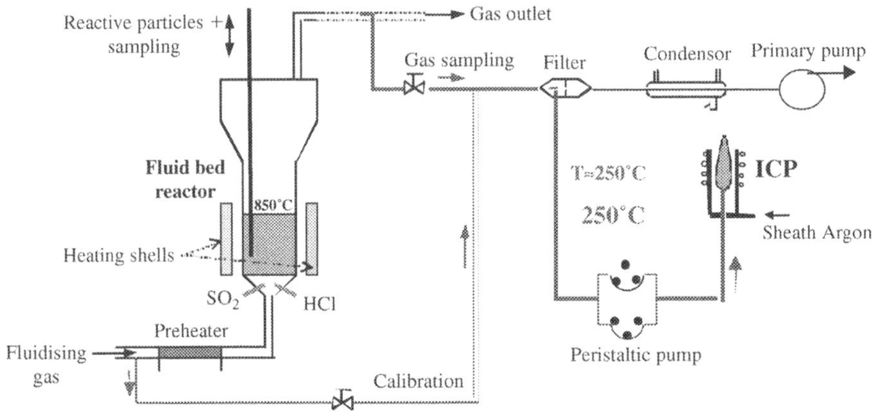

Figure 10.8 Scheme of the experimental set-up.

(Jobin Yvon JY 38S) was adapted for direct gas injection into the argon plasma (Hassaine et al., 2001; Trassy & Diemiaszonek, 1994). The gas sample is aspired through a 5-m long line, heated to avoid any condensation. Then, a small gaseous flow (100 mL/min) is injected into the plasma torch by using a secondary sampler (peristaltic pump), as shown in Figure 10.8. The accuracy of concentration measurements depends on the plasma temperature. The limits of detection of this method depend on the experimental conditions, the flow rate and the spectrometer itself; they can be as low as a few $\mu g/m^3$ (Hassaine et al., 2001).

Until now, without operating calibration procedure, only profiles of relative intensity, which are proportional to concentration profiles, were obtained. At a given wavelength characteristic of the metal, the intensity of the signal is proportional to the metal's concentration, as long as the gas composition and the flow rate of sampling are constant. Research is currently undertaken on the calibration to obtain absolute metal concentrations from the ICP-OES intensity measurements.

10.4.1.3 Model wastes

In order to simplify the method development and validation, experiments were done with model wastes, which were prepared by matrix impregnation. Two matrix types were considered so far. The mineral matrix (porous alumina particles: $d_p = 1.6-2.0$ mm, $\rho_p = 1348$ kg/m^3, initial metal concentration q_0), permits solid sampling during the experiments. A more RAW made from real MSW (Pierotti, 1999) was also used.

A special procedure was developed to make and spike the particles of RAW. The waste was mixed with sand ($d_p = 0.5$ mm, MSW/sand weight ratio = 2:1) and wallpaper glue in an aqueous solution of the metal chlorides ($CdCl_2$, $PbCl_2$ and $ZnCl_2$). Then the mixture was shredded, pressed into small cylinders (diameter: 10 mm; length: 6–12 mm; density: 600 kg/m^3) with a hydraulic press, and dried in an oven. Spike samples had Pb and Zn concentrations of the same order of magnitude than in real MSW. For Cd, the concentration was about 500-fold, because of the detection limits of the ICP (Table 10.5).

Table 10.5 Cd, Pb and Zn concentrations of typical MSW and spiked samples.

Concentration (mg/kg)	Cd	Pb	Zn
Typical MSW (Chandler et al., 1997)	5–15	400–1000	500–2000
Impregnated alumina matrix	700–5000	2500–5500	5000
Impregnated RAW matrix	4000	3000	3000

10.4.2 Theoretical study

10.4.2.1 Method

The method intends to determine the vaporisation rate of HM from the profile of emission intensity of metallic spectral lines in the outlet gas. It is presented here for transient conditions, on the basis of the steady-state model (Abanades et al., 2003b). The method involves three steps:

1. First, two theoretical models were developed:
   - The direct model that calculates the HM concentration at the gas outlet $C_o(t)$ from the vaporisation rate $r = \mathrm{d}q/\mathrm{d}t$.
   - The inverse model that calculates the metal's vaporisation rate $r = \mathrm{d}q/\mathrm{d}t$ from its concentration profile in the outlet gas $C_o(t)$. So far, it is in dimensionless form because only relative signals could be obtained by the ICP.
2. The method was applied and validated in the case of the spiked mineral model waste:
   - The metal's vaporisation rate $r = \mathrm{d}q/\mathrm{d}t$ is obtained experimentally by solid sampling. Then, the HM concentration at the gas outlet (C_o) is calculated by applying the direct model to the previously measured rate r, thus leading to the metal's vaporisation flux f (s^{-1}).
   - The dimensionless vaporisation flux f is obtained from the normalised profile of concentration at the outlet (C_o/C_{omax}), that is the normalised profile of emission intensity (I/I_{max}) since $I = \propto C_o$.
   - Comparing the results validates the method.
3. The method was applied to spiked RAW, when the only information is the metal concentration in the outlet gas.

10.4.2.2 Direct model

The direct model is based on a modified version of Kunii and Levenspiel's model (1991), developed for beds with very thick clouds (Geldart group B particles), and adapted to the case of HM vaporisation. The adsorption and condensation of metals on particles are neglected, they will be considered when data are available.

The main assumptions of the model are:

- the two-phase bed is composed of the bubbles and the emulsion
- the bed is isothermal
- bubbles have a uniform diameter
- the gas flow in the emulsion phase is at minimum fluidisation
- the gas is in plug flow everywhere

- the gas velocity (U) is constant
- there is a mass transfer between the two phases, and the global mass transfer coefficient K_{BE} is:

$$K_{BE} = 4.5\frac{U_{mf}}{D_B} \tag{10.1}$$

where U_{mf} is the minimum fluidisation velocity and D_B is the bubble diameter
- chemical and physical reactions have no effect on the volumetric flow rate.

This macroscopic approach uses only the global flux of the generation of metallic species at the particles' surface. The direct model predicts the metal's concentration in the bubble phase and in the emulsion phase as a function of the bed height, that are combined to give the concentration at the outlet.

10.4.2.3 Inverse model

The inverse model predicts the vaporisation rate of the metal from its concentration in the outlet gas. At a given wavelength where the metal emits, the analysis gives on-line the temporal profile of the emission intensity. The normalised intensity (I/I_{max}) is identical to the normalised concentration ($C_o^* = C_o/C_{omax}$) since intensity and concentration are proportional. In addition, the dimensionless vaporisation flux $f\,(s^{-1})$ is related to the mass flow rate of vaporisation r (mg/kg/s) by:

$$f = \frac{\gamma_B \theta\, r \rho_p}{C_{omax}/F_B} \tag{10.2}$$

with γ_B: volume fraction of solid particles dispersed in bubbles; θ: volume fraction of reactive metal-spiked particles in the bed; $r(t)$: mass flow rate of vaporising metal from reactive particles; ρ_p: density of reactive metal-spiked particles; C_{omax}: maximum concentration of metallic species at the bed surface; F_B: volume fraction of the bed consisting of bubbles.

The unsteady model's equations can be written in dimensionless form as:

$$\partial C^*_{Bi}/\partial C^*_{Bi}\ \partial h + (1/U^*_B)\ \partial C^*_{Bi}/\partial t = f/U^*_B - (K_{BE}/U^*_B)(C^*_{Bi} - C^*_{Ei}) \tag{10.3}$$

$$\begin{aligned}\partial C^*_{Ei}/\partial h + (1/U_{mf})\ \partial C^*_{Ei}/\partial t = {} & (1-\varepsilon_{mf})f/\gamma_B U_{mf} \\ & + (F_B/(1-F_B))(K_{BE}/U_{mf})(C^*_{Bi} - C^*_{Ei})\end{aligned} \tag{10.4}$$

$$C^*_{0i} = \left[C^*_{Bi} + \frac{(1-F_B)}{F_B} \cdot C^*_{Ei} \right]_{h=H} \tag{10.5}$$

where the dimensionless concentrations C^*_E (emulsion phase) and C^*_B (bubble phase) are defined as:

$$C^*_E = \frac{C_E}{C_{omax}/F_E} \qquad C^*_B = \frac{C_B}{C_{omax}/F_B} \tag{10.6}$$

10.4.3 Determination of the metal vaporisation kinetics

10.4.3.1 Validation of the method with alumina matrix

Direct modelling of $CdCl_2$ vaporisation In the case of Cd-spiked alumina matrix at 850°C, the experimental rate profile $r(t)$ of Cd vaporisation was obtained by sampling particles in the reactor at given times and analysing them. This profile was then used as inlet data in the direct model, which calculates the time course of the metal concentration in the gas $C_0(t)$ (Figure 10.9). The model's results agree well with reality: $CdCl_2$ vaporises very rapidly in the first moments, then slower and slower because of the formation of the binary oxide CdO-Al_2O_3 layer.

On-line measurement of $CdCl_2$ vaporisation The normalised intensity (I/I_{max}) can be deduced from the measurement of the net intensity of emission for a characteristic wavelength of Cd. The intensity, thus the metal's concentration in the gas, exhibited a peak almost instantaneously after the injection of the reacting sample into the fluid bed, whatever the matrix (Abanades et al., 2002b, 2003a).

Figure 10.10 plots the kinetics of the gas (obtained from the emission intensity) and of the spiked alumina particles (obtained from solid sampling) relative to Cd vaporisation from alumina matrix. The Cd vaporisation rate (left ordinate) decreases always. The Cd concentration in gas (right ordinate) increases sharply in the first moments, with a 3–4-min lag due to both the ICP measurement technique and to the 5-m long sampling line, then it decreases until it becomes undetectable.

$CdCl_2$ vaporisation flux: method validation The rate of vaporisation of Cd is obtained by applying the inverse model to the Cd normalised intensity profile (deduced from Figure 10.10). As shown in Figure 10.11, the calculated flux of vaporisation compares fairly well with the experimental profile, obtained from the analysis of solid samples, for the alumina matrix. The gap is about 25% in the first moments, then it

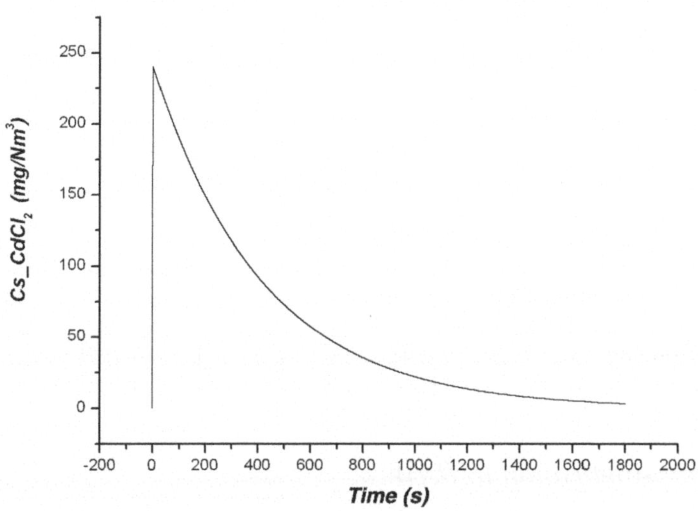

Figure 10.9 $CdCl_2$ concentration in off-gas versus time (from direct model).

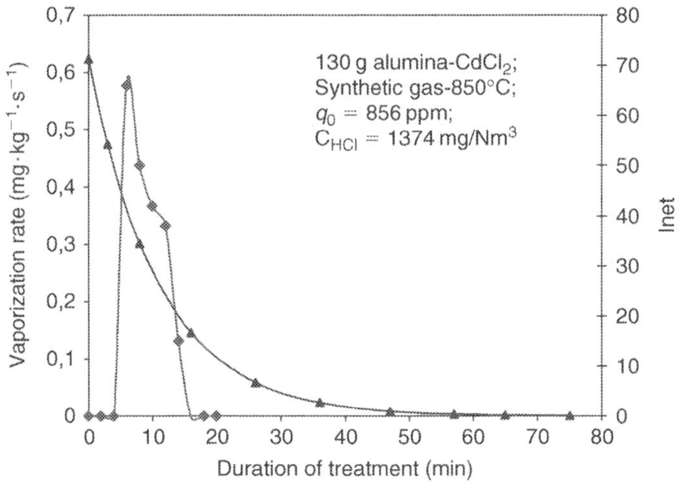

Figure 10.10 Cd vaporization rate from alumina, and emission intensity versus time (experimental results).

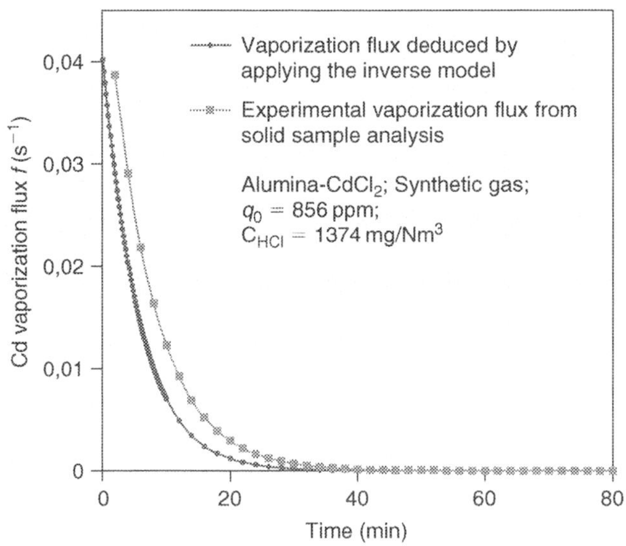

Figure 10.11 Comparison between the Cd vaporization flux (f) obtained from the inverse model and the experimental one.

decreases with time. This good fit validates the inverse method: by applying the inverse model, the rate of vaporization of HMs can be estimated from the on-line analysis of the outlet gas.

10.4.3.2 Application to RAW

The RAW contains organic species since they are made from real MSW. Therefore, nothing can be measured but the concentration profiles of metals in the exhaust gas.

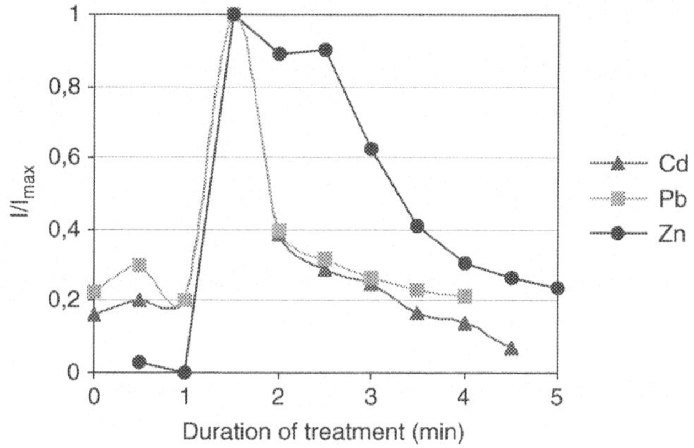

Figure 10.12 Relative concentration of Cd, Pb and Zn versus time.

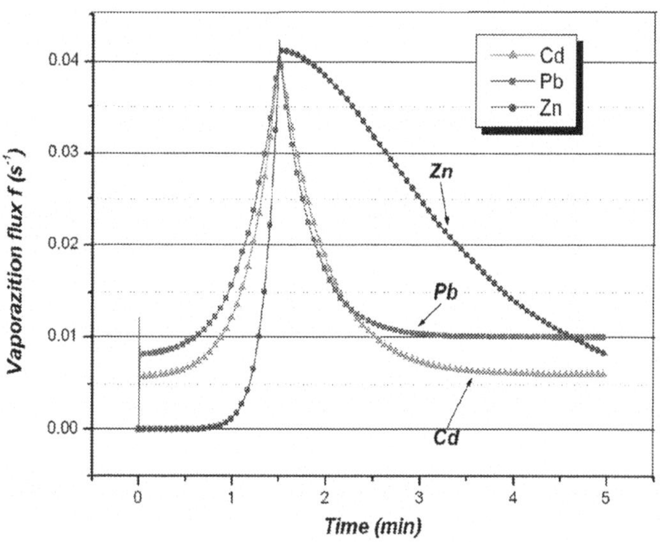

Figure 10.13 Results of inverse model: metals' vaporization flux versus time.

Experiments were run with RAW model waste spiked with Cd (q_0 = 4160 ppmw), Pb (q_0 = 2980 ppmw) or Zn (q_0 = 2700 ppmw). They all involved 10 g of RAW sample, and were run with synthetic gas without HCl at 850°C. Figure 10.12 plots the normalised profiles of the metal concentrations at the outlet, measured on-line, which exhibit a maximum value at time 1.5 min. All three profiles can be fitted, with different fits for time <1.5 min and >1.5 min. For each metal, the transient vaporization flux can be calculated by applying the inverse model to experimental data (Figure 10.13). This gives discretised transient results (Abanades et al., 2005). Current work concerns the deduction of kinetic laws from the relations between the vaporization flux and the metal concentration profiles in the solid waste.

10.5 CONCLUSION

The fate of HMs in waste incineration processes is an important problem: their accumulation in solid residues and their possible release in fumes are of main concern, slackening the implementation of incineration. A complete thermodynamic study was developed, and the ultimate residues were characterised physico-chemically, pointing out the influence of operating conditions and gas composition. An inverse method to determine the kinetics of HM vaporisation and to study the effect of operating conditions on these kinetics was developed. It was validated in a lab-scale fluid bed incinerator with Cd, Pb and Zn-spiked mineral matrices, then implemented with spiked RAW. The metal vaporisation dynamics were obtained from the on-line gas analysis, leading to kinetic laws that will be introduced in a general model simulating a waste incinerator. In addition, vaporised HMs may be stabilised into ultimate residues by adsorbing them on orthophosphates.

ACKNOWLEDGEMENTS

ADEME (French Environment and Energy Agency) is acknowledged for its partial financial support of this work. Authors are grateful to R. Flamand and B. Gagnepain for assistance in characterisation studies, and to J. Liu for participation to the kinetic study.

REFERENCES

Abanades, S. (2001) Behaviour of heavy metals speciation in solid household waste incineration processes (in French). Thèse de doctorat en Sciences de l'Ingénieur, Université de Perpignan.

Abanades, S., Flamant, G. and Gauthier, D. (2001) Modelling of heavy metal vaporization from a mineral matrix. *J. Hazard. Mater.* **88**(1), 75–94.

Abanades, S., Flamant, G., Gagnepain, B. and Gauthier D. (2002a) Fate of heavy metals during municipal solid waste incineration. *Waste Manage. Res.* **20**(1), 55–68.

Abanades, S., Flamant, G. and Gauthier D. (2002b) Kinetics of heavy metal vaporization from model wastes in fluidized bed. *Environ. Sci. Technol.* **36**(17), 3879–3884.

Abanades, S., Gauthier, D., Flamant, G. and Trassy, C. (2003a) On-line analysis of heavy metal in fumes from a fluid bed waste incinerator. *Combust. Sci. Technol.* **175**(4), 759–774.

Abanades, S., Flamant, G. and Gauthier, D. (2003b) The kinetics of vaporization of a heavy metal from a fluidized waste by an inverse method. *Combust. Flame* **134**, 315–326.

Abanades, S., Flamant, G., Gauthier, D., Tomas, S. and Huang, L. (2005) Development of an inverse method to identify the kinetics of vaporization of heavy metal during waste incineration in fluidized bed. *J. Hazard. Mater.* **A124**, 19–26.

ADEME, La composition des ordures ménagères en France (1998) *Données et Références*.

Barton, R., Clark, W. and Seeker, W. (1990) Fate of metals in waste combustion systems. *Combust. Sci. Technol.* **74**, 327–342.

Brunner, P.H. and Mönch, H. (1986) The flux of metals through municipal solid waste incinerators. *Waste Manage. Res.* **4**, 105–119.

Chandler, A.J., Eighmy, T.T., Hartlén, J., Hjelmar, O., Kosson, D.S., Sawell, S.E., Van der Sloot, H.A. and Vehlow, J. (1997) *Municipal Solid Waste Incinerator Residues*. The International Ash Working Group (IAWG). Elsevier Science B.V., Amsterdam.

Cheynet, B. (1988a) Complex chemical equilibria calculation with the THERMODATA system. *Computer Software in Chemical and Extractive Metallurgy. Proc. Metal. Soc. CIM* **11**, 31–44, Pergamon Press, New York.

Cheynet, B. (1988b) THERMODATA: on-line integrated information system for inorganic and metallurgical thermodynamics. In *Computerized Metallurgical Databases*, pp. 29–40. The Metallurgical Society, Inc., Warrendale, PA, USA.

Durlak, S.K., Biswas, P. and Shi, J. (1997) Equilibrium analysis of the affect of temperature, moisture and sodium content on heavy metal emissions from municipal solid waste incinerators. *J. Hazard. Mater.* **56**, 1–20.

Frandsen, F., Dam-Johansen, K. and Rasmussen, P. (1994) Trace elements from combustion and gasification of coal – an equilibrium approach. *Prog. Energ. Combust. Sci.* **20**, 115–138.

Gagnepain, B. (1998) Heavy metal speciation in solid household waste incineration residues and contribution to the interpretation of volatilisation processes (in French). Thèse de doctorat en Sciences de l'Ingénieur, Université de Perpignan.

Gauthier, D., Kribi, S., Bailliez, S., Abanades, S., Nzihou, A. and Flamant, G. (2005) Vaporisation and/or stabilisation of heavy metals in thermal treatment of waste: case studies of municipal wastes and dredged materials. *Proceedings of the 1st International Conference on Engineering for Waste Treatment*, May 17–19, Albi, France.

Hassaine, S., Trassy, C. and Proulx, P. (2001) Continuous emission monitoring of metals in flue gases by ICP-OES: role of calibration and sample gas. *High Temp. Mater. Process.* **5**, 313–331.

Kirby, C.S. and Rimstidt, J.D. (1993) Mineralogy and surface properties of municipal solid waste ash. *Environ. Sci. Technol.* **27**, 652–660.

Kunii, D. and Levenspiel, O. (1991) *Fluidization Engineering*, 2nd edn. Butterworth-Heinemann Series in Chemical Engineering, Boston.

Menard, Y., Patisson, F., Sessiecq, P., Ablitzer, D., Merz, R. and Seifert, H. (2002) Transient combustion of municipal solid waste in a grate furnace: modelling and experiments. *European Conference on the Future of Waste Management in Europe*, Proceedings, Strasbourg, VDI, pp. 239–242.

Morf, L., Brunner, P. and Spaun, S. (2000) Effect of operating conditions and input variations on the partitioning of metals in a municipal solid waste incinerator. *Waste Manage. Res.* **18**, 4–15.

Owens, T., Wu, C. and Biswas, P. (1995) An equilibrium analysis for reaction of metal compounds with sorbents in high temperature systems. *Chem. Eng. Commun.* **133**, 31–52.

Pfrang-Stotz, G. and Schneider, J. (1995) Comparative studies of waste incineration bottom ashes from various grate and firing systems, conducted with respect to mineralogical and geochemical methods of examination. *Waste Manage. Res.* **13**, 273–292.

Pierotti, G. (1999) Société TIRU Paris, Personal Communication.

Tessier, A., Campbell, P. and Bisson, M. (1979) Sequential extraction procedure for the speciation of particulate trace metals. *Anal. Chem.* **51**(7), 844–851.

Trassy, C. and Diemiaszonek, R. (1994) Heavy elements in gaseous effluents: study of an on-line analytical method. *High Temp. Chem. Process.* **3**, 449–458.

Verhulst, D., Buekens, A., Spencer, P. and Eriksson, G. (1996) Thermodynamic behavior of metal chlorides and sulfates under the conditions of incineration furnaces. *Environ. Sci. Technol.* **30**, 50–56.

Wey, M., Hwang, J. and Chen, J. (1996) The behavior of heavy metal Cr, Pb and Cd during waste incineration in fluidized bed under various chlorine additives. *J. Chem. Eng. Jpn.* **29**(3), 494–500.

Wu, C. and Biswas, P. (1993) An equilibrium analysis to determine the speciation of metals in an incinerator. *Combust. Flame* **93**, 31–41.

Yan, R., Gauthier, D., Flamant, G. and Badie, J.M. (1999) Thermodynamic study of the behaviour of minor coal elements and their affinities to sulphur during coal combustion. *Fuel* **78**(15), 1817–1829.

11

Functional interpretation of gas composition

P. Hobbs

11.1 INTRODUCTION

At first, the interpretation of the composition of a gas brings a range of impressions, but when put in the context of, "has that cheese matured?", or "has that beer fermented well?", then the process in hand is evaluated if it has been successful or not. Cheese- and beer-making processes have been successfully monitored by using multi-sensors or electronic noses to monitor the non-methane volatile organic compounds (NMVOCs) in the headspace of the food process (Gardner et al., 1994; Gibson et al., 1995; Schaller et al., 1998).

To place NMVOC emissions from mostly organic decay process in the social context requires an understanding of the perceptions of the public and also the employees of the plants involved with managing such fermentation processes. These provide an insight into how an odour problem evolves from an acceptable background odour into an odour nuisance. NMVOC emissions and especially those that are odorous can cause considerable difficulties for the local population and subsequently the plant operators. There is increasingly less tolerance to odour problems from the general population. There are underlying causes for this. At the individual psychological level, malodour has been

 Waste Gas Treatment for Resource Recovery edited by Piet N.L. Lens, Christian Kennes, Pierre Le Cloirec and Marc Deshusses. ISBN: 1843391279. Published by IWA Publishing, London, UK.

associated with decay and often disease, thus avoidance of the area or source is encouraged. Moreover, more houses are being built on brownfield sites that encroach onto the perimeter of industrial plants and waste treatment processes. This was also the case for greenfield areas where dwellings are located closer to composting sites. In the UK several composting business have been forced to close by the local authorities due to odour emissions that are a nuisance. The formation of local pressure groups has led to more social pressure that often centres on health or property prices.

The perceptions of the plant employees are, perhaps, more important, because they are responsible for preventing or minimising emissions. Proactive communications policy with the local population helps to reduce odour problems and any subsequent pressure group activity. With good communications the local population can provide the plant operators with more information that they may not acquire on site, because of wind or climatic factors. Also, the perception of plant operators to observe the trace gas emissions during fermentation processes is important, but this was only so for those VOCs that give an olfactory response. Although processes can be monitored using the analysis of volatile fatty acids (VFAs) from the fermentation liquor (Bjornsson et al., 2000) that contribute to the VOCs in the headspace, there are good arguments for monitoring them in the gaseous phase. The arguments include that there is a relationship of the liquor concentration to the headspace concentration according to Henry's law (Braida & Ong, 1998). Further the headspace concentration is subject to the effects of the chemistry at the liquid surface. These effects include pH which can affect the emissions of ammonia, for example when long chain fatty acids form a layer on the surface (Sommer & Sherlock, 1996). A more reliable profile of the chemical components (whether odorous or not) or trace NMVOCs can be obtained by chemical analysis which best involves collecting the trace NMVOCs and analysing them by gas chromatography mass spectrometry (GC-MS). Such observations are process and plant site dependent and are more likely to involve production of an organic output or waste management. Inorganic sources of gases can also be monitored to observe the process, but these are often not odorous, for example carbon dioxide gas in combustion processes.

A major objective of this chapter is to outline both the chemical and the physical rationale behind the functional interpretation of gas composition, as applied to industrial "wastes" or industrial resources. We can therefore ask such questions as: Does a process emitting unpleasant odours, whether as a periodic, very offensive source or a continuous monotonous source, provide an indication that the process is not operated effectively? This has implications regarding sustainability in terms of wasting energy, where the operators are not considering the economic or the environmental consequences. This chapter will only consider emissions of NMVOCs and does not consider inorganic compounds.

11.2 SOURCES OF VOLATILE ORGANIC COMPOUNDS

11.2.1 NMVOCs emissions

There are numerous sources of NMVOCs, but the amounts emitted and if they contribute to air pollution need to be considered to develop a perspective of sustainability of the process. Such NMVOCs range from sources that maybe part or totally anaerobic.

Often as the process becomes more anaerobic or aerobic, the NMVOC profile changes (Hobbs et al., 1999). In this context, the NMVOCs originating from decomposition processes can then be identified and how such emissions can be reduced or minimised. The NMVOCs comprise a wide range of chemical compounds that have been mentioned, but also include hydrocarbons as well as oxygenated and halogen-containing species. The environmental impact of NMVOCs includes their involvement in the formation of ground level ozone, but they may also cause a range of ill-health effects. NMVOCs include benzene and 1,3-butadiene, which are carcinogens.

Figure 11.1 shows the fraction of the 545 kt $year^{-1}$ of NMVOCs emissions in 2001 for the European countries. There appears no clear indication why NMVOCs are about twice as high from the UK and France, but it is possibly a result of the way the emissions data are collated. Figure 11.2 shows the breakdown of the sources for the whole of Europe. The oil industry contributes the largest amount of NMVOCs emitted at 39% of the total. Surface treatments, which includes sewage works and products requiring the use of organic solvents contributes 17% of the total. In addition, there are biogenic sources that originate from vegetation such as trees and agriculture, which includes emissions from livestock and manure sources (Hobbs et al., 2004). At present, these include sulfides, VFAs, phenols and indoles.

Of the NMVOCs mentioned above, those generated from industrial production processes offer us the best opportunities to reduce their emissions at source. However, the capacity to monitor a process can be limited by the ability to measure the emissions.

11.2.2 Biochemistry of the anaerobic environment

Generally, NMVOCs are produced in an anaerobic environment that is a consequence of a high oxygen demand. The oxygen demand cannot be satisfied due to

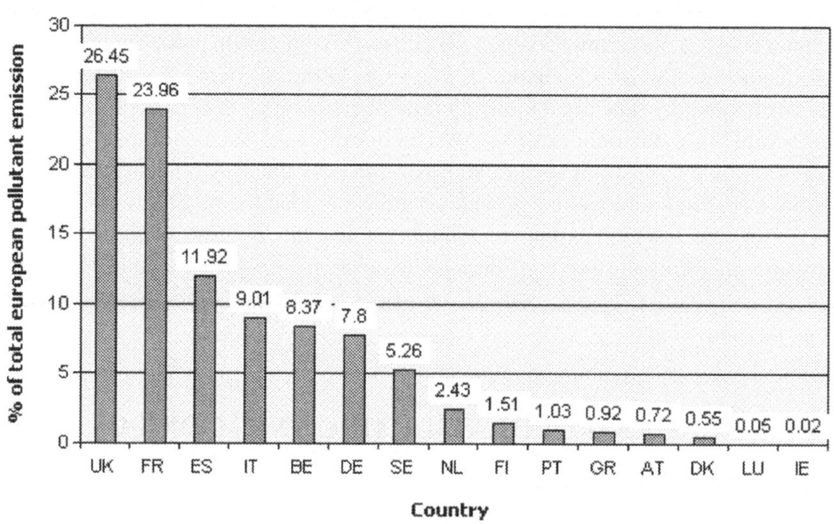

Figure 11.1 NMVOCs from European countries 2001. (*Source*: European Pollutant Emission Register http://www.eper.cec.eu.int/eper/)

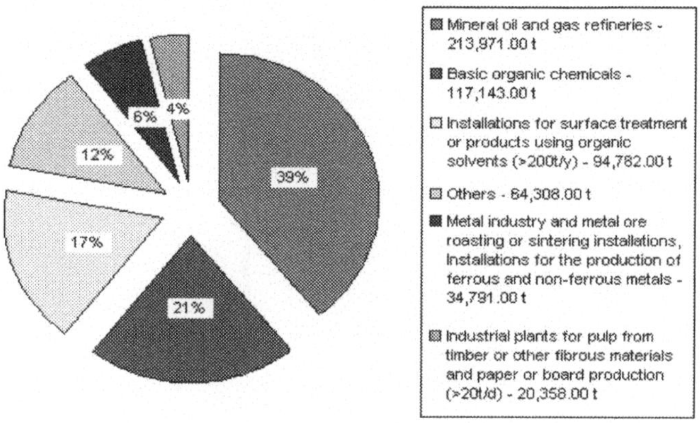

Figure 11.2 Proportions of NMVOCs emitted from Europe. (*Source*: European Pollutant Emission Register http://www.eper.cec.eu.int/eper/)

the high organic content of the waste is being oxidised. Sewage treatment works (Hvitved-Jacobsen & Vollertsen, 2001), composting sites (Noble et al., 2001) and livestock housing and manure storage tanks (Hartung, 1992) all show similar AD decomposition products of sulfides, VFAs, amines and phenols, but not similar NMVOC profiles in terms of the concentration profile.

Four stages of anaerobic digestion (AD) are distinguished (Gujer & Zehnder, 1983), the first stage involves hydrolysis of natural polymers, which are often insoluble, into smaller molecules that are soluble. For example, polysaccharides decay to sugars, proteins to amino and fatty acids (Spoelstra, 1980). The process is catalysed by enzymes such as cellulase, amylase, lipase or protease that are produced outside the cell wall of microbes. The second stage concerns acidogenesis, where the acid-forming bacteria form VFAs, CO_2 and hydrogen. In the third stage acetogenesis converts fatty acids into acetic acid and the final stage methanogenesis occurs mostly from the actions of acetoclastic and hydrogenotrophic organisms to produce methane from acetic acid and H_2/CO_2 (Gujer & Zehnder, 1983).

Microbes involved in the decomposition processes are influenced by their substrate composition and environment, for example thermophilic organisms normally perform (at 55°C) the same processes faster because of the higher temperature than the mesophilic organisms that operate at about 35°C. In some AD processes, the oxygen content can be controlled to produce a different profile of NMVOCs. Figure 11.3 shows that the NMVOC emission profile may change with decreasing oxygen concentrations, or depth of the storage tank. In an anaerobic environment, decay products normally use the most readily available source of electron acceptors. Therefore, instead of carbon dioxide and water, odorous molecules such as hydrogen sulfide, methanethiol, butanoic acid, phenols and indoles are produced from amino acids or organo-sulfur compounds. The majority of these compounds originate from the decomposition of protein (Mackie et al., 1998; Spoelstra, 1980) some of which contain sulfur. Some NMVOCs are exclusive to specific protein sources (Table 11.1) and some are partially associated with protein decay (Table 11.2). For example, up to 50% of the nitrogen in livestock

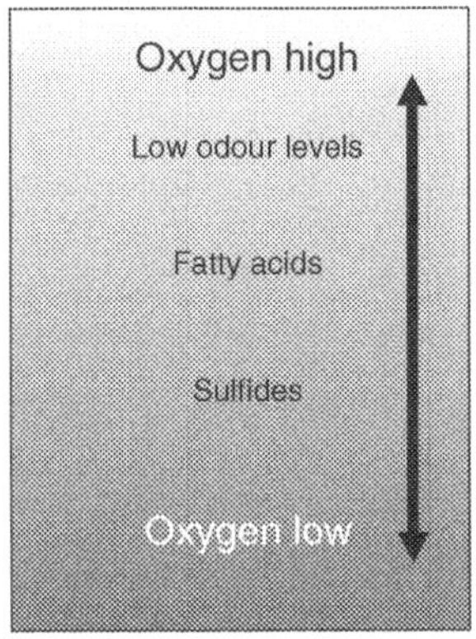

Figure 11.3 Changes in oxygen concentration, or redox potential with depth, in an anaerobic storage tank, showing the most probable location of odour production.

Table 11.1 Source of NMVOCs from specific amino acids.

NMVOC	Source
Methyl sulfides	Methionine and cysteine
4-methyl phenol and 4-ethyl phenol	Tyrosine
Phenols and phenylpropionic acid	Phenyalanine
Dihydroskatole	Histidine
Indole and skatole	Tryptophan
2-methyl propanoic	Valine
2-methyl butanoic	Isoleucine
3-methyl butanoic	Leucine

After Spoelstra et al. (1980); Mackie et al. (1998).

Table 11.2 Source of NMVOCs from less specific amino acid decay pathways.

NMVOC	Source
Acetic acid	Glycine, alanine and serine
Butyric acid	Histidine
Propanoic acid	Threonine (and oils)

faeces can be present as undigested protein and can be available for decomposition. Generally, straight chain fatty acids originate from a range of natural polymers that include polysaccharides, fibre and lipids as well as protein.

11.3 ANALYSIS OF NMVOCS

Sampling the headspace of an anaerobic decomposition process has the major advantage of not having to perform any extraction technique from such a difficult analytical matrix. There are numerous sensors available to monitor NMVOCs in the headspace, however, before selecting a rapid or continuous means of monitoring, the most complete profile possible is needed, as some sensors can bias the results as most do not provide adequate concentration or identification information. GC offers the best means of separating odorous components. Sampling and the choice of detector for the GC and its limitations can impose upon chromatographic configurations. Often NMVOCs which are malodorous can be in the parts per billion (v/v) range and at such low concentrations, surface absorption and chemical activity become major concerns regarding sample integrity. Sample preconcentration onto an adsorbent is then necessary prior to introduction into the GC system for better detection.

11.3.1 Concentrating the sample

Numerous adsorbents are available that have a range of specific properties to collect NMVOCs. Recognised choices of adsorbents for odour trapping NMVOCs are present in the literature (Adib et al., 1999; Devai & DeLaune, 1997; Huxham & Thomas, 2000; Mccaffrey et al., 1994). Their use and selection are available on the web sites presented in Table 11.3. Most are recognised methods, designated for groups of compounds generally having similar chemical activity.

The selectivity to NMVOCs with chemical groups, and the mass that an adsorbent can adsorb depends on their pore size and surface attraction properties. Adsorbents have a range of different surface areas (per unit mass) and different particle sizes appropriate to different airflow conditions. Generally adsorbents with larger surface areas will have a greater capacity to adsorb smaller molecules. If too much NMVOC is passed through the adsorbent, then a steady state between the mass on the adsorbent and the concentration in the eluted sample will be reached. To ensure this breakthrough volume is not reached, manufacturers recommend the largest volume that can be sampled without NMVOC loss. Such volumes have been determined and are available from manufacturers, the scientific literature and electronic media. For example, Tenax, a common hydrophobic adsorbent has a web page (www.sisweb.com/index/referenc/tenaxtam.htm) that provides details, such as breakthrough volumes (the point at which the trapped NMVOC elutes from the adsorbent) for a range of polar compounds showing the dependency on temperature (www.tu-harburg.de/et1/private/gk/break/break.htm).

Table 11.3 Web sites with application notes for the analysis of NMVOCs.

- http://www.markes.com/
- http://www.oico.com/appnvsv.htm
- http://water.wr.usgs.gov/pnsp/pest.rep/voc.html
- http://www.restekcorp.com/voa/voa.htm
- http://www.cdc.gov/niosh/homepage.html
- http://www.jandw.com/GCAppnotes.htm
- http://www.sisweb.com/index/referenc/apnotetd.htm
- http://www.battelle.org/environment/ASAT/canister.html
- http://www.restekcorp.com/voa/gcconfig.htm
- http://www.gerstel.com/solutions/indexe.htm
- http://www.skcinc.com/guides.html

Although adsorbents have proved a successful means to trap volatile compounds, there are some problems. First, moisture inhibits and prejudices the NMVOC profile adsorbed. Second, often no one adsorbent can trap the complete range of compounds, so different adsorbents are placed in series to one another in the tubes. The adsorbents are placed in an order such that the sample migrates through the tube to the increasing likelihood of being trapped. The largest molecules will be trapped first and the more volatile species will be trapped further down the tube. Desorption from the multi-bed tube should be performed with the carrier gas flow in the opposite direction to the adsorption mode. This is because less volatile compounds are more likely to produce broader chromatographic peaks if they pass through the stronger adsorbent for the more volatile compounds. In this event, the limit of detection will be reduced and this is a major problem for analytical methods with such limiting amounts present for detection.

11.3.2 Gas chromatography

GC is ideally suited for the rapid separation of complex NMVOC components that contribute to odour formation, especially with the use of capillary columns that offer a higher peak resolution. The technique principally involves a carrier gas that passes over a stationary phase for which the volatile components have a differential affinity to effect separation. The efficiency of the chromatography can be improved by precision temperature control of the column and constant flow of carrier gas. Generally, a range of detectors can be used with the MS proving the most effective for the identification of unknown NMVOC components.

As NMVOCs are nearly all polar, a non-polar column can be recommended for separation (Hobbs et al., 1999). Capillary columns have an internal thin film with dimensions of between 0.05 or 5 μm and an internal diameter between 0.01 and 0.5 mm and operate with gas flows lower than 1 mL min^{-1}, so only a small volume of sample originating from the thermal desorption is required. The small volume, of about 0.5 mL, can be a problem and cryofocusing of the NMVOCs can be performed to reduce the volume and increase the sensitivity of the detector. Sulfur-containing compounds will readily adsorb into silica or metal or porous polymer surfaces, to overcome this difficulty coating the internal surface of the gas chromatograph is recommended.

11.3.3 Mass spectrometry

The MS has been the detector of choice because of the capacity to identify unknown compounds. Co-eluting peaks can often be identified because there are mutually exclusive ions present from each compound, so that they can be quantified. Normally MS systems operate in the nanogram range when scanning the full mass scale of 10–600 mass units but this can be improved by single ion monitoring. Single ion monitoring with modern quadrupole MS systems means that the appropriate masses for each peak can be more sensitively detected as it elutes.

The inflow of oxygen into the ion source from a gas sample will react with the hot surfaces of the ion lens to reduce the sensitivity by the build up of electrostatic charges on these charged surfaces. Chemical ionisation rather than electronic ionisation may be used to increase sensitivity as less fragmentation occurs.

11.4 NMVOC EMISSION RATES

11.4.1 Determination of emission rates

When sampling emissions, it needs to be considered whether to use discrete or continuous monitoring techniques and how they relate to the NMVOC generating process. This may involve determination of gas flow rates as well as considering the factors that affect the sample such as moisture, temperature and the presence of particles. NMVOCs emissions are affected by the age of the waste, the waste type and the degree to which oxygen can diffuse into the bulk of the material.

It is necessary to take more exacting measurements to determine the emission rate from a source to obtain data for modelling or understand processes that affect the emission rates. This may require an open or a closed chamber system (Cumby et al., 1995). Results can be expressed as mass emitted per unit area or mass emitted per unit volume per unit area, which is a necessary for anaerobic digestion or decaying processes.

Emission rates can be determined in an enclosed chamber from the headspace concentration based on Henry's law behaviour, from an inverse exponential curve with time, limited at the saturated vapour pressure (SVP) of a volatile compound. The slope is the kinetic rate of emission. However, Henry's law applies to pure solutions rather than the complex mixture present in wastes.

In open chambers, the airflows at the surface should be well characterised and emissions can be observed for non-interfering (laminar) or turbulent events. Clean air has to be introduced at one end of the open system for sampling. However, in a closed system the concentration of the volatile compounds increases, which makes them easier to analyse, but more complex mathematics are necessary as the increasing concentrations suppress emissions. For compounds with increasing molecular weights (MWts) their time to reach saturation concentrations can be of the order of months, for example the explosive trinitrotoluene (MWt 261). Generally NMVOCs compounds from decay or fermentation processes have molecular weights between about 30 and about 150.

11.4.2 Physical factors affecting emission rates

11.4.2.1 Effects of wind speed

There are general rules concerning the emission rates from a surface as a function of wind speed. Laminar flows are present at low air speeds and surface emissions will be diffusion limited. At greater wind speeds, turbulence will increase the emission rate because mixing reduces the headspace concentration. Should this occur, then emission rates will be limited by the diffusion of the volatile compounds to the liquid surface. Therefore, above a threshold wind speed the emission rate will not increase and, of course, the NMVOC concentration will decrease in the air.

For a liquid or a semi-liquid waste, the calculation of emission rates and threshold wind speeds will be complex. The emission profile changes with temperature and airspeed, because of different kinetic rates of release from the waste surface. Different VOCs will produce a changing emission profile which will add to the complexity of determining the olfactory offensiveness and odour concentration (OC) emitted.

Zahn et al. (1997) positively correlated wind velocity of between 0.2 and 9.4 $m\,s^{-1}$ with the concentration of NMVOC emissions from a pig waste lagoon. The maximum concentration of total NMVOCs of 9 $\mu g\,L^{-1}$ occurred at a wind velocity of 3.6 $m\,s^{-1}$. Zahn et al. (1997) also recognised differential processes occurring in the emissions plume: some NMVOCs did not travel as far. Whether this was due to NMVOCs dissolving in the moisture in the air or attachment to particles was not identified. Chemical processes such as photolytic reaction with sunlight may also play a role.

11.4.2.2 Henry's law

The emission rates of VOCs, originating from a liquid surface, depend on several variables but they mostly depend on the concentration in the liquid and the chemical or driving energy available for release of VOCs into the air above the surface. William Henry first published his law (Henry's law) in 1803 that expresses the ratio of the concentration between air and liquid at equilibrium concentrations for a volatile compound, providing no chemical reactions occur. Henry's law is a thermodynamic expression of the potential for emission. Henry's law constant can be approximated from the ratio of vapour pressure to solubility.

The implications are that those compounds that are virtually insoluble in water, such as the hydrocarbons and the alkyl sulfides have a greater presence in the headspace above the source than those that are soluble (Vitenberg, 1991). Those that react with water and are pH dependent, such as the phenols and VFAs make the calculation more complicated, but Henry's law addresses unionised gas species. Therefore, the effect of pH needs to be determined before the unionised concentration can be calculated.

11.4.2.3 Effect of pH on gas concentration in the headspace

Before a gas can be formed from an ion in solution, the molecule has to normally lose or attach a hydrogen ion as in the case for ammonium and acetate, for example:

$$NH_4^+ + OH^- = NH_3 + H_2O \tag{11.1}$$

$$CH_3COO^- + H^+ = CH_3COOH \tag{11.2}$$

This is graphically illustrated in Figure 11.4 where the dissociation constant (pKa) for each volatile compound is used to show the fraction of unionised NMVOC and ammonia in pure water as the pH changes. The majority of NMVOCs from AD are principally related to the hydrogen ion concentration, i.e. the pH. So for the ionisation or solubilisation of phenols, a pH of 9.2 or greater is required and for the VFAs a pH of 4.2 or greater. The non-linear interaction of concentration of unionised species with pH is shown in Figure 11.5.

11.4.2.4 Non-ideal Henry's law conditions

Electrolytes and surfactants in the form of polymeric proteins, saccharides and lipids all cause potential deviations from Henry's law (Thomas, 1990). This is because lipid layers can form at the air/liquid interface and prohibit emissions of NMVOCs (Sommer & Sherlock, 1996). Also because many organic acids are polar at one end

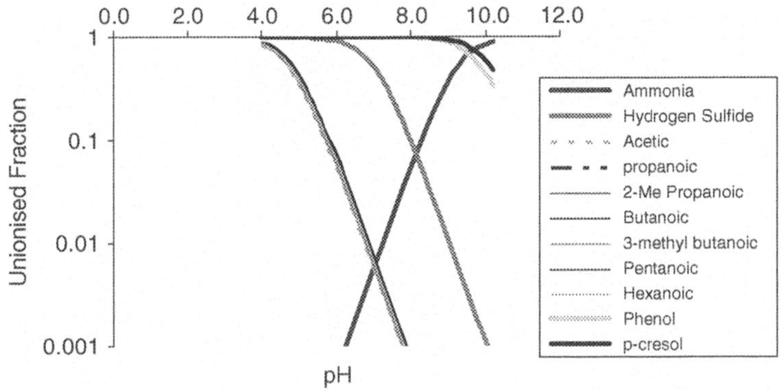

Figure 11.4 Graphical representation of the theoretical fraction of unionised compounds that can volatilise as a function of pH.

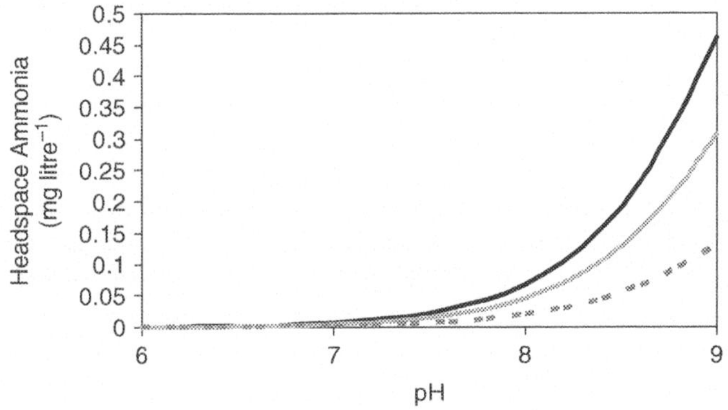

Figure 11.5 Variation of ammonia headspace concentration from three different ammonium concentrations at equilibrium. (-- 1 mg/l; — 4.2 mg/l; ▬ 5.8 mg/l).

and hydrophobic at the other, they can increase the acidity of the surface layer of water. As a consequence ammonia for example may reform to ammonium and remain in solution (Cumby et al., 1995). During AD many other gases are erupting from the bulk volume, so CO_2 may dissolve and spurge other gases to the surface.

11.5 DIFFERENTIATING PROCESSES USING THE NMVOC PROFILE

There are several feedback techniques to control processes using single sensors. These are in use mainly for liquid media, for example pH, redox or conductivity measurements which are used extensively in the wastewater industry and other fermentation processes (Bjornsson et al., 2000; Cheng et al., 1999; Cimander & Mandenius, 2002; Eldem et al., 2004; Jantsch & Mattiasson, 2004; Sasaki et al., 2003). Increasingly, algorithms that are more complex are necessary to deal with the interpretation of data from a larger number of sensors, such as an "electronic nose". Such data can be

Figure 11.6 Odour emissions from compost produced for mushroom growing. This picture shows a site with Phase I stage of composting in the foreground, the straw and in the background the effluent storage tank.

processed by more complex numerical analysis, for example fuzzy logic (MarsiliLibelli & Muller, 1996; Murnleitner et al., 2002; Polit et al., 2001), but this has not yet been used for process control, so far only for process monitoring. In the first instance, NMVOCs were collected using adsorption processes to concentrate to a sufficient mass for analysis by GC-MS as described earlier to ensure that no NMVOCs are missed that a sensor system may not identify. The investigations involve: (1) assessing composting processes and the impact of the substrate, (2) reducing the odour emissions from stacked sewage sludge cake (SSC) and (3) a more speculative investigation involving monitoring the NMVOCs emitted from their sheep faeces to assess dietary evaluation.

11.5.1 Composting

11.5.1.1 Composting processes

NMVOCs can be used to diagnose the status of compost produced to grow mushrooms and to reduce odorous emissions (Noble et al., 2001). Odour emissions are a major waste management problem and consequently produce a financial burden on compost production for mushroom growers in many EU countries (Duns et al., 1997; Noble & Gaze, 1994). Mushroom composting involves wetting and mixing straw and animal manure in windrows (pre-wetting) and leaving this mixture in long windrow stacks for Phase I. In addition to straw and manure (often from horses or chicken houses) as a nitrogen source, gypsum can be added as a sulfur source. During these stages, fermentation can be uncontrolled, producing anaerobic zones that are a source of odorous NMVOCs that cause unacceptable emissions (Figure 11.6). To reduce emissions during pre-wetting and Phase I, new management techniques were introduced (Noble et al., 2002). The trace volatile emission profiles were recognised as indicators of the complex biological processes that occur and were used to investigate composting production decisions. This was especially necessary with large outputs of about 1000 tons of compost per week per site.

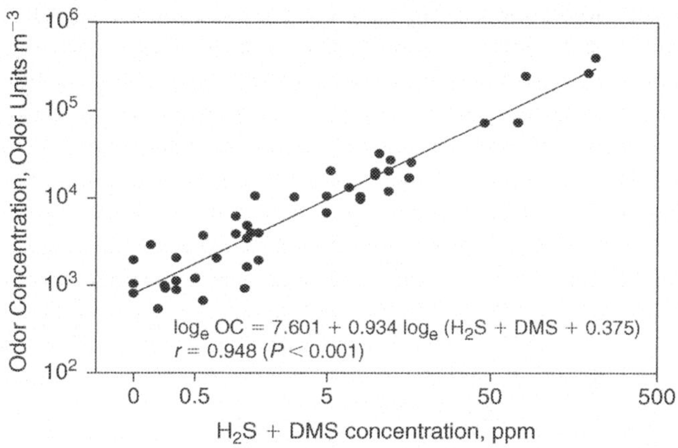

Figure 11.7 Relationship between the combined on-site H_2S and DMS concentrations and the 24 h OC on mushroom composting sites. Each point was the mean of two sample determinations.

Thermal desorption-GC-MS was used in order to detect both odorous and non-odorous NMVOCs at concentrations as low as parts per trillion. Twenty-six VOCs are related to the different stages of the composting process. A relatively small number of chemical compounds were identified as responsible for the offensive composting odours and were systematically recognised by producing the highest number or multiple above their odour threshold when the concentration found was divided by the odour threshold concentration. Gas detector tubes were used for on-site measurement of the most concentrated NMVOCs. Odorants that exceeded their published olfactory detection thresholds (the point at which the odour is just perceived and has an OC of 1) by the greatest order of magnitude, were (in decreasing order): H_2S, dimethyl sulfide (DMS), butanoic acid, methanethiol and trimethylamine. Concentrations of ammonia were not significantly correlated with OC, and they were not significantly affected by the use of aeration. In contrast, there was a close correlation ($r = 0.918$, $P < 0.001$) between the OC of air samples from the composting materials and the combined H_2S + DMS concentrations (Figure 11.7), measured on site with colorimetric gas detector tubes.

In many cases, local authorities were threatening to enforce compulsory closures of the mushroom composting sites because of odour complaints. By adopting different composting turning regimes, composting processes were improved and odour nuisance complaints reduced by 90% to help prevent the enforcement of site closure. Such an approach also ensured that the finished compost was of the highest quality.

11.5.1.2 The effect of different substrates

The traditional ingredient of mushroom (*Agaricus bisporus*) compost, straw bedding horse manure, provided both a source of carbon (C) and nitrogen (N) to the composting organisms. Due to difficulties in availability and variability in the composting substrate, different formulations with straw and various N sources have been developed.

In most countries, wheat straw has been preferred to other cereal straw types that lose their structure integrity more rapidly. N sources include animal manures, such as poultry, pig and cattle, and other organic wastes such as dried blood, cottonseed meal, brewery wastes, horn, whey powder, molasses and sewage-based products. Inorganic N sources include ammonium nitrate and sulfate, calcium nitrate, urea and urea formaldehyde. Due to its low cost and high N content, broiler poultry manure is currently accepted for most mushroom composting activities. However, poultry manure has a serious odour problem, both on its own and when incorporated into compost.

The effects of different straw types and organic and inorganic N sources on the chemical composition and OC of composting emissions, compost analysis and mushroom yield were examined using bench-scale and large-scale (windrows and aerated tunnels) experimental systems (Noble et al., 2002). There was a close correlation between the butanol or combined H_2S + DMS concentration and logarithm of the OC of air samples taken from different composting ingredients ($r = 0.83$ and 0.87, $P < 0.001$). Replacement of poultry manure with the other N sources at 50–100% or wheat straw with rape straw in aerated tunnel or windrow composts reduced the composting OC and emission of odorous sulfur-containing compounds by less than 10%, but also reduced the mushroom yield. Urea and cocoa meal may be suitable for low odour pre-wetting of straw, with addition of poultry manure immediately before aerated tunnel composting. A proportion of rape straw in compost may reduce the formation of anaerobic zones since it maintains its structure and porosity better than wheat straw.

Poultry manure composts and hop waste compost (2.3% N rate) had higher OCs and ammonia concentrations than most formulations and the N sources at the 1.8% N rate produced a higher OC than the 2.3% N rate. Volatiles found in the compost odour samples at concentrations exceeding their detection thresholds were NH_3, butanol and sulfides (Table 11.4). Several other compounds (mainly organic acids) were found at concentrations just above their detection thresholds, but were not significantly different between treatments. The sulfide mainly responsible for exceeding its detection threshold was DMS, followed by H_2S, except for ammonium sulfate composts where methanethiol predominated. Little or no S-containing compounds were detected in the air from cocoa meal, hop waste or urea composts. Other odorants identified with GC-MS but not significantly different between treatments were: acetic acid (0.37–1.043 $mg\,m^{-3}$), butanoic acid (0.70–2.68 $mg\,m^{-3}$), dimethyl disulfide (0.10–0.19 $mg\,m^{-3}$), 4-ethyl phenol (0.06–0.35 $mg\,m^{-3}$), isopropyl alcohol (0.28–0.79 $mg\,m^{-3}$), 3-methyl butanoic acid (0.66–4.92 $mg\,m^{-3}$), methyl ethyl ketone (1.76–4.68 $mg\,m^{-3}$), 4-methyl phenol (0.14–0.33 $mg\,m^{-3}$), methyl propanoic acid (0.33–1.07 $mg\,m^{-3}$), pentanoic acid (0.63–2.48 $mg\,m^{-3}$), phenol (0.29–0.51 $mg\,m^{-3}$) and propanoic acid (0.14–0.74 $mg\,m^{-3}$).

11.5.2 Sewage sludge cake

Also other fermentation processes that cause odour problems, such as those that inadvertently occur in stored SSC are likely to benefit from enabling air to permeate through the bulk of the material. Sewage treatment plants generate nuisance odours, but those from the SSC are known to produce greater problems especially when they

Table 11.4 Odour and NMVOC concentrations of air samples from different aerated tunnel compost formulations.

Compost formulation	N rate[a]	Odour concentration ($OU\,m^{-3}$)	Concentration ($mg\,m^{-3}$)							
			H_2S	DMS	MeSH[b]	NH_3	Acetone	Ethanol	Butanol	Propanol[c]
Wheat straw +	1.8	22 [7.70][b]	0.09	0.26	0.07	7 [1.97][d]	0.74 [0.11][d]	3.5 [1.23][d]	2.8 [1.08][d]	1.03
Poultry manure	2.3	6342 [8.76]	0.73	2.06	0.46	80 [4.38]	1.38 [0.56]	10.26 [2.36]	13.60 [2.64]	3.84
Rape straw +	1.8	2619 [7.87]	0.07	0.34	0.15	41 [3.73]	0.29 [−0.41]	2.92 [1.19]	1.94 [0.84]	0.28
Poultry manure	2.3	3584 [8.18]	0.49	0.52	0.33	10 [4.68]	1.25 [0.49]	12.6 [2.57]	13.52 [2.63]	0.15
Wheat straw +	1.8	488 [6.19]	0	0	0	9 [2.24]	0.21 [−0.53]	0.86 [0.21]	1.08 [0.38]	0.29
Cocoa meal	2.3	977 [6.89]	0	0	0	23 [3.16]	0.63 [0.01]	0.08 [−0.79]	0.83 [0.19]	0.10
Wheat straw +	1.8	854 [6.70]	0	0	0	2 [0.91]	1.07 [0.37]	2.07 [0.90]	0.23 [−0.51]	0.63
Hop waste	2.3	2970 [8.00]	0.05	0.13	0.09	50 [3.93]	1.60 [0.68]	1.66 [0.71]	8.16 [2.14]	0.45
Wheat straw +	1.8	699 [6.55]	0	0	0.12	12 [2.49]	0.34 [−0.34]	3.28 [1.30]	2.44 [1.04]	0
Ammonium sulfate	2.3	1045 [6.95]	0.02	0.12	0.27	20 [3.00]	1.44 [0.59]	6.04 [1.86]	1.89 [0.82]	0.37
Wheat straw +	1.8	670 [6.51]	0	0	0	5 [1.71]	0.65 [0.03]	1.16 [0.43]	1.18 [0.44]	0
Urea	2.3	843 [6.74]	0	0.02	0	41 [3.11]	0.70 [0.07]	0.55 [−0.07]	0.88 [0.23]	0.15
LSD, $P < 0.05$		[1.50]	0.11	1.15	0.35	[3.11]	[0.99]	[2.49]	[1.77]	2.47
Detection threshold [8]			0.03	0.006	0.002	4	34.67	55	1.51	6.03

[a]N content of DM of blended compost ingredients.
[b]Methanethiol.
[d]Figures in square parentheses are $\log_e$ transformations, shown next to back-transformed values.

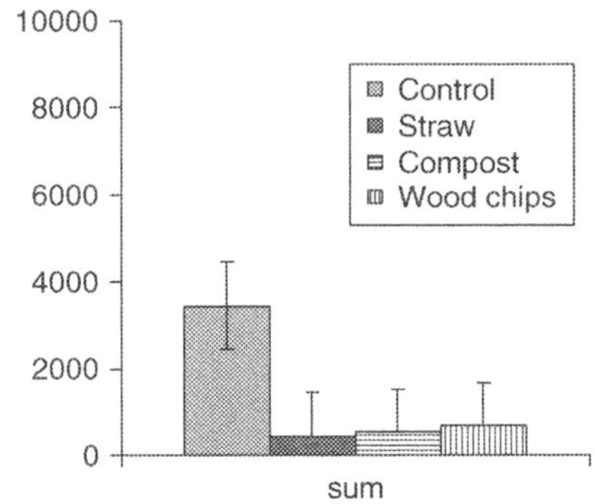

Figure 11.8 Fivefold odour emission reduction after treating SSC with additions of compost, straw and wood chippings.

are disturbed prior to spreading or removal. Earlier researchers have identified H_2S as a marker compound to determine olfactory response to sewage odours, but this was found not to be the case in a recent study (Stuetz et al., 1999). SSC odours have been identified as containing mostly sulfide compounds (Bonnin et al., 1990). There has been a continuing need to ensure that odour emissions are minimised to meet legislative requirements for the water industry. For example, Severn Trent Water produces 160,000 tons (dry solids) of digested sludge per year and approximately 60% was recycled to agriculture in recent years (personnel communication). This has been recognised by the UK government as the Best Practicable Environmental Option (Lovell et al., 2005). Increasing sludge odour annoyance from operational handling and land application of stored sludge was a problem odour source and effective treatment was necessary to avoid enforcement action form local authorities. However, the long-term effectiveness of these chemical products to reduce odours was not well known and they have only been shown to be effective for only 1 or 2 days of storage. Some chemical additives appeared to increase sulfide emissions and did not adsorb or transform these problem odours. Some organic materials such as compost, wood chips and straw were used at the 5 m^3 scale. They were found to have the capacity to adsorb odours and aid oxygen transfer. Figure 11.8 shows the fivefold decrease of odour emissions achieved when treating the same volume of sewage sludge. The sulfides that were responsible for the odours were those other than H_2S and methanethiol (Lovell et al., 2005). They were DMS, dimethyl disulfide and dimethyl trisulfide. These were the main odorous compounds associated with the sludge cake (Lovell et al., 2005). These results prompted further investigations on a larger scale.

11.5.3 Differentiating dietary intake

To illustrate that other processes may be investigated by monitoring NMVOCs, emissions from sheep faeces were monitored to assess dietary intake and digestive variation that occur in sheep. Differences in the NMVOC profile were identified between sheep fed ensiled kale, rye grass and lucerne. Interestingly, the datasets reveal a range of effects, for example, the metabolic profile from the faeces also

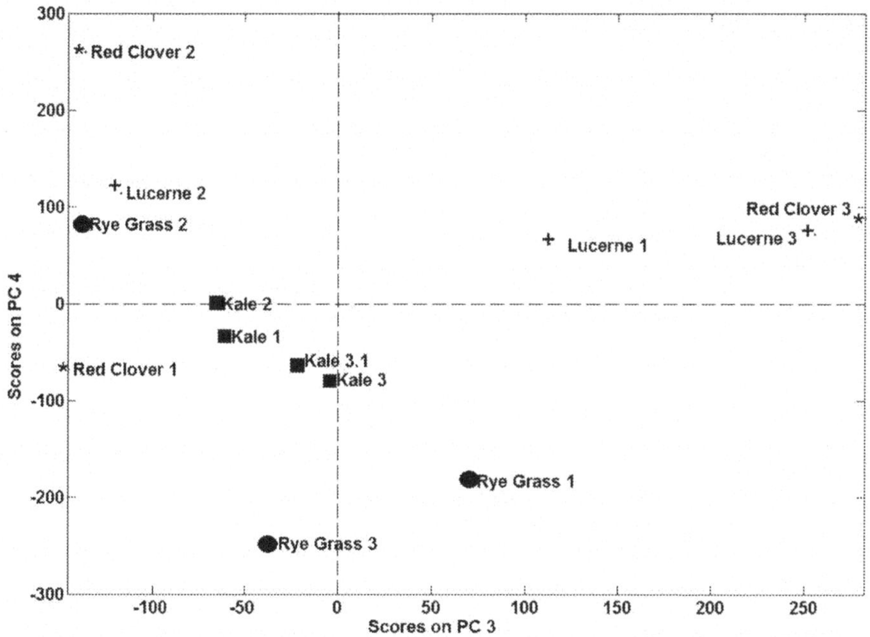

Figure 11.9 Clustering of sampling periods produced by principal component analysis from the headspace samples of faeces from sheep fed different ensiled diets.

distinguished the dietary intake: some of the metabolic compounds are the precursors to odour emissions. Principal component analysis revealed that there were differences in the NMVOCs profile that provide a window on the differences due to the sheep's digestion (Figure 11.9). Some of the major NMVOCs that elucidate dietary intake differences include DMS, a range of phenols including 4-methyl phenol (that is a decomposition product of tyrosine and tryptophan), methyl ethyl ketone and some primary amines. The data interpretation is, however, in the early stages and it is anticipated that the NMVOC profile maybe more sensitive to the dietary intake than the metabolic profile.

11.6 CONCLUSION

This chapter describes new approaches to treating and dealing with odorous problem wastes or resource materials using information from understanding both chemical and physical processes. This information enables managers of "waste streams" that are involved with handling compost, SSC and manure to consider the NMVOC profile as providing an insight into decay processes that may occur. The studies presented here show that there is a requirement for improving analytical capabilities. These should be portable for rapid assessment of process control failure or quantification of the odour emission potential. Recently, there has been considerable use of the multi-sensor systems, so-called electronic noses, that may be a real option for process control following detailed chemical investigation of the processes involved.

REFERENCES

Adib, F., Bagreev, A. and Bandosz, T.J. (1999) Effect of surface characteristics of wood-based activated carbons on adsorption of hydrogen sulfide. *J. Colloid Interf. Sci.* **214**, 407–415.

Bjornsson, L., Murto, M. and Mattiasson, B. (2000) Evaluation of parameters for monitoring an anaerobic co-digestion process. *1990.* **54**, 844–849.

Bonnin, C., Laborie, A. and Paillard, H. (1990) Odor nuisances created by sludge treatment – problems and solutions. *Water Sci. Technol.* **22**, 65–74.

Braida, W.J. and Ong, S.K. (1998) Air sparging: air–water mass transfer coefficients. *Water Resour. Res.* **34**, 3245–3253.

Cheng, N., Lo, K.V. and Yip, K.H. (1999) pH as a real-time control parameter in swine wastewater treatment process. *Adv. Environ. Res.* **3**, 116–178.

Cimander, C. and Mandenius, C.F. (2002) Online monitoring of a bioprocess based on a multi-analyser system and multivariate statistical process modelling. *J. Chem. Technol. Biotechnol.* **77**, 1157–1168.

Cumby, T.R., Moses, B.S.O. and Nigro, E. (1995) Gases from livestock slurries: emission kinetics. *Seventh International Symposium on Agricultural and Food Processing Wastes (Isafpw95)* **95**, 230–240.

Devai, I. and DeLaune, R.D. (1997) Field sampling of trace levels of hydrogen sulfide with the use of solid adsorbent preconcentration. *Field Anal. Chem. Technol.* **1**, 145–149.

Duns, G.J., Ripley, B.D. and Rinker, D.L. (1997) The analysis of mushroom composting odours. *Mushroom World* 58–70.

Eldem, N.O., Ozturk, I., Soyer, E., Calli, B. and Akgiray, O. (2004) Ammonia and pH inhibition in anaerobic treatment of wastewaters, Part I: Experimental. *J. Environ. Sci. Health A-Toxic/Hazard Subst Environ. Eng.* **39**, 2405–2420.

Gardner, J.W., Pearce, T.C., Friel, S., Bartlett, P.N. and Blair, N. (1994) A multisensor system for beer flavor monitoring using an array of conducting polymers and predictive classifiers. *Sensor. Actuator B-Chem.* **18**, 240–243.

Gibson, N.B., Costigan, G.T., Swannell, R.J. and Woodfield, M.J. (1995) Volatile organic-compound (Voc) emissions during malting and beer manufacture. *Atmos. Environ.* **29**, 2661–2672.

Gujer, W. and Zehnder, A.J.B. (1983) Conversion processes in anaerobic-digestion. *Water Sci. Technol.* **15**, 127–167.

Hartung, J. (1992) Emission and control of gases and odorous substances from animal housing and manure stores. *ZBL. Hyg. Umweltmed.* **192**, 389–418.

Hobbs, P.J., Misselbrook, T.H. and Cumby, T.R. (1999) Production and emission of odours and gases from ageing pig waste. *J. Agricul. Eng. Res.* **72**, 291–298.

Hobbs, P.J., Webb, J., Mottram, T.T., Grant, B. and Misselbrook, T.M. (2004) Emissions of volatile organic compounds originating from UK livestock agriculture. *J. Sci. Food Agricul.* **84**, 1414–1420.

Huxham, M. and Thomas, C.P. (2000) Sampling procedures for intrinsically valid volatile organic compound measurements. *Analyst* **125**, 825–832.

Hvitved-Jacobsen, T. and Vollertsen, J. (2001) Odour formation in sewer networks. In *Odours in Wastewater Treatment* (eds. R. Stuetz and F.B. Frechen), pp. 33–68. IWA publishing, London.

Jantsch, T.G. and Mattiasson, B. (2004) An automated spectrophotometric system for monitoring buffer capacity in anaerobic digestion processes. *Water Res.* **38**, 3645–3650.

Lovell, M.J., Hobbs, P.J., Williams, J.S., Chadwick, D.R. and Smith, R. (2005) Investigating additives for controlling odour from stored centrifuged digested sludge (biosolids). *Environ. Odour Manage.*

Mackie, R.I., Stroot, P.G. and Varel, V.H. (1998) Biochemical identification and biological origin of key odor components in livestock waste. *J. Anim. Sci.* **76**, 1331–1342.

MarsiliLibelli, S. and Muller, A. (1996) Adaptive fuzzy pattern recognition in the anaerobic digestion process. *Pattern Recogn. Lett.* **17**, 651–659.

Mccaffrey, C.A., Maclachlan, J. and Brookes, B.I. (1994) Adsorbent tube evaluation for the preconcentration of volatile organic-compounds in air for analysis by gas-chromatography mass spectrometry. *Analyst* **119**, 897–902.

Murnleitner, E., Becker, T.M. and Delgado, A. (2002) State detection and control of overloads in the anaerobic wastewater treatment using fuzzy logic. *Water Res.* **36**, 201–211.

Noble, R. and Gaze, R.H. (1994) Controlled environment composting for mushroom cultivation – substrates based on wheat and barley straw and deep litter poultry manure. *J. Agricul. Sci.* **123**, 71–79.

Noble, R., Hobbs, P.J., Dobrovin-Pennington, A., Misselbrook, T.H. and Mead, A. (2001) Olfactory response to mushroom composting emissions as a function of chemical concentration. *J. Environ. Qual.* **30**, 760–767.

Noble, R., Hobbs, P.J., Mead, A. and Dobrovin-Pennington, A. (2002) Influence of straw types and nitrogen sources on mushroom composting emissions and compost productivity. *J. Ind. Microbiol. Biotechnol.* **29**, 99–110.

Polit, M., Genovesi, A. and Claudet, B. (2001) Fuzzy logic observers for a biological wastewater treatment process. *Appl. Numer. Math.* **39**, 173–180.

Sasaki, N., Suehara, K.I., Kohda, J., Nakano, Y. and Yano, T. (2003) Effects of C/N ratio and pH of raw materials on oil degradation efficiency in a compost fermentation process. *J. Biosci. Bioeng.* **96**, 47–52.

Schaller, E., Bosset, J.O. and Escher, F. (1998) "Electronic noses" and their application to food. *Food Sci. Technol.-Leb.-Wiss. Technol.* **31**, 305–316.

Sommer, S.G. and Sherlock, R.R. (1996) pH and buffer component dynamics in the surface-layers of animal slurries. *J. Agricul. Sci.* **127**, 109–116.

Spoelstra, S.F. (1980) Origin of objectionable odorous components in piggery wastes and the possibility of applying indicator components for the study of odour development. *Agricul. Environ.* **5**, 241–260.

Stuetz, R.M., Fenner, R.A. and Engin, G. (1999) Assessment of odours from sewage treatment works by an electronic nose, H2S analysis and olfactometry. *Water Res.* **33**, 453–461.

Thomas, R.G. (1990) Volatilization from water. In *Handbook of Chemical Property Estimation Methods* (ed. W.J. Lyman), pp. 15–34. American Chemical Society, Washington, DC.

Vitenberg, A.G. (1991) Equilibrium accumulation methods in determining volatile trace components by gas-chromatography. *J. Anal. Chem. USSR* **46**, 1548–1565.

12

Biomarkers of air pollution

C. de Burbure and A. Bernard

12.1 INTRODUCTION

12.1.1 Air pollution and health

Recording and monitoring of air pollution locally, nationally and globally have become established procedures over the past decades. However, it is also necessary to be able to assess any acute or chronic impact of this air pollution on the various populations on earth, and this led to the development of the field of biomarkers and biomonitoring. Of course, air pollution at a given site will affect not only the local environment, its various ecosystems, plant, animal and human populations via air, water and soil contamination, it can also have effects on a much broader and international scale (Figure 12.1). This has led in the past two decades to extensive research into the development and application of complex geographical mathematical models of wind dispersion (Gulliver & Briggs, 2005; Harshadeep et al., 2004) and to an impressive flurry of various biomonitoring tools with ever-increasing sensitivity and specificity in man as well as in plant and animal species. The description of biomarkers of pollution in the various species is absolutely fascinating and ranges from the study of single-cell algae via lichens and mosses, fungi and higher plants, invertebrates such

 Waste Gas Treatment for Resource Recovery edited by Piet N.L. Lens, Christian Kennes, Pierre Le Cloirec and Marc Deshusses. ISBN: 1843391279. Published by IWA Publishing, London, UK.

Figure 12.1 Air pollution and its global impact on nature along the water cycle.

as molluscs and mussels, fish, frogs, birds and mammals through to decomposers (Hauck, 2005; Regoli et al., 2006). However, this chapter aims to focus on biomarkers used to assess the impact of air pollution in man.

Probably one of the very first accounts of the effects of air pollution on man and his environment and on the importance of the winds was written in 400 BC by Hippocrates (460–370 BC). In his book called *Airs, Waters and Places*, he recorded the following observations: "But such cities as lie to the west, and which are sheltered from winds blowing from the east, and which the hot winds and the cold winds of the north scarcely touch, must necessarily be in a very unhealthy situation: in the first place the waters are not clear" … "Their voices are rough and hoarse owing to the state of the air, which in such a situation is generally impure and unwholesome". Hippocrates urged his followers to always examine patients within their whole environmental context, and can rightly be considered not only as the father of medicine but also the founder of environmental medicine: "Whoever wishes to investigate medicine properly, should proceed thus: in the first place to consider the seasons of the year, … Then the winds, the hot and the cold, … the qualities of the waters … These things one ought to consider most attentively" (Baum, 2000).

So air pollution and its effects on man have been recognised for some time, and indeed the whole concept of biomonitoring is not new either. The first renal biomarker was probably glucosuria reported as far back as the 5th and 6th century AD

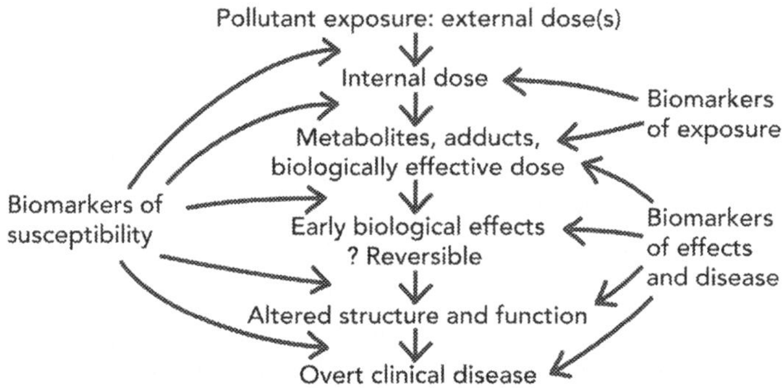

Figure 12.2 Biomarkers of susceptibility, exposure, effects and disease in the continuum between pollutant exposure and overt clinical disease. Ideally, exposure biomarkers reflecting either the parent compound or its derivatives should be good dosimeters of the pollutant studied. Note that susceptibility factors can influence each step of the biomarker spectrum (adapted from DeCaprio, 1997 and Madden & Gallagher, 1999).

in the Sanskrit literature by two notable Indian physicians Susruta and Charuka (Frank, 1957).

12.1.2 Biomarker categories

In 1989, the National Research Council defined biomarkers as indicators of events in biological systems or samples. They classified biomarkers into three categories (Figure 12.2): biomarkers of exposure, effect and susceptibility, based on their relationship to the development of disease following exposure to a given toxicant (NRC, 1989a, b; 1992a, b).

Biomarkers of exposure enable the identification of an exogenous substance within a given biologic system, of an interactive product between a xenobiotic compound and the endogenous components, or other events in the biologic system related to exposure. *Biomarkers of effect* correspond to any changes that are qualitatively or quantitatively predictive of health impairment or potential impairment resulting from exposure. *Biomarkers of susceptibility* are indicators that the health of an organism is particularly sensitive to the challenge of exposure to a xenobiotic compound. However, owing to fast-developing new laboratory techniques using gene chips, immunoassays and polymerase chain reactions for the detection of somatically mutated DNA, unusual RNA or protein expression in tissues or body fluids, a chapter on biomarkers, and more specifically biomarkers of effects, would nowadays not be complete without including some *biomarkers of disease*, in particular cancer biomarkers.

12.2 THE IDEAL BIOMARKER

12.2.1 Criteria for biomarkers

Biomarkers should ideally be easily applicable, measured in a minimally invasive way, allowing for repeated measurements over time. Moreover, they should identify early stages of disease, be indicative of disease prognosis and correlate well with progression and response to therapy.

This is particularly important to consider when applying them to children, who are uniquely vulnerable to many environmental pollutants because of their different behaviour patterns – notably playing outdoors and being more prone to hand-to-mouth contamination – and because of their smaller size, which can lead them to be exposed to relatively higher pollution levels than adults (Baum, 2000; Wickre et al., 2004). In addition, their body is still growing and developing, so that key developmental processes can be disturbed or even adversely affected by exposures that would have little effect on adults. With regard to the lungs for example, the number of respiratory bronchioles and alveoli increase during childhood, 80% of the alveoli arise following birth (Pinkerton & Joad, 2006; Roth-Kleiner & Post, 2003). The lungs undergo a phase of rapid growth along with the whole body between the ages of 10 and 18 years and are then particularly sensitive to ambient air pollutants such as nitrogen dioxide, acid vapour, particulate matter (PM) with a diameter $<2.5\,\mu m$ ($PM_{2.5}$) and elemental carbon (Gauderman et al., 2004).

12.2.2 Biomarker validation

One of the major difficulties is to apply the right biomarkers or the right battery of biomarkers in order to detect both the early biological response to air pollutants and the right pollutant causing the effect. Yet biomonitoring exposure to air pollutants can be a real challenge, as ambient air is far from being a simple gas formula, its complex mixture often misleads scientists into pointing the finger to the wrong culprit or only one of many toxicants involved. For example, the "smelter disease" in workmen replacing pipes in a Japanese sulfuric acid plant was originally attributed to SO_2 fumes. However, it was eventually linked to skin absorption of mercuric fumes (Koizumi et al., 1994). Guidelines on air pollution have therefore to be continuously updated in accordance with evolving scientific knowledge. In turn, discovery and use of new and better biomarkers become more relevant in a society conscious of its responsibilities towards the environment. In terms of application to human health, biomarker validation criteria should include specificity, sensitivity, dose–response correlations, degree of inter- and intra-individual variability, data concerning kinetics, confounding and/or modifying factors, as well as relevant sampling and analytical procedure properties. This all takes time, well-designed studies, as well as national and global coordination.

12.2.3 Legal implications

The Clean Air Act, last amended in 1990, is a good example of the interrelationship between the fast-developing field of biomonitoring and regulations requiring periodic review of the science upon which the standards are based and the standards themselves. The United States Environmental Protection Agency (US EPA) was required to set National Ambient Air Quality Standards (NAAQS) that is acceptable concentrations for widespread pollutants considered harmful to public health and the environment. Two types of national air quality standards were established: primary standards meant to protect the health of the general population including the more susceptible ones such as asthmatics, children and the elderly, and secondary standards to protect public welfare, including protection against visibility impairment, damage to animals, crops, vegetation and buildings.

The US EPA set NAAQS in 1990 for six most common air pollutants, also referred to as "criteria" pollutants, for which every major metropolitan area in the US is required to have at least one monitoring site: ozone (O_3), carbon monoxide (CO), sulfur dioxide (SO_2), lead (Pb), nitrogen dioxide (NO_2) and particulate matter ($PM_{2.5}$). Only 6 years later, in November 1996, the US EPA already proposed tighter regulation standards for PM and ozone, and in 2004 issued new Clean Air Rules concerning Clean Air Interstate, Mercury, Non-road Diesel, Ozone and Fine Particles (US EPA, 2006).

Thus biomarkers criteria also evolve with time, in accordance with new environmental and scientific circumstances. For instance, the gold standard for assessing renal effects of pollutants on the glomerular filtration rate is inulin clearance. However, it is cumbersome and several alternative methods are used in daily clinical practice, such as creatinine or cystatin C measurements in serum and in urine. The latter compounds serve as biomarkers of effects for chemicals known to affect the kidney such as heavy metals (Martini et al., 2003).

The decline in glomerular filtration rate observed with increasing body burden and urinary excretion of lead, cadmium and mercury for example are well documented, often potentiated by arsenic exposure (Akesson et al., 2005; Hong et al., 2003; Jarup, 2003; Liu et al., 2000). However, as emission levels decrease with the application of more stringent regulations, a new correlation pattern is emerging (Figure 12.3): lower levels of exposure to heavy metals such as lead correlate to an increase in glomerular filtration rate, as a likely early response of the kidney (de Burbure et al., 2006; Roels et al., 1994; Weaver et al., 2003). Indeed, because of a renal functional reserve, a rise in serum creatinine only becomes clinically pathological once 75% of the kidney glomeruli are destroyed, which is far from an early response. More sensitive markers of cellular damage are therefore used in conjunction with creatinine, such as *N*-acetyl β-D-glucosaminidase, for example. The latter glucosaminidase is a proximal tubule lysosomal enzyme, which shows a clear dose–response increase with cadmium for example, both in adults and in children, at environmental exposure levels (Akesson et al., 2005; de Burbure et al., 2006).

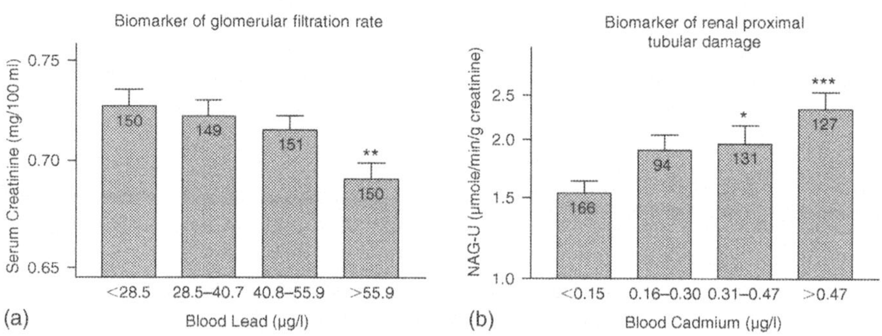

Figure 12.3 Environmental exposure study in 800 children (age 8–12; France, Poland, Czech Republic). (A) Decreased serum creatinine hence increased glomerular filtration rate in higher lead quartile levels (means + SE). (B) Lysosomal enzyme *N*-acetyl β-D-glucosaminidase in urine (NAG-U): dose–response increase with blood cadmium, reflecting tubular damage (de Burbure et al., 2006). *: $P < 0.5$; **: $P < 0.01$; ***: $P < 0.001$.

12.3 DIVERSITY OF POLLUTANTS AND BIOMARKERS

The main pollutants of concern contaminating the 10,000–20,000 L of ambient air that reach the 40–120 m^2 gas exchange area of our lungs daily (Salvi & Holgate, 1999) can be seen in Figure 12.4. Indeed, far from being simply 20% oxygen and 80% nitrogen, the air we breathe at home, at work and during our various journeys contains about 1% of very complex mixtures of fumes and fragrances.

Cigarette smoke alone contains over 4000 various components, whereas there are 18,000 different high-molecular-weight compounds on a carbonaceous core described in diesel exhaust (Grigg, 2002). Over 40 of these compounds in diesel exhaust are recognised as toxic components. There are currently around 190 compounds listed as hazardous air pollutants by the EPA (US EPA, 2006), over 750 compounds with Threshold Limit Values are published annually by the American Conference of Governmental Industrial Hygienists (ACGIH), and a further 109 substances are listed by them as Chemical Substances Under Study (ACGIH, 2006).

Biomarkers of exposure, effect, susceptibility and disease need to be evaluated for each pollutant, and there are daily new ones being actively developed. A single chapter on biomarkers cannot be exhaustive, particularly when looking at the literature. Indeed, interest in biomarkers has literally exploded in the past 20 years, with 330,000 papers mentioning "biomarkers" to date, including more than 16,500 papers concerning "lung biomarkers" alone (Figure 12.5).

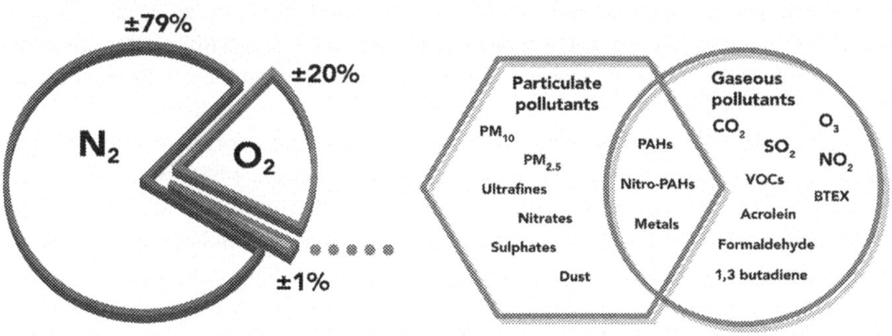

Figure 12.4 Composition of ambient air. PAHs: polycyclic aromatic hydrocarbons; CO_2: carbon dioxide; BTEX: benzene, toluene, ethylbenzene, xylene (adapted from Grigg, 2002 and Health Effects Institute, 2000).

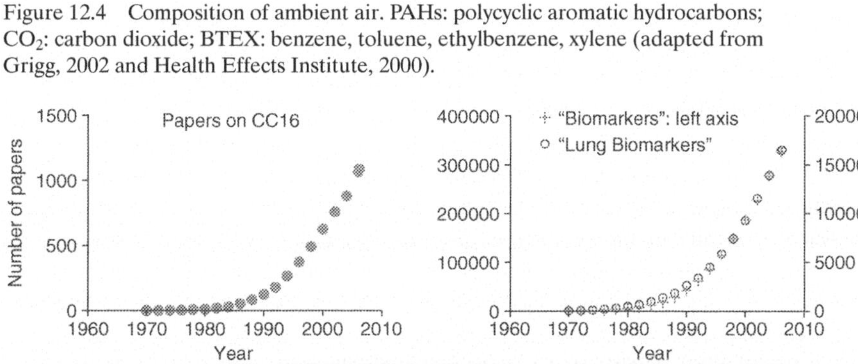

Figure 12.5 PubMed search on Clara cell protein (1085 papers to date), Biomarkers (°), and Lung Biomarkers (+), indicating the almost identical exponential rise in publications.

Therefore, this chapter presents only those biomarkers of main contributors to air pollution that have shown to be most valuable in daily environmental health practice and research, because of their great reliability, ease of application and preferably non-invasive methodology that makes them particularly handy to use in susceptible populations such as the frail, elderly and children in particular.

12.4 EXPOSURE BIOMARKERS

12.4.1 Exposure biomarkers in environmental medicine

Biomarkers of exposure should by definition enable the detection and measurement of internal exposure to a given pollutant within a biologic system. Historically, the performance by James Marsh of a test detecting the presence of arsenic during a murder trial in 1832 was probably one of the very early demonstrations of the practical use of a biomarker of exposure. James Marsh further developed his technique, enabling the detection of up to 1/50th of a milligram of arsenic, and published it in 1838 in *The Edinburgh Philosophical Journal*. Arsenic is a good example illustrating the importance of taking into consideration the time and duration of exposure, the half-life of the pollutant, its physico-chemical properties as well as possible confounders when using biomarkers of exposure.

Because of its high affinity for sulfydryl groups, arsenic accumulates in keratin-rich tissues such as skin, hair and nails. Its levels in hair and nails will be elevated within 1 week of acute poisoning and return to background levels within a few months (Lauwerys & Hoet, 2001). As hair grows at a rate of about 1 cm/month, the segmental distribution of arsenic along a hair shaft can estimate the exposure time (Koons & Peters, 1994). Fingernail arsenic levels have been shown to correlate well with air arsenic levels in occupational exposure cases (Agahian et al., 1990). Since inorganic arsenic is rapidly cleared from blood, its blood levels therefore indicate either high-level exposure or chronic stable exposure due for example to drinking water, although there are other confounding sources such as food (NRC, 1999). After methylation by arsenic methyltransferase AS3MT, for which 26 single nucleotide polymorphisms have just recently been described (Wood et al., 2006), inorganic arsenic is excreted into urine, where it can be measured along with its metabolites monomethylarsonate (MMA) and dimethylarsinate (DMA). However, ingestion of seafood such as bivalves, which contain high quantities of DMA in particular, are a confounder of total urinary arsenic levels (Buchet et al., 1996).

Biomarkers of exposure are thus informative with regard to actual toxicant content of various "biological compartments". Even if the history of biomarkers of exposure started with the investigation of murder cases, autopsy is clearly not an option in environmental medicine. Exposure biomarkers can be used to dose pollutant levels in whole organs from autopsy specimens, biopsy or surgical material under extreme circumstances, or when investigating or biomonitoring background levels of exposure to a given air pollutant, for example dioxins or polychlorinated biphenyls (PCBs) (De Saeger et al., 2005). Scientists and public health officers alike will generally preferably look for biomarkers of exposure in easily attainable sample compartments, such as exhaled breath (see Figure 12.6), sputum and saliva, blood and urine specimens, although as we've seen with

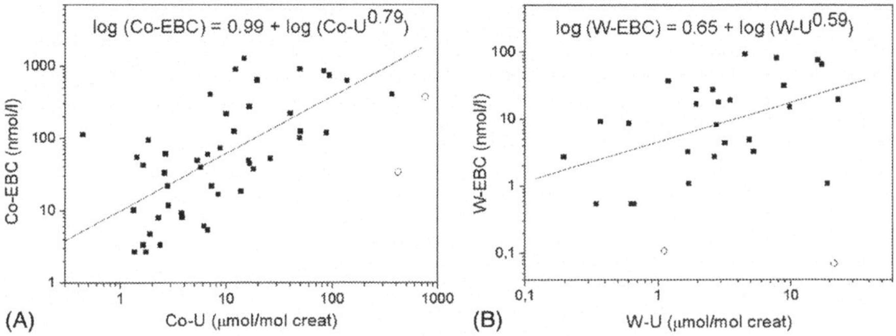

Figure 12.6 Correlations between urinary (U) and exhaled breath condensate (EBC) levels of (A) Cobalt (Co) and (B) Tungsten (W) in workers producing hard-metal alloys and tools. Courtesy of Dr M. Corradi and Prof. A. Mutti (adapted from Goldoni et al., 2004).

arsenic, hair and nails also prove useful for assessing exposure to various metals such as cadmium and lead, in children for example (Anwar, 2005).

12.4.2 Illustrative examples

12.4.2.1 Carbon monoxide and tobacco smoke

We can't speak about biomarkers of exposure and air pollution without mentioning the classic carboxyhaemoglobin (COHb) measurements for monitoring carbon monoxide (CO) exposure. After reaching the lungs, CO diffuses rapidly across the alveolo-capillary barrier and 80–90% binds reversibly to haemoglobin (Hb), for which its affinity is about 250 times that of oxygen, 245 to be precise (Maynard & Waller, 1999). The COHb levels depend on inhaled CO concentration, duration of exposure and alveolar ventilation (exercise). The most well-known model of CO uptake is the Coburn–Foster–Kane exponential equation (Gallagher & Mason, 2004): levels of COHb increase rapidly at the onset of exposure, start to level off after 3 h, and reach a steady state after 6–8 h, when alveolar breath and ambient air CO levels are almost equal. The elimination half-life of COHb while breathing room air is 2–6.5 h, depends on the initial COHb levels. As CO is eliminated unchanged via the lungs, it can be measured in exhaled breath. Hand-held CO monitoring devices have therefore been developed and adapted for estimation of COHb levels by exhaled breath analysis, and successfully used in the field, for example on the fireground to assess CO exposure in firefighters (Cone et al., 2005).

The CO concentration in breath can also serve as estimator of true prevalence of smoking, since cigarette smoke contains about 5% CO (cigars and pipe: 10–15%). Exhaust fumes, gas cooking or heating and endogenous CO formation are other important sources of CO exposure. A recent large-scale study in two British populations using a cutoff value of 10 ppm of exhaled CO has shown the low sensitivity of this biomarker, as 30% of self-reported smoking Europeans and more than 50% smokers of South Asian origin (who generally smoked fewer cigarettes per day) had levels below 10 ppm. Even using a cut-off at 6 ppm would potentially miss a large number of smokers, so that this method is reportedly still a long way from replacing nicotine, cotinine (classic marker notably in saliva) or thiocyanate (from hydrogen cyanide in tobacco smoke) measurements in body fluids (Pearce & Hayes, 2005; Scherer, 2005). Reviewing the great multitude of studies concerning tobacco smoke biomarkers and their various confounders, the only ones so far that do not suffer from other confounders are cotinine and the tobacco-specific nitrosamine metabolite 4-(methylnitrosamino)-1-(3-pyridyl)-1-butanol (NNAL) with its detoxified NNAL-glucuronide (Scherer, 2005).

12.4.2.2 Benzene

Primarily derived from gasoline release and therefore an ubiquitous pollutant, indoor concentrations of this no-threshold carcinogen can actually be many times higher than those of a busy street, particularly in the homes of smokers and users of air fresheners. The latter indeed release a cocktail of irritants such as formaldehyde, acrolein and volatile organic compounds (VOCs), including benzene levels up to 75 times the US National Institute of Occupational Safety and Health's recommended short-term exposure limit (15 min) of 3.2 mg/m^3 (SCHER, 2006). At least 50% of inhaled benzene is absorbed through the lungs, but dermal absorption could contribute 20–40% of the total dose of absorbed benzene. Benzene can be measured either as unmetabolised benzene in blood, serum, exhaled breath and urine, or as metabolites. Measurement of unmetabolised benzene in urine is the choice biomarker at low environmental exposure (Fustinoni et al., 2005). There are major metabolites of benzene such as phenol, trans, trans-muconic acid, hydroquinone and catechol, and minor metabolites, such as *S*-phenylmercapturic acid (SPMA) for example. In a recent study in 250 benzene-exposed workers and 139 control workers in China, Kim et al. (2005) describe an interesting dose-related decline in four major metabolites and total metabolites produced (μmol/L per ppm benzene), ranging between 2.5- and 26-fold, as group median benzene exposures increased between 0.027 and 15.4 ppm. Reductions in metabolite production below 1 ppm were most pronounced for cathecol and phenol, indicating that metabolism at low exposure favoured the formation of toxic metabolites, hydroquinone and muconic acid (Kim et al., 2005).

12.4.2.3 1,3-Butadiene

Studies concerning personal exposures to air pollutants have shown these can often be twice as high as ambient measurements, particularly in the case of volatile organic components (Serrano-Trespalacios et al., 2004). Validated biomarkers of exposure are thus particularly important when studying known carcinogens such as 1,3-butadiene, a chemical widely used in the manufacture of synthetic rubber, that has been linked to childhood cancers and lymphoproliferative disorders (Graff et al., 2005; Knox, 2006). Recent studies in three population groups exposed to significantly different median levels of 1,3-butadiene (2.38, 1.62 and 0.88 μg/m^3, for toll collectors, urban weekday group and suburban weekend group, respectively) indicated higher levels of two urinary mercapturic metabolites in the toll collectors ($P = 0.06$, Sapkota et al., 2006). Adducts in Hb have also been shown to be a sensitive method for monitoring cumulative exposures to 1,3-butadiene at or above 0.35 ppm (van Sittert et al., 2000).

12.4.2.4 Polycyclic aromatic hydrocarbons

Biological monitoring of polycyclic aromatic hydrocarbons (PAHs), formed by incomplete combustion of fuel and pyrolysis of organic materials (roasting), is substantially confounded by diet, tobacco smoke and drugs containing coal tar, as used in psoriasis. There are hundreds of known PAHs, several of which are classified as probable carcinogens. The non-carcinogenic PAH pyrene, which is always present in PAH mixtures, produces a metabolite so far considered the most suitable and widely used biomarker of exposure to PAHs: urinary 1-hydroxypyrene (Sorensen et al., 2003). Other biomarkers are also widely used such as DNA adducts in white blood cells, Hb or albumin adducts of benzo-a-pyrene, and urinary 1 hydroxypyrine–glucuronide. The latter is considered with urinary 1-hydroxypyrene to be markers of recent exposure (Scherer, 2005).

12.4.2.5 Ozone

Indoor ozone concentrations are typically only 10–30% of outdoor values, as ozone is a highly reactive gas (Ozkaynak, 1999). This contrasts with levels of volatile organic compound, which can be 100 times higher than outdoor levels, particularly in new buildings. Very little pure O_3 exists in the nasal or lung lining fluid after deposition as O_3 is highly oxidative and over 90% of the inhaled O_3 is extracted, leading to a cascade of peroxidation reactions with the formation of ozonides, carbonyls, hydrogen peroxide (H_2O_2) and reactive oxygen species (ROS) derived from both O_3 and activated macrophages and neutrophils. Reaction products such as increased H_2O_2, ortho-tyrosine and 8-epi-prostaglandin $F_{2\alpha}$ are used in exhaled breath, nasal and bronchoalveolar lavage fluid respectively, although these are arguably already biomarkers of effect (Madden & Gallagher, 1999).

12.4.2.6 Sulfur dioxide

Sulfur dioxide (SO_2) is known only too well along with nitric oxides as one of the main culprits of acid rain and for its major role in the severe pollution episodes in Donora (Pennsylvania), London and New York in the 1940s, 1950s and 1960s. Although its emission levels have reduced by 50% compared to 1980 as most industrial combustion plants are nowadays fitted with flue gas desulfurisation technology, it is still a major contributor to the $PM_{2.5}$ fraction of PM. High ground levels of SO_2 can build up very quickly locally, depending on weather and living conditions, and it remains a major pollutant in many areas of the globe. Its harmful effects on skin, as it changes to sulfuric acid, and lungs have been described and studied since 1949 (Dunner et al., 1949). Epidemiologic studies consistently provide evidence of adverse health effects of SO_2, yet among the published literature to date there are few studies using specific exposure biomarkers for SO_2, even though 80–90% of it gets excreted as sulfate in urine (Shy et al., 1973). *S*-sulfonate in nasal lavage samples has been described as an effective biomarker of short-term (24 h) exposure, as levels are about three-fold those in plasma (Bechtold et al., 1993).

12.4.2.7 Nitrogen dioxide

Nitrogen dioxide (NO_2) is one of the major components of air pollution in densely populated areas. It is highly reactive, poorly water soluble and just like SO_2 appears to have only very few exposure biomarkers described among the published literature to date. Some authors have proposed the dosage of NO binding to the iron of haemoproteins to form a complex detectable by electron spin resonance, without much generalisation of its use (Maples et al., 1991).

12.4.2.8 Particulate matter

Because of their complex composition, various approaches have been developed aimed at monitoring specific constituents of PM, such as biological, organic, metal and acid components. For example, PAH-derived DNA and protein adducts enable the assessment of particulate-associated organic exposure, whereas metal components can be dosed individually in blood, urine, exhaled breath (see Figure 12.6) or even nails and hair as seen above. The classification into inhalable (enter during breathing) and respirable (reach alveoli) particles, and the gravimetric division into the coarse PM_{10} (aerodynamic diameter $<10\,\mu m$), fine $PM_{2.5}$ ($<2.5\,\mu m$) and ultrafine fractions (<100 nm) are mainly linked to their potential health effects.

12.5 BIOMARKERS OF EFFECTS

Biomarkers of effects are indicators of inflammation, oxidative stress, immunotoxicity, direct cellular toxicity or mutagenicity. They can, just like exposure biomarkers, be measured in various ways and in various biological compartments. For years, clinicians have been relying on physiological and biological parameters such as spirometry and respiratory symptoms to assess lung toxicity of pollutants. However, lung function tests are notoriously unreliable, particularly with small children, the sick or the elderly (Borm, 1994). Besides, they do not adhere to the strict definition of biomarkers as a measurement of response of an endogenous component to a xenobiotic, although this point is hotly debated by both clinicians and some toxicologists.

Of course there are always some specific effects that can be linked to a given toxicant. For example, the competition effects of lead with calcium or zinc are well known. The latter effect gives rise to the inhibition of aminolevulinic acid dehydratase (ALAD) enzyme and hence to the accumulation of zinc protoporphyrin IX in urine. This test has been shown to have low sensitivity at environmental exposure levels (Chalevelakis, 1995). Also ferrochelatase inhibition due to lead, decreased protein transcription and hence a decrease in cytochrome P450 are well known (Wagner & Tephly, 1975). However, new developments in the field of non-invasive biomarkers of early pulmonary effects are particularly promising, as detailed below.

12.5.1 The pneumoproteinaemia concept

The lining fluid covering the respiratory epithelium of the nasal and conducting airway passages is indeed the very first line of defence to inhaled components, as it enables the highly efficient mucociliary apparatus to transport both dust particles above $3\,\mu m$ in diameter and macrophages full of ingested material back up to the laryngopharynx at a speed of 13 mm/min. Beside macrophages and anti-bacterial enzymes, lining fluid also contains several anti-oxidants such as glutathione peroxidase and *S*-transferase, urate,

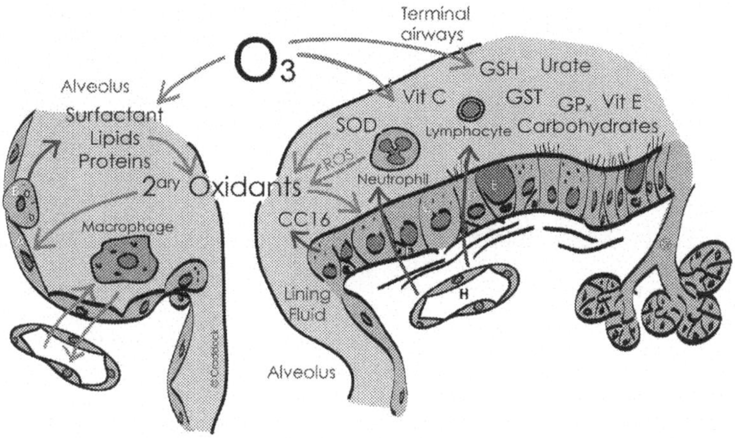

Figure 12.7 Ozone interactions with the terminal airways lining fluid and cells. The lining fluid contains 12 times more lipids than proteins (over 150 proteins identified) and acts as first line of defence against oxidative stress or infection. A: type I pneumocyte; B: type II pneumocyte; C: Clara cell; D: ciliated airway cell; E: goblet cell; F: basal regenerative cell; G: bronchial gland producing proteins and mucus; H: blood vessel; GPx: glutathione peroxidase; GST: glutathione *S*-transferase; SOD: sodium dismutase.

sodium dismutase, vitamin C, as well as several lung-specific proteins, among which the 16-kD Clara cell protein, surfactant proteins A, B, C, D and various mucin-associated antigens (see Figure 12.7) (Hermans & Bernard, 1999).

These various compounds, cells, enzymes, lipids and proteins can in theory all be measured in lavage fluid, sputum or even exhaled breath. Although nasal lavage is relatively non-invasive and easy to apply, it only relates to a very local though first line of response to toxic insult. Bronchial or bronchoalveolar lavage on the other hand are commonly used for dosing proteins secreted by the respiratory epithelium and thereby assessing response to pollutants, but these techniques are much more cumbersome and cannot be advocated for studying the effects of air pollution in children or in field conditions. Hence the discovery that small doses of lung-specific proteins, secreted mainly if not exclusively in the airways, could also be measured in serum in various amounts depending on the toxic insult, led to the development of the brand new pneumoproteinaemia concept.

Based on the assumption of an increased epithelial permeability allowing leakage of both lung-specific proteins into serum and of albumin from plasma into surfactant, this concept has been used in animals and humans to assess both acute and chronic effects of air pollution. Exposure to tobacco smoke for example (Figure 12.8) has been shown to lead over time to a decline in the Clara cell numbers, and hence to an overall reduction in CC16, and to an increased leakage of surfactant proteins such as surfactant protein B (SPB) into serum (Robin et al., 2002).

The actual mechanisms responsible for the increased permeability following oxidative stress and inflammation of the airways are only beginning to be understood. Recent studies demonstrate notably the role played by sphingosine-1-phosphate S1P, which disrupts tight junctions within an hour upon liberation by mast cells, platelets

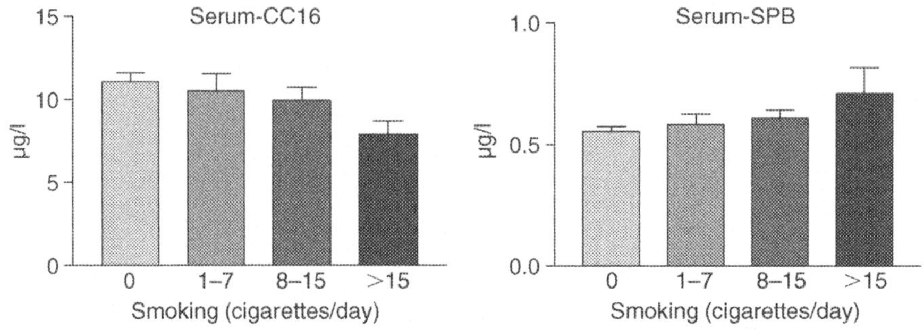

Figure 12.8 Levels of Clara cell protein (CC16) and surfactant protein B (SPB) in serum according to number of cigarettes smoked per day in a Ukrainian mixed population. Toxicity to the Clara cell is indicated by a decline in CC16 with increased smoking, even though airway permeability is increased as demonstrated by a rise in SPB levels in serum (courtesy of Dr S. Fierens).

Table 12.1 Blood biomarkers of lung disease.

Lung epithelium-specific proteins	Surfactant proteins	SPA, SPB, SPD
	Mucin antigens	KL-6/MUC1
	Clara cell protein	CC16
	Epithelial markers	CK19, Ca19.9, SLX
Cytokines and other parameters	Chemokines/cytokines	MCP-1, MIP-1a, ITAC/CXCL-11, TNF
	Anti-oxidant enzymes/ collagen peptides	GST, GPx, SOD, type III procollagen
	T-cell activation	sIL-2R
	Macrophage/monocyte activity	ACE, neopterin, β-glucuronidase, LDH

MCP-1: monocyte chemoattractant protein; MIP-1a: monocyte inflammatory proteins; ITAC/CXCL-11: IFN-inducible T-cell-a chemoattractant; TNF: tumour necrosis factor; CK19: cytokeratin 19, Ca19.9: carbohydrate antigen sialyl Lewis (a); SLX: sialyl Lewis (x); LDH: lactate dehydrogenase; ACE: angiotensin converting enzyme; sIL-2R: soluble IL-2 receptor (adapted from review by Tzouvelekis et al., 2005).

and epithelial cells stimulated by proinflammatory cytokines IL-1 and tumour necrosis factor (TNF). S1P induces rapid loss of ZO-1 reactivity, an essential component of the cytoplasmic plaque associated with tight junctions, as well as of the tetraspannin Claudin-18, an integral membrane organiser of tight junctions (Gon et al., 2005). This concept has led to the development of a number of novel blood biomarkers of lung disease included in Table 12.1.

12.5.2 Markers of airway inflammation such as exhaled NO

Clearly, the interactions of air pollutants with the many components of airway lining fluid are key to their further health effects. Depending on their physico-chemical properties, airway toxicants will first react with anti-oxidants present in lining fluid before they directly or indirectly affect epithelial cells. Epithelial cells are indeed the

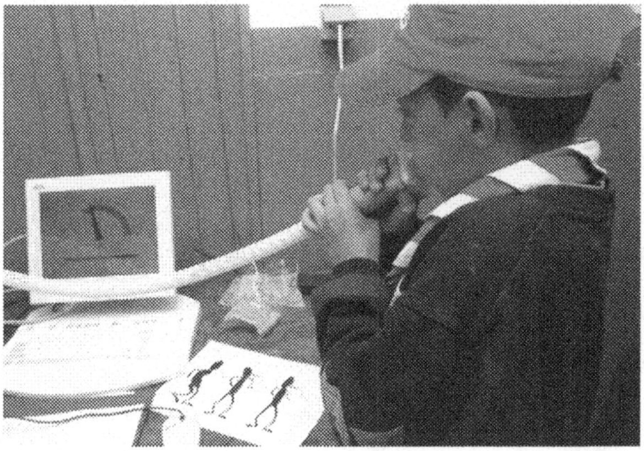

Figure 12.9 Demonstration of exhaled NO measurement by a school-age boy.

main producers of nitric oxide (NO), which they secrete along with cytokines such as the granulocyte-macrophage-colony stimulating factor and eotaxin in response to oxidative stress, various cytokines, endotoxins or viruses. Inducible type II NO synthase (iNOS) stimulation occurs via NF-κB, itself downregulated by glucocorticoids. Since NO affinity for Hb is 100 times that of CO (hence 24,500 that of O_2), Hb is an excellent scavenger for NO, ensuring it is very unlikely that NO enters alveolar air from pulmonary blood vessels. Thus, exhaled NO is a good biomarker of epithelial response to airway toxicants, and its measurement is furthermore relatively easy even in children (see Figure 12.9). Excretion of NO in normal nasal airways is about 500 nL/min, five times that of the lower airways. Nasal NO levels attain means of 450 ppb whereas bronchial exhaled NO averages 10–20 ppb in health and higher levels in asthma (Bromberg, 1999; Struben et al., 2005).

12.5.3 Markers of oxidative stress

12.5.3.1 Oxidative stress using PM as an example

This is probably one of the favourite areas of research in terms of health effects, as most air pollutants will lead to some oxidative stress-related injury, as seen in Figure 12.10 which highlights some of its major complications. Exposure to ambient air PM is associated with pulmonary and cardiovascular diseases and cancer. The mechanisms of PM-induced health effects are believed to involve inflammation and oxidative stress (see Figure 12.10). The oxidative stress mediated by PM may arise from direct generation of ROS from the surface of particles, soluble compounds such as transition metals or organic compounds, altered function of mitochondria or NADPH-oxidase and activation of inflammatory cells capable of generating ROS and reactive nitrogen species. Resulting oxidative DNA damage is likely implicated in cancer risk and may serve as marker for oxidative stress relevant for other ailments caused by particulate air pollution. Studies with various model/surrogate particle preparations, such as carbon black, suggest that the surface area is the most important determinant of the effect of ultrafine particles (UFP, diameter less than 100 nm), whereas chemical composition may be more

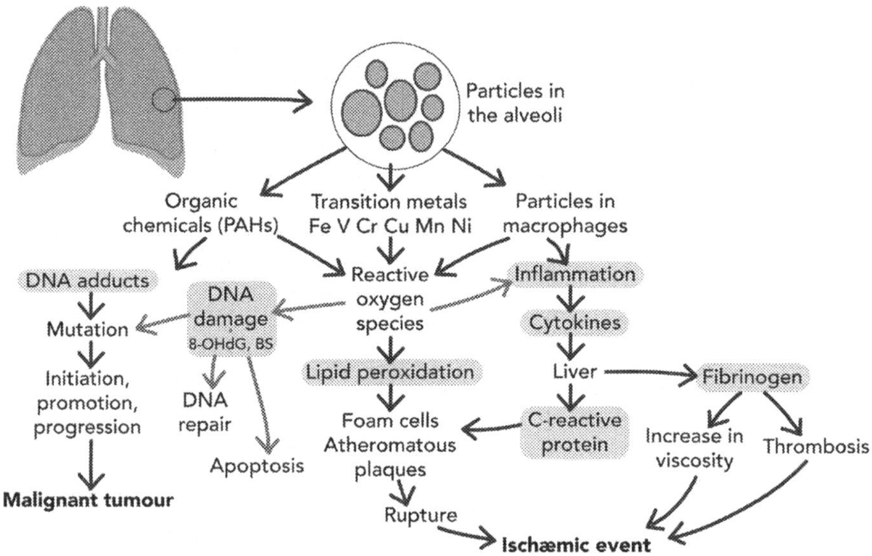

Figure 12.10 Illustration of various effects attributed to both surface and organic components of particulate matter. Highlighted biomarkers of effects, inflammatory mediators, evidence of DNA damage (8-OHdG: 8-hydroxy-2′-deoxyguanosine; BS: base substitutions), lipid peroxidation products can be measured in various biological fluids, including exhaled breath (adapted from Risom et al., 2005 and Sorensen et al., 2003).

important for larger particles. Guanine oxidation in DNA has been correlated with exposure to $PM_{2.5}$ and UFP outdoor and indoor. Oxidative stress-induced DNA damage appears to be an important mechanism of action of urban particulate air pollution (Risom et al., 2005). This is of particular concern when studying air pollution and its effects on human health. Searching Medline for papers on lung biomarkers ($n = 16{,}484$ at the time of writing) and cancer indicates 54.1% of the papers to be concerned with this particular issue. When it comes to studying biomarkers, cancer far outruns other important health issues linked to air pollution such as asthma (4.7%), or chronic obstructive pulmonary disease (1.6%).

12.5.3.2 Biomarkers of oxidative stress

The great variety of oxidative stress biomarkers can be measured in bronchoalveolar lavage fluid, nasal lavage, sputum (induced or not), serum and urine, as well as in exhaled breath condensate (EBC). The latter technique uses a cooling system allowing water vapour to condensate and has emerged as a new non-invasive tool that may prove useful in air pollution biomonitoring. Indeed, whereas exhaled NO as a marker of inflammation was rather non-specific, EBC contains a number of mediators relating to the NO pathway, including nitrite as a metabolite of nitric oxide, 3-nitrotyrosine, nitrosothiols plus small molecular mediators associated with oxidative stress such as hydrogen ions and hydrogen peroxide (Liu & Thomas, 2005). It also enables the detection of leukotriene B4, 8-isoprostane (oxidation product of tissue phospholipids), malondialdehyde (commonly measured in serum and tissues as thiobarbituric acid reactive substances (TBARS), used as gold standard for oxidative stress assay), acrolein, other aldehydes produced during lipid peroxidation, prostaglandins, albumin and other proteins, for example cytokines, and DNA. Exciting developments occurring in the realm of proteomics, presently used in the analysis of small proteins or peptides such as neutrophil defensins 1 and 2, S100A8 and S100A9 (calgranulins A, B), salivary proline-rich peptide P-C and P-D or lysozyme C in lung lavage fluids (Merkel et al., 2005) bring hope that with evolving technology less invasive methods (EBC, sputum analysis) will soon be sensitive enough for the detection of early effects of air pollutants on the respiratory tract.

12.5.4 Other cellular effects and immunotoxicity

Direct physico-chemical interaction with cellular components, be they lipids, carbohydrates, nuclear components or proteins such as enzymes, can along with oxidative stress influence cell function to varying degrees, maybe leading it to death via the apoptotic pathway. Ozone inhalation for example induces ciliated cell loss, an increased epithelial mitotic index and neutrophilic inflammation, damage to alveolar type I cells and disturbed alveolar macrophage function. Evidence of stress or injury to the airway epithelial cells is found in their liberation of large amounts of arachidonate from the sn-2 position of phospholipids, as well as several cytokine products and various enzymes such as lactate dehydrogenase and *N*-acetyl β-D-glucosaminidase. Free arachidonate and residual lysophospholipids produce PGE_2 and $PGF_{2\alpha}$ (cyclooxygenase pathway), 5-, 12- and 15-hydroxyeicosatetraenoates (HETE) (lipoxygenase pathways) and platelet activating factor (PAF). The various cytokines belong to different groups: chemotactic factors for inflammatory cells, colony stimulating factors promoting survival, activation and differentiation of mobile inflammatory cells and certain resident cells (e.g. subepithelial dendritic cell network), growth factors such as TGF-β and finally pleiotropic cytokines IL-6, IL-11, TNF-α and IL-6. The latter cytokines regulate lymphocytes and hepatic synthesis of acute phase proteins (Bromberg, 1999). Various cellular enzymes can be either induced, such as cytochrome CyP1A1 in lung cells by PAHs and 8-OHdG repair enzymes in leukocytes, or inhibited like monoamine oxidase A in brain, heart, lung and kidneys by tobacco smoke (Fowler et al., 2005).

Other effects can be directly mediated via Ah receptors. Organic compounds of dust and diesel particles and 2,3,7,8-tetrachlorodibenzo-p-dioxin (TCDD), the most potent AhR ligand, have recently been shown to enable the differentiation of the human macrophage cell line U937 into foam cells, precursors of atheroma. These cells have raised mRNA levels of cyclooxygenase-2, IL-1β, TNF-α as well as elevated levels of matrix-degrading metalloproteinases (MMP)-1, MMP-3, MMP-12 and MMP-13 (Vogel et al., 2005).

The immunotoxic properties of cigarette smoke, acid aerosols, metals and organic compounds such as benzene, trichloroethylene, dioxins, phenols, organotonins and diester phorbol compounds are also under full investigation. This is because they cause a multitude of effects: several enzyme disfunction and base substitutions have been demonstrated in lymphocytes and leukocytes, including chromosomal aberrations, sister chromatid exchanges, micronuclei, comets and somatic mutations (Gilmour et al., 2001; Scherer, 2005).

12.5.5 Biomarkers of air pollution in organs other than lung

12.5.5.1 The kidney

As the principal excretion pathway of xenobiotics, the kidney is particularly vulnerable to pollutants. Acute lead exposure for example is classically characterised by Fanconi's syndrome, with aminoaciduria, hypophosphataemia, relative hyperphosphaturia and glucosuria, the first biomarker of renal effect described above (see Section 12.1.1). In view of the stricter regulations applied in most industrial countries, chronic environmental exposure to historic pollution is nowadays more common, leading to the need for reliable biomarkers of early effects, such as the lysosomal enzyme *N*-acetyl β-D-glucosaminidase and retinol-binding protein for example (Figures 12.3B and 12.11).

Interestingly, Japan, Korea, Malaysia and Taiwan carry out nowadays routine urinary screening programmes looking for haematuria and proteinuria in school children. This results in a net decreased incidence in the diagnosis of childhood end-stage renal disease (Yap et al., 2005).

Table 12.2 indicates the most commonly used biomarkers of renal effects among the many discovered and developed over the past 20 years.

12.5.5.2 The central nervous system

Interestingly, although it is easily understood how inhaled particles are likely to be deposited in the lungs, it appears from the new emerging discipline of nanotoxicology that the central nervous system (CNS) could also be targeted, notably by airborne solid UFP. Observations in rodents indicate indeed that deposits on the olfactory mucosa of the nasopharyngeal region of the respiratory tract can subsequently translocate via the olfactory nerve. It is estimated that over 50% of inhaled UFP can be deposited in the nasopharyngeal region during nasal breathing depending on particle size. Since approximately 20% of the UFP deposited on the olfactory mucosa of the rat can be translocated to the olfactory bulb, the significance of this up to now unrecognised clearance pathway in humans remains to be investigated (Oberdorster et al., 2004, 2005).

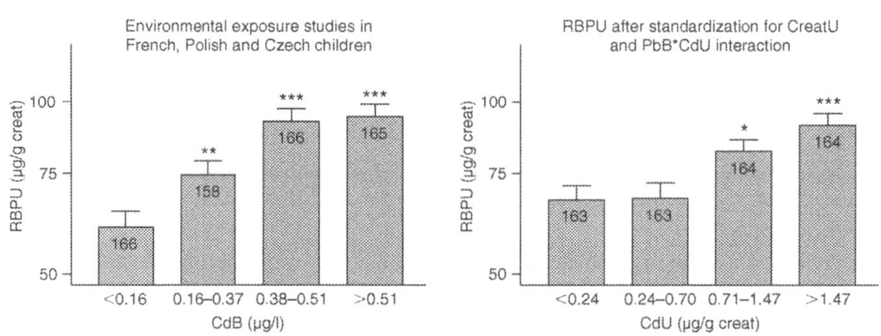

Figure 12.11 Positive correlations between cadmium levels in blood (CdB) and urine (CdU), and urinary retinal-binding protein (RBPU), indicating a decrease in proximal tubular reabsorption of small molecular weight proteins at levels much lower than the recommended WHO limits (de Burbure et al., 2006). *: $P < 0.5$; **: $P < 0.01$; ***: $P < 0.001$.

Table 12.2 Renal biomarkers useful in environmental toxicology studies.

	Biomarker origin	Kidney biomarker name
Serum	GFR markers	Creatinine, β2-microglobulin, cystatin C
	GBM integrity markers	Laminin, anti-GBM antibodies
Urine	*Plasma-derived proteins*	
	High molecular weight	Albumin, transferrin
	Low molecular weight	β2-microglobulin, retinol-binding protein, α1-microglobulin, Clara cell protein, α-amylase
	Kidney derived	
	Antigens	
	Glomerulus	Fibronectin, laminin
	Proximal tubule	Brush border antigens, alkaline phosphatase
	Loop of Henle	Tamm–Horsfall glycoprotein
	Enzymes	Glutathiones-*S*-transferase, *N*-acetyl-β-D-glucosaminidase (A, B)
	Others	Glycosaminoglycans, prostanoids
	Others	Amino acids, total proteins, glucose, calcium

Effects of air pollutants such as lead have long been studied on neurotransmission of gamma amino-butyric acid, serotonin, adrenalin and noradrenalin, acetyl choline, glutamate, dopamine or other peptides. Lead as a cation competes with calcium for numerous cellular functions at pre- and post-synaptic levels and can have direct neurotoxic effects by inducing apoptosis. Using specific biomarkers to the neurotransmitter studied, for example serum prolactin and urinary homovanillic acid in relation to the dopaminergic system, it can be shown that even low environmental exposure to heavy metals such as lead, cadmium, arsenic and mercury can lead to complex interactive effects on the CNS (de Burbure et al., 2006).

12.5.5.3 The liver

Once air pollutants are in the body, just like any xenobiotic, they will be distributed via the circulation, often via transport proteins, deposited in various organs depending on their chemical and physical properties, metabolised notably by the liver before they eventually are or not excreted via different routes. SO_2 for example can induce apoptosis not only in the lung but also in the liver, by increasing mRNA and protein levels of apoptotic markers such as p53 and bax and decreasing bcl2, while increasing caspase 3 activity (Bai & Meng, 2005a, b). Rats exposed to SO_2 have decreased cytochrome P4501-A1 and A2 mRNA levels (Qin & Meng, 2005). Disturbances of the cytochrome P450-aided oxygenation of xenobiotics were indeed studied in 35 kindergarten children's urine by measuring [^{15}N]methacetin (Rehwagen et al., 2001), demonstrating both the confounding impact of tri- and tetrachloroethylene exposure on SO_2 alteration of liver function in the 23 affected children and the reversibility of the effect 3 years later (Herbarth et al., 2004).

Other liver enzyme biomarkers are the caffeine breath test, which monitors CYP1A2 activity by following the appearance of $^{13}CO_2$ in exhaled breath resulting from the oxidative demethylation of [3-^{13}C-methyl]caffeine, or urinary D-glucaric acid, since glucuronidation is a major biotransformation pathway for a wide variety of xenobiotics and drugs. Pesticides, PCBs and TCDD indeed all increase urinary D-glucaric acid, and they also lead to excess total porphyrin excretion, with skin symptoms in the final stage. Despite relatively high body burden of PCBs and dioxin-like compounds in the Canadians from the Lower North Shore region of the St Lawrence river, only smoking had a significant correlation with these three biomarkers of hepatic enzyme induction. This supports the hypothesis that smoking-induced liver CYP1A2 activity alters haem metabolism and increases the biotransformation of mono-ortho PCB congeners (Ayotte et al., 2005).

12.6 BIOMARKERS OF SUSCEPTIBILITY

There are marked inter- and intra-species sensitivity differences to air pollutants (Ichinose et al., 1988; Roux et al., 1996). Ozone is a striking example. A 3-hour exposure to 400 ppb O_3 in rats leads to transient bacterial inactivation impairment against streptococcal infection, whereas mice show bacterial proliferation, polymorphonuclear cell influx into the lungs and excess mortality. This clearly indicates caution is needed in extrapolating animal data to humans (Gilmour & Selgrade, 1993).

Children are particularly vulnerable to toxic pollutants, even before they are born, as lipophilic components pass the placental barrier and are stored into the foetal adipose tissue. A Dutch cohort study for example reminds us that even background perinatal exposure to dioxins can be linked to a reduction in lung function, increased bronchial obstruction, reduced platelets yet increased thrombopoietin levels. This indicates production problems at stem cell level and increased CD45RA and T-helper CD4 + cells implying interference with the immune system (ten Tusscher et al., 2003). With regard to benzene toxicity, DNA damage measured by 8-hydroxy-2′-deoxyguanosine in lymphocyte DNA and urine and by comet assay was shown to be positively correlated to levels of SPMA. Since higher excretion levels of benzene's metabolites trans, trans-muconic acid and SPMA were observed in subjects with the NAD(P)H:quinone oxidoreductase with or without genotype compared to the wild type, these results suggest that benzene exposure could have genotoxic effects in susceptible people even at ambient concentrations (Sorensen et al., 2003).

Dietary factors, notably anti-oxidant intake, have to be borne in mind in populations at risk such as the elderly, as they are also well-known determinants of individual response to air pollutants. Oxidation of lung lipids has been shown to be prevented by both vitamin C and E. Vitamin C is more effective against NO_2 oxidation whereas vitamin E is more effective against O_3 damage. Vitamin C supplementation also protects against the effects of ambient air pollution in adults when testing them by histamine challenge (Menzel, 1992). Interestingly, folate facilitates both the first and second methylation steps of arsenic, resulting in excretion of lower proportions of inorganic arsenic and MMA but higher DMA, thereby decreasing arsenic toxicity. But individuals with hyperhomocysteinaemia, that is about 30% of European elderly and over 60% of men in Bangladesh, have a reduced ability to methylate MMA into DMA and are therefore at greater risk (Gamble et al., 2005).

Studying gene polymorphism is particularly fashionable today, variants being discovered almost daily for various enzymes, receptor proteins or protein precursors.

Epoxide metabolites of 1,3-butadiene, probably the initiators of its carcinogenic effects, are detoxified predominantly by hydrolysis mediated by the microsomal epoxide hydrolase (mEH; EPHX1) enzyme. Beside the well-studied coding region polymorphisms of the mEH gene encoding Tyr113His and His139Arg substitutions, seven other polymorphic sites in the 5′-flanking region of the mEH gene are described. They differentially affect mEH gene transcriptional activities, hence total mEH enzymatic activity. Recent studies in 49 non-smoking workers from two styrene–butadiene rubber facilities in southeast Texas recently confirmed genotoxic effects of 1,3-butadiene to be inversely correlated with predicted mEH activity (Abdel-Rahman et al., 2005). Polymorphisms in the genes encoding for delta-ALAD and the vitamin D receptor have recently been linked to modified associations between lead biomarkers of exposure and renal outcome (Weaver et al., 2006).

Asthmatic subjects may also be more susceptible to the adverse effects of pollutants such as O_3 and NO_2, as shown by several epidemiological studies and demonstrated by studies on bronchial epithelial cells of non-atopic non-asthmatic and atopic asthmatic subjects. Following a 24-hour incubation to 50–100 ppb O_3 and 200–400 ppb NO_2, asthmatic epithelial cells showed an increased release of inflammatory mediators such as IL-8, GM-CSF, RANTES and soluble intercellular adhesion molecule 1 (sICAM-1), whereas non-asthmatic epithelium showed increases in IL-8 and sICAM-1 only (Bayram et al., 2001). Asthmatic susceptibility to SO_2 has also been linked to a wild-type allele of the TNF-alpha promoter polymorphism (Winterton et al., 2001).

12.7 CONCLUSION

Biomarkers of air pollution are probably as varied if not more than the pollutants themselves. There are exciting new developments tending towards abandoning invasive techniques such as bronchoalveolar lavage to the profit of non-invasive technologies. Proteomics were briefly mentioned, but there are also transcriptomic and metabonomic technologies that through their continued development will hopefully completely change the field of biomarkers in the years to come, pressing towards the everlasting goal of increased prevention and better health for all.

REFERENCES

Abdel-Rahman, S.Z., Ammenheuser, M.M., Omiecinski, C.J., Wickliffe, J.K., Rosenblatt, J.I. and Ward Jr, J.B. (2005) Variability in human sensitivity to 1,3-butadiene: influence of polymorphisms in the 5′-flanking region of the microsomal epoxide hydrolase gene (EPHX1). *Toxicol. Sci.* **85**(1), 624–631.

ACGIH©. (2006) *TLVs® and BEIs®*. ISBN: 1-882417-62-3 Copyright © 2006; 227 pp. http://www.acgih.org

Agahian, B., Lee, J.S., Nelson, J.H. and Johns, R.E. (1990) Arsenic levels in fingernails as a biological indicator of exposure to arsenic. *Am. Ind. Hyg. Assoc. J.* **51**, 646–651.

Akesson, A., Lundh, T., Vahter, M., Bjellerup, P., Lidfeldt, J., Nerbrand, C., Samsioe, G., Stromberg, U. and Skerfving, S. (2005) Tubular and glomerular kidney effects in Swedish women with low environmental cadmium exposure. *Environ. Health Perspect.* **113**(11), 1627–1631.

Anwar, M. (2005) Arsenic, cadmium and lead levels in hair and toenail samples in Pakistan. *Environ. Sci.* **12**(2), 71–86.

Ayotte, P., Dewailly, E., Lambert, G.H., Perkins, S.L., Poon, R., Feeley, M., Larochelle, C. and Pereg, D. (2005) Biomarker measurements in a coastal fish-eating population environmentally exposed to organochlorines. *Environ. Health Perspect.* **113**(10), 1318–1324.

Bai, J. and Meng, Z. (2005a) Effects of sulfur dioxide on apoptosis-related gene expressions in lungs from rats. *Regul. Toxicol. Pharmacol.* **43**(3), 272–279.

Bai, J. and Meng, Z. (2005b) Expression of apoptosis-related genes in livers from rats exposed to sulphur dioxide. *Toxicology* **216**(2–3), 253–260.

Baum, C. (2000) *Air Quality Assessments: Implications for Pediatric Environmental Health.* A Project of the American College of Medical Toxicology funded by a Cooperative Agreement with the US Agency for Toxic Substances and Disease Registry. https://www.acmt.net/content/resources/monographs/AirQuality.pdf

Bayram, H., Sapsford, R.J., Abdelaziz, M.M. and Khair, O.A. (2001) Effect of ozone and nitrogen dioxide on the release of proinflammatory mediators from bronchial epithelial cells of nonatopic nonasthmatic subjects and atopic asthmatic patients in vitro. *J. Allergy Clin. Immunol.* **107**(2), 287–294.

Bechtold, W.E., Waide, J.J., Sandstrom, T., Stjernberg, N., McBride, D., Koenig, J., Chang, I.Y. and Henderson, R.F. (1993) Biological markers of exposure to SO_2: S-sulfonates in nasal lavage. *J. Expo. Anal. Environ. Epidemiol.* **3**(4), 371–382.

Borm, P.J. (1994) Biological markers and occupational lung disease: mineral dust-induced respiratory disorders. *Exp. Lung Res.* **20**(5), 457–470.

Bromberg, P.A. (1999) Structure-function relationships. In *Air Pollution and Health* (eds. S.T. Holgate, J.M. Samet, K.S. Koren and R.L. Maynard), pp. 269–294. Academic Press, London.

Buchet, J.P., Lison, D., Ruggeri, M., Foa, V. and Elia, G. (1996) Assessment of exposure to inorganic arsenic, a human carcinogen, due to the consumption of seafood. *Arch. Toxicol.* **70**, 773–778.

Chalevelakis, G., Bouronikou, H., Yalouris, A.G., Economopoulos, T., Athanaselis, S. and Raptis, S. (1995) Delta-aminolaevulinic acid dehydratase as an index of lead toxicity. Time for a reappraisal? *Eur. J. Clin. Invest.* **25**(1), 53–58.

Cone, D.C., MacMillann, D.S., Van Gelder, C., Brown, D.J., Weir, S.D. and Bogucki, S. (2005) Noninvasive fireground assessment of carboxyhemoglobin levels in firefighters. *Prehosp. Emerg. Care* **9**(1), 8–13.

de Burbure, C., Buchet, J.P., Leroyer, A., Nisse, C., Haguenoer, J.M., Mutti, A., Smerhovsky, Z., Cikrt, M., Trzcinka-Ochocka, M., Razniewska, G., Jakubowski, M. and Bernard, A. (2006) Renal and neurological effects of cadmium, lead, mercury and arsenic in children: Evidence of early effects and multiple interactions at environmental exposure levels. *Environ. Health Perspect.* **114**, 584–590.

De Saeger, S., Sergeant, H., Piette, M., Bruneel, N., Van de Voorde, W. and Van Peteghem, C. (2005) Monitoring of polychlorinated biphenyls in Belgian human adipose tissue samples. *Chemosphere* **58**(7), 953–960.

DeCaprio, A.P. (1997) Biomarkers: coming of age for environmental health and risk assessment. *Environ. Sci. Technol.* **31**(7), 1837–1847.

Dunner, L., Hardy, R. and Bagnall, D.J. (1949) Pneumoconiosis after exposure to sulphur dioxide fumes and dust from coke fires. *Lancet* **2**(27), 1214–1216.

US EPA (US Environmental Protection Agency) (2004). *Clean Air Rules of 2004*. Last updated on Thursday, March 2nd, 2006. URL: http://www.epa.gov/cleanair2004

US EPA (2006). *The original list of hazardous air pollutants*. Last updated on Thursday, March 9th, 2006. URL: http://www.epa.gov/ttn/atw/188polls.html

Fowler, J.S., Logan, J., Wang, G.J., Volkow, N.D., Telang, F., Zhu, W., Franceschi, D., Shea, C., Garza, V., Xu, Y., Ding, Y.S., Alexoff, D., Warner, D., Netusil, N., Carter, P., Jayne, M., King, P. and Vaska, P. (2005) Comparison of monoamine oxidase a in peripheral organs in nonsmokers and smokers. *J. Nucl. Med.* **46**(9), 1414–1420.

Frank, L.L. (1957) Diabetes mellitus in the texts of old Hindu medicine (Charaka, Susruta, Vagbhata). *Am. J. Gastroenterol.* **27**(1), 76–95.

Fustinoni, S., Consonni, D., Campo, L., Buratti, M., Colombi, A., Pesatori, A.C., Bonzini, M., Bertazzi, P.A., Foa, V., Garte, S., Farmer, P.B., Levy, L.S., Pala, M., Valerio, F., Fontana, V., Desideri, A. and Merlo, D.F. (2005) Monitoring low benzene exposure: comparative evaluation of urinary biomarkers, influence of cigarette smoking, and genetic polymorphisms. *Cancer Epidemiol. Biomarker. Prev.* **14**(9), 2237–2244.

Gallagher, F. and Mason, H.J. (2004) Carbon monoxide poisoning in two workers using an LPG forklift truck within a coldstore. *Occup. Med. (London)* **54**(7), 483–488.

Gamble, M.V., Liu, X., Ahsan, H., Pilsner, R., Ilievski, V., Slavkovich, V., Parvez, F., Levy, D., Factor-Litvak, P. and Graziano, J.H. (2005) Folate, homocysteine, and arsenic metabolism in arsenic-exposed individuals in Bangladesh. *Environ. Health Perspect.* **113**(12), 1683–1688.

Gauderman, W.J., Avol, E., Gilliland, F., Vora, H., Thomas, D., Berhane, K., McConnell, R., Kuenzli, N., Lurmann, F., Rappaport, E., Margolis, H., Bates, D., and Peters, J. (2004) The effect of air pollution on lung development from 10 to 18 years of age. *New Engl. J. Med.* **351**(11), 1057–1067.

Gilmour, M.I. and Selgrade, M.K. (1993) A comparison of the pulmonary defenses against streptococcal infection in rats and mice following O_3 exposure: differences in disease susceptibility and neutrophil recruitment. *Toxicol. Appl. Pharmacol.* **123**(2), 211–218.

Gilmour, M.I., Daniels, M., McCrillis, R.C., Winsett, D. and Selgrade, M.K. (2001) Air pollutant-enhanced respiratory disease in experimental animals. *Environ. Health Perspect.* **109**(Suppl. 4), 619–622.

Goldoni, M., Catalani, S., De Palma, G., Manini, P., Acampa, O., Corradi, M., Bergonzi, R., Apostoli, P. and Mutti, A. (2004) Exhaled breath condensate as a suitable matrix to assess lung dose and effects in workers exposed to cobalt and tungsten. *Environ. Health Perspect.* **112**(13), 1293–1298.

Gon, Y., Wood, M.R., Kiosses, W.B., Jo, E., Sanna, M.G., Chun, J. and Rosen, H. (2005) S1P3 receptor-induced reorganization of epithelial tight junctions compromises lung barrier integrity and is potentiated by TNF. *Proc. Natl. Acad. Sci. USA* **102**(26), 9270–9275.

Graff, J.J., Sathiakumar, N., Macaluso, M., Maldonado, G., Matthews, R. and Delzell, E. (2005) Chemical exposures in the synthetic rubber industry and lymphohematopoietic cancer mortality. *J. Occup. Environ. Med.* **47**(9), 916–932.

Grigg, J. (2002) The health effects of fossil fuel derived particles. *Arch. Dis. Child.* **86**(2), 79–83.

Gulliver, J. and Briggs, D.J. (2005) Time-space modeling of journey-time exposure to traffic-related air pollution using GIS. *Environ. Res.* **97**(1), 10–25.

Harshadeep, N., Guttikunda, S. and Shah, J.J. (2004) Session on simple interactive databases and models to support air quality management: brief introduction to the session. *Better Air Quality Workshop 2004*. Clean Air Initiative. http://www.cleanairnet.org/cai/1403/articles-59386_background_paper.pdf

Hauck, M. (2005) Epiphytic lichen diversity on dead and dying conifers under different levels of atmospheric pollution. *Environ. Pollut.* **135**(1), 111–119.

Health Effects Institute (2000) *HEI Strategic Plan for the Health Effects of Air Pollution 2000–2005*. March 2000. Health Effects Institute, Cambridge, MA.

Herbarth, O., Franck, U., Krumbiegel, P., Rehwagen, M., Rolle-Kampczyk, U. and Weiss, H. (2004) Noninvasive assessment of liver detoxification capacity of children, observed in children from heavily polluted industrial and clean control areas, together with assessments of air pollution and chloro-organic body burden. *Environ. Toxicol.* **19**(2), 103–108.

Hermans, C. and Bernard, A. (1999) Lung epithelium-specific proteins: characteristics and potential applications as markers. *Am. J. Resp. Crit. Care Med.* **159**(2), 646–678.

Hong, F., Jin, T.Y., Lu, G.D. and Yin, Z.Y. (2003) Renal dysfunction in workers exposed to arsenic and cadmium [in Chinese]. *Zhonghua Lao Dong Wei Sheng Zhi Ye Bing Za Zhi* **21**(6), 432–436.

Ichinose, T., Arakawa, K., Shimojo, N. and Sagai, M. (1988) Biochemical effects of combined gases of nitrogen dioxide and ozone. II. Species differences in lipid peroxides and antioxidative protective enzymes in the lungs. *Toxicol. Lett.* **42**(2), 167–176.

Jarup, L. (2003) Hazards of heavy metal contamination. *Br. Med. Bull.* **68**, 167–182.

Kim, S., Vermeulen, R., Waidyanatha, S., Johnson, B.A., Lan, Q., Rothman, N., Smith, M.T., Zhang, L., Li, G., Shen, M., Yin, S. and Rappaport, S.M. (2005) Using urinary biomarkers to elucidate dose-related patterns of human benzene metabolism. *Carcinogenesis* (in press).

Knox, E.G. (2006) Roads, railways, and childhood cancers. *J. Epidemiol. Community Health* **60**(2), 136–141.

Koizumi, A., Aoki, T., Tsukada, M., Naruse, M. and Saitoh, N. (1994) Mercury, not sulphur dioxide, poisoning as cause of smelter disease in industrial plants producing sulphuric acid. *Lancet* **343**(8910), 1411–1412.

Koons, R.D. and Peters, C.A. (1994) Axial distribution of arsenic in individual human hairs by solid sampling graphite furnace AAS. *J. Anal. Toxicol.* **18**, 36.

Lauwerys, R.R. and Hoet, P. (2001) *Industrial Chemical Exposure – Guidelines for Biological Monitoring*, 3rd edn, 638 pp. Lewis Publishers, CRC Press, Boca Raton, Florida.

Liu, J. and Thomas, P.S. (2005) Exhaled breath condensate as a method of sampling airway nitric oxide and other markers of inflammation. *Med. Sci. Monit.* **11**(8), MT53–MT62.

Liu, J., Liu, Y., Habeebu, S.M., Waalkes, M.P. and Klaassen, C.D. (2000) Chronic combined exposure to cadmium and arsenic exacerbates nephrotoxicity, particularly in metallothionein-I/II null mice. *Toxicology* **147**(3), 157–166.

Madden, M.C. and Gallagher, J.E. (1999) Biomarkers of exposure. In *Air Pollution and Health* (eds. S.T. Holgate, J.M. Samet, K.S. Koren and R.L. Maynard), pp. 417–430. Academic Press, London.

Maples, K.R., Sandstrom, T., Su, Y.F. and Henderson, R.F. (1991) The nitric oxide/heme protein complex as a biologic marker of exposure to nitrogen dioxide in humans, rats, and in vitro models. *Am. J. Resp. Cell. Mol. Biol.* **4**(6), 538–543.

Martini, S., Prevot, A., Mosig, D., Werner, D., van Melle, G. and Guignard, J.P. (2003) Glomerular filtration rate: measure creatinine and height rather than cystatin C! *Acta Paediatr.* **92**(9), 1052–1057.

Maynard, R.L. and Waller, R. (1999) Carbon monoxide. In *Air Pollution and Health* (eds. S.T. Holgate, J.M. Samet, K.S. Koren and R.L. Maynard), pp. 749–796. Academic Press, London.

Menzel, D.B. (1992) Antioxidant vitamins and prevention of lung disease. *Ann. NY. Acad. Sci.* **669**, 141–155.

Merkel, D., Rist, W., Seither, P., Weith, A. and Lenter, M.C. (2005) Proteomic study of human bronchoalveolar lavage fluids from smokers with chronic obstructive pulmonary disease by combining surface-enhanced laser desorption/ionization-mass spectrometry profiling with mass spectrometric protein identification. *Proteomics* **5**(11), 2972–2980.

NRC (US National Research Council) (1989a) *Biologic Markers in Pulmonary Toxicology*, 179 pp. National Academy Press, Washington, DC.

NRC (US National Research Council) (1989b) *Biologic Markers in Reproductive Toxicology*, 395 pp. National Academy Press, Washington, DC.

NRC (US National Research Council) (1992a) *Biologic Markers in Immunotoxicology*, 206 pp. National Academy Press, Washington, DC.

NRC (US National Research Council) (1992b) *Environmental Neurotoxicology. Report of Committee on Neurotoxicology and Models for Assessing Risk*, 154 pp. National Academy Press, Washington, DC.

NRC (National Research Council) (1999) *Arsenic in Drinking Water. Subcommittee on Arsenic in Drinking Water, Committee on Toxicology, Board on Environmental Studies and Toxicology Commission on Life Science*. National Academy Press, Washington, DC.

Oberdorster, G., Sharp, Z., Atudorei, V., Elder, A., Gelein, R., Kreyling, W. and Cox, C. (2004) Translocation of inhaled ultrafine particles to the brain. *Inhal. Toxicol.* **16**(6–7), 437–445.

Oberdorster, G., Oberdorster, E. and Oberdorster, J. (2005) Nanotoxicology: an emerging discipline evolving from studies of ultrafine particles. *Environ. Health Perspect.* **113**(7), 823–839.

Ozkaynak, H. (1999) Exposure assessment. In *Air Pollution and Health* (eds. S.T. Holgate, J.M. Samet, K.S. Koren and R.L. Maynard), pp. 149–164. Academic Press, London.

Pearce, M.S. and Hayes, L. (2005) Newcastle Heart Project. Newcastle Thousand Families Study. Self-reported smoking status and exhaled carbon monoxide: results from two population-based epidemiologic studies in the North of England. *Chest* **128**(3), 1233–1238.

Pinkerton, K.E. and Joad, J.P. (2006) Influence of air pollution on respiratory health during perinatal development. *Clin. Exp. Pharmacol. Physiol.* **33**(3), 269–272.

Qin, G. and Meng, Z. (2005) Effect of sulfur dioxide inhalation on CYP1A1 and CYP1A2 in rat liver and lung. *Toxicol. Lett.* **160**(1), 34–42 [Epub 2005 July 14].

Regoli, F., Gorbi, S., Fattorini, D., Tedesco, S., Notti, A., Machella, N., Bocchetti, R., Benedetti, M. and Piva, F. (2006) Use of the land snail Helix aspersa as sentinel organism for monitoring ecotoxicologic effects of urban pollution: an integrated approach. *Environ. Health Perspect.* **114**(1), 63–69.

Rehwagen, M., Krumbiegel, P., Koschny, I., Rolle-Kampczyk, U., Richter, M. and Herbarth, O. (2001) The [^{15}N]methacetin liver function test characterizes multicomponent exposure of children in industrially polluted regions. *Isot. Environ. Health Stud.* **37**(2), 167–174.

Risom, L., Moller, P. and Loft, S. (2005) Oxidative stress-induced DNA damage by particulate air pollution. *Mutat. Res.* **592**(1–2), 119–137.

Robin, M., Dong, P., Hermans, C., Bernard, A., Bersten, A.D. and Doyle, I.R. (2002) Serum levels of CC16, SP-A and SP-B reflect tobacco-smoke exposure in asymptomatic subjects. *Eur. Resp. J.* **20**(5), 1152–1161.

Roels, H., Lauwerys, R., Konings, J., Buchet, J.P., Bernard, A., Green, S., Bradley, D., Morgan, W. and Chettle, D. (1994) Renal function and hyperfiltration capacity in lead smelter workers with high bone lead. *Occup. Environ. Med.* **51**, 505–512.

Roth-Kleiner, M. and Post, M. (2003) Genetic control of lung development. *Biol. Neonate.* **84**(1), 83–88.

Roux, E., Guibert, C., Crevel, H., Savineau, J.P. and Marthan, R. (1996) Human and rat airway smooth muscle responsiveness after ozone exposure in vitro. *Am. J. Physiol.* **271**(4 Pt 1), L631–L636.

Salvi, S. and Holgate, S.T. (1999) Mechanisms of particulate matter toxicity. *Clin. Exp. Allergy.* **29**(9), 1187–1194.

Sapkota, A., Halden, R.U., Dominici, F., Groopman, J.D. and Buckley, T.J. (2006) Urinary biomarkers of 1,3-butadiene in environmental settings using liquid chromatography isotope dilution tandem mass spectrometry. *Chem. Biol. Interact.* **160**(1), 70–79.

SCHER (Scientific Committee on Health and Environmental Risks) (2006) *Opinion on the report "Emission of chemicals by air fresheners Tests on 74 consumer products sold in Europe" (BEUC report January 2005) Adopted by the SCHER during the 9th plenary of 27 January 2006.* Available online at http:// europa.eu.int/comm/health/ph_risk/committees/04_scher/docs/scher_o_026.pdf

Scherer, G. (2005) Biomonitoring of inhaled complex mixtures-ambient air, diesel exhaust and cigarette smoke. *Exp. Toxicol. Pathol.* **57**(Suppl. 1), 75–110.

Serrano-Trespalacios, P.I., Ryan, L. and Spengler, J.D. (2004) Ambient, indoor and personal exposure relationships of volatile organic compounds in Mexico City Metropolitan Area. *J. Expo. Anal. Environ. Epidemiol.* **14**(Suppl. 1), S118–S132.

Shy, C.M., Hasselblad, V., Burton, R.M., Nelson, C.J. and Cohen, A.A. (1973) Air pollution effects on ventilatory function of US schoolchildren. Results of studies in Cincinnati, Chattanooga, and New York. *Arch. Environ. Health* **27**(3), 124–128.

Sorensen, M., Skov, H., Autrup, H., Hertel, O. and Loft, S. (2003) Urban benzene exposure and oxidative DNA damage: influence of genetic polymorphisms in metabolism genes. *Sci. Total Environ.* **309**(1–3), 69–80.

Struben, V.M., Wieringa, M.H., Mantingh, C.J., Bruinsma, S.M., de Jongste, J.C. and Feenstra, L. (2005) Silent and humming nasal NO measurements in adults aged 18–70 years. *Eur. J. Clin. Invest.* **35**(10), 653–657.

ten Tusscher, G.W., Steerenberg, P.A., van Loveren, H., Vos, J.G., von dem Borne, A.E., Westra, M., van der Slikke, J.W., Olie, K., Pluim, H.J. and Koppe, J.G. (2003) Persistent hematologic and immunologic disturbances in 8-year-old Dutch children associated with perinatal dioxin exposure. *Environ. Health Perspect.* **111**(12), 1519–1523.

Tzouvelekis, A., Kouliatsis, G., Anevlavis, S. and Bouros, D. (2005) Serum biomarkers in interstitial lung diseases. *Respir. Res.* **6**, 78.

van Sittert, N.J., Megens, H.J., Watson, W.P. and Boogaard, P.J. (2000) Biomarkers of exposure to 1,3-butadiene as a basis for cancer risk assessment. *Toxicol. Sci.* **56**(1), 189–202.

Vogel, C.F., Sciullo, E., Wong, P., Kuzmicky, P., Kado, N. and Matsumura, F. (2005) Induction of proinflammatory cytokines and C-reactive protein in human macrophage cell line U937 exposed to air pollution particulates. *Environ. Health Perspect.* **113**(11), 1536–1541.

Wagner, G.S. and Tephly, T.R. (1975) A possible role of copper in the regulation of heme biosynthesis through ferrochelatase. *Adv. Exp. Med. Biol.* **58**, 343–354.

Weaver, V.M., Lee, B.K., Ahn, K.D., Lee, G.S., Todd, A.C., Stewart, W.F., Wen, J., Simon, D.J., Parsons, P.J. and Schwartz, B.S. (2003) Associations of lead biomarkers with renal function in Korean lead workers. *Occup. Environ. Med.* **60**, 551–562.

Weaver, V.M., Lee, B.K., Todd, A.C., Ahn, K.D., Shi, W., Jaar, B.G., Kelsey, K.T., Lustberg, M.E., Silbergeld, E.K., Parsons, P.J., Wen, J. and Schwartz, B.S. (2006) Effect modification by delta-aminolevulinic acid dehydratase, vitamin D receptor, and nitric oxide synthase gene polymorphisms on associations between patella lead and renal function in lead workers. *Environ. Res.* (in press).

Wickre, J.B., Folt, C.L., Sturup, S. and Karagas, M.R. (2004) Environmental exposure and fingernail analysis of arsenic and mercury in children and adults in a Nicaraguan gold mining community. *Arch. Environ. Health* **59**(8), 400–409.

Winterton, D.L., Kaufman, J., Keener, C.V., Quigley, S., Farin, F.M., Williams, P.V. and Koenig, J.Q. (2001) Genetic polymorphisms as biomarkers of sensitivity to inhaled sulfur dioxide in subjects with asthma. *Ann. Allergy Asthma Immunol.* **86**(2), 232–238.

Wood, T.C., Salavaggione, O.E., Mukherjee, B., Wang, L., Klumpp, A.F., Thomae, B.A., Eckloff, B.W., Schaid, D.J., Wieben, E.D. and Weinshilboum, R.M. (2006) Human arsenic methyltransferase (AS3MT) pharmacogenetics: gene resequencing and functional genomics studies. *J. Biol. Chem.* (in press).

Yap, H.K., Quek, C.M., Shen, Q., Joshi, V. and Chia, K.S. (2005) Role of urinary screening programmes in children in the prevention of chronic kidney disease. *Ann. Acad. Med. Singapore.* **34**(1), 3–7.

PART III: Bioreactors for waste gas cleaning

13

Gas purification by heterogeneous photocatalysis

B. Sánchez, A.I. Cardona, J. Peral and M.I. Litter

13.1 INTRODUCTION

Environmental legislation and actions are placing more and more emphasis on the elimination of undesirable airborne organic pollutants. Volatile organic compounds (VOCs) can chemically pollute the troposphere both by themselves as well as in combination with other compounds – basically NO_x – generating even more dangerous secondary pollutants, such as ozone. The need for reducing the formation and concentrations of ozone is the basis of new ever-stricter regulations for VOCs (Directive 1999/13/CE). These substances can be chemically destroyed (mineralized) or treated by adsorption, absorption incineration or condensation. In this context, gas-phase photocatalysis is one of the Advanced Oxidation Technologies that has been studied for the last 20 years for the treatment of air pollution from various sources.

Heterogeneous photocatalysis for the removal of VOCs is based on the oxidation of these undesirable airborne compounds, using a semiconductor catalyst activated by light from a certain wavelength. Therefore, the components necessary for

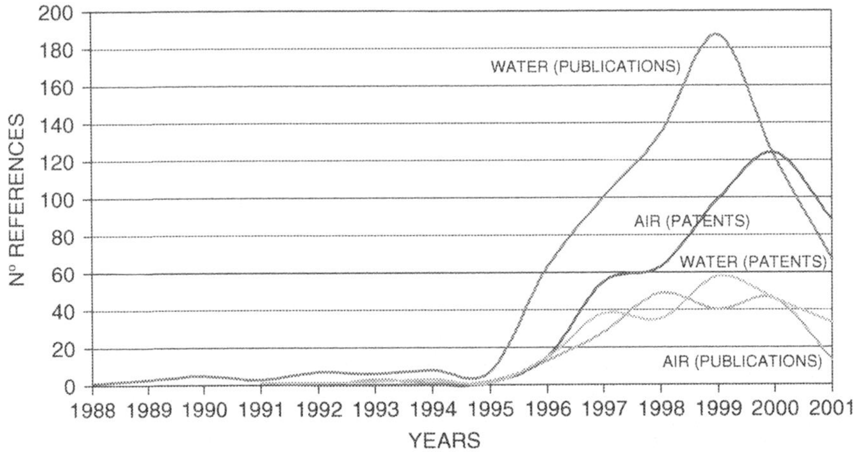

Figure 13.1 Comparison between the number of international publications and patents presented from 1989 to 2001 related to photocatalysis in air and water. (By author, with data from Blake 1997).

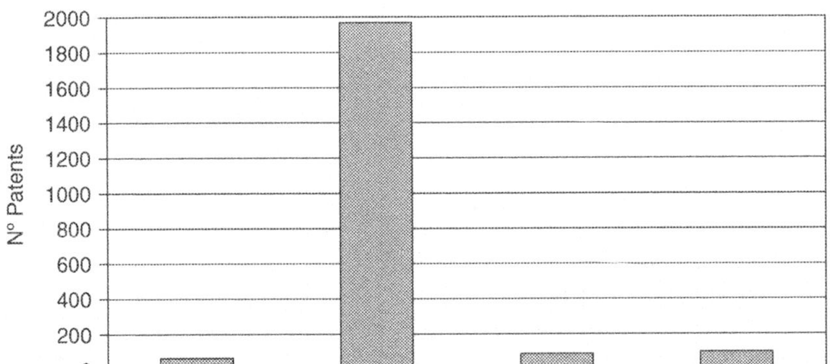

	Europe	Asia	America	Undefined
N°	67	1965	84	96

Figure 13.2 Number of worldwide patents related to photocatalysis in water and gas by continent up to 2001. (By author, with data from Blake 1997).

a photocatalytic reaction to take place are a compound to be degraded, an oxidizing compound such as oxygen contained in the air, a medium where the reaction takes place (in this case air itself), a photocatalyst (such as titanium dioxide (TiO_2) or other semiconductor compound), and a source of ultraviolet (UV) light whether natural (sunlight, in photoreactors configured as solar collectors) or artificial (lamps: mercury, xenon, black lights, UVA, etc.).

The potential environments for application of this technology are industrial and residential spaces where significant concentrations of VOCs are detected. The effectiveness of this process for the vapor-phase destruction of various substances, including alcohols, ketones, aromatic compounds, nitrogenated substances and halogenated hydrocarbons, has been shown.

Numerous research activities have concentrated on finding the ideal catalyst for the process, on the analysis of susceptibility of various types of organic compounds to oxidization and on the potential applications of the technology. However, already from the beginning, results have been contradictory or inconclusive which has slowed down their practical application.

The complexity of such studies and the difficulty in comparing the results in the many studies undertaken is understandable when the different configurations and compositions of photoreactors and catalysts, the intrinsic characteristics of the compounds to be treated, and the different light sources employed are also taken into consideration. The majority of the studies on photocatalytic processes have been carried out in experimental laboratory reactors and some prototypes have been developed from them into commercial purification systems. However, scale-up from laboratory conditions to commercial systems is complicated because the design and start-up of such systems demands macro-kinetic calculations based on micro-kinetic data and assumptions on reaction mechanisms which are not always available.

Among the advantages of gas-phase reaction mechanisms, which are different from aqueous systems, is the relatively low UV-energy levels required for it in contrast to reactions in water. Fox and Dulay (1993) and Jardim et al. (1994) have indicated other advantages of this technology: (1) diffusion is favored; (2) $HO^{\bullet}$ trappers such as bicarbonates or carbonates do not interfere; (3) electron trappers such as O_2 are rarely speed-limiting and (4) few photons are absorbed by water. Nevertheless, more attention has been given to water research, at least from the point of view of the number of publications in international journals. However, the ratio of patents for air are almost double compared to those related to water (Figure 13.1). Japan is the outstanding country in the world in the number of photocatalysis patents (Figure 13.2).

13.2 GAS-PHASE PHOTOCATALYTIC DEGRADATION OF POLLUTANTS

Gas-phase photocatalytic degradation of pollutants is influenced by a series of parameters that go from the type of catalyst, configuration of the photoreactor and light source employed to pollutant adsorption on the catalyst, which is affected by the catalyst properties and the temperature at which the process is carried out. The amount of air to be treated for a particular photoreactor-catalyst configuration defines the residence time, the flow rate and mass transfer processes.

Some materials have ideal properties for developing photocatalytic reactions, for example, TiO_2, ZnO, CdS, Fe_2O_3, FeOOH, ZnS, etc. Many of these materials are excited by light with a moderate energy content, absorbing part of the radiation from the solar spectrum incident on the terrestrial surface ($\lambda > 310\,nm$). This makes them of great interest because of the possibility of making use of sunlight. Moreover, these materials are economically accessible and many of them are present in nature and are involved in environmental chemical processes.

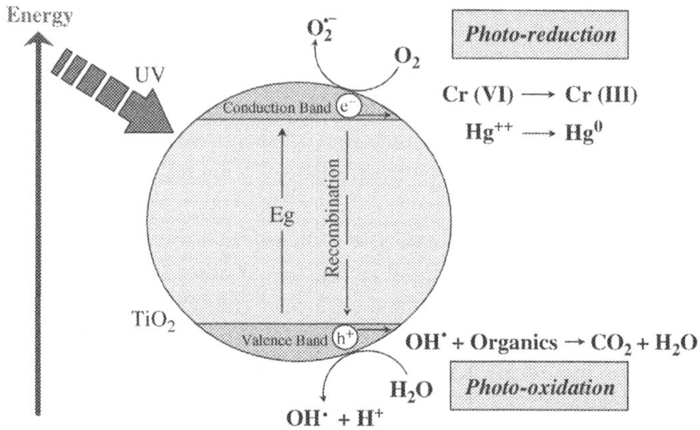

Figure 13.3 Solar photocatalytic detox process. Schematic representation of a TiO_2 catalyst particle.

The photocatalysts studied most to date are the semiconductor oxides and among them, TiO_2 in different forms and preparations. Titanium dioxide is able to produce electron hole pairs by light absorption in the near ultraviolet (UVA) with wavelengths under 387 nm, thus favoring oxide–reduction reactions of organic compounds adsorbed on the surface of the oxide (Figure 13.3). The majority of this chapter refers to these photochemical processes, using this material as the photocatalyst.

Several laboratory-scale reactors have been built to test and develop the UV/TiO_2 technology with artificial light from UV lamps (Blanco et al., 2004), that can differ in geometry, power, lifetime and type (Figure 13.4). Different configurations of reactors described include annular reactors with a fine layer of TiO_2 impregnated on the internal surface of the reactor walls; TiO_2 supported on a porous fiber mesh; monoliths; TiO_2 supported on particles and TiO_2 supported on optic fiber bundles. Optic fibers have the advantage of direct radiation transfer to the photocatalyst and a large activated surface area compared to the reactor volume. However, the potential deactivation of the catalyst due to heating of the optic fiber bundle must be taken into consideration. The main advantage of monoliths is the low-pressure drop that develops when the gas stream passes through it.

13.2.1 Background and controversy

The study of photocatalytic treatment in gas-phase systems began simultaneously with aqueous systems. The work of Gravelle et al. (1971); Formenti et al. (1971); Djeghri et al. (1974); Djeghri and Teichner (1980) on selective oxidation of organic molecules is the first study of the use of TiO_2 for the oxidation of organic molecules in the gas phase. Surprisingly, the purpose of these authors was to use photocatalysis as a method of synthesis, although they soon desisted in their efforts, as results showed that it was impossible to avoid oxidation of the organic species into CO_2. In general, oxidation of an alkene (C_nH_{2n+2}) with three or more carbon atoms gave rise to the formation of ketones ($C_mH_{2m}O$) and aldehydes ($C_mH_{2m+2}O$) with $2 < m < n$. The reactivity of

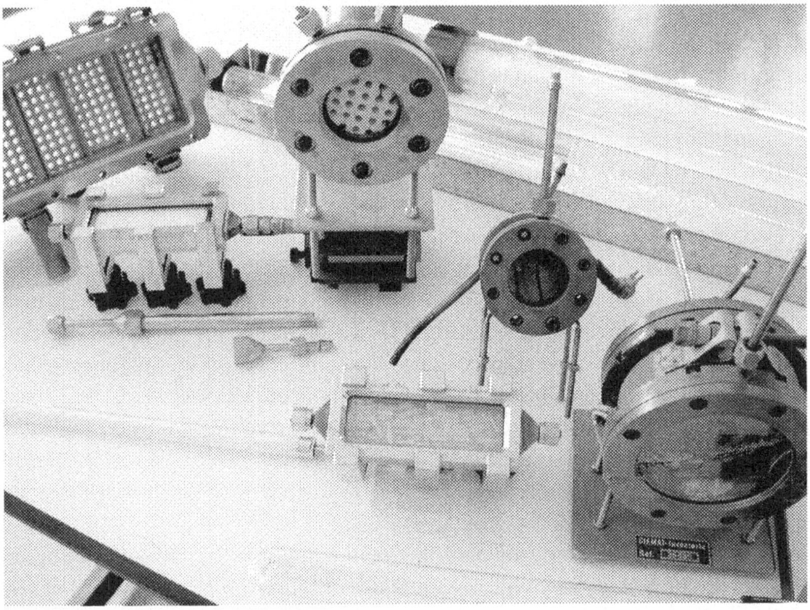

Figure 13.4 Different gas-phase photoreactors designed and tested at CIEMAT facilities (B. Sánchez).

different types of alkenes follows a $C_{tertiary} > C_{quaternary} > C_{secondary} > C_{primary}$ sequence. The carbon atom preferentially attacked is the one with the greatest electronic density and the least steric impedance, although any of the paraffin carbon atoms could be subject to the oxidative attack. These authors proposed reaction mechanisms that go through the formation of alcohols, which, in the case of secondary and tertiary carbons, undergo dehydration forming intermediate unsaturated species. Nevertheless, the simplest hydrocarbons, such as methane, ethane, ethylene, propylene, 1-butylene and 2-butylene reacted directly into CO_2 and H_2O without passing through any detectable intermediate.

During the same period, other authors studied the adsorption and desorption of oxygen on the catalyst surface (Hermann et al., 1981; Courbon et al., 1977, 1984; Munuera et al., 1979). The role played by oxygen in heterogeneous photocatalysis has been amply discussed. Teichner himself was the first to detect the existence of O_2^- and O^- species on the surface of illuminated TiO_2 and in contact with O_2 gas (Formenti et al., 1971). In photoconductivity experiments (electrical conductivity of an irradiated semiconductor electrode), Courbon et al. (1977, 1984) and Hermann et al. (1981) confirmed that these species are formed, although the relative proportions in which they appear depends on the semiconductor employed. The studies by these authors further showed that photocatalysis and photoconductance are parallel, both phenomena showing the same dependence on the incident light intensity. Moreover, González-Elipe et al. (1979) and Munuera et al. (1979, 1980) found that the adsorption of oxygen on the surface of TiO_2 did not take place in darkness. Under irradiation, the Ti^{4+} centers in the crystalline network capture electrons from the conduction band to form Ti^{3+}. In the absence of oxygen, these species are stable and cause the TiO_2 to appear bluish, but this color rapidly disappears in the presence of oxygen. The same authors demonstrated that photo-adsorption of oxygen is favored on highly hydroxylated surfaces, a circumstance which strengthens the capture of photogenerated holes in the valence band.

Cunningham et al. (1981) studied the oxidation of alcohols in the gas phase, and described two possible reactive pathways for their degradation: dehydrogenation with formation of an aldehyde or ketone and dehydration with appearance of an olefin. The reaction mechanism depends on the partial pressure of the alcohol in gas phase: high pressures favor the physisorption of the alcohol and the Elay–Rideal mechanism prevails, while at low pressure, the systems obey to Langmuir–Hinshelwood kinetics. It also indicates that dissociated and adsorbed O_2 is responsible for the dehydrogenation.

According to some authors, the existence of adsorbed water on the semiconductor surface is necessary for gas-phase activity to be sustainable (Bickley et al., 1973; Dibble et al., 1990). According to this idea, if the gas treated contains no water, the catalyst uses surface water to generate the $^{\bullet}OH$ radicals (species considered responsible for the direct attack of the organic material) until they are used up and reactivity disappears.

In photocatalytic reactions, the photogenerated holes on the TiO_2 surface are able to oxidize any organic compound. In aqueous solution, the production of hydroxyl radicals is favored due to the abundance of hydroxyl groups on the TiO_2 surface. However, in the gas phase, the organic substrates can capture the photogenerated holes themselves (Alberici & Jardim, 1997), since the number of water molecules adsorbed on the catalyst is much smaller. Although in the presence of water steam, there are always $^{\bullet}OH$ groups on the TiO_2 surface and their contribution to photooxidation is considerable, the adsorption of organic compounds in the gas phase is a basic prerequisite for an efficient detoxification process (Alberici & Jardim, 1997).

The controversy about the role of the oxidants in the gas phase is greater than in the liquid phase, because of the contradictory results obtained by different authors, with different operating parameters, types of contaminants and concentrations tested. The controversy is mainly about some organo-chlorinated compounds and the possible action of chlorine radicals as triggers of the reaction (Nimlos et al., 1996; Romero et al., 1999). In a study of correlations between the photocatalytic oxidation of airborne contaminants by chlorine radicals, hydroxyl radicals and photogenerated holes on the semiconductor (Luo & Ollis, 1996), it was found that in the presence of chlorine atoms, the active radical on the surface is apparently $Cl^{\bullet}$. Other authors point out that either $Cl^{\bullet}$ or $^{\bullet}OH$ can initiate the oxidation pathway depending on the catalyst characteristics, such as porosity and internal surface area (Yamazaki et al., 1996).

It is widely accepted that the oxygen molecule is an efficient captor of electrons from the semiconductor conduction band, inhibiting the undesired process of recombination of electron hole pairs. Furthermore, oxygen has a small number of several very reactive species. It has been demonstrated that the presence of oxygen is required for the photo-destruction of many organic and inorganic compounds. However, Alberici et al. (2001) question the need for oxygen as an oxidant, they observe similar propylamine and diethylamine oxidation rates in the absence and in the presence of oxygen but the influence of the partial pressure of oxygen was negligible. On the other hand, pyridine is not destroyed efficiently in the absence of oxygen, observing only

6% conversion. According to Fox and Chen (1981) absorbed amines can capture the photogenerated gap, inhibiting electron hole recombination. The decomposition of amines takes place due to an electrophilic attack and weakening of the C—N bond, which may be due to the higher electronic density of the nitrogen atom. In pyridine, the electrons located in the nitrogen atom are less available for trapping hole (orbital sp^2) than those of the aliphatic amines (orbital sp^3), explaining why the C—N bond in pyridine does not react efficiently in the absence of oxygen.

13.2.2 Photocatalysis of selected compounds

13.2.2.1 Alcohols, aldehydes, ketones, acids and olefins

Many of these compounds are typical airborne pollutants in indoor air. Among the alcohols studied, the most notable is isopropanol (IPA). Bickley et al. (1973) used TiO_2 in its rutile form to study the reaction under two different experimental conditions: (a) with an adsorbed layer of IPA, but without the alcohol gas and (b) with pre-adsorbed IPA and the alcohol gas. In both cases, acetone, which is the main oxidation product, is displaced from the surface of the catalyst by water, and IPA mineralization yields are modest. Acetone gas without IPA generates noticeably larger amounts of CO_2. Furthermore, the photo-adsorption of oxygen increases with the degree of surface coating by IPA, and the photocatalytic activity was greater the gentler degasification of the catalyst pretreatment was. This indicates that the activity is closely associated with the surface hydroxyls concentration. The authors conclude that the progress of acetone and IPA gas is the result of a complex equilibrium of the species between the two phases (gas and surface), an equilibrium that is gradually modified as the reaction progresses. Hiromichi and Hiroaki (1999) studied the photoactivity of IPA on different TiO_2 films and Changrani and Raupp (2000) modeled an annular reactor for treatment of IPA.

Heterogeneous photooxidation of methyl butanol (Figure 13.5) leads to dehydrogenation with formation of aldehydes and ketones and dehydration with appearance of olefins (Walker et al., 1977; Cunningham et al., 1981). Note that dehydration is not an oxidation reaction. The tendency of these alcohols to oxidize goes in the order of secondary > tertiary > primary.

Blake and Griffin (1988) found that catalytic photooxidation of *n*-butanol in the presence of TiO_2 follows first-order kinetics with regard to the alcohol concentration, zero order for the O_2 concentration and negative order for the water concentration. The main intermediates in the reaction are butanal (89%) and 1-butene (11%). This selective reactivity was maintained in very diverse experimental situations. The catalytic reaction in the dark (thermal activation), which only begins to be important above 470 K, produces larger proportions of alkenes and combustion products. Infrared (IR) analysis of the catalyst surface shows absorption bands of carboxylated species which the authors associate with secondary oxidation of butanal.

Peral and Ollis (1992) studied photooxidation with low concentrations of TiO_2 (50–250 ppmv) of organic compounds with different organic functional groups typical of closed atmospheres. Acetone and 1-butanol reactions obey Langmuir–Hinshelwood kinetics. Equation (13.1) is for a continuous reactor operating with differential conversion, in which K is the adsorption equilibrium constant of the organic compound, k_0 the surface rate kinetic constant, ε the coefficient of molar absorption of TiO_2, ν the

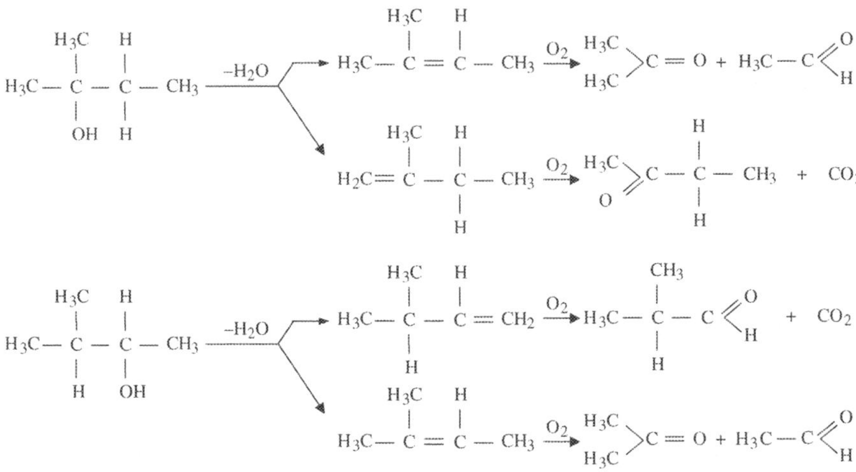

Figure 13.5 Some of the initial methyl butanol reaction steps subjected to the heterogeneous photocatalytic processes described by Walker et al. (1997).

linear velocity of gas through the TiO_2 bed and α is an exponential reflecting the dependence between the reaction rate and the incident light intensity:

$$\frac{\ln \frac{C}{C_0}}{(C - C_0)} = -\frac{k_0 K}{\alpha \varepsilon \nu} \frac{1}{(C - C_0)} - K \tag{13.1}$$

No intermediate reaction was detected for acetone, while 1-butanol generated butanal and other unidentified compounds. The increase in steam pressure inhibited photooxidation of acetone, but affected the reactivity of 1-butanol very little. The oxidation of acetone turned out to be proportional to $I_0^{0.7}$ ($I_0 \equiv$ intensity of incident radiation), which indicates that part of the electron-gap pairs recombined. In the aqueous phase at low irradiation intensities (under a near solar UV equivalent), the rate order was one, while for higher intensities at a solar equivalent, the order was 0.5 (D'Oliveira et al., 1990; Egerton & King, 1979; Kormann et al., 1991; Okamoto et al., 1985).

El-Maazawi et al. (2000) determined that the reaction rate was dependent on oxygen pressure, acetone coating and water adsorption, and proposed three potential mechanisms for photooxidation of acetone adsorbed on TiO_2. In the absence of pre-adsorbed water, O^- forms and reacts with the adsorbed acetone molecules. Another mechanism recurs to oxygen from the TiO_2 network and, in the presence of adsorbed water, the hydroxyl radicals trigger the photooxidation of acetone.

Nimlos et al. (1996) detected acetaldehyde, acetic acid, formaldehyde, formic acid and traces of methyl formiate, ethyl formiate and ethyl acetate as intermediates of the photodegradation of ethanol. The adsorption parameters of ethanol and of some of the intermediates on TiO_2 suggest the existence of two types of reactive surface sites and changes in the surface properties of the catalyst under illumination. Pikenton et al. (1999) showed that the photooxidation of ethanol on an optical microfiber impregnated with TiO_2 is highly effective, with formation of acetic acid and water; water inhibits the reaction.

Sato (1983) found kinetics and reaction products of photo-decarboxylation of formic and acetic acids (photo-Kolbe) on TiO_2, describing the influence of the partial water pressure. It was possible to revert the observed catalyst deactivation by treatment in an H_2 atmosphere at 200°C. The detailed reaction mechanism proposed explains the formation of CO_2, H_2 and chains of more than two carbon atoms.

Muggli and Falconer (1999) studied the photocatalytic decomposition of acetic acid. Acetic acid is photocatalytically decomposed on TiO_2 at ambient temperature in an inert atmosphere by two parallel routes. In one route, the acetic acid is decomposed into CO_2 and apparently forms hydrogen and methyl groups that combine on the TiO_2 surface to form CH_4. In the other route, the acetic acid extracts oxygen from the TiO_2 network to form adsorbed water, CO_2 and C_2H_6. In the primary route, the formation of CH_4 and CO_2 does not consume oxygen from the network. The first step in the photocatalytic decomposition of acetic acid seems to be the dissociation of the O—H bond with production of surface acetates. However, the acetic acid adsorbed molecularly reacts at the same rate and with the same selectivity as the surface acetates. In the presence of gas-phase oxygen, the methyl groups adsorbed are oxidized before they are able to generate CH_4.

Cao et al. (1999a) studied the photocatalytic oxidation of 1-butene in the gas phase with TiO_2 and SnO_2. Under illumination with UV at a major wavelength of 300 nm and a maximum of 352 nm, 5 nm diameter SnO_2 particles, showed an initial photoactivity three times greater than 30 nm-diameter Degussa P-25 TiO_2 particles in the absence of steam, while 22 nm SnO_2 particles turned out to be inactive because of the absence of active surface hydroxyl groups and small surface areas.

Noguchi et al. (1998) studied the photocatalytic degradation of formaldehyde on TiO_2, one of the major causes of the sick building syndrome. In the dark, TiO_2 has a high capacity for adsorption of formaldehyde. Under illumination, total mineralization into CO_2 and H_2O is obtained in a process that responds to the Langmuir–Hinshelwood kinetic model. In synthesis, TiO_2 is a good absorbent and photocatalyst for the elimination of formaldehyde in the gas phase. At high formaldehyde concentrations, Nimlos et al. (1993) identified formic acid as a subproduct. The same authors, and Sopvan et al. (1996) working with acetaldehyde, typical pollutant from cigarette smoke, identified formaldehyde, acetic acid, methyl formiate and CO_2 as byproducts. Matsubara et al. (1995) indicated quantum yields of nearly 100% in the decomposition of gaseous acetaldehyde using paper containing TiO_2.

Falconer and Magrini (1998) studied the thermal catalytic and photocatalytic oxidation of low concentrations of acetaldehyde in air (60 ppmv) on TiO_2 (Degussa P-25) with 0.5% Pt between 24°C and 200°C. On Pt/TiO_2 the contribution of the photocatalytic oxidation is maximum at 140°C, where the conversion is 2.8 times greater than at 24°C. At high temperatures, the TiO_2 catalyst without Pt deposition is deactivated.

13.2.2.2 Aromatic compounds

Benzene is an indoor pollutant known to be carcinogenic to humans. Jacoby et al. (1996) found the amount of CO formed in the photocatalytic oxidation of benzene to be negligible. However, with large amounts of the compound ($\geqslant$10 ppmv) and short reaction times, polymers containing aldehyde and conjugated carboxylic acid functions were formed. Einaga et al. (1999) pointed out its complete oxidation in air on Pt (1.0% by weight)/TiO_2 catalysts at ambient temperature and 65% relative humidity. When the amount of Pt on TiO_2 was increased, the selectivity for the formation of CO_2 increased and the formation of CO decreased.

Toluene is another majority aromatic substance in industrial and motor vehicle emissions and indoor air. Ibusuki and Takeuchi (1986) and Obee and Brown (1995) studying the degradation of toluene and Peral and Ollis (1992) studying *m*-xylene degradation demonstrated a noticeable acceleration of the CO_2 formation reaction in the presence of a photocatalyst. The reactions require water; although the total absence of water does not imply disappearance of reactivity, it does mean that the intermediates as benzaldehyde, nitrotoluene, HCN, etc., stagnate. In contrast, Sánchez et al. (1992), Romero et al. (1995), Avila et al. (1998) and Blanco et al. (1996) incorporated massic TiO_2 in a monolithic support structure. Using residence times of less than 2 s with no recirculation, they were able to discriminate the thermal – between 473 and 523 K – from the photon effect, and concluded that under those conditions the photoreaction of toluene and xylene was irrelevant.

Magrini et al. (1999) made two pilot-scale field tests to determine the effectiveness of photocatalytic oxidation treatment units with sunlight as the light source to destroy emissions of volatile chlorinated compounds, toluene and methyl ethyl ketone (MEK). Lew et al. (1999) also studied the photocatalytic degradation of MEK and xylenes on TiO_2 anatase. These compounds are present in many industrial applications and are usually found in industrial emissions. The tests were carried out in an annular photoreactor and a thin layer of TiO_2 immobilized on the reactor surface with 0.1–75% oxygen in volume, pollutant concentrations between 10 and 40 ppmv, relative humidity between 0% and 84% and residence times between 13 and 20 s. The primary photocatalytic MEK degradation product was CO_2 with acetaldehyde as an intermediate. In dry air, a destruction efficiency of 96% was attained for an initial 10 ppm_v. CO_2 and H_2O make up approximately 85% of degraded MEK; 2% corresponds to acetaldehyde. The destruction efficiency dropped when the initial concentration increased to 40 ppmv.

Ibrahim and Lasa (1999) degraded toluene in an innovative reactor with recirculation (*Photo-CREC-Air*), in which TiO_2 is supported on a mesh that allows good contact with the UV radiation, TiO_2 and polluted air. Although adsorption of toluene is low under the test conditions used, initial photodegradation rates per gram of catalyst were between 0.005 and 0.05 μmol/s. The apparent quantum yield is often over 100%, and sometimes even 450%. Maira et al. (2001) photo-oxidized gas-phase toluene using controlled nanometric TiO_2 prepared by a modified sol–gel method. Under both wet and dry conditions, the dominant products were CO_2 and H_2O. Electronic and structural effects were responsible for the excellent behavior of 6 nm TiO_2.

13.2.2.3 Compounds with heteroatoms (halogens, nitrogen and sulfur)

Halogens Photocatalytic oxidation of halogenated organic compounds has been well studied, in part because of the relative ease with which they react and in part due to the importance this group of substances has in gas-phase pollution. One of the most relevant articles is by Dibble et al. (1990) on oxidation of trichloroethylene (TCE), a compound used reiteratively as a model substrate for photocatalytic oxidation, both in gas phase and in aqueous dissolution. In a continuous reactor with TiO_2, the total stoichiometry of the oxidation reaction is as represented by:

$$Cl_2C = CHCl + \tfrac{3}{2}O_2 + H_2O \xrightarrow{h\nu,\ TiO_2} 2CO_2 + 3HCl \qquad (13.2)$$

The kinetics for the process follows the Langmuir–Hinshelwood–Hougen–Watson equation with positive reaction orders related to TCE and O_2, but a negative reaction order for water. The presence of steam, although necessary to maintain the degree of catalyst surface hydroxylation and to meet the reaction stoichiometry, strongly inhibits the rate in the early stages.

Nimlos et al. (1993) and Jacoby et al. (1994) detected the formation of significant amounts of toxic intermediates (phosgene, dichloroacetyl chloride (DCAC), CO_2, Cl_2, etc.) in mass spectrometry and IR spectroscopy studies of rapid TCE degradation. Nimlos et al. (1993) suggest a DCAC chain formation mechanism, in which the TCE molecules are attacked by chlorine atoms. The appearance of dichloroethylene and perchloroethylene (PCE) is justified by the same mechanism, while the formation of phosgene requires photocatalytic oxidation of DCAC. For 98% degradation of TCE at concentrations of 100 ppmv, 75 ppmv of phosgene are detected. This amount is much higher than the acceptable limits for phosgene in breathable atmospheres (0.1 ppmv, according to criteria of the US Environmental Protection Agency (USEPA)), and these results generated a debate on the possible limitations of heterogeneous photocatalysis as an air purification method.

Nimlos' experiments were made with residence times on the order of milliseconds, since the purpose was not to produce complete oxidation, but to study reaction intermediates. Residence times on the order of seconds or minutes (the most common for the reactions described to date) should be sufficient for these toxic intermediates, as they are more reactive than TCE itself. Blanco et al. (1998), Romero et al. (1999) and Cardona et al. (1999a) using a tube reactor with a concentric UV lamp and a sepiolite/TiO_2 catalyst found that for the conditions employed, an increase in residence time to over 1.5 s is sufficient to destroy these intermediates and to achieve complete mineralization of TCE (Figure 13.6).

Yamazaki-Nishida et al. (1993) studied the degradation of TCE in a continuous reactor with a bed of granulated TiO_2 prepared by sol–gel, which enabled a large specific surface. The results corroborated that formation of phosgene can be avoided by selecting a suitable residence time. Upadhya and Ollis (1998) developed a simple model to explain the light intensity dependence of TCE degradation under various experimental conditions, that is, square root dependence at low TCE concentrations and linear dependence at high concentrations. This model is based on a mechanism involving direct attack on gaps and trapping of electrons by oxygen (without involving $Cl^\bullet$ or other species) with two limiting cases: (a) when recombination dominates, the dependency on light intensity is on the order of 0.5 and (b) when the reaction rate dominates, the dependency is first order. The validity of the model is considered important to set the reactor design conditions, as they must work at intensities that guarantee low recombination rates. For

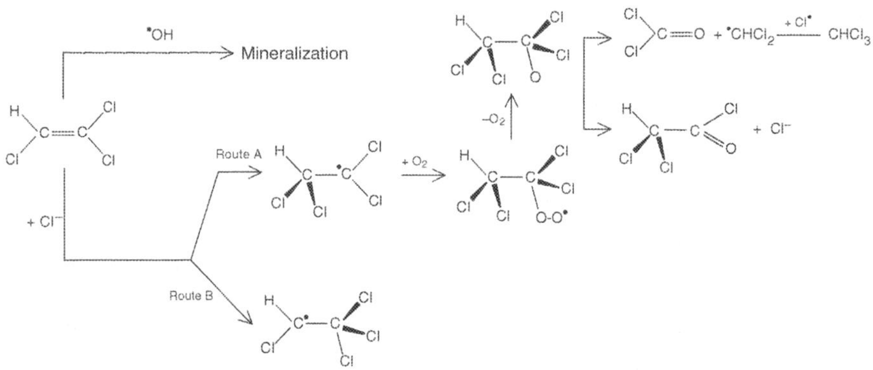

Figure 13.6 Degradation route proposed for TCE using monolithic catalysts (Romero et al., 1999; Sánchez et al., 1996). The chlorine radicals trigger the process. Residence times of 1.5 s allow complete degradation of TCE, obtaining simple traces of DCAC and chloroform.

further information on the photooxidation of TCE, see: Annapragada et al. (1997); Chung-Hsuang & Mariñas (1997); d'Hennezel & Ollis (1997); Dressen et al. (1998); Luo & Ollis (1996) and Phillips & Raupp (1992).

Yuan et al. (1999) studied the photocatalytic oxidation of PCE gas on glass impregnated with Degussa P-25 TiO_2 using a tubular Pyrex-glass reactor illuminated by four 10 W UV lamps. The reaction products detected were chloroform, carbon tetrachloride pentachloroethane, hexachloroethane, phosgene, chlorine, hydrochloric acid, carbon monoxide and carbon dioxide. Hung and Yuan (1998) used the same experimental system but with glass beads impregnated with TiO_2 to study the influence of PCE concentration (between 10 and 50 ppmv), oxygen concentration (0.31–21%), residence time (between 0.55 and 1.83 s) and light intensity (0.4–1 mW/cm^2). The conversion was 99.8% with the highest residence times and the presence of water inhibited the reaction.

Magrini et al. (1999) analyzed the effectiveness of photocatalytic oxidation treatment units with sunlight to destroy emissions of TCE and dichloroethylene. Martin et al. (1998) studied the photocatalytic degradation of chlorobenzene on TiO_2, while Sánchez et al. (2000) made preliminary studies of photocatalytic oxidation of orthodichlorobenzene. Alberici et al. (1998) photocatalytically degraded tetrachloroethylene, chloroform and dichloromethane.

Chlorofluorocarbons (CFC) have attracted much attention. Kutsuna et al. (1992), Tanaka and Isanaga (1994), Raupp (1995) and Tajima et al. (1997) have proven that heterogeneous photocatalysis with TiO_2 is able to efficiently destroy these little-reactive species and, therefore, serve as the basis for their emission treatment technologies. Ibusuki and Takeuchi (1994) also suggest that semiconductor metal oxide particles found suspended in the terrestrial atmosphere can provide one of the few natural routes for destruction of CFCs in the troposphere. Filby et al. (1981) used ZnO as the photocatalyst to destroy CFC. Kutsuna et al. (1999) studied adsorption in the dark, and photocatalytic degradation of some methyl perfluoroalkyl ethers on TiO_2 particles in air, potential substitutes for the hydrochlorofluorocarbons and hydrofluorocarbons.

Table 13.1 Catalytic photooxidation rates of some odors in the presence of TiO_2 (Canela et al., 1999).

Odor	k (min^{-1})
CH_3CHO	0.11
$(CH_3)_2CHCOOH$	0.0078
$CH_3C_6H_5$	0.059
CH_3SH	0.13
H_2S	0.13
$(CH_3)_3N$	0.16

Nitrogenated compounds In contrast to the abundant bibliography on the photocatalytic destruction of hydrocarbons and chlorated hydrocarbons, very little work has been published on the destruction of nitrogenated compounds. These toxic pollutants, present in indoor atmospheres, are important due to their impact on public health and because their ability to deactivate catalysts is well known. Increasing attention is devoted to the elimination of amines and pyridines, compounds that cause foul odors. Huang et al. (1999) studied the photooxidation of triethylamine (TEA) in a flat photoreactor, nitrogen and water on titanium dioxide in presence of oxygen. TEA is a compound used widely in industry and found in many commercial applications (including hair care products). It has a foul odor and may be cancerous. TEA oxidized into CO_2, leaving some byproducts on the thin catalyst layer. The intrinsic oxidation reaction rate depends on the concentration of TEA, humidity and light intensity.

Alberici et al. (2001) studied the photocatalytic TiO_2/UVA process for the destruction of nitrogenated compounds responsible for foul odors (pyridine, propylamine and diethylamine) with and without oxygen. They confirmed the degradation by a mass balance that includes the concentrations measured with GC/FID at the photoreactor inlet and outlet, reporting 90% degradation of pyridine (99 ppmv), 90% of diethylamine (235 ppmv) and 89% of propylamine (99 ppmv). The odor of propylamine disappeared after irradiation. However, diethylamine produced a foul odor even after 30 min of irradiation. The experiments were carried out in a once-through-flow reactor at a rate between 180 and 250 mL/min using synthetic air (21% oxygen and 0.8% relative humidity) or nitrogen (0.8% relative humidity) as the carrier gas at ambient temperature. In all cases, deactivation of the catalyst was observed, attributed to remaining byproducts.

Suzuki et al. (1999) used a multi-channel monolithic reactor with the walls coated with TiO_2 to study the destruction of organic compounds containing nitrogen or sulfur that cause foul odors in various automobile industry environments. Table 13.1 shows the first-order rate constant for organic and inorganic odor removal and demonstrates that the technique is highly efficient for this type of application.

Sulfur compounds The bibliography on destruction by photocatalysis of sulfurated compounds is limited. Canela et al. (1999) carried out a preliminary study on the possible application of the TiO_2/UVA photocatalytic process for the destruction of odorous compounds such as trimethylene sulfide (C_3H_6S), propylene sulfide (C_3H_6S), thiophene

(C_4H_4S) and methyl disulfide ($C_2H_6S_2$) in a once-through annular reactor, with TiO_2 supported on glass with a UV ($\lambda_{max} = 365$ nm) black light at ambient temperature. The flow rate was 250–500 mL/min, which corresponds to a residence time of 1.6–0.8 min. The results indicated that trimethylene sulfide and propylene present at low concentrations (20 and 86 ppmv in synthetic air with 23% relative humidity) can be mineralized almost completely (99%) with generation of CO_2 and SO_4^{2-}. The intensity of the odor after treatment with TiO_2/UV is below the threshold of smell. Peral and Ollis (1997) achieved 10% destruction of dimethyl sulfide with a TiO_2/UV process with no sulfur compounds detected on TiO_2, attributed to the formation of SO_2 and SO_3 or some other mercaptans.

13.2.2.4 Inorganic species

Much fewer studies on the heterogeneous photocatalytic oxidation of gas-phase inorganic species have been carried out. The process studied most is the oxidation of ammonia. Mozzanega et al. (1979) observed that the electrical photoconductivity of TiO_2 was unaffected by the presence of oxidation products as N_2 or N_2O. They concluded that these species do not compete with O_2 for adsorption or in capturing electrons. The N_2 and N_2O formation rates depend on the partial pressure of O_2, but the partial pressure of NH_3 only affects the formation of N_2, from which it may be concluded that the N_2 formation results from a surface reaction between NH_3 and O_2 (with Langmuir–Hinshelwood kinetics) while the reaction of O_2 with intermediate nitrogen species yields N_2O. It is postulated that two types of O_2 are adsorbed on the TiO_2 surface.

The possibility of replacing the Haber process by the more economical and gentler photocatalytic reduction of nitrogen to ammonia, according to:

$$N_2\ (g) + 3H_2O\ (l) \xrightarrow{h\nu} 2NH_3\ (g) + 1.5O_2\ (g) \qquad (13.3)$$

was viewed as economically important. A brief review of this reaction was published by Litter and Navío (1996).

Cant and Cole (1992) observed the formation of N_2 and N_2O, in a 2:1 ratio, in the photocatalytic reaction between NH_3 and NO. From $^{15}NH_3$ and ^{14}NO, $^{14}N^{15}N$ and $^{14}N\ ^{14}NO$ are formed. The ratio between the reaction rates of NH_3 and ND_3 is 1:3. The overall reaction obeys the Langmuir–Hinshelwood equation and depends on the partial pressure of O_2 and NH_3, while N_2O formation kinetics are independent of the latter's partial pressure.

A series of studies carried out in Japan point out the feasibility of treating low concentrations of nitrogen oxides with photocatalysis. Murata (2000) considered this technology a solution for the elimination of NO_x pollution produced by traffic in large urban areas. Yumoto et al. (1999) studied the destruction of NO_2 and Takeuchi (1998) the elimination of NO, NO_2 and SO_2 in concentrations between 0.01 and 10 ppmv. Fukaya et al. (1997) studied the elimination of NO_x in ambient air using ceramic blocks impregnated with a film of TiO_2; 94% and 98% NO_x was eliminated using several blocks with single and triple impregnation, respectively. Hashimoto et al. (2001) studied the photocatalytic oxidation of nitrogen oxide on a zeolite and TiO_2 composite catalyst.

Canela et al. (1998) found destruction of up to 99% of hydrogen sulfide in the presence of O_2 with TiO_2/UV–VIS in a concentration range between 33 and 855 ppmv. The catalyst is not deactivated in the short-time tested when the H_2S concentration is 217 ppmv, but is at high concentrations such as 600 ppmv. The main subproduct of the reaction is the sulfide ion, which is mostly deposited on the catalyst. The process seems feasible for the elimination of odors in wastewater treatment plants. Suzuki (1993) calculates 0.13/min for H_2S and methyl mercaptan catalytic photooxidation reaction rates under his experimental conditions (Table 13.1).

The photo-reduction of CO_2 has also recently attracted attention due to the urgency of abating the amount of carbon dioxide produced by man and the resulting greenhouse effect. Hemminger et al. (1978) reported for the first time the photocatalytic production of methane from CO_2 and steam at 42°C using an $SrTiO_3$ crystal and a Pt contact sheet. Yamashita et al. (1994) described the photo-induced reduction of CO_2 by H_2O using TiO_2 where the main product was methane.

13.2.3 Process improvements

13.2.3.1 Use of modified catalysts

The most widely used catalysts in gas-phase photocatalytic reactions are impregnated or supported, including pellets and massic, although powdered catalysts have been used as well. Table 13.2 summarizes the characteristics of some modified catalysts employed for the destruction of gas pollutants.

Kim and Park (1998) sought better photocatalytic activity in anatase TiO_2 by controlling the size of the particle. Vigil et al. (1999) prepared TiO_2 films synthesized by activation with microwaves, seeking a simple photocatalyst preparation method with a view to make the product cheaper for its commercialization. Different supports influence the photodecomposition rate of propanal gas markedly (Takeda et al., 1997), with activity varying in the order mordenite > activated carbon > zeolites.

Watanabe et al. (1993) and Matsubara et al. (1995) first pointed out different types of TiO_2 thin films covering paper, mosaics, alumina, glass and paint under low illumination with higher photocatalytic activity than P-25.

Table 13.2 Some modified catalyst systems studied.

Catalyst	Photoreactions studied	Reference
TiO_2 + Fe_2O_3 combinations	Destruction of $Cl_2HCCOOH$, oligo-carboxylic acids and 4-nitrophenol; $CHCl_3$ oxidation; N_2 reduction	Pal et al. (1999)
Iron oxides	CO_2 reduction	Matsumoto (1996)
TiO_2/Pt	Miscellaneous organic compounds	Vorontsov et al. (2000)
Ceramic pastes, especially based on TiO_2 + WO_3	TCE	Cardona et al. (1999b)
TiO_2 on porous supports (alumina, glass)	Ácido acético	Sawunyama et al. (1998)
Based on nanoparticulate TiO_2	1-butene toluene	Cao et al. (1999b) and Maira et al. (2001)

13.2.3.2 Periodic illumination

Gas-phase photocatalytic oxidation with TiO_2 has some limitations common to the aqueous process and has some important characteristics of its own. Its low quantum yield, associated with rapid electron-gap recombination, is one of the common disadvantages. Sczechowski et al. (1993) suggest that the low quantum yield is because the rate-limiting step does not require photons. They tested their ideas in experiments in which the catalyst was illuminated periodically, so that the rate-limiting step can take place before injecting more photons into the system. Upadhya and Ollis (1998) propose a model that qualitatively fits Sczechowki's data and that assumes that the rate-limiting step in the dark is the adsorption of O_2 on the surface and/or transfer of the electrons from the conduction band to adsorbed O_2. On this basis, Buechler et al. (1999) used controlled periodic illumination (CPI) in the photocatalytic oxidation of gas-phase TCE. A versatile reactor was designed for the purpose with an opaque cylinder and adjustable grill (allowing light to go through) inserted between the reactor and the lamps, which is rotated by a variable-speed motor. The reactor operates in either of two possible geometric configurations, tubular and annular, which allow the study of different flow systems. In these experiments, the illuminated area of the catalyst is constant, since the rotation does not change the number of photons injected into the system. However, during rotation, the region illuminated and the distribution of light over the entire surface of the TiO_2 film changes. Each individual TiO_2 particle has a short period of light and a longer period of darkness. At low flow rates, a positive effect is observed when the reaction extends over the whole catalyst area. This effect disappears when the flow of reagents was greater. It was assumed that the rotation provides each site on the catalyst with enough time to desorb the products and intermediates and adsorb TCE and O_2 before light returns to the same place.

13.2.3.3 Addition of ozone

Ozone has been used to reinforce the photocatalytic oxidation reaction. This oxidant is particularly attractive for gas-phase reactions. Ozone is used in various industrial processes, such as bleaching paper. In these cases, photocatalysis with TiO_2 in the presence of O_3 can be of interest. In processes that do not employ or generate ozone, the cost of its production can be both economically and environmentally prohibitive.

Pichat et al. (2000) used ozone to mineralize 130 ppmv of *n*-octane in O_2 in stationary state at flow rates of between 5 and 25 L/h and achieved elimination efficiencies of 97–100%. The reactor length is insufficient to achieve mineralization of the intermediate products, which react with each other and with the *n*-octane, which enter continuously into the reactor. The ozone is not absorbed at the wavelengths used in the tests and is stable in the absence of O_2. Using only O_2 without UV irradiation at a flow rate of 14 L/h and 3.5% (v/v) of O_3 in O_2, almost complete elimination of *n*-octane is achieved (130 ppmv), but only a small percentage of mineralization. Under the same conditions, the percentage of mineralization is 75% when lamps are turned on and goes up to 90% when the concentration of O_3 is also increased to 5.5% (v/v).

Alberici and Jardim (1997) proposed the use of heterogeneous photocatalysis combined with ozonization in order to minimize catalyst deactivation. This combination was successfully tested for the photocatalytic destruction of pyridine, which was poorly degraded with only UV/TiO_2.

13.2.4 Effect of water and temperature

13.2.4.1 Water

The effect of water on the process rate and efficiency were presented in previous sections (Bickley et al., 1973; Blake & Griffin, 1988; Cao et al., 1999b; Dibble et al., 1990; El-Maazawi et al., 2000; Hung & Yuan, 1998; Peral & Ollis, 1992; Pilkenton et al., 1999; Sato, 1983). In gas-phase photocatalytic reactions, the effect of water is sometimes clearly beneficial and in others, prejudicial to photoactivity. In general, it is accepted that water has two different functions. Low humidity helps to maintain oxidation rates constant due to replacement of hydroxyl groups consumed in the reaction. On the other hand, high concentrations of water impede photoactivity because the competitive adsorption of water decreases the surface concentration on the substrate at the active sites decrease. These phenomena are not only observed on TiO_2, but also on SnO_2 (Alberici & Jardim, 1997). Its relative importance varies from system to system, depending on the role played by the radicals generated from water and substrate. For example, Park et al. (1999) found that the addition of water favors complete oxidation of ethylene (C_2H_4) into CO_2 and H_2O with O_2 and a finely powdered TiO_2 photocatalyst. The addition of a constant amount of steam leads to an increase in the production of CO_2, while it has no effect on the production of CO. Adsorbed water and O_2 are activated by the photocatalyst irradiated to generate, respectively, $^{\bullet}OH$ and $O_2^{-\bullet} + O_3^{-\bullet}$ radicals, which play a significant role as an active species in the complete photocatalytic oxidation of ethylene.

13.2.4.2 Temperature

The possibility of using high temperatures introduces two aspects typical for gas-phase processes. On the one hand, the possibility of deactivation/reactivation of the catalyst (which will be discussed in the following section, since it can occur even at low temperatures), and on the other hand, the possibility of heterogeneous catalysis in the dark. Vorontsov et al. (1999) studied the effect of temperature and composition of the pollution stream on the photocatalytic oxidation of acetone with TiO_2. They worked at a controlled temperature of between 40°C and 140°C, and with the oxygen in air as the oxidant. The stationary state was typically reached in 30 min below 80°C, but took longer at higher temperatures. The oxidation rate was significantly higher at 80°C than at 40°C in acetone concentrations between 25 and 3000 ppmv. As with the dependency on steam, the dependency on the acetone concentration changed in the high temperature range. Near 115°C, the highest concentration had the lowest reaction rate. The oxidation rates are maximum when the temperature increases. No CO_2 formation was detected in the dark at any of the temperatures tested.

Falconer and Magrini (1998) studied the catalytic and photocatalytic oxidation of low concentrations of acetaldehyde in air (60 ppmv) with Degussa P-25 TiO_2 with 0.5% of Pt in a temperature range from 24°C to 200°C. The contribution of photocatalytic oxidation is maximum at 140°C, where the conversion was 2.8 times that observed at 24°C.

Sánchez et al. (1996) and Cardona et al. (1999b) studied the conversion and mineralization of gas-phase TCE, toluene and xylene between 60°C and 200°C, using the same monolithic catalyst made of a mass mixture of magnesium silicate and TiO_2 on a flat reactor with front illumination. The destruction efficiency curve for aromatic compounds as a function of temperature follows purely catalytic behavior, practically unaffected by photons. The photocatalytic destruction efficiency of TCE is maximum between 60°C and 125°C. At higher temperatures there is a sharp drop in photoactivity due to the desorption of TCE. Between 60°C and 125°C, the catalytic destruction (thermal catalytic) of TCE is negligible. In this sense, Magrini et al. (1999) also point out the treatment of chlorated compounds (TCE and dichloroethylene) at ambient temperature, and toluene and MEK at higher temperatures.

Sánchez et al. (1999) and Cardona et al. (1999b) studied the conversion of TCE in a tubular reactor at constant temperatures between 60°C and 100°C. Conversion was greater at 60°C than at 100°C, but the percentage of mineralization was higher at 100°C. This result seems to indicate two stages: a first step converting TCE into byproducts favored at lower temperatures and a second step converting byproducts into CO_2 and HCl favored by the increase in temperature.

13.2.5 Behavior of mixtures

Luo and Ollis (1996) and d'Hennezel and Ollis (1997) have indicated that the destruction rates for pollutants with low quantum efficiencies can be promoted substantially by the addition of a more reactive pollutant, such as a chlorinated olefin (TCE, PCE or trichlorophenol (TCP)). Thus, for example, the addition of TCE promoted very quick conversion of toluene (90–100% in 5.6 ms). On the other hand, high levels of toluene impeded this promotion and, at the same time, very strongly depressed TCE conversion. The photocatalytic degradation rates of compounds such as isooctane, methylene chloride and chloroform were also found to increase dramatically in the presence of TCE (Berman & Dong, 1993; Sauer et al., 1995). The conversion of a mixture of TCE and PCE yields a higher conversion of both reagents in comparison to that of isolated compounds (Berman & Dong, 1993). The authors believe that the added sensitizer function provides the radicals required to trigger the chain reaction leading to the destruction of the pollutant. In the case of TCE or PCE, the chlorine atoms are suggested to trigger the reaction.

13.2.6 Deactivation of the photocatalyst

Deactivation and regeneration of the catalyst is a basic issue in photocatalytic reactions. Many authors have observed catalyst deactivation in their gas-phase reactions. In catalysis, it is always desirable for the catalytic activity to be long lasting, since the possibility of the catalyst surface to be poisoned, with the consequent loss of its activity negatively impacts on the commercial competitiveness of this technology. It is believed that some reaction intermediates, such as carboxylic acids formed during the incomplete oxidation of organic compounds, may be chemically adsorbed on the active catalyst sites provoking their deactivation.

Although the destruction of pollutants is generally found to be very efficient in the gas phase, low reaction rates or deactivation of the catalyst are observed with changes in color of the catalyst surface, especially in the case of aromatic compounds. To explain the deactivation and drop in photocatalytic activity, three possible causes have been proposed: blocking of the active sites on the catalyst by CO_2 or other carbon deposits,

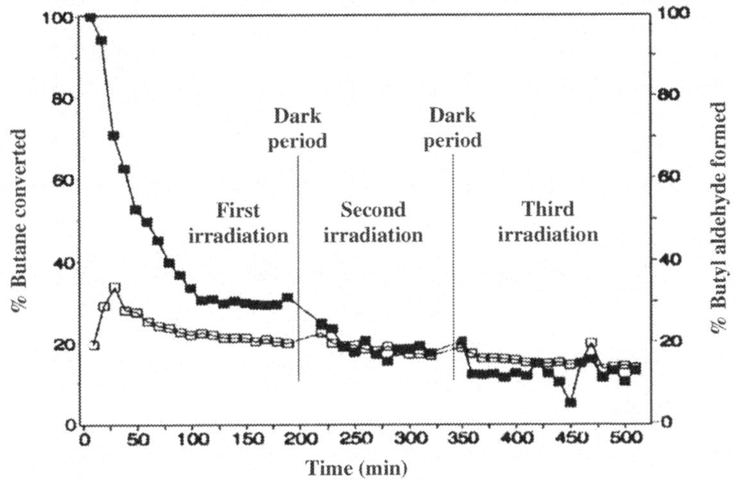

the over-consumption of surface oxygen species, and irreversible surface dehydroxylation (Alberici & Jardim, 1997; Cunningham et al., 1981; Fu et al., 1997).

Peral and Ollis (1992), in the work cited above, found that the percentage of 1-butanol degraded diminishes in successive irradiations (Figure 13.7). The activity of TiO_2 cannot be recovered with only one passage through unpolluted air in the dark and radiation must be applied simultaneously. These observations point to the formation of organic deposits (and therefore, potentially oxidizable), not very volatile, and with noticeably slower degradation kinetics than the starting product.

The same authors (1993, 1997) have studied the photooxidation of various organic compounds containing heteroatoms: decamethyl tetrasiloxane (DMTS), indol, pyrrol and dimethyl sulfide (Figure 13.8). With the exception of dimethyl sulfide, a continued decrease in percentage degraded over time is referred in the rest of the compounds. Equation (13.4) has been proposed to describe the deactivation kinetics with v on degradation rate for time t, v_0 the initial rate and b a parameter that depends on the substance being degraded:

$$\frac{v}{v_0} = \frac{1}{1 + t^b} \tag{13.4}$$

DMTS heteroatoms in indol and pyrrol were detected on the surface of the TiO_2 used for photooxidation by Auger electronic spectroscopy. Figure 13.9 shows the nitrogen peak in the Auger electronic spectrum of the TiO_2 sample used for pyrrol oxidation. The authors have tentatively assigned the peak to nitrates formed by complete oxidation of the starting compound which, in absence of aqueous dissolution, are impossible to eliminate. A considerable carbon peak, indicative of organic deposits, is also observed. All attempts to recover the catalyst activity were unfruitful.

Sauer and Ollis (1996) in their review, indicate deactivation of the catalyst in continuous fixed-bed reactors with no recirculation, while they found no evidence of

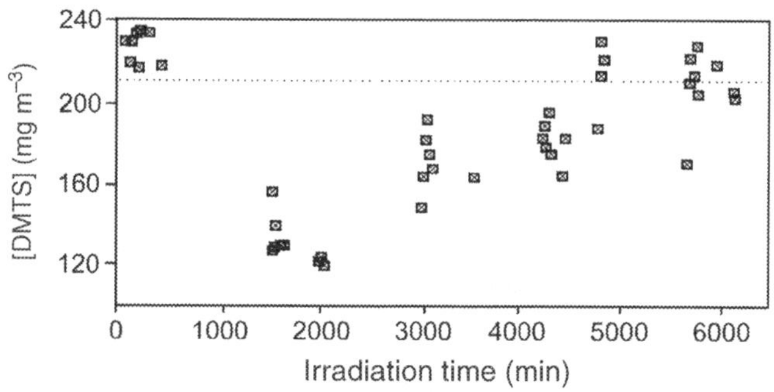

Figure 13.8 Decrease in photoactivity of a sample of TiO_2 used to treat dimethyl sulfide in a continuous reactor. The concentration of dimethyl sulfide is measured over time at the reactor outlet (Peral & Ollis, 1997).

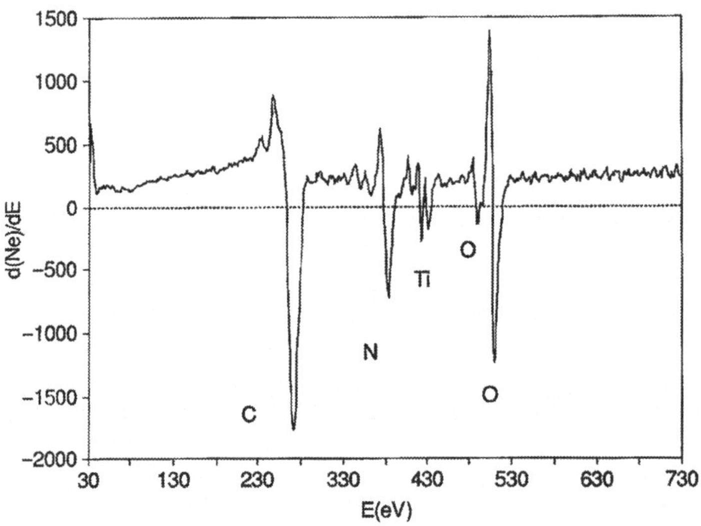

Figure 13.9 Auger spectrum of the TiO_2 surface deactivated by photocatalytic treatment of pyrrol, in which the different elements detected on the catalyst surface are observed (Peral & Ollis, 1997).

deactivation in experiments in discontinuous systems. The use of discontinuous reactors requires systematic repetition of experiments (with the same semiconductor sample) so that if deactivation phenomena occur, they are observable.

Temperature changes significantly affect adsorption/desorption equilibrium of reagents and products. If the temperature is sufficiently high, the semiconductor can also act as a conventional heterogeneous catalyst, improving the process kinetics. For these two reasons, an increase in temperature could enable the formation of species formed at ambient temperature that poison the semiconductor surface to be oxidized or desorbed, thus avoiding the deactivation process.

Photocatalytic oxidation of acetone is faster as the temperature increases to 80°C (Vorontsov et al., 1999). At higher temperatures, the process slows down. The rate at 163°C is nearly five times lower than at 80°C. This decrease is explained by deactivation of TiO_2: the catalyst turns brown after the reaction at high temperature and gradually recovers activity during the process at 72°C. However, the nature of this deactivation is unclear.

Photocatalytic oxidation of acetaldehyde on TiO_2 catalysts deactivates quickly at high temperatures due to a parallel thermal catalytic reaction (Falconer & Magrini, 1998). The addition of Pt slows down the deactivation. Apparently, Pt provides oxygen to the TiO_2 that oxidizes the decomposition products of acetaldehyde in a reaction in the dark. TiO_2 deactivated without Pt was regenerated by photocatalytic oxidation at ambient temperatures.

Alberici and Jardim (1997) studied the photocatalytic oxidation of different types of VOCs (17 compounds) present in the atmosphere. They only detected deactivation with toluene, but activity was recovered by illuminating the catalyst in the presence of hydrogen peroxide.

Canela et al. (1999) found that sulfate ions adsorbed on the catalyst surface during photocatalytic oxidation of sulfurated compounds do not cause deactivation of the catalyst, except when processing high concentrations of H_2S (600 ppmv) At 217 ppmv of H_2S, photocatalytic activity is maintained for 24 h.

13.3 POTENTIAL APPLICATIONS

13.3.1 Treatment of air in building interiors

Heterogeneous photocatalysis could become the best method for treating indoor environments containing pollutants, that have even at low levels severe consequences for human health (Peral et al., 1997). Adsorption of pollutants on activated carbon, among other techniques, has been used for treatment of this type of atmosphere. In this non-destructive process, the pollutants are accumulated on the surface of the adsorbent until it is used up. The carbon cassettes must be regenerated periodically, which produces a new pollution phase with high concentrations that still requires to be treated. By employing heterogeneous photocatalysis, indoor air can be decontaminated, purified and deodorized at ambient temperature, using molecular oxygen as the only oxidant and ambient light or sunlight. For example, Pichat et al. (2000), Peral and Ollis (1993, 1997) and Sánchez et al. (1998, 2000) report on the use of gas-phase heterogeneous photocatalysis for deodorization of gas streams and closed atmospheres. The studies carried out on treatment of indoor air include residences, offices and industrial bays. There are also studies financed by the NASA on the air quality in spacecraft and space stations. Research on the problem of ethylene in greenhouses and food storage cold rooms (CELSS – Controlled Ecological Life Support Systems). In this context, see the study of Park et al. (1999) on the photocatalytic oxidation of ethylene into CO_2 and H_2O with powdered TiO_2, in the presence of O_2 and H_2O.

Pichat et al. (2000) verified the ability of TiO_2 to purify and deodorize indoor air and industrial effluent gases using glass fiber impregnated with TiO_2 made by the Matrix Photocatalytic company (Canada). In a laboratory photoreactor with a lamp emitting at

365 nm, the authors determined the elimination efficiency of three pollutants with very different physical–chemical characteristics (CO, *n*-octane and pyridine). They obtained 5–10 μmol/Wh consumed by the lamp for 50–2000 ppmv concentration and flow rates of 25–50 L/h (dry air or O_2). These values allowed them to view the usefulness of a system of continuous renewal of reasonable size for the destruction of pollutants in indoor air, except for CO and CH_4, which were too highly concentrated. Using a prototype individual air purifier, they were able to reduce the average concentrations of benzene, toluene and xylene in a reference room by a factor 2 or 3.

13.3.2 Industry

Photocatalysis does not seem to be a suitable technique for the treatment of concentrated gas effluents of industrial gas emissions. The catalyst tends to become quickly saturated, and its performance is low. The olfactory limits of many compounds causing foul odor are in a ppb range, which poses severe requirements for effective photocatalytic odorizing. Pichat et al. (2000), in a presumably over-dimensioned extrapolation, calculated that 10 kW UV lamps would be required to eliminate a pollutant present in an industrial gas effluent in concentrations of around 25 ppm_v at a gas flow rate of 100 m^3/h. This can be used as a guide to estimate whether a gas effluent can be treated by photocatalysis with TiO_2. Related to the treatment of industrial emissions, the work by Kemme et al. (1999) on the treatment of an effluents from a propellant manufacturer merits to be mentioned.

13.3.3 Exteriors

The remediation of outdoor air is relevant for cities or highly polluted areas, where pollution comes from automobile exhaust or industrial processes. As the air cannot be kept near the source of emission or in a single place, the pollutants are diluted and transported by the wind. The majority of methods for decontaminating large spaces have high operating and power costs. In this sense, heterogeneous photocatalysis may be a good candidate for treating volumes of air in such large areas.

Ibusuki et al. (1993), Ibusuki and Takeuchi (1994) and Takeuchi (1998) proposed an ingenious method of photocatalytically reducing low levels of NO_x and SO_2 on Japanese highways. The system uses sunlight and 1 m^2 sheets of fluorocarbonated polymer film coated with TiO_2, with a filler such as activated carbon, cement or siliconized paint. The co-products of the photocatalytic reaction, mainly nitric and sulfuric acids, are dissolved by the frequent heavy rains in Japan and carried to the sewer system. The authors report that a 1 m^2 purifier sheet can eliminate more than 0.02 moles of NO_x gas (which can be found in the atmosphere in concentrations of from 0.01 to 5 ppm).

13.3.4 Disinfection

Fujishima et al. (1999) described the bactericidal power of photocatalysis. The issue was also studied in water by Ohgaki et al. (2000) and Kersters et al. (1998), among others.

In air, the basic strategy consists of fixing the bacteria on a photocatalytically active surface for their irradiation. Horie et al. (1998) evaluated the photocatalytic sterilization rates of *Escherichia coli* on TiO_2 irradiated with several UV light sources, black light fluorescent, actinic blue fluorescent and daylight fluorescent. With 1×10^{-2} kg/m^3 of TiO_2 and an initial cell concentration of 1×10^{11} cells/m^3, and variable intensities in the reactor (in the range between 0–22 W/m^2, 0–14 W/m^2 y 0–1.1 W/m^2, respectively, for each type of lamp), the apparent *E. coli* sterilization rate constants were determined for these lamps. Values obtained were compared with those derived from experiments with high-pressure mercury lamps. The constants can be correlated with the amount of light absorbed by the TiO_2. Therefore, the spectral distribution of the lamps defines its effectiveness. The authors used the results to predict *E. coli* sterilization rate constants with sunlight. The experimental data are in agreement with the results predicted in a reactor illuminated by sunlight at average light intensities of 14–23 W/m^2.

Jacoby et al. (1998) deposited *E. coli* on surfaces impregnated with TiO_2 and achieved their photooxidation in air into CO_2, providing proof of total oxidation of organic matter in the cells. They used scanning electron microscopy techniques and isotopic marking with ^{14}C to prove that the carbon content in *E. coli* oxidizes in the presence of TiO_2 and suggested that a photocatalytically active surface for disinfection be hooked up to an air purification system.

13.3.5 Odor treatment

Urban and/or industrial sludge and wastewater treatment plants emit odors that are intensely annoying to local residents. Such odors can be caused by low concentrations of compounds containing sulfur, such as mercaptans, organic sulfides, disulfides and hydrogen sulfide, and nitrogenated compounds such as amines and pyridines. There are several well-established conventional odor control technologies, including adsorption by activated carbon, biofiltration, thermal oxidation and prevention of formation of these compounds. Each one of these techniques has a series of advantages and disadvantages and varied application costs.

Huang et al. (1999) and Suzuki et al. (1999) studied photocatalytic oxidation of TEA and a mixture of acetaldehyde, isobutyric acid, toluene, methyl mercaptan, hydrogen sulfide and trimethylamine at concentrations of around 500 ppmv. The work of Canela et al. (1998, 1999) on photocatalytic oxidation of sulfurated compounds was presented in Section 13.2.2.4.

13.3.6 Passive treatment of architectural elements: self-cleaning materials

In recent years, mainly in Japan, much attention has been given to self-cleaning materials, such as windows and tiles impregnated with TiO_2 used in traffic tunnels, bathrooms, etc. Coated surfaces, films, paints or papers containing TiO_2 and ceramic tiles coated with TiO_2 are currently used in sanitary environments such as bathrooms and hospitals, and even in special bays. Thus, stains can be eliminated by keeping the windows and vehicle windshields clean, removing finger prints or grease from walls and reducing colonies of pathogenic microorganisms, as well as deodorizing public and private spaces.

Heller (1995), Paz et al. (1995) and Sitkiewitz and Heller (1996) developed self-cleaning windows able to oxidize solid organic material deposited on it and abate the organic load in the steam phase. To be efficient, a photocatalytic self-cleaning system must meet certain conditions. The glass coated with TiO_2 must be perfectly clear, with no light dispersion, which requires very small particles with a very narrow size distribution of between 30 and 50 nm. Furthermore, it must have a significant oxidation rate, with negligible loss of activity and be abrasion resistant. A suitable material was obtained by partial hydrolysis of titanium triisopropoxide acetylacetonate spread on glass or quartz and calcinated at 500°C. With this material, a layer of 2000 Å of stearic acid on glass was completely eliminated at a rate of 20 nM h^{-1}. These authors also reported the decomposition of oily materials such as vegetable oil or cigarette tar using ordinary fluorescent illumination.

There are a multitude of patents on photocatalytic systems (Kameda et al., JP6262187; Nishimoto et al., JP6411622), see Figures 13.1 and 13.2. Suzuki (1993, JP60187322) along with a Japanese automobile company, patented a compact purifier for cars that uses a photocatalyst made of cordierite coated with TiO_2.The purifier can be operated with a 12 V DC battery, consumes only 10 W and reduces the concentration of pollutants to almost half after 2 h of operation. The design can be easily applied to other domestic habitats polluted by cooking smells, pets or bathrooms.

Fluoranthene is one of the most abundant polyaromatic hydrocarbons (PAHs) among atmospheric aerosols originating from fossil fuel combustion processes. It is photolyzed relatively slowly and, due to its low vapor pressure, is not significantly eliminated by evaporation on surfaces where it is deposited. Romeas et al. (1999) studied the elimination of fluoranthene on window panes or any glass object in general, and suggested that self-cleaning glass can be obtained by depositing a thin transparent layer of TiO_2 nanoparticles on glass by patented methods. To test its self-cleaning properties, he sprayed the glass with fluoranthene. Under UV light, fluoranthene was eliminated at a rate of 0.73 nmol/h/cm^2 of glass, along with all of its degradation products. Contrary to unimpregnated glass, the transparency of the glass recovered. The percentage of fluoranthene converted into volatile carbonylic products released into the ambient air was lower than without TiO_2 impregnation. The reaction is complex, and 40 intermediate products were detected.

Murata (2000), Yumoto et al. (1999), Fukaya et al. (1997) and Hashimoto et al. (2001) demonstrated that photocatalytically active surfaces can eliminate nitrogen oxide. New self-cleaning materials based on TiO_2 have been created, that are already used in traffic tunnels, in paving tiles in interiors and exteriors of buildings and in glass covers of city illumination systems. Fujishima (2000) used the TiO_2 photocatalyst under solar irradiation to prevent contagion in buildings, and Murata (1998) treated paving stones for the same purpose.

13.3.7 Treatment of polluted soil and water

Polluted soil can be treated by vapor extraction followed by photocatalytic treatment of the air stream. The same principle is applied to pollution of liquids. The selection of the phase in which the photocatalytic decontamination is to be done depends on the cost of the process. The work done by Changrani et al. (1998) on integration of photocatalytic oxidation for treatment of groundwater should be mentioned. It has been estimated that the cost of detoxifying TCE in groundwater by combining air stripping and photocatalytic oxidation is significantly lower than the cost of direct water treatment by any advanced oxidation method. In particular, treatment of gas VOCs should be preferred to direct treatment of aqueous effluents for several reasons (Berman & Dong, 1993):

(1) air streams are cleaner and vary less because the non-volatile compounds have already been removed from them
(2) the gas-phase photocatalytic processes are more efficient because the competition from water for catalyst sites is low and there is more interaction between the catalyst and the organic compound
(3) gas-phase rates are less likely to be limited by diffusion
(4) treatment design is simpler.

Brunet et al. (1999) present a photocatalytic system for the control of toxic air emissions containing compounds such as TCE and PCE found in soil, in groundwater and in industrial effluents. They affirm that the cost of the system is acceptable. This technology uses UV light and a suitable catalyst to absorb and destroy pollution in ambient conditions, even with high humidity. Air stripping and soil extraction-vapor processes efficiently transfer gas pollution to where they can be economically treated by photocatalysis without the risk of formation of hazardous byproducts. A commercial-size unit can treat $20\,m^3$/min and up to 1000 ppmv of TCE, mixed with PCE, dichloroethylene, trichloroethane and vinyl chloride. A system, called AIR 2000, has been installed in Stamina Mills Superfund en Rhode Island. System monitoring as part of the USEPA program has shown total destruction efficiency of 99–99.99%. The cost of system operation is approximately 20% of a corresponding activated carbon adsorption process.

REFERENCES

Alberici, R.M. and Jardim, W.F. (1997) Photocatalytic destruction of VOCs in the gas-phase using titanium dioxide. *Appl. Catal. B: Environ.* **14**(1–2), 55–68.

Alberici, R.M., Mendes, M.A., Jardim, W.F. and Eberlin, M. (1998) Mass spectrometry on-line monitoring and MS2 product characterization of TiO_2/UV photocatalytic degradation of chlorinated volatile organic compounds. *J. Am. Soc. Mass Spectrom.* **9**, 1321–1327.

Alberici, R.M., Canela, M.C., Eberlin, M.N. and Jardim, W.F. (2001) Catalyst deactivation in the gas phase destruction of nitrogen-containing organic compounds using TiO_2/UV–VIS. *Appl. Catal. B.* **30**, 389–397.

Annapragada, R., Leet, R., Changrani, R. and Raupp, G.B. (1997) Vacuum photocatalytic oxidation of trichloroethylene. *Environ. Sci. Technol.* **31**, 1898–1901.

Avila, P., Bahamonde, A., Blanco, J., Sánchez, B., Cardona, A.I. and Romero, M. (1998) Gas-phase photo-assisted mineralization of volatile organic compounds by monolithic titania catalysts. *Appl. Catal. B: Environ.* **17**, 75–88.

Berman, E. and Dong, J. (1993) Photocatalytic decomposition of organic pollutants in gas streams. "In *The Third International Symposium Chemical Oxidation: Technology for the Nineties*" (eds. W.W. Echenfelder, A.R. Bowers and J.A. Roth), pp. 183–189. Technomic Publishers, Chicago.

Bickley, R.I., Munuera, G. and Stone, F.S. (1973) Photoadsorption and photocatalysis at rutile surfaces. II. Photocatalytic oxidation of isopropanol. *J. Catal.* **31**, 389–407.

Blake, D.M. (1997) *Bibliography of Work on the Heterogeneous Photocatalytic Removal of Hazardous Compounds from Water and Air.* Number 2 to October 1996. NREL/TP-430-22197. 1997. National Renewable Energy Laboratory, Golden, CO (United States).

Blake, N.R. and Griffin, G.L. (1988). Selectivity control during the photoassisted oxidation of 1-butanol on titanium-dioxide. *J. Phys. Chem.* **92**(20), 5697–5701.

Blanco, J., Avila, P., Bahamonde, A., Alvarez, E., Sánchez, B. and Romero, M. (1996) Photocatalytic destruction of toluene and xylene at gas phase on a titania based monolithic catalyst. *Catal. Today* **29**, 437–442.

Blanco, J., Bahamonde, A., Avila, P., Sánchez, M., Romero, M., Cardona, I., Funken, K.H., Reichert, M., Rovatti, M. and Nicolella, C. (1998) Utilization of Monoliths for Photocatalytic Destruction of Organic Pollutants in Gaseous Emissions. *Waste Water Treatment and Monitoring, Abatement of Emissions to the Atmosphere, Cleaner Technologies*, pp. 179–187. Cranfield University Press, UK.

Blanco, J., Malato, S., Peral, J., Sánchez, B. and Cardona, A. (2004) Diseño de reactores para fotocatálisis: evaluación comparativa de las distintas opciones. Eliminación de contaminantes por fotocatálisis heterogénea. Colección Documentos CIEMAT, pp. 305–328.

Brunet, R.A.H., Pearcey, R., Kittrell, J.R., Macking, G. and Wise, C.A. (1999) Site remediation using photocatalytic VOC destruction of chlorinated hydrocarbons. *Air and Waste 92nd Annual Meeting and Exhibition Proceedings*. St. Louis, MO (United States), Pittsburgh, PA (United States) Air and Waste Management Association Publication Date: 1999 Page: vp., *Paper 99.826* ([9500] p), pp. 20–26.

Buechler, K.J., Noble, R.D., Koval, C.A. and Jacoby, W.A. (1999) Investigation of the effects of controlled periodic illumination on the oxidation of gaseous trichloroethylene using a thin film of TiO_2. *Ind. Eng. Chem. Res.* **38**(3), 892–896.

Canela, M.C., Alberici, R.M. and Jardim, W.F. (1998) Gas-phase destruction of H2S using TiO_2/UV-VIS. *J. Photochem. Photobiol. A: Chem.* **112**, 73–80.

Canela, M.C., Alberici, R.M., Sofía, R.C.R., Eberlin, M.N. and Jardim, W.F. (1999) Destruction of malodorous compounds using heterogeneous photocatalysis. *Environ. Sci. Technol.* **33**, 2788–2792.

Cant, N.W. and Cole, J.R. (1992) Photocatalysis of the reaction between ammonia and nitric oxide on TiO_2 surfaces. *J. Catal.* **134**, 317.

Cao, L., Spiess, F.J., Huang, A., Suib, S.L., Obee, T.N., Hay, S.O. and Freihaut, J.D. (1999) Heterogeneous photocatalytic oxidation of 1-butene on SnO_2 and TiO_2 films. *J. Phys. Chem. B: Mater. Surf. Interf. Biophys.* **103**(15), 2912–2917.

Cao, L., Huang, A., Spiess, F.J. and Suib, S.L. (1999) Gas-phase oxidation of 1-butene using nanoscale TiO_2 photocatalysts. *J. Catal.* **188**(1), 48–57.

Cardona, A.I., Sánchez, B., Romero, M., Fabrellas, B., García, E., Blanco, J., Avila, P. and Bahamonde, A. (1999) Solarized photoreactors for degradation of chlorinated organics in air. *J. Phys. IV France* **9**, 271–276.

Cardona, A.I., Sánchez, B., Bahamonde, A. and Avila, P. (1999) Estudio comparativo de las eficiencias de destrucción de TCE en fase gas por vía fotocatalítica utilizando diferentes catalizadores basados en TiO_2. (eds. Ginesa Blanco and César Mira), *Actas SECAT'99.*

Changrani, R., Raupp, G.B. and Turchi, C. (1998) Integration of photocatalytic oxidation with air stripping of contaminated aquifers. Report for 1 October 1993–30 June 1997. *PB-99-127920/XAB*. Arizona State University, Tempe, AZ (United States); Zentox Corporation, Ocala, FL (United States); National Risk Management Research Laboratory, Sustainable Technology Division, Cincinnati, OH (United States).

Changrani, R.G. and Raupp, G.B. (2000) Two-dimensional heterogeneous model for a reticulated-foam photocatalytic. *Amer. Inst. Chem. Eng.* **46**(4), 829–842.

Chung-Hsuang, H. and Mariñas, B.J. (1997) Role of chlorine and oxygen in the photocatalytic degradation of trichloroethylene vapor on TiO_2 films. *Environ. Sci. Technol.* **331**, 562.

Courbon, H., Formenti, M. and Pichat, P. (1977) Study of oxygen isotopic exchange over ultraviolet irradiated anatase samples and comparison with the photooxidation of isobutane into acetone. *J. Phys. Chem.* **81**, 550.

Courbon, H., Hermann, J. and Pichat, P. (1984) Effect of platinum deposits on oxygen adsorption and oxygen isotope exchange over variously pretreated, ultraviolet-illuminated powder titanium dioxide. *J. Phys. Chem.* **88**, 5210.

Cunningham, J., Hodnett, B.K., Ilyas, M., Tobin, J., Leahy, E.L. and Fierro, J.L.G. (1981) Dehydrogenation versus dehydration of aliphatic alcohols on metal oxides. *Faraday Discuss. Chem. Soc.* **72**, 283.

d'Hennezel, O. and Ollis, D.F. (1997) Trichloroethylene-promoted photocatalytic oxidation of air contaminants. *J. Catal.* **167**, 118–126.

D'Oliveira, J., Al-Sayyed, G. and Pichat, P. (1990) Photodegradation of 2- and 3-chlorophenol in titanium dioxide aqueous suspensions. *Environ. Sci. Technol.* **24**, 990–996.

Dibble, L.A., Raupp, L.A. and G.B. (1990) Kinetics of the gas-solid heterogeneous photocatalytic oxidation of trichloroethylene by near UV illuminated titanium dioxide. *Catal. Lett.* **4**, 345–354.

Directive 1999/13/CE Volatile Organic Compounds (VOCs) resulting from certain industrial activities (1999).

Djeghri, N. and Teichner, S.J. (1980) Heterogeneous photocatalysis: the photooxidation of 2-methylbutane. *J. Catal.* **62**, 99.

Djeghri, N., Formenti, M., Juillet, F. and Teichner, S.J. (1974) Photointeraction on the surface of titanium dioxide between oxygen and alkanes. *Faraday Discuss. Chem. Soc.* **58**, 85.

Driessen, M.D., Goodman, A.L., Miller, T.M., Zaharias, G.A. and Grassian, V. (1998) Gas-phase photooxidation of trichloroethylene on TiO_2 and ZnO: influence of trichloroethylene pressure, oxygen pressure, and the photocatalyst surface on the product distribution. *J. Phys. Chem.* **102**, 549.

Egerton, T.A. and King, C.J. (1979) The influence of light intensity on photoactivity in TiO_2 pigmented systems. *J. Oil Col. Chem. Assoc.* **62**, 386.

Einaga, H., Futamura, S. and Ibusuki, T. (1999) Complete oxidation of benzene in air over Pt/TiO_2 photocatalysts. *Sekiyu Gakkai-Shi (J. Jpn. Pet. Inst.)*, **42**(5), 363–364.

El-Maazawi, M., Finken, A.N. and Nair, A.B. (2000) Adsorption and photocatalytic oxidation of acetona on TiO_2: an in situ transmission FT-IR study. *J. Catal.* **119**(1), 138–146.

Falconer, J.L. and Magrini-Bair, K.A. (1998) Photocatalytic and thermal catalytic oxidation of acetaldehyde on Pt/TiO_2. *J. Catal.* **179**(1), 171–178.

Filby, W.G., Mintas, M. and Gusten, H. (1981) *Ber. Bunsenges. Phys. Chem.* **85**, 189.

Formenti, M., Juillet, F., Meriaudeau, P. and Teichner, S.J. (1971). Heterogeneous photocatalysis for partial oxidation of paraffins. *Chem. Technol.* **1**, 680.

Fox, M.A. and Chen, C.C. (1981) Mechanistic features of the semiconductor photocatalyzed olefin-to-carbonyl oxidative cleavage. *J. Am. Chem. Soc.* **103**, 6757.

Fox, M.A. and Dulay, M.T. (1993) Heterogeneous photocatalysis. *Chem. Rev.* **93**, 341–357.

Fu, X., Zeltner, W.A. and Anderson, M.A. (1997) The gas-phase photocatalytic mineralization of benzene on porous titania-based catalysts. *Appl. Catal. B: Environ.* **6**, 209–224.

Fujishima, A. (2000) Application of titanium oxides photocatalysts for architecture prevention. *Taiyo Enerugi* **26**(2), 2–6.

Fujishima, A., Hashimoto, K. and Watanabe, T. (1999) *Photocatal. Fund. Appl. BKC, Inc.* 24–37.

Fukaya, M., Taoda, H., Watanabe, E., Nonami, T., Iseda, K., Kato, K., Kunieda, S. and Kato, S. (1997) Photocatalytic removal of nitrogen oxides from ambient air using solar energy. *Aichi (Japan). Proceedings of JSES/JWEA Joint Conference* **28**, 11.

González-Elipe, A., Munuera, G. and Soria, J. (1979) Photo-adsorption and photo-desorption of oxygen on highly hydroxylated TiO_2 surfaces. Part 2. Study of radical intermediates by electron paramagnetic resonance. *J. Chem. Soc. Faraday Trans. 1* **75**, 748.

Gravelle, P.C., Juillet, F., Meriaudeau, P. and Teichner, S.J. (1971) Surface reactivity of reduced titanium dioxide. *Discuss. Faraday Soc.* **52**, 140.

Hashimoto, K., Wasada, K., Osaki, M., Shono, E., Adachi, K., Toukai, N., Kominami, H. and Kera, Y. (2001) Photocatalytic oxidation of nitrogen oxide over titania-zeolite composite catalyst to remove nitrogen oxides in the atmosphere. *Appl. Catal. B: Environ.* **30**, 429–436.

Heller, A. (1995) Chemistry and applications of photocatalytic oxidation of thin organic films. *Acc. Chem. Res.* **28**, 503–508.

Hemminger, J.C., Carr, R. and Somorjai, G.A. (1978) The photoassisted reaction of gaseous water and carbon dioxide adsorbed on the $SrTiO_3$ (111) crystal face to form methane. *Chem. Phys. Lett.* **57**, 100–104.

Hermann, J., Disdier, J. and Pichat, P. (1981) Oxygen species ionosorbed on powder photocatalyst oxides from room-temperature photoconductivity as a function of oxygen pressure. *J. Chem. Soc. Faraday Trans. 1* **77**, 2815.

Hiromichi, I. and Hiroaki, K. (1999) Photocatalytic activities films prepared from peroxo-modified anatase sol with added TiO_2 powder. Nippon Seramikkusu Kyokai Gakujutsu Ronbunshi. *J. Ceram. Soc. Jpn.* **107**, 1241, 73–77.

Horie, Y., Taya, M. and Tone, S. (1998) Evaluation of photocatalytic sterilization rates of *Escherichia coli* cells in titanium dioxide slurry irradiated with various light sources. *J. Chem. Eng. Jpn.* **31**(4), 577–584.

Huang, A., Cao, L., Chen, J., Spiess, F.J., Suib, S.L., Obee, T.N., Hay, S.O. and Freihaut, J.D. (1999) Photocatalytic degradation of triethylamine on titanium oxide thin films. *J. Catal.* **188**(1), 40–47.

Hung, C.H. and Yuan, C.S. (1998) Gas-phase photocatalytic degradation of perchloroethylene on glass pellets immobilized with TiO_2. p. *Anonymous Proceedings of the 91 Annual Meeting and Exhibition*. Bridging International Boundaries: clean production for environmental stewardship. Pittsburgh, PA (United States). Air and Waste Management Association.

Ibrahim, H. and Lasa, H. (1999) Novel photocatalytic reactor for the destruction of airborne pollutants reaction kinetics and quantum yields. *Ind. Eng. Chem. Res.* **38**(9), 3211–3217.

Ibusuki, T. and Takeuchi, K. (1986) Toluene oxidation on UV – irradiated titanium dioxide with and without O_2, NO_2 or H_2O at ambient temperature. *Atmos. Environ.* **20**, 1711–1715.

Ibusuki, T. and Takeuchi, K. (1994) Removal of low concentration nitrogen oxides through photoassisted heterogeneous catalysis. *J. Molecul. Catal.* **88**, 93–102.

Ibusuki, T., Kutsuna, S. and Takeuchi, K. (1993) Removal of low concentration air pollutants through photoassisted heterogeneous catalysis. In *Photocatalytic Purification of Water and Air* (eds. D.F. Ollis and H. Al-Ekabi), pp. 375–386. Elsevier Science Publisher, B.V. Amsterdam.

Jacoby, D.M., Nimlos, M.R., Blake, D.M., Noble, R.D. and Koval, C.A. (1994) Products, intermediates, mass balances, and reaction pathways for the oxidation of trichloroethylene in air via heterogeneous photocatalysis. *Environ. Sci. Technol.* **28**, 1661.

Jacoby, W.A., Blake, D.M. Fennell, J.A., Boulter, J.E., Vargo, L.M., George, M.C. and Dolberg, S.K. (1996) Heterogeneous photocatalysis for control of volatile organic compounds in indoor air. *J. Air Waste Manage. Assoc.* **46**, 891–898.

Jacoby, W.A., Maness, P.C., Wolfrum, E.J. and Blake, D.M. (1998) Mineralization of bacterial cell mass on a photocatalytic surface in air. *Environ. Sci. Technol.* **32**(17) 2650–2654.

Jardim, W.F., Alberici, R.M., Takiyama, M.M.K. and Huang, C.P. (1994) Gas-Phase photocatalytic destruction of trichloro-ethylene (TCE) using UV/TiO_2. *Hazard. Ind. Wastes* **26**, 230–238.

Kameda, M., Naka, R., Sumida, T. and Kawashima, M. *Japan Kokai Jisuyoshinan Kcho*, JP6262187.

Kemme, M.R., Lateulere, M. and Maloney, S.W. (1999) Reducing air pollutant emissions from solvent multi-base propellant production. *Final Report. AD-A-367284/XAB; CERL-TR-99/71*, p. 85. Army Construction Engineering Research Laboratory, Champaign, IL (United States).

Kersters, I., De Keyser, T. and Verstraete, W. (1998) Sensitivity of bacteria to photoactivated titanium dioxide in comparison with UV irradiation. *Indian J. Eng. Mater. Sci.* **5**(4), 211–216.

Kim, K.Y. and Park, S.B. (1998) Photocatalytic activity of anatase titania particles of controlled size prepared in an aerosol reactor. *Hwahak Konghak J. Korean Inst. Chem. Eng.* **36**(1), 116–119.

Kormann, C., Bahnemann, D.W. and Hoffmann, M.R. (1991) Photolysis of chloroform and other organic molecules in aqueous TiO_2 suspensions. *Environ. Sci. Technol.* **25**(3), 494–500.

Kutsuna, S., Takeuchi, K. and Ibusuki, T. (1992) Adsorption and reaction of trichlorofluoromethane on various particles. *J. Atmos. Chem.* **14**, 1.

Kutsuna, S., Takeuchi, K., Ibusuki, T. and Toma, M. (1999) Photocatalytic degradation of some methyl perfluoroalkyl ethers on TiO_2 particles in air: the dependence on the dark adsorption, the products, and the implication for a possible tropospheric sink. *Environ. Sci. Technol.* **33**(7), 1071–1076.

Lew, K.K., Herrera, M. and Marinas, B.J. (1999) Photocatalytic degradation of methyl ethyl ketone (MEK) and xylenes on anatase titanium dioxie. p. *Anonymous Air and Waste 92nd Annual Meeting and Exhibition*, Pittsburgh, PA (United States). Air and Waste Management Association.

Litter, M.I. and Navío, J.A. (1996) Photocatalytic properties of iron-doped titania semiconductors. *A: Chem.* **98**, 171–181.

Luo, Y. and Ollis, D.F. (1996) Heterogeneous photocatalytic oxidation of trichloroethylene and toluene mixtures in air: kinetic promotion and inhibition, time-dependent catalyst activity. *J. Catal.* **163**, 1–11.

Magrini, K., Watt, A., Boyd, L., Wolfrum, E.J., Larson, S.A., Roth, C. and Glatzmaier (1999) Application of solar photocatalytic oxidation to VOC-containing airstreams. *Paper RAES 99.7691. New York, NY (United States) American Society of Mechanical Engineers. RAES 99.*

Maira, A.J., Yeung, K.L., Soria, J., Coronado, J.M., Belver, C., Lee, C.Y. and Augugliaro, V. (2001) Gas-phase photo-oxidation of toluene using nanometer-size TiO_2 catalysts. *Appl. Catal. B. Environ.* **29**, 327–336.

Martin, R.S., Richardson, C.P., Fan, T. and Halbert, C.P. (1998) Control of gas-phase chlorobenzene using TiO_2-mediated photocatalysis and packed-bed absorption. p. *Anonymous 91st Annual Meeting and Exhibition of the Air and WMA*, Pittsburgh, PA (United States). Air and Waste Management Association.

Matsubara, H., Takada, M., Koyama, S., Hashimoto, K. and Fujishima, A. (1995) Photoactive TiO_2 containing paper: preparation and its photocatalytic activity under weak UV light illumination. *Chem. Lett.* 767–768.

Matsumoto, Y. (1996) Energy positions of oxide semiconductors and photocatalysis with iron complex oxides. *J. Solid State Chem.* **126**(2), 227–234.

Mozzanega, H., Herrmann, J.M. and Pichat, P. (1979) Ammonia oxidation over UV-irradiated titanium dioxide at room temperature. *J. Phys. Chem.* **83**, 2251.

Muggli, D.S. and Falconer, J.L. (1999) Parallel pathways for photocatalytic decomposition of acetic acid on TiO_2. *J. Catal.* **187**, 230–237.

Munuera, G., Rives-Arnau, V. and Saucedo, A. (1979) Photo-adsorption and photo-desorption of oxygen on highly hydroxylated TiO_2 surfaces. Part 1. Role of hydroxyl groups in photo-adsorption. *J. Chem. Soc. Faraday Trans. 1* **75**, 736–747.

Munuera, G., Gonzalez-Elipe, A.R., Soria, J. and Sanz, J. (1980) Photo-adsorption and photo-desorption of oxygen on highly hydroxylated TiO_2 surfaces. Part 3. Role of H_2O_2 in photo-desorption of O_2. *J. Chem. Soc. Faraday Trans. 1* **76**, 1535.

Murata, Y. (1998) *Semento Konkurito (Cement and Concrete)*, **622**, 32–37.

Murata, Y. (2000) Technology for air purification on photocatalytic titanium oxide. *J. Soc. Inorg. Mater., Jpn.* **7**(284), 50–56.

Nimlos, M.R., Jacoby, W.A., Blake, D.M. and Milne, T.A. (1993) Direct mass spectroscopic studies of the destruction of hazardous wastes. 2. Gas-phase photocatalytic oxidation of trichloroethylene over TiO_2: products and mechanisms. *Environ. Sci. Technol.* **27**, 723.

Nimlos, M.R., Wolfrum, E.J., Brewer, M.L. and Fennell, J.A. (1996) Gas-phase heterogeneous photocatalytic oxidation of ethanol: pathways and kinetic modeling. *Environ. Sci. Technol.* **30**, 3102–3110.

Nishimoto, M., Watanabe, R. and Moriya, K. *Japan Kokai Tokkyo Koho*, JP6411622.

Noguchi, T., Fujishima, A., Hashimoto, K. and Sawunyama, P. (1998) Photocatalytic degradation of gaseous formaldehyde using TiO_2 film. *Environ. Sci. Technol.* **32**(23), 3831–3833.

Obee, T.N. and Brown, R.T. (1995) TiO_2 Photocatalysis for indoor air applications: effects of humidity and trace contaminant levels on the oxidations rates of formaldehyde, toluene, and 1,3-butadiene. *Environ. Sci. Technol.* **29**, 1223–1231.

Ohgaki, S., Otaki, M., Katayama, H., Oguma, K. and Hirata, T. (2000) Inactivation of cryptosporidium and *E. coli* by UV-irradiation and by photocatalytic reaction. *Tokyo Daigaku Kogakubu Sogo Shikenjo Nenpo*, **58**, 65–70.

Okamoto, K., Ysamamoto, Y., Tanaka, H. and Itaya, A. (1985) Kinetics of heterogeneous photocatalytic decomposition of phenol over anatase TiO_2 powder. *Bull. Chem. Soc. Jpn.* **58**, 2023.

Pal, B., Nogami, G. and Sharon, M. (1999) Preparation and characterization of TiO_2/Fe_2O_3 binary mixed oxides and its photocatalytic properties. *Mate. Chem. Phys.* **59**(3), 254–261.

Park, D.R., Zhang, J., Ikeue, K., Yamashita, H. and Anpo, M. (1999) Photocatalytic oxidation of ethylene to CO_2 and H_2O on ultrafine powdered TiO_2 photocatalysts in the presence of O_2 and H_2O. *J. Catal.* **185**(1), 114–119.

Paz, Y., Luo, Z., Rabenberg, L. and Heller, A. (1995) Photooxidative self-cleaning transparent titanium dioxide films on glass. *J. Mater. Res.* **10**, 2842.

Peral, J. and Ollis, D.F. (1992) Heterogeneous photocatalytic oxidation of gas-phase organics for air purification: Acetone, 1-butanol, butyraldehyde, formaldehyde, and m-xylene oxidation. *J. Catal.* **136**, 554–565.

Peral, J. and Ollis, D.F. (1993) "In *Photocatalytic Purification and Treatment of Water and Air*" (eds. D.F. Ollis and H. Al-Ekabi), p. 741. Editores, Elsevier.

Peral, J. and Ollis, D.F. (1997) TiO_2 photocatalyst deactivation by gas-phase oxidation of heteroatom organics. *J. Molecul. Catal. A: Chem.* **115**, 347–354.

Peral, J., Domènech, X. and Ollis, D.F. (1997) Heterogeneous photocatalytic oxidation of gas-phase organics for air purification: acetone, 1-butanol, butyraldehyde, formaldehyde, and m-xylene oxidation. *J. Catal.* **70**, 117.

Phillips, L.A. and Raupp, G.B. (1992) Infrared spectroscopic investigation of gas-solid heterogeneous photocatalytic oxidation of trichloroethylene. *J. Molecul. Catal.* **77**, 297.

Pichat, P., Disdier, J., Hoang-Van, C., Mas, D., Goutailler, G. and Gaysse, C. (2000) Purification/deodorization of indoor air and gaseous effluents by TiO_2 photocatalysis *Catal. Today* **63**, 363–369.

Pilkenton, S., Hwang, S.J. and Raftery, D. (1999) Ethanol photocatalysis on TiO_2 coated optical microfiber, supported monolayer, and powdered catalysts: An in situ NMR study. *J. Phys. Chem. B: Mater. Surf. Interf. Biophys.* **103**(50), 11152–11160.

Raupp, G.B. (1995) Photocatalytic oxidation for point-of-use abatement of volatile organic compounds in microelectronics manufacturing. *J. Vacuum Sci. Technol. B, Microelect. Processing and Phenomena*, **13**, 1883–1887.

Romeas, V., Pichat, P., Guillard, C., Chopin, T. and Lehaut, C. (1999) Testing the efficacy and the potential effect on indoor air quality of a transparent self-cleaning TiO_2-coated glass through the degradation of a fluoranthene layer. *Ind. Eng. Chem. Res.* **38**(10), 3878–3885.

Romero, M., Sánchez, B. and Fabrellas, B. (1995) Photocatalytic destruction of toluene and xylene in air streams with a monolithic receiver. *European Workshop on Technologies for Environmental Protection*. Bilbao. España.

Romero, M., Blanco, J., Sánchez, B., Vidal, A., Malato, S. and Cardona, A.I. and García, E. (1999) Solar Photocatalytic degradation of water and air pollutants: challenges and perspectives. *Solar Energy* **66**(2), 169–182.

Sánchez, B., Romero, M., Sánchez, M. and Vidal, A. (1992) *Solar Destruction of Hazardous Wastes 6th International Symposium on Solar Thermal Concentrating Technologies*, Vol II Ed, Mojacar, *Ciemat*.

Sánchez, B., Romero, M., Cardona, A., Fabrellas, B., Avila, P. and Bahamonde, A. (1996). Photoassisted catalytic oxidation of trichloroethylene, toluene and xylene by using TiO_2 based monoliths in gaseous emissions. *Solar Thermal Concentrating Technologies 8th International Symposium Köln*, Germany. Vol. 3, 1489–11498.

Sánchez, B., Cardona, A.I. and Romero, M. (1998) Tratamiento fotocatalítico de compuestos orgánicos volátiles en fase gas. In *CYTED Ciencia y Tecnología para el desarrollo*. (ed.), pp. 73–79. Catalizadores y Adsorbentes para la Protección Ambiental en la Región Iberoamericana. *CYTED Subprog. V*.

Sánchez, B., Cardona, A.I., Romero, M., Avila, P. and Bahamonde, A. (1999) Influence of temperature on gas-phase photo-assisted mineralization of TCE using tubular and monolithic catalysts. *Catal. Today* **54**, 369–377.

Sánchez, B., Cardona, A.I. and Avila, P. (2000) Fotocatálisis, una tecnología ambientalmente correcta para la depuración de gases. Colegio Oficial de Físicos, Aproma, and Instituto de la Ingeniería de España. *V Congreso Nacional del Medio Ambiente*. Madrid.

Sato, S. (1983) Photo-Kolbe reaction at gas-solid interfaces. *J. Phys. Chem.* **87**, 3531.

Sauer, M.L. and Ollis, D.F. (1996) Catalyst deactivation in gas-solid photocatalysis. *J. Catal.* **163**, 215–217.

Sauer, M.L., Hale, M.A. and Ollis, D.F. (1995) Heterogeneous photocatalytic oxidation of dilute toluene-chlorocarbon mixture in air. *J. Photochem. Photobiol. A: Chem.* **88**, 169–178.

Sawunyama, P., Yasumori, A. and Okada, K. (1998) The nature of multilayered TiO_2-based photocatalytic films prepared by a sol-gel process. *Mater. Res. Bull.* **33**(5), 795–801.

Sczechowski, J.G., Koval, C.A. and Noble, R.D. (1993) Evidence of critical illumination and dark recovery times for increasing the photoefficiency of aqueous heterogeneous photocatalysis. *J. Photochem. Photobiol. A.* **74**, 273.

Sitkiewitz, S. and Heller. A. (1996). Photocatalytic oxidation of benzene and stearic acid on sol-gel derived TiO_2 thin films attached to glass. *New J. Chem.* **20**(2), 233–241.

Sopyan, I., Watanabe, M., Murasawa, S., Hashimoto, K. and Fujishima, A. (1996) Efficient TiO_2 powder and film photocatalysts with rutile crystal structure. *Chem. Lett.* 69–70.

Suzuki, K. (1993) Photocatalytic air purification on TiO_2 coated honeycomb support. In *Photocatalytic Purification and Treatment of Water and Air* **3**, 421–434.

Suzuki, K., Satoh, S. and Yoshida, T. (1999) *Oenki Kagaku*, **59**, 521.

Suzuki, K., Yoshimoto, T., Honda, K., Fujishima, A. and Baba, R. *Japan Kokai Tokkyo Koho*, JP60187322.

Tajima, M., Niwa, M., Fujii, Y., Koinuma, Y., Aizawa, R., Kushiyama, S., Kobayashi, S., Mizuno, K. and Ohuchi, H. (1997) Decomposition of chlorofluorocarbons on TiO_2—ZrO_2. *Appl. Catal. B: Environ.* **12**, 263.

Takeda, N., Ohtani, M., Torimoto, T., Kuwataba, S. and Yoneyama, H. (1997) Evaluation of diffusibility of adsorbed propionaldehyde on titanium dioxide-loaded adsorbent photocatalyst films from its photodecomposition rate *J. Phys. Chem. B.* **101**(14), 2644–2649.

Takeuchi, K. (1998) Purification and remediation of ambient air by photocatalysis. *Taiki Kankyo Gakkaishi*, **33**(3), 139–150.

Tanaka, T. and Isanaga, T. (1994) Photodegradation of chlorofluorocarbon alternatives on metal oxide. *Solar Energy* **52**, 447–450.

Upadhya, S. and Ollis, D.F. (1998) A simple kinetic model for the simultaneous concentration and intensity dependence of TCE photocatalyzed destruction. *J. Adv. Oxid. Technol.* **3**, 199–202.

Vigil, E., Saadoun, L., Rodriguez-Clemente, R., Ayllón, J.A. and Domenech, X. (1999) TiO_2 thin films grown using microwave-activated solutions. *J. Mater. Sci. Lett.* **18**, 1067–1069.

Vorontsov, A.V., Savinov, E.N. and Kurkin, E.N. (1999) Study of TiO_2 deactivation during gaseous acetona photocatalytic oxidation. *J. Catal.* **186**(2), 318–324.

Vorontsov, A.V., Stoyanova, I.V., Kozlov, D.V., Simagina, V.I. and Savinov, E.N. (2000) Kinetics of the photocatalytic oxidation of gaseous acetona over platinized titanium dioxide. *J. Catal.* **189**(2), 360–369.

Walker, A., Formenti, M. and Meriaudeau, P. Teichner S.J. (1977) Heterogeneous photocatalysis: photo-oxidation of methyl butanols. *J. Catal.* **50**, 237.

Watanabe, T., Kitamura, A., Kojima, E., Nakayama, C., Hashimoto, K. and Fujishima, A. (1993) Photocatalytic activity of TiO_2 thin film under room light. *Photocatal. Purificat. Treat. Water Air* **3**, 747–751.

Yamashita, H., Nishiguchi, H., Kamada, N., Anpo, M., Teraoka, Y., Hatano, H., Ehara, S., Kikui, K., Palmisano, L., Sclafani, A., Schiavello, M. and Fox, M.A. (1994) Photocatalytic reduction of CO_2 with H_2O on TiO_2 and Cu/TiO_2 catalysts. *Res. Chem. Interm.* **20**, 815–823.

Yamazaki-Nishida, S., Nagano, K.J., Phillips, L.A., Cervera-March, S. and Anderson, M.A. (1993) Photocatalytic degradation of trichloroethylene in the gas phase using titanium dioxide pellets. *J. Photochem. Photobiol. A: Chem.* **70**, 95.

Yamazaki-Nishida, S., Fu, X., Anderson, M.A. and Hori, K. (1996) Chlorinated byproducts from the photoassisted catalytic oxidation of trichloroethylene and tetrachloroethylene in the gas phase using porous TiO_2 pellets. *J. Photochem. Photobiol. A: Chem.* **97**, 175.

Yuan, C.S., Hsu, B.C., Wu, J.F. and Hung, C.H. (1999) Reaction products of gas-phase photocatalytic degradation of perchloroethylene over titanium dioxide (UV/TiO_2). p. *Anonymous Air and Waste 92nd Annual Meeting and Exhibition Proceedings*, Pittsburgh, PA (United States). Air and Waste Management Association.

Yumoto, H., Inoue, T., Li, S.J., Sako, T. and Nishiyama, K. (1999) Application of ITO films to photocatalysis. *Thin Solid Films* **345**(1), 38–41.

14

Removal of airborne particulate matter

L. Le Coq

14.1 INTRODUCTION

Air pollution comprises the presence of gaseous molecules and particulate matter in indoor or outdoor air at such concentrations that they have a noticeable negative effect on health (Bascom et al., 1996; Pope et al., 1995) or the environment (Boubel et al., 1994). Among air pollutants, particulate matter can have a wide range of anthropogenic or natural sources. Natural events generate primary particles usually greater than 10 μm in diameter (rock erosion, volcanic emissions, sea spray, ...). Anthropogenic activities generate smaller secondary particles which arise mainly from reaction sources (atmospheric reactions of gaseous emissions, industrial chemical reactions, ...) and from combustion sources (engines, power plants, ...). Primary or secondary particles exhibit very different physico-chemical and structural properties (Cachier et al., 2004; Marinoni et al., 2005).

Considering the different industrial activities, there is a variety of particle treatment systems available (Dickenson, 1997; Fayed & Otten, 1997). Thus, a collection process can be designed taking into account the clean gas specification, the operating conditions (such as gas flow), the effluent particle concentration and the gas properties (temperature, pressure, humidity). The air treatment process must respect the existing regulations in terms of particle emission, human exposure and explosion hazard.

14.2 DEFINITION – PARTICULATE POLLUTANT SOURCES AND IMPACTS

As far as environmental engineering is concerned, a particle can be defined as a solid or liquid material suspended in air or able to be stirred up upon exposure to an air flow. Particles can be classified according to their characteristic size, which can vary from 0.005 to approximately 100 μm (USEPA, 1996). For a given size, the particles of a class can have different chemical composition, depending on their source. As a result, different words exist to name, define and classify pollutant particles:

- *Total suspended particulates (TSP)*: According to the European Committee for Standardisation (CEN), TSP represent all airborne particles. The US definition takes into account all the particles with an aerodynamic diameter less than 40–50 μm, whereas the European standard considers aerodynamic diameters up to approximately 100 μm. TSP include particles entering the human body by the nose or the mouth, that is to say the inhalable fraction from a toxicological approach. Nevertheless, a large part of the particles constituting TSP are too big to enter the human respiratory system. As a result, this particle classification is not suitable to characterise the impact of particles on human health.
- *Particulate matter of aerodynamic diameter less than 10 μm (PM_{10})*: Based on the ability of PM_{10} to enter the human respiratory system, the US Environmental Protection Agency (USEPA, 1996) defined PM_{10} as the standard to characterise air quality. In this context, the European Community issued the 1999/30/CE Directive, which replaced TSP by PM_{10} measurements. It is not easy to compare TSP with PM_{10} measurements but USEPA considers that the PM_{10} weight fraction represents approximately 40–70% of TSP. The European Directive proposes another conversion relation: $TSP = 1.2\,PM_{10}$.
- *Particulate matter of aerodynamic diameter less than 2.5 μm ($PM_{2.5}$)*: The $PM_{2.5}$ classification contains those particles that penetrate human lung cells. Consequently, it is a relevant factor to describe air quality from a medical point of view (ISO, 1992).
- *Particulate matter of aerodynamic diameter less than 1 μm (PM_1)*: Although this particle fraction is measured and its sampling and effects on health are studied, PM_1 classification is not currently recommended by any legal standard or decree.
- *Dust*: Dust defines all the solid particles that have been stirred up in air as a result of various kinds of activity or disturbance. In fact, this definition represents the solid fraction of TSP.
- *Smog, smoke and mist*: Mist is formed by airborne liquid droplets. The droplet size is usually greater than 1 μm. Smoke is the results of aerosols from combustion source. Smog is a dense mixture of smoke, mist and other atmospheric pollutants (nitrogen oxides, tropospheric ozone, volatile organic compounds (VOCs), …) (Baron & Willeke, 1992).

- *Aerosol*: Aerosol defines solid or liquid airborne particles whose settling velocity is negligible. Aerosol and TSP lead to equivalent particle classification in ambient air at low wind velocity. Aerosols can contain dust, mist, smog and microbial particles.

14.2.1 Origin of air pollutant particles

Figure 14.1 shows that air pollutant particles can have very different particle size distributions, depending on their source (Dickenson, 1997). $PM_{2.5}$ are mainly secondary particles formed by the condensation and coalescence of gas and vapours. These fine particles are formed by the incomplete combustion in vehicle engines, wood fires, industrial activities and cigarettes. PM_{10} are mainly produced by physical phenomena such as erosion. Indoor coarse particles originate from floor dust, abrasive wear of surfaces and domestic human activities (movement, textile friction, cleaning activities). Nevertheless, a small fraction of coarse particles is measured in indoor air because of fast sedimentation. In an urban environment, the outdoor particles are mostly $PM_{2.5}$ and stem from diesel and fuel-injected engines.

Particles from industrial activities can be separated into two classes (Marinoni et al., 2005):

(1) Dust from engineering activities, also called primary particles. They are generated by operation units such as milling, crushing or bursting. They also stem from solid attrition during their supply or their use. Primary particles are mostly bigger than $PM_{2.5}$ and a very small fraction is smaller than 1 μm.
(2) Particles generated by chemical reaction or a change in thermal conditions are called secondary particles. They are formed by a change of phase occurring during a chemical reaction, gas condensation or liquid solidification. Secondary particles are mainly PM_1 but agglomeration or coalescence mechanisms can lead to an increase in particle size to form $PM_{2.5}$.

Particle size has a direct influence on indoor–outdoor mass transfer. Indeed, PM_{10} and $PM_{2.5}$ have very different residence times in air and travelling distances. $PM_{2.5}$ can flow

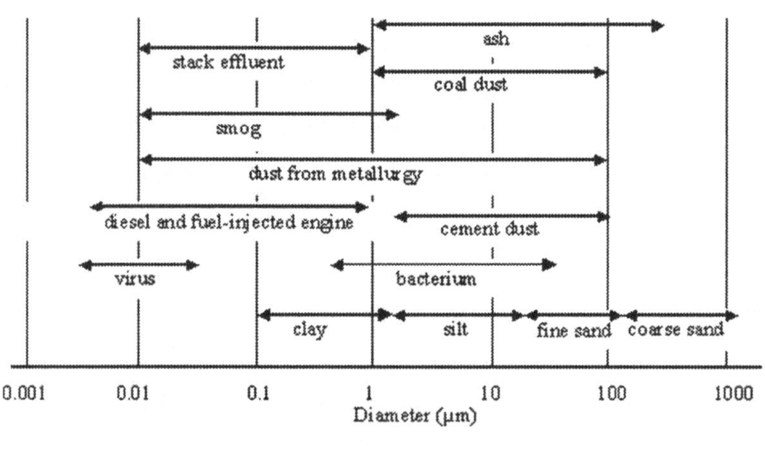

Figure 14.1 Particle size distribution according to source.

over 100–1000 km and can reach residence times in air of several months (Baron & Willeke, 1992). On the contrary, PM_{10} only remain suspended in air for a few hours. Fine particles flow in air according to draft and penetrate indoor air through heating ventilation and air conditioning (HVAC) systems. The dispersion of coarse particles is comparatively limited due to their high sedimentation velocity and so they show only a weak infiltration of indoor air.

14.2.2 Structure and physico-chemical properties

Particle classes depend on their structure and physico-chemical properties. These influence particle residence time in air as well as medical and environmental effects and thus treatment process design.

14.2.2.1 Size, shape and density of particles

In most cases, particle size is defined by its aerodynamic equivalent diameter. The definition of particle aerodynamic diameter is the sphere diameter of density $10^3\,kg\,m^{-3}$ having the same fall velocity as the considered particle for equal air conditions (temperature, pressure, relative humidity and velocity). The aerodynamic diameter depends strongly on particle shape (Kaye, 1999) and density.

14.2.2.2 Influence of particle composition

The chemical composition of particles varies according to the disposal source. Indeed, urban $PM_{2.5}$ (power generation and road traffic) are mainly composed of sulfate (SO_4^{2-}), nitrate (NO_3^-), organic compounds and carbon. Natural particles are mainly composed of metallic oxides (alumina, iron, silicon, titanium), pollens and oceanic compounds. Particles from industrial emissions such as coal combustion are mostly composed of three chemical classes:

(1) Polynuclear aromatic hydrocarbons (PAH) from unburnt organic carbon (1–90 wt% of particles according to combustion conditions).
(2) Mineral matter of coal that can escape with airborne particles (Si, Al, Ca, Fe).
(3) Heavy metals contained in coal can be vaporised in the burner to form metallic vapour. The vapour is nucleated in very fine aerosol particles (10–30 nm), which coagulate with each other or with dust to form $PM_{2.5}$.

Indoor airborne particles are composed of organic and inorganic solids and of chemical compounds in fractions depending on several factors, such as building location, room function, ventilation and heating system.

14.2.2.3 Influence of electrical resistivity and surface charge

Particle electrostatic charges have an influence on aggregation mechanisms and on particle capture in treatment systems involving electrostatic phenomena (Peukert & Wadenpohl, 2001).

In the same way, it is important to consider the particle electrical resistivity in the design of an electrostatic precipitator (ESP). Indeed, this parameter corresponds to particle resistance divided by particle size. ESPs are commonly used for dust capture when electrical resistivity varies between approximately 10^4 and $10^{11}\,\Omega$cm (Peukert & Wadenpohl, 2001).

Electrical resistivity and surface charge depend on particle chemical composition, moisture content and temperature. For example, the electrical resistivity of cement dust varies between 10^{11} and $10^{13}\,\Omega$cm according to the cement kiln process and design (Bapat, 2001). Fly ash electrical resistivity of power generation plant is related to burned fuel properties and vary in the range 10^8 up to $10^{14}\,\Omega$cm. An other example from glass industry leads to electrical resistivity for dust generated within the furnace ranging from 10^6 to $10^{13}\,\Omega$cm (Caputo et al., 1996).

14.2.3 Noxious effects on health and the environment

The characterisation of particle emission from industrial activities is regulated due to their noxious effects on health and the environment.

14.2.3.1 Effect on human health

Considering the toxicological aspects, pollutant particles have a different behaviour from gas due to their selective inhalation and the way they are deposited in the airway (Hester & Harrison, 1998). Consequently, particles have two possible noxious effects on human health:

(1) A first effect related to particle size:
- Coarse particles larger than 10 μm are stopped in the throat (pharynx) or the voice box (larynx) and are rejected.
- Particles between 3 and 10 μm can reach the trachea and bronchi.
- Fine particles between 0.01 and 3 μm become clogged in the lung alveoli and are responsible for most of the noxious effects of particles on health (Hester & Harrison, 1998).
- Particles smaller than 0.001 μm have the same behaviour as molecules so they can leave the airway during breathing.

(2) A second effect related to the toxicological, physico-chemical and microbiological properties:
- Particles are composed of hazardous chemical compounds such as heavy metals or PAH.
- Noxious effects depend on the crystallographic state and the surface of particles as well as the compound solubility.
- Particles can harbour pathogenic microorganisms, such as mildew, bacteria or viruses, as well as volatile organic compounds of microbial origin (VOCm).

14.2.3.2 Impact on the environment

In normal atmospheric conditions, particles can be transported through several tens to some hundreds of kilometres. As a result, airborne particles produced by human activities are not concentrated in industrial or urban areas, but can pollute residential areas (Boubel et al., 1994). The flight time and corresponding flight distance of particles depends on their size and their density as well as on meteorological conditions (wind, pressure and rain).

Acidic dust is produced by SO_2, NO_2 and chemical acid emissions (Fangmeijer et al., 1994; Krupa & Arndt, 1990). These particles are partly responsible for acid rain, which has a direct impact on the chemical equilibrium of an ecosystem: changes of metal solubility or induction of eutrophication.

Particles can contain heavy metals (such as Cd, Hg and Pb) as they can be adsorbed on solid particle surfaces (Fang et al., 2005).

Also, persistent organic pollutants (POP), which are micropollutants of high toxicity and recalcitrance, can be adsorbed onto dust surfaces (Boubel et al., 1994). The POP chemical class contains PAH and polychlorinated biphenyls (PCB), which are produced by organic matter or urban waste combustion.

14.3 MEASUREMENT OF PARTICLE SIZE DISTRIBUTION AND CONCENTRATION

Different systems for measuring particle concentration are available and can be used according to the particle characteristics to be determined and the use of the data: pollutant emission measurement and regulation, control of air treatment systems or security. Table 14.1 summarises the different systems for particle concentration measurement and their conditions of use.

Two classes of device for measuring particle size distribution and concentration can be distinguished (Kaye, 1999):

(1) Indirect measurement systems based on particle sampling onto a filter or impaction plate. Concentration measurement is then performed by weight, counting, or light obscuration. Particle size and shape can be characterised by microscopic examination coupled with image analysis. The major drawback of these systems is the measurement time and cost. On the other hand, they do allow accurate and non-destructive measurement of samples that can be stored.

(2) Direct measurement or continuous measurement systems, based on different physical principles such as piezoelectric gauge, beta gauge and electrical or optical systems. These allow a continuous and rapid characterisation of the particle concentration.

14.4 AIRBORNE PARTICLE TRANSPORT – THEORETICAL APPROACHES

Particle transport in air flow depends on different forces, which can change the particle flow path and velocity. Moreover, the particle treatment systems described in Part 14.5 are based on the action of some of these forces to capture particles. Thus, theoretical equations of forces are proposed as a basis for the design and optimisation of treatment systems.

Table 14.1 Particle concentration measurement systems.

Methodology	Conditions of use	Principle
Optical systems		
Opacimeter	High smoke and smog concentration.	Deviation measurement of the emitted and transmitted light intensity when passing through a given volume containing particles.
Photometer	Weight concentration.	Intensity measurement of diffused light inside a polluted air volume (monochromatic light).
Optical counter	Number concentration and size distribution for low concentration aerosol (0.1–40 μm).	Intensity measurement of light diffused by one particle (white or monochromatic light).
Condensation nucleus counter	Number concentration of submicron aerosol.	Optical counter with condensation device to allow size increase of particles formed by vapour condensation.
Continuous systems		
Piezoelectric gauge counter	Weight concentration of solid aerosol.	Oscillation frequency measurement of a particle collector charged with a mass of PM_{10}.
Beta gauge method	Weight concentration of solid aerosol.	β ray attenuation measurement when penetrating a mass of deposited particle.
Electrical sensor	Weight or number concentration.	Electrostatic classification.
Aerodynamic systems	Weight or number concentration and size distribution.	Measurement based on sedimentation, centrifugal process, cyclonic precipitation,impaction or particle acceleration.

14.4.1 Particle falling velocity

Considering a spherical solid particle in a non-moving fluid, the particle equation of motion (Equation (14.3)) depends on the drag force (Equation (14.1)) exerted by the fluid on the particle and the force of acceleration (Equation (14.4)) due to gravity and buoyancy. Buoyancy is the upward force acting on an object immersed in a fluid, enabling it to float or at least to appear lighter.

The fluid drag force is thus defined as (Baron & Willeke, 1992):

$$\vec{F}_{t} = C_{d}\frac{\rho_{f}}{2}\vec{U}_{p}^{2}\frac{\pi d_{p}^{2}}{4} \tag{14.1}$$

with:

U_p: particle velocity ($m\,s^{-1}$)
d_p: particle aerodynamic diameter (m)
F_t: fluid drag force ($m\,kg\,s^{-2}$)
ρ_f: fluid density ($kg\,m^{-3}$)
C_d: drag coefficient.

C_d depends on the particle Reynolds number (Re_r) for a relative particle velocity, which is defined as the particle to fluid velocity difference. The drag coefficient of the particle can be expressed thanks to the Brauer equation for a solid sphere and a flow regime ranging from Stokes viscous flow to Newton turbulent flow ($Re_r = 5 \cdot 10^5$) (Midoux, 1993):

$$C_{d} = \frac{24}{Re_{r}} + \frac{3.73}{\sqrt{Re_{r}}} - \frac{4.83 \cdot 10^{-3}\sqrt{Re_{r}}}{1 + 3 \cdot 10^{-6} Re_{r}^{3/2}} + 0.49 \tag{14.2}$$

The equation of motion for an isolated particle with diameter d_p in a non-moving fluid is then:

$$\frac{\pi d_{p}^{3}}{6}\rho_{p}\frac{d\vec{U}_{p}}{dt} = \vec{F}_{gA} + \vec{F}_{t} \tag{14.3}$$

$$\vec{F}_{gA} = \frac{\pi d_{p}^{3}}{6}(\rho_{p} - \rho_{f})\vec{g} \tag{14.4}$$

with:

ρ_p: particle density ($kg\,m^{-3}$)
F_{gA}: resultant force of gravity force and buoyancy ($m\,kg\,s^{-2}$)
g: acceleration of gravity ($m\,s^{-2}$).

The particle fall velocity is obtained for $\frac{dU_p}{dt} = 0$. For Stokes flow ($Re_r < 1$), it is equal to:

$$U_{p}^{\lim} = \frac{(\rho_{p} - \rho_{f})d_{p}^{2}}{18\mu} \tag{14.5}$$

with:

μ: fluid dynamic viscosity (Pa s).

14.4.2 Forces acting on particles in a moving fluid

Particle flow paths in an air flow can be calculated according to the Basset–Boussinesq–Oseen (BBO) equation (Hjelmfelt & Mockros, 1966):

$$\frac{\pi d_p^3}{6}\rho_p\frac{d\vec{U}_p}{dt} = \vec{F}_{gA} + \vec{F}_t + \vec{F}_{ma} + \vec{F}_p + \vec{F}_m \tag{14.6}$$

with:
F_{ma}: virtual mass force (m kg s^{-2})
F_p : pressure gradient force (m kg s^{-2})
F_m : Basset force (m kg s^{-2}).

The drag force F_t represents the viscous friction of the moving fluid on the moving particle:

$$\vec{F}_t = C_d\frac{\rho_f}{2}(\vec{U}_f - \vec{U}_p)^2\frac{\pi d_p^2}{4} \tag{14.7}$$

with:
U_f: fluid velocity (m s^{-1})

The virtual mass force F_{ma} corresponds to the fluid mass that is moving due to particle motion:

$$\vec{F}_{ma} = \frac{\pi d_p^3}{12}\rho_f\left(\frac{d\vec{U}_f}{dt} - \frac{d\vec{U}_p}{dt}\right) \tag{14.8}$$

The pressure gradient force F_P is the result of pressure change (P) at the particle surface due to fluid acceleration:

$$\vec{F}_p = \frac{\pi d^3}{6}\,\vec{\text{grad}}\,P \tag{14.9}$$

The Basset history force represents the damping of fluid motion considering a particle acceleration horizon (after a particle movement):

$$\vec{F}_m = -\frac{3}{2}d^2\sqrt{\pi\rho_f\mu}\int_0^t\frac{dU_p - dU_f}{d\tau}\frac{d\tau}{\sqrt{t-\tau}} \tag{14.10}$$

This equation of particle motion neglects particle collisions with other particles or a wall, particle rotation effects as well as Brownian diffusion and electrostatic effects.

It is possible to add to the BBO equation of motion an electrostatic force, which cannot be neglected when considering particle capture on an electret (dielectric insulating polymer) fibrous filter or ESP. This force can take into account all the electrostatic forces acting on a particle, such as Coulomb force or dielectric forces (Bottner,

2003; Wang, 2001). In the specific case of a charged particle in the vicinity of a polarised collector (fibre or electrode), the electrostatic force (F_e) is:

$$\vec{F}_e = q\vec{E} \tag{14.11}$$

with:
F_e: Coulomb electrostatic force ($m\,kg\,s^{-2}$)
q: charge of the particle ($A\,s\,m^{-2}$)
E: electrical field ($kg\,m\,A^{-1}s^{-3}$).

14.4.3 Sphericity correction

The equation of particle motion (Equation (14.6)) is obtained for spherical particles. When the considered particles are not spheres, particle shape orientation in the air flow has an influence on drag force. This can be done via the particle sphericity index, that is the ratio of the surface area of the sphere having the same volume as the particle to the surface area of the particle. This index is a correction term that can be added into the equation of motion (Equation (14.6)). The smaller the index is, the smaller the sedimentation velocity is.

14.4.4 Influence of particle concentration

When the distance between two particles is of the same magnitude as the aerodynamic diameter ($<10\,d_p$), particle interactions cannot be neglected. In most real situations, the equation of particle motion (Equation (14.6)) has to be modified for particle concentrations higher than $5\%_{vol.}$. To take into account particle–particle interactions due to electrostatic and hydrodynamic phenomenon, an interparticle force can be added to the equation of motion (Kim & Zydney, 2005).

14.5 AIRBORNE PARTICLE TREATMENT SYSTEMS

The system specifications for airborne particle treatment can be very different depending on the area of application: industrial or domestic, indoor air or gaseous effluent. In the following sections, the most commonly used systems for airborne particle collection are presented in terms of implementation, design and operating conditions.

14.5.1 Fibrous filter media

Porous media for airborne particle filtration can be divided into three classes:

(1) Packed beds of granular matter such as sand. These are reliable, robust and cheap but provide a low filtration efficiency for fine particle collection.
(2) Consolidated porous media such as sintered ceramic. These are only used for specific applications due to their prohibitive pressure drops.
(3) Fibrous filter media are widely used in different filtration devices as explained in the following paragraphs.

Figure 14.2 Scanning electron microscopic photographs of different fibrous filter media: woven fibrous cloth (A), melt-blown (B) and glass fibre paper (C).

14.5.1.1 Operating conditions and principles

The two major filtration devices using fibrous media are bag filters and cassette or cartridge filters. These systems allow batch or continuous filtration. Concerning continuous filtration, several filtration devices are used in parallel configuration to allow dust unloading of some of the filters (by vibration/back blowing or compressed air) during air treatment.

Fibrous filter media have different structures (Figure 14.2):

- Woven fibrous media such as cloths or knitted media
- Non-wovens such as felts or melt-blowns
- Technical papers.

According to the performance of the fibrous media treatment, the filtration device can cover a wide range of filtration efficiencies and areas of application:

- *ULPA/HEPA filters (ultra low penetration air/high efficiency particulate air)*: Used in pharmacy, electronics, hospitals, handling of hazardous substances and foodstuff industry.
- *Fine filters*: Used in HVAC and industry as main filters or as intermediate filters for higher-class filters.
- *Coarse filters*: Used in common HVAC applications and industrial gaseous effluent as main filters or pre-filters.

The different classes of fibrous media can lead to two major filtration mechanisms: surface filtration, with the building of a filtration cake during clogging and deep filtration. For surface filtration (coarse and fine filters), clogged fibrous media can easily be cleaned by cake removal. In the case of deep filtration (ULPA/HEPA and some fine filters), captured particles are eliminated with the clogged fibrous media.

Bag filters Polluted air flows through fibrous media contained in a cylinder or a cube. Particles are captured on the fibrous media, so the pores are slowly clogged (Figure 14.3). This clogging improves the global filtration efficiency but increases pressure drops. When a maximal pressure drop is reached, the fibrous media must be

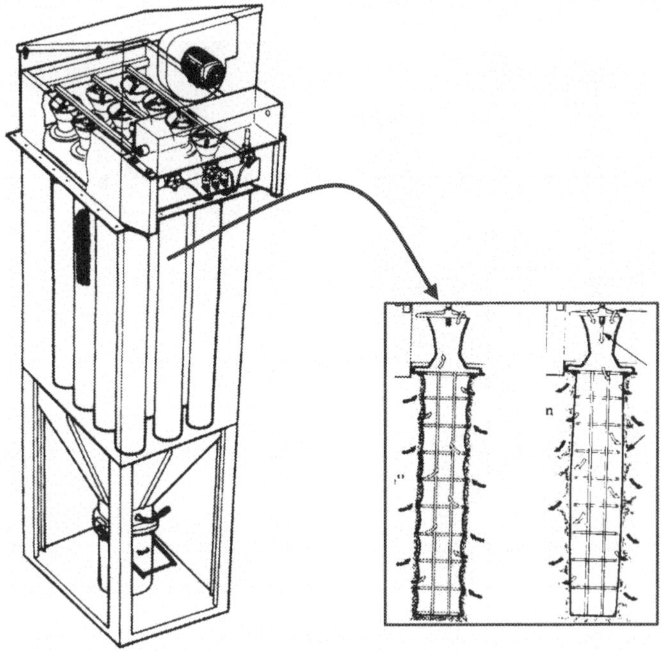

Figure 14.3 Bag filters: operation, clogging and dust unloading.

unloaded by back blowing or pressure surge (Calle et al., 2002). These filtration devices are widely used for dust collection during powder conveying and handling.

Cartridge filters Cartridge filters are formed by pleated fibrous media in a cylindrical or planar configuration (Figure 14.4). This filtration device allows a high filtration surface to be achieved for a small required volume and for lower pressure drops than bag filters. Indeed, the pleats enable an increase of the filtering surface in a given volume and a decrease of the air frontal velocity, which leads to a pressure drop decrease. Pleating parameters (height, angle and spacing of pleats) must be calculated in order to optimise the pressure drop and the particle capture efficiency (Del Fabbro et al., 2002).

14.5.1.2 Design of filters

For filter design, pressure drop and filtration efficiency models are used. Among the different works dealing with pressure drop versus filter clogging, captured particles are considered to change the structural parameters of the fibrous media. Thus, fibrous media can be modelled as a serial association of underlayers, each one homogeneously clogged with a given amount of particles (Thomas et al., 1999). Most filtration efficiency models are based on these capture mechanisms (Figure 14.5):

- Brownian diffusion
- Direct interception
- Inertial impaction

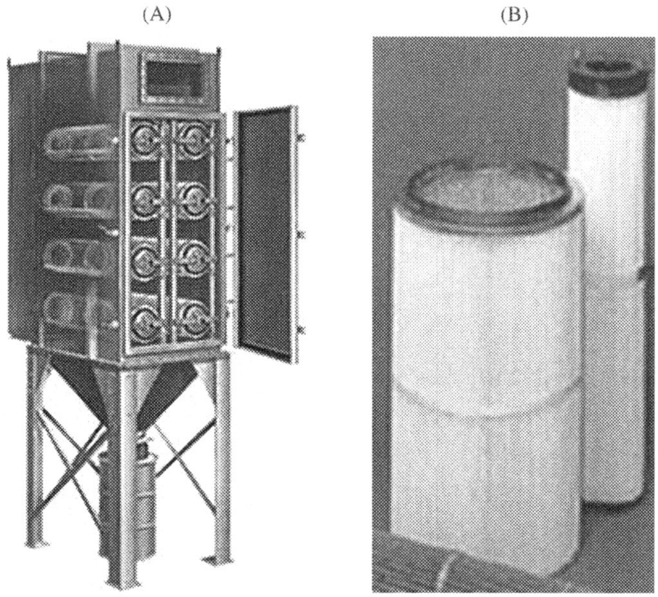

Figure 14.4 (A) Cartridge filter device (Met-Pro Corporation) and (B) cartridges (SF-Filtres).

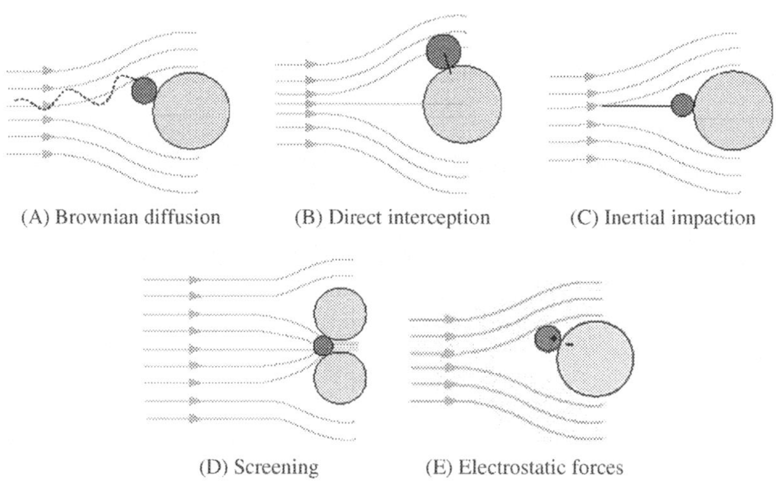

Figure 14.5 Particle capture mechanisms onto a fibre (Brown, 1993).

- Screening
- Electrostatic forces.

Brownian diffusion Airborne particles smaller than the gas mean free path (λ = 0.067 μm for air) can collide with gas molecules and consequently can randomly change their flow path. This phenomenon results in an increased probability of particle collision with a collector (Figure 14.5A). Particle capture due to Brownian diffusion is characterised by the Peclet number (*Pe*), which is the particle flow ratio between convection and diffusion:

$$Pe = \frac{d_p U_0}{D_{diff}} \tag{14.12}$$

For a sphere:

$$D_{diff} = \frac{kTCu}{3\pi\mu d_p} \tag{14.13}$$

with:

U_0: fluid superficial velocity ($m\,s^{-1}$)
D_{diff}: Brownian diffusion coefficient ($m^2 s^{-1}$)
k: Boltzmann constant ($J\,K^{-1}$)
T: temperature (K)
Cu: Cunningham correction factor
Kn_p: particle Knudsen number, $Kn_p = \frac{2\lambda}{d_p}$.

Cu considers flow unsteadiness due to the same magnitude of particle size and gas mean free path:

$$Cu = 1 + Kn_p \left(1.207 + 0.44 \exp\left(\frac{-0.78}{Kn_p}\right)\right) \tag{14.14}$$

Direct interception Fine particles are usually light and so inertia forces can be neglected. As a result, fine particles follow the air-stream lines while passing round collectors. Direct interception of a particle occurs if the distance between the particle centre and the collector is less than the particle radius (Figure 14.5B). The characteristic dimensionless number (*R*) is the diameter ratio of particle (d_p) to fibre (d_f):

$$R = \frac{d_p}{d_f} \tag{14.15}$$

Inertial impaction Coarse particles are heavy and their flow path is influenced by inertia forces. These act on them, and induce sharp acceleration or change of direction of the particle. Then, particles can pass in the vicinity of a fibre surface and can be captured by impaction (Figure 14.5C). The inertial impaction mechanism is characterised by the Stokes number (*St*), which is defined as the ratio of the stopping distance of a particle to a characteristic dimension of the obstacle (d_f):

$$St = \frac{Cu\rho_p d_p^2 U_0}{18\mu_g d_f} \tag{14.16}$$

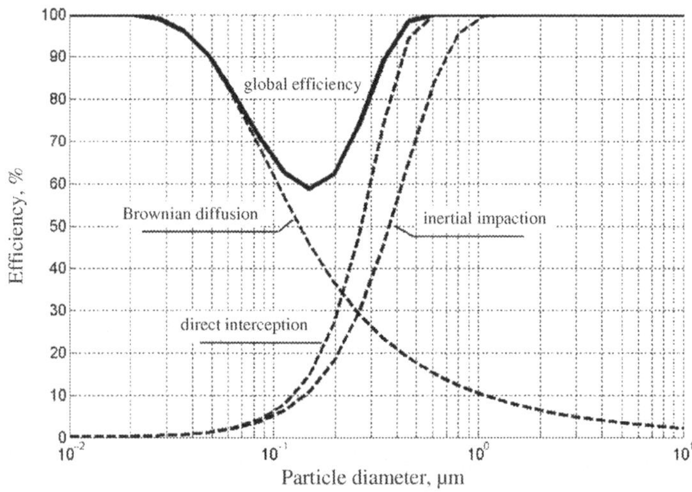

Figure 14.6 Single fibre efficiencies versus particle size.

Screening Screening is the major capture mechanism for particles greater than the fibrous medium pore size.

Electrostatic forces Particles with a surface charge can be captured on fibres of opposite charge. The electrostatic force mechanism is widely used to improve the filtration efficiency of particles too small to be captured by inertial impaction and too big to collide under Brownian diffusion effects. These particles correspond to the minimum of fractional collection efficiency curve and their corresponding size is called most penetrating particle size (MPPS). To do so, electret filter media made of charged synthetic fibres allow an increasing initial filtration efficiency of an unused filter. The main drawback of this filtration system concerns the rapid decrease in filtration efficiency with media clogging.

14.5.1.3 Filtration efficiency

As shown in Figure 14.6, the main efficient particle capture mechanisms are Brownian diffusion, direct interception and inertial impaction (Brown, 1993). Operating conditions such as gas velocity, particle size and particle density have an influence on these capture mechanisms (Table 14.2).

Filtration efficiencies at the level of one fibre, and considering each capture mechanism separately, are called single fibre efficiencies. The global single fibre efficiency (η) is calculated from theoretical or semi-empirical models of single fibre efficiencies for each capture mechanism (Brown, 1993; Davies, 1973). It is considered that a particle that collides with a fibre is captured under Van Der Waals, electrostatic or intermolecular forces. As far as deep filtration onto non-woven fibrous media is concerned, the initial filtration efficiency (E) is calculated as follows:

$$E = 1 - \exp\left(-\frac{4\alpha Z\eta}{\pi(1-\alpha)d_f}\right) \tag{14.17}$$

Table 14.2 Influence of operating conditions on capture mechanisms.

	Air velocity ↗	Particles	
		Size ↗	Density ↗
Diffusion	↘	↘	0
Interception	0	↗	↘
Inertia	↗	↗	↗

↗ Increasing efficiency for an operating parameter increase.
↘ Decreasing efficiency for an operating parameter increase.
0 No effect on efficiency for an operating parameter increase.

with:
E: initial filtration efficiency
α: medium packing fraction
Z: medium thickness (m)
η: single fibre efficiency
d_f: fibre diameter (m).

For the specific case of a woven fabric made of multi-fibre yarns, the initial filtration efficiency can be estimated from a model considering two continuous pore size levels: the inter-yarn pore level and the intra-yarn pore level (Figure 14.7) (Benesse et al., 2006).

As multi-fibre yarn is considered as a homogeneous porous medium, yarn efficiency ($E_{\text{intra-yarn}}$) depends on the single fibre efficiency (η_f) (Equation (14.20)). η_f represents the efficiency of one single fibre contained in a yarn and is calculated from Equation (14.21) considering the particle capture mechanisms by diffusion ($\eta_{d,f}$), impaction ($\eta_{i,f}$) and interception ($\eta_{r,f}$). At the filter level, the cloth is formed by a continuous weaving of yarns, which are considered as monofilament collectors. In the same way as for multi-fibre yarn, cloth efficiency (E) depends on single monofilament yarn efficiency (η_{yarn}) (Equation (14.18)). η_{yarn} is calculated from single monofilament yarn efficiencies for each capture mechanism added to $E_{\text{intra-yarn}}$ (Equation (14.19)) (Benesse et al., 2006):

$$E = 1 - \exp\left(-\frac{4\alpha_{\text{filter}} Z \eta_{\text{yarn}}}{\pi(1 - \alpha_{\text{filter}}) d_{\text{yarn}}}\right) \quad (14.18)$$

$$1 - \eta_{\text{yarn}} = (1 - \eta_{\text{d,monofil}})(1 - \eta_{\text{i,monofil}})(1 - \eta_{\text{r,monofil}})(1 - \eta_{\text{intra-yarn}}) \quad (14.19)$$

$$E_{\text{intra-yarn}} = 1 - \exp\left(-\frac{4\alpha_{\text{intra-yarn}} d_{\text{yarn}} \eta_f}{\pi(1 - \alpha_{\text{intra-yarn}}) d_f}\right) \quad (14.20)$$

$$1 - \eta_f = (1 - \eta_{d,f})(1 - \eta_{i,f})(1 - \eta_{r,f}) \quad (14.21)$$

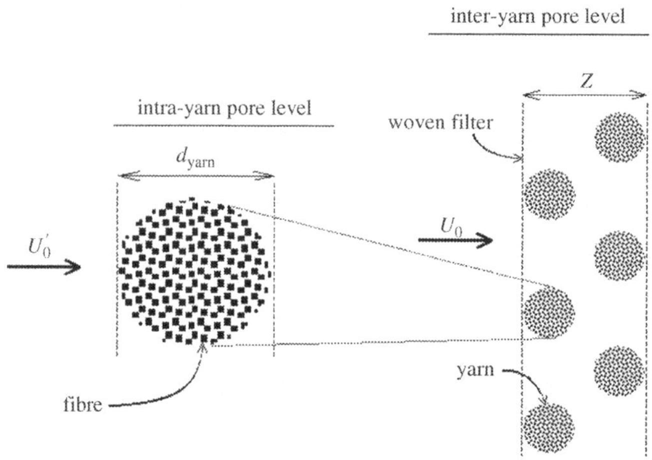

Figure 14.7 Transversal cross-section of woven media made of multi-fibre yarns.

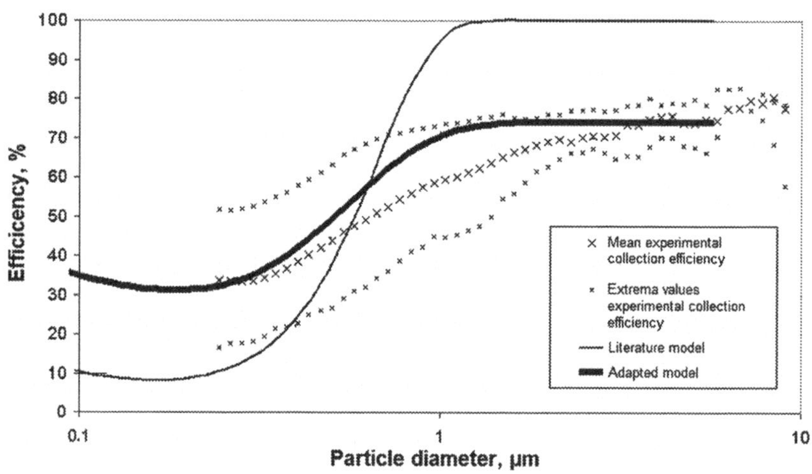

Figure 14.8 Woven cloth filtration efficiency versus PM size (Benesse et al., 2006).

Figure 14.8 shows the experimental and predicted initial filtration efficiencies versus particle size of a mat cloth made of activated carbon fibre twisted into yarn (d_f = 14 µm, d_{yarn} = 200 µm and U_0 = 0.5 m s^{-1}). The filtration efficiency calculated with the present model allows a good prediction of experimental values compared with classical model from literature developed for homogeneous fibrous media.

14.5.1.4 Qualification of filter

The European Committee for Normalisation (CEN) has proposed a standard for filter qualification (Eurovent 4/9, 1997).

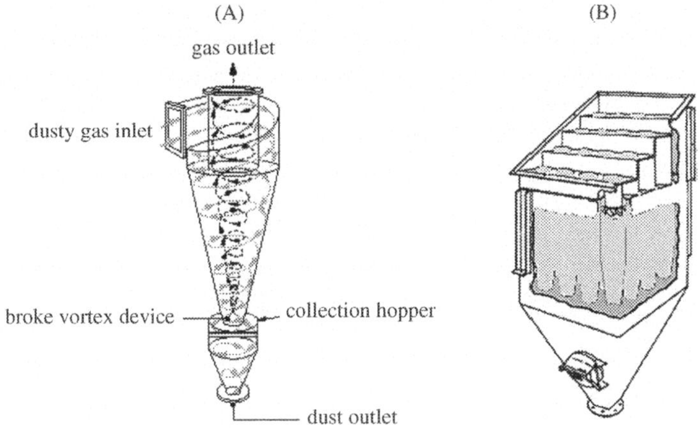

Figure 14.9 Lateral entry cyclone (A) and multi-cyclone (B).

14.5.2 Cyclones

14.5.2.1 Operation principle

Cyclones are used to remove mist or particles from air by centrifugal, inertia and gravity forces. The gas/particle mixture flows along the wall and, at the bottom of the cyclone, gas and coarse particle flow paths separate. Gas streams along the cyclone spindle and fine dust flows away through the chimney while the main part of the coarser dust is dragged off the particle collecting vent hole. Cyclones are efficient (>95%) for capturing particles bigger than about 10 μm. For dust streams with particles greater than several hundred microns, a settling chamber must be preferred due to the lower installation and operating costs compared to cyclones. The main advantages of this particle separation process for dust sizes ranging from 5 to 200 μm is robustness, low cost and easy implementation and handling for a wide range of operating conditions, that is temperature, gas pressure and particle concentration (Fayed & Otten, 1997). In favourable operating conditions (high gas velocity and particles greater than 5 μm), cyclones can reach 99% filtration efficiency. Nevertheless, for high gas velocity conditions, special precautionary measures must be taken to avoid cyclone wall erosion and particle attrition, which are responsible for decreases in efficiency and wear of process equipment (Da Silva et al., 2003).

14.5.2.2 Operating conditions

Although many different types of cyclone have been built for industrial gas cleaning, two main classes can be distinguished: reverse-flow and multi-flow cyclones.

Reverse flow and straight flow cyclones Figure 14.9A shows a tangential entry cyclone, which is the most common reverse flow system. After entering the cyclone, the gas forms a vortex with high tangential velocity, which gives particles high centrifugal forces for low pressure drops. The shape of the entry causes the gas to spin. Optional shapes of the entry such as swirl vane or scroll can be applied (Fayed & Otten, 1997).

Straight-through cyclones have an axial entry. The dusty gas enters at one end and begins to spin due to a swirl vane entry shape. Within the cyclone, centrifugal forces push the particles to the wall. Gas and separated particles exit separately at the opposite side of the entry. The swirl vane entry shape causes higher pressure drops than the tangential entry cyclone for the same separation performance. Values for the cyclone pressure drop range from approximately 0.1–2 kPa. There is a trade-off between separation efficiency and pressure drop, high efficiency being associated with a high pressure drop.

Multi-cyclones Multi-cyclones are composed of parallel cyclones of small sizes (Figure 14.9B). Small diameters enable higher separation efficiencies and smaller threshold particle sizes to be separated (minimum size of 1 μm) than mono-cyclones. Multi-cyclones must be carefully implemented to ensure a good flow distribution in the cyclone chambers.

14.5.2.3 Design

Cyclone design is based on complex equations for material and force balances. Nevertheless, two major simplified design approaches can be used:

(1) The first one gives standard design specifications based on tables listing gas inlet velocity and cyclone diameter as a function of particle size range and separation efficiency. The other design parameters (entry shape size, cyclone cylinder and cone heights) are then calculated from the cyclone diameter (Fayed & Otten, 1997; Perry, 1997). Among the different standard design methods, the Stairmand one is widely used (Stairmand, 1951).
(2) The second design method uses dedicated software for the design and optimisation of chemical engineering operation units. Among the available products, ASPEN PLUS (Aspen Tech) or PROSIM (Prosim SA) software can be used to perform cyclone design and design parameter optimisation for known gas operating conditions and for a specified separation efficiency.

14.5.3 Electrostatic precipitators

14.5.3.1 ESP operation

Airborne particles are characterised by an electrical charge, partly due to their chemical composition and the electrostatic phenomena related to frictional forces generated by the air flow. This electrical charge can have a negative impact on particle capture performance. Indeed, charged particles can lead to uncontrolled bonding on the operation unit walls or ducts of an industrial plant. It can result in contamination of the finished product (like in an agro-business plant) or give explosive dust problems.

To avoid this, the electrostatic precipitator concept is based on a controlled electrical charge of the particle. For this, dust is charged by gas-controlled ionisation under a corona discharge (high negative voltage of several tens of kV). Charged particles are then captured on a collector surface passing through the electrical field generated by the discharge electrodes. Dust unloading is performed by vibration, beating or washing of the collector plates. The ESP process enables a collection efficiency of more than 99.5% to be reached for all particle size ranges (Fayed & Otten, 1997).

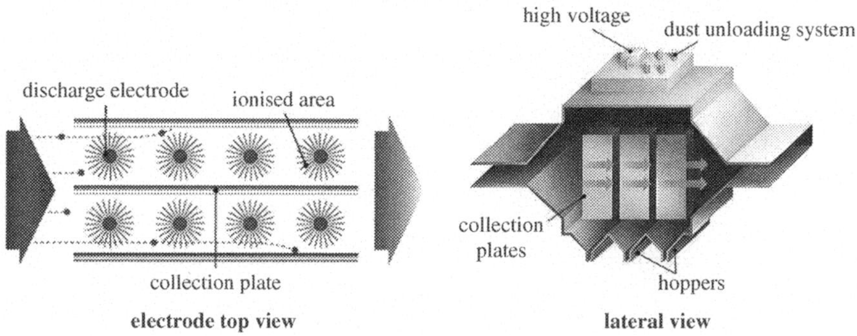

Figure 14.10 Electrostatic filter (Powerspan Corporation).

The ESP process can take various shapes depending on the discharge electrode shape, the unloading techniques or a possible combination with other particle capture mechanisms such as a venturi scrubber (Figure 14.10).

The ESP efficiency depends on the particle size and presents a minimum for 0.2–0.5 μm aerodynamic diameter range. Indeed, for these particle sizes, Brownian diffusion and inertial impaction are negligible mechanisms of particle–ion impacts.

14.5.3.2 ESP design

ESP design can be performed using the simple Deutsch–Anderson equation, which gives a relationship between the collection efficiency (E) and the collector surface for the given operating conditions (Fayed & Otten, 1997):

$$E = 1 - \exp\left(-\frac{A}{U_0}\right)^{W} \tag{14.22}$$

with:

E: collection efficiency

A: collector surface (m^2)

U_0: gas flow rate through the ESP ($cm\,s^{-1}$)

W: migration velocity ($m\,s^{-1}$) represents the particle velocity in the vicinity of the collector surface and under the influence of the electrical fields.

Another design approach is to use the design software for chemical engineering already presented in 14.5.2.3. In this case, operating conditions, particle properties (size and dielectric constant) and required collection efficiency must be fixed. Some empirical rules based on expertise are also useful for ESP design purposes.

14.5.4 Wet scrubbers

Wet scrubbers are used for dust separation as well as absorption of chemical compounds (SO_2 or NH_3) and VOC/odour biosorption. Among the different existing technologies, packed column and spraying washers are mostly used (Fayed & Otten, 1997).

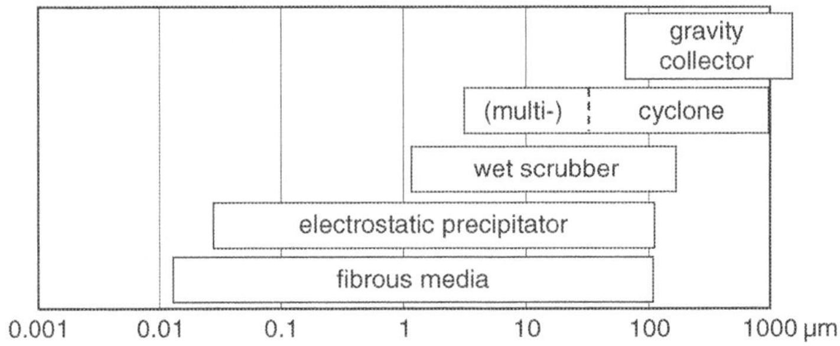

Figure 14.11 Separation process efficiency versus particle size.

The collection efficiency of a packed column increases when the packing is structured. Spray scrubbers include open spray towers and cyclone spray towers. The choice of the scrubber type depends on the available space and the dust/scrubber combination, which has an effect on plugging and corrosion, as well as liquid and solid handling. For a specific application, such as a paint workshop, a water curtain allows workers to be efficiently protected from paint aerosols.

14.5.5 Comparison

Figure 14.11 presents a classification of separation processes according to the airborne particle size (Dickenson, 1997; Fayed & Otten, 1997). Settling chambers are often used because of their low cost for dust diameters of more than some hundred microns. Cyclones are unsuitable for gas de-dusting when the particle size is less than about 5 μm, as efficiency sharply falls off for smaller sizes. When dust is sticky or hygroscopic, a wet scrubber is often chosen as these particles can stick to the cyclone walls and are thus not discharge in the collection hopper. Airborne particles of a diameter less than 1 μm are efficiently removed from air using fibrous media filters or electrostatic precipitators.

14.6 EXAMPLES OF INDUSTRIAL APPLICATIONS

Airborne particle collection technologies differ according to the industrial activity and the treatment process application (effluent treatment, industrial air de-dusting, ventilation and air-conditioning systems). This is due to historical reasons as well as to particular specifications such as temperature, pressure, chemical resistance and space available.

14.6.1 Industrial effluent gas treatment

Table 14.3 presents a classification of air treatment processes according to industrial activity.

14.6.2 Indoor air treatment

The indoor air of industrial premises, offices, public places or vehicles is commonly ventilated and air-conditioned. Air treatment processes are used in order to obtain

Table 14.3 Particle separation processes classified according to industrial activity.

	Fibrous media	Cyclone	ESP	Wet scrubber
Siderurgy, cement plant	++		++	
Waste incineration plant	++		++	
Paper, wood and glass plant		++	++	
Agro-business plant	++			+
Chemical plant	++			
Petrochemical plant		+	++	
Paint, varnish workshop	++			++

Table 14.4 Filter media classified according to indoor air application.

	Coarse filter	Fine filter	HEPA/ULPA
Indoor comfort, office, standard pollution workshop	++		
Microengineering, pharmaceutical plant, hospital, indoor air of vehicle, public place, office		++	
Clean room, operating block, nuclear plant			++

clean industrial surrounding air, ensure people's safety, improve air quality for people's comfort and to reduce operating costs. Air quality influences the cleanliness of HVAC systems and consequently thermal and aeraulic performances.

14.6.2.1 Airborne particle treatment processes classified according to application

For indoor air treatment applications, filter media are widely used and a classification of the efficiency of filter media is presented in Table 14.4.

14.6.2.2 Process intensification: combined treatment of inert particles and microbial aerosols/VOCs and odours

Considering building and set materials (suite, wall and floor furniture), insulation improvement and mean time spent indoors (80–90%), HVAC systems must combine particle and molecule treatment. The filtration processes for particulate pollutants must be efficient with regard to microbial aerosols (fungus, spore, bacterial and virus) as allergy diseases and virus and microbial transmission in indoor air increase. Thus, low cost treatment systems able to eliminate efficiently all these pollutants have to be developed. Among the possible solutions, an interesting way forward is to consider complex fibrous media to filter and treat these pollutants (Lorimier et al., 2004).

14.7 CONCLUSION

Airborne particulate matter produced by industrial activities can have noxious effects on human health due to their size and their chemical composition. In order to reduce the aerosol content in industrial gas or in indoor air (industrial or public), fibrous filter media and electrostatic precipitators are effective processes to remove secondary particles of small size. Wet scrubbers and cyclones are low cost processes which can be easily implement to collect particles bigger than 5 μm.

REFERENCES

Bapat, J.D. (2001) Application of ESP for gas cleaning in cement industry – with reference to India. *J. Hazard. Mater.* **81**(3), 285–308.

Baron, P.A. and Willeke, K. (1992) *Aerosol Measurement: Principles, Techniques, and Applications*, 2nd edn, John Wiley & Sons, ISBN 0471284068.

Bascom, R., Bromberg, P.A., Costa, D.A., Devlin, R., Dockery, D.W., Frampton, M.W., Lambert, W., Samet, J.M., Speizer, F.E. and Utell, M. (1996) State of the art: health effects of outdoor air pollution (Parts 1 and 2). *Am. J. Resp. Crit. Care* **153**, 3–50, 477–498.

Benesse, M., Le Coq, L. and Solliec, C. (2006) Collection efficiency of a woven filter made of multifiber yarn: experimental characterization during loading and clean filter modeling based on a two-tier single fiber approach. *J. Aerosol Sci.* **37** (in press).

Bottner, C.U. (2003) The role of the space charge density in particulate processes in the example of the electrostatic precipitator. *Powder Technol.* **135–136**, 285–294.

Boubel, R.W., Fox, D.L., Turner, D.B. and Stern, A.C. (1994) *Fundamentals of Air Pollution*, 3rd edn, Academic Press. ISBN 0121189309.

Brown, R.C. (1993) *Air Filtration, An Integrated Approach to the Theory and Applications of Fibrous Filters.* Pergamon Press. ISBN 0080412742.

Cachier, H., Aulagnier, F., Sarda, R., Masclet, P., Besombes, J.L., Marchand, N., Despiau, S., Croci, D., Mallet, M., Laj, P., Marinoni, A., Deveau, P.A., Roger, J.C., Putaud, J.P., Van Dingenen, R., Dell'Acqua, A., Viidanoja, J., Martins-Dos Santos, S., Liousse, C., Cousin, F., Rosset, R., Gardat, E. and Galy-Lacaux, C. (2004) Aerosol studies during the ESCOMPTE Experiment: an overview. *Atmos. Res.* **74**, 547–563.

Calle, S., Bemer, D., Vaxelaire, X., Thomas, D., Contal, P. and Leclerc, D. (2002) Change in filtration efficiency during dust cake removal from fabric filters. *World Congress on Particle Technology 4*, Sydney, Paper 40 CD-Rom.

Caputo, C., Giacchetta, G. and Pelagagge, P.M. (1996) Economical comparison of conventional and pulsed electrostatic precipitators in industrial applications. *Proceedings of the 6th International Conference on Electrostatic Precipitation*, Budapest.

Da Silva, P., Briens, C. and Bernis, A. (2003) Development of a new rapid method to measure erosion rates in laboratory and pilot plant cyclones. *Powder Technol.* **131**, 111–119.

Davies, C.N. (1973) *Air Filtration*. Academic Press. ISBN 0122056604.

Del Fabbro, L., Laborde, J.C., Merlin, P. and Ricciardi, L. (2002) Air flows and pressure drop modelling for different pleated industrial filters. *Filtr. Separat.* **39**(1), 34–40.

Dickenson, T.C. (1997) *Filters and Filtration Handbook*, 4th edn. Elsevier, Oxford. ISBN 1856173224.

Directive 1999/30/EC of 22 April 1999, relating to limit values for sulphur dioxide, nitrogen dioxide and oxides of nitrogen, particulate matter and lead in ambient air.

Eurovent 4/9 (1997) Method of testing air filters used in general ventilation for determination of fractional efficiency. Eurovent/Cecomaf, Paris.

Fang, G.C., Wu, Y.S., Huang, S.H. and Rau Hester, J.Y. (2005) Review of atmospheric metallic elements in Asia during 2000–2004. *Atmos. Environ.* **39**, 3003–3013.

Fangmeijer, A., Hadwiger-Fangmeier, A., Van der Eerden, L. and Jäger, H.J. (1994) Effects of atmospheric ammonia on vegetation – a review. *Environ. Pollut.* **86**, 43–82.

Fayed, M. and Otten, L. (1997) *Handbook of Powder Science and Technology.* Chapman & Hall. ISBN 0412996219.

Hester, R.E. and Harrison, R.M. (1998) *Air Pollution and Health*, Royal Society of Chemistry. ISBN 0854042458.

Hjelmfelt, A.T. and Mockros, L.F. (1966) Motion of discrete particles in a turbulent, fluid. *Appl. Sci. Res.* **16**, 149–161.

ISO CD7708 (1992) *Air Quality – Particle Size Fraction Definition for Health-Related Sampling*. International Organization for Standardization, Geneva.

Kaye, B.H. (1999) *Characterization of Powders and Aerosols.* Wiley-VCH, New York. ISBN 3527288538.

Kim, M. and Zydney, A.L. (2005) Particle–particle interactions during normal flow filtration: model simulations. *Chem. Eng. Sci.* **60**, 4073–4082.

Krupa, S.G. and Arndt, U. (1990) Special issue on the Hohenheim long term experiment. *Environ. Pollut.* **68**, 193–478.

Lorimier, C., Le Coq, L., Subrenat, A. and Le Cloirec, P. (2004) Combined filtration VOCs/particles on fibrous filtering medias. *World Filtration Congress*, New Orleans.

Marinoni, A., Laj, P., Deveau, P.A., Marino, F., Ghermandi, G., Aulagnier, F. and Cachier, H. (2005) Physicochemical properties of fine aerosols at Plan d'Aups during ESCOMPTE. *Atmos. Res.* **74**, 565–580.

Midoux, M. (1993) *Mécanique et rhéologie des fluides en génie chimique*, Ed. Lavoisier Tec et Doc, ISBN 2852069288.

Perry, R.H. (1997) *Perry's Chemical Engineers' Handbook.* McGraw-Hill. ISBN 0070498415.

Peukert, W. and Wadenpohl, C. (2001) Industrial separation of fine particles with difficult dust properties. *Powder Technol.* **118**, 136–148.

Pope, C.A., Thun, M.J., Namboodiri, M.M., Dockery, D.W., Evans, J.S., Speizer, F.E. and Heath, C.W. (1995) Particulate air pollution as a predictor of mortality in a prospective study of U.S. adults *Am. J. Resp. Crit. Care Med.* **151**, 669–674.

Stairmand, C.J. (1951) The design and performance of cyclone separators. *T. I. Chem. Eng.* **29**, 356–383.

Thomas, D., Contal, P., Renaudin, V., Penicot, P., Leclerc, D. and Vendel, J. (1999) Modelling pressure drop in hepa filters during dynamic filtration. *J. Aerosol Sci.* **30**(2), 235–246.

USEPA (1996) Air quality criteria for particulate matter, **I**. EPA/600/P-95/001aF. US Environmental Protection Agency, Washington, DC.

Wang, C.S. (2001) Electrostatic forces in fibrous filters – a review. *Powder Technol.* **118**, 166–170.

15

Bioprocess engineering aspects of biological waste gas treatments

P. Le Cloirec

15.1 INTRODUCTION

Due to public awareness of the need for human health and environmental protection, a strong technological base has been developed for air quality control in recent years. Consequently, environmental legislations and regulations have been proposed and carried out to restrain gaseous polluted emissions. Removing air pollutant such as volatile organic compounds (VOC) or odorous molecules is achieved by different ways (Le Cloirec, 1998):

- Transfer and/or concentration into a liquid phase (absorption–condensation) or onto a solid (adsorption).
- Destruction by thermal oxidation, catalytic treatment, flaring or degradation by microorganisms.

It is well known that biological processes are the most competitive systems for the treatment or deodorization of waste gases characterized by high flow rates and low concentrations of the contaminants (Kosteltz et al., 1996; Le Cloirec et al., 1999).

Biological units commonly used for air treatment are based on three different technologies: biofilter, biotrickling filter and bioscrubber. More recently, two other types of bioreactors have been developed: membrane bioreactors and two-phase bioreactors. But theses systems have no industrial applications at this time.

The aim of this chapter is to present the engineering aspects of biological air treatment processes, especially the removal of VOC and odorous molecules. This presentation is divided into four parts: (1) an approach of air treatments. The biological processes are shown in a general context (2) biofilters, (3) biotrickling filters and (4) bioscrubbers. For each treatment technology, a simple presentation of the system will be proposed but the design and modeling of the process will be much more developed.

15.2 WASTE GAS TREATMENT PROCESSES

15.2.1 Air treatment processes

Due to the complex mixture of pollutants in air, several processes are available to control gaseous emissions. Figure 15.1 shows an overview of the most important treatment processes. As a function of the size of the pollutant (particles or molecules), a first classification shows two sets of processes. Specifically, for removal of molecules such as acid gases (SO_x, HCl, ...), nitrogen oxides (NO_x), carbon dioxide and organics (VOC, odors, dioxins, furans, ...), several treatment processes can be useful. In this general presentation, biological processes are dedicated to the elimination by degradation of organic compounds present in air.

15.2.2 Biological treatments of waste gases (Le Cloirec et al., 2005)

Biological treatment can be distinguished either by the mobility of the microorganisms and the liquid phase (as in bioscrubber) or only the liquid phase (as in trickling filter). In biofilters and trickling filters, the microorganisms are fixed in a support or a packing material (Devinny et al., 1999; Ottengraf, 1987). Figure 15.2 and Table 15.1 summarize the different technologies. Performances of such processes are interesting when operating parameters, as waste gas loads and washing solution loads for bioscrubbing and trickling filtering, are well controlled (Herrygers et al., 2000; Kennes & Thalasso, 1998).

15.3 BIOFILTERS

15.3.1 General presentation of the process

Figure 15.3 presents a schematic biofilter, the most basic system used in waste gas treatment. Air loaded with pollutant flows through a humidified packing material on which microorganisms (i.e. the biocatalyst) grow as a biofilm. The pollutants are transferred into the liquid film and then into the biofilm to be degraded by microorganisms. Additional substrate (nitrogen and phosphate sources) used as nutrients can be introduced through humidification water.

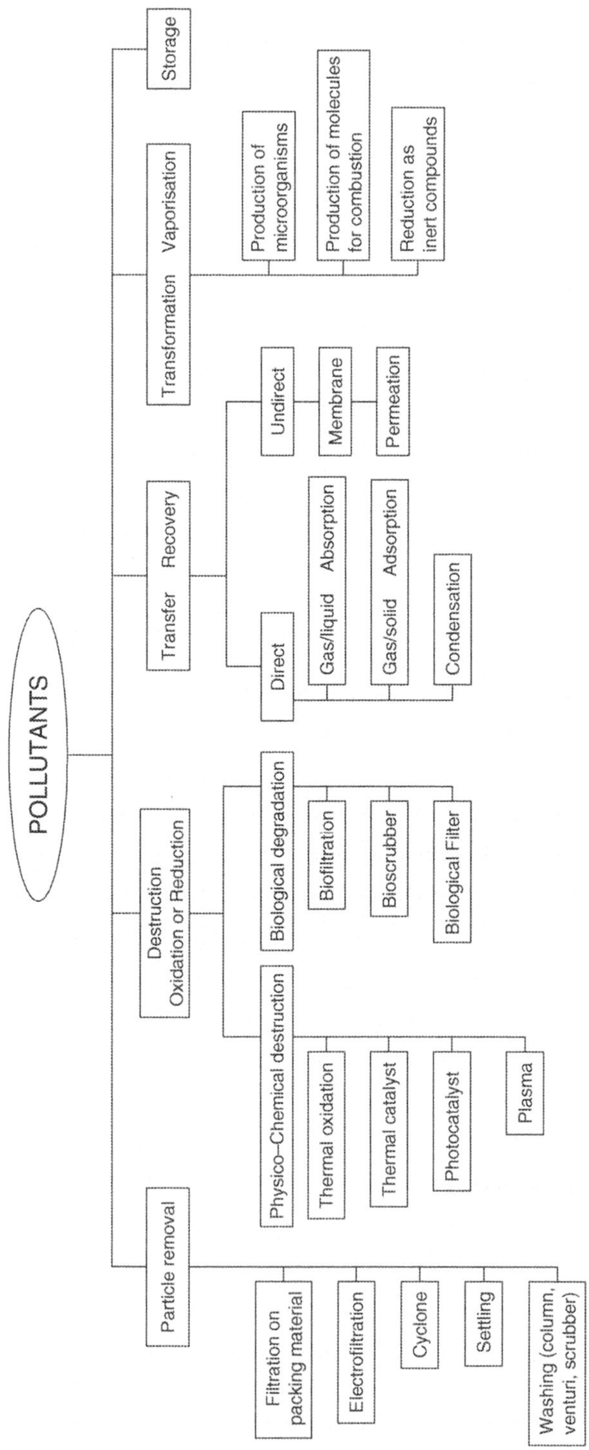

Figure 15.1 Overview of air treatment processes.

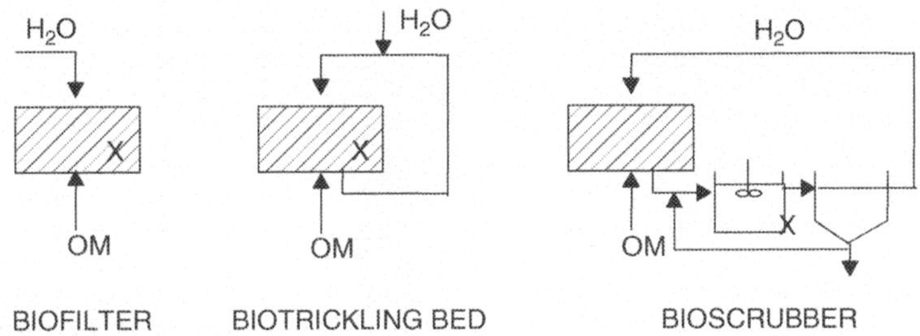

Figure 15.2 Schematic presentation of biological processes in air treatment. OM: organic matter (substrate), X: biomass.

Table 15.1 Classification of biological treatment as a function of the mobility of the liquid phase and microorganisms.

Biomass	Aqueous phase	
	Mobile	Immobile
Activated sludge	Bioscrubber	–
Attached biomass	Biotrickling filter	Biofilter

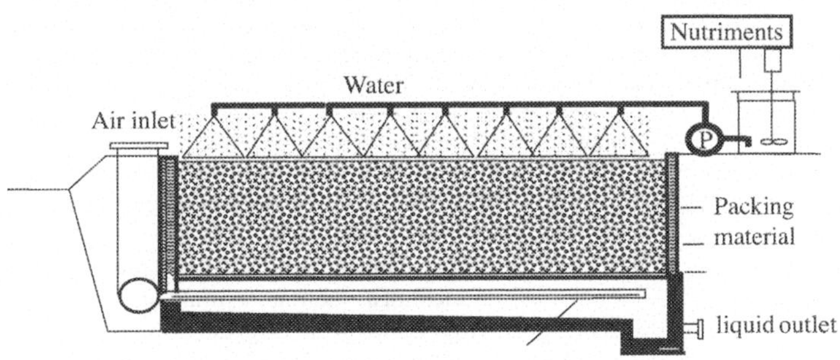

Figure 15.3 Schematic presentation of a classical open biofilter.

Systems are usually working in counter- (open biofilters) or co-current flow (closed reactor). An example of a biofilter treating odorous waste gas is given in Figure 15.4. The biofilters are covered and the purified air is sent to a chimney. The flow is 50,000 $m^3 h^{-1}$, the filter surface is $2 \times 250\,m^2$ and the depth is close to 2.0 m. The system is packed with a mineral solid waste (mussel shells). This material is a microorganism support, but reacts also with, and thus neutralizes, acid volatile compounds (fatty acid, H_2S, ...) present in the odorous emission.

15.3.2 Practical operating conditions and performances

Table 15.2 gives a range of general operating conditions and performances of biofilters. The polluted air velocities are generally low due to the slow degradation kinetics.

Figure 15.4 Example of a biofilter used for odorous waste gas treatment.

Table 15.2 General operating conditions applied in biofilters.

Parameter	Value	Remarks
Gas phase velocity (U_G) ($m\,h^{-1}$)	100–500	Low values due to low kinetics of degradation
Air residence time (s)	15–90	Depending of the molecule degradation kinetic alcohols > ketones > *n*-alkanes > aromatics
Bed porosity (–)	0.4–0.95	High values avoid clogging
Specific surface area (S) ($m^2 m^{-3}$)	100–400	High values give a better mass transfer and a higher biomass concentration (biotrickling filter)
Filter depth (H) (m)	0.5–2.5	Compromise between the residence time and the pressure drop
Pressure drop (m H_2O)	0.1–0.5	Depending of the clogging and support compressing
Air humidity (%)	60–100	High values are better
Water pH	5–9	Depending of the solubility of pollutants
Temperature (T) (°C)	10–40	Mesophilic microorganisms Thermophilic microorganisms at temperature ranging from 50°C to 70°C
Acclimation time (day)	8–30	Function of the biodegradability of pollutants Utilization of an inoculum (diluted activated sludge)
Pollutant concentration ($mg\,m^{-3}$)	1–1000	Possible inhibition at higher concentration by molecules or degradation by-products
Efficiencies (%)	90–99	Depending of molecules
Life time (year)	2–5	Inorganic packing materials increase the life time

Therefore, a large biofilter surface is required for such as a system. The depth is close to 2 m. This value is a compromise between the necessary residence time and the pressure drop. Biodegradation, and thus performance, is a function of the pollutant. Fass et al. (1999) give a relative aerobic biodegradability order of various classes of compounds:

aliphatics > aromatics > chlorinated compounds > cyclic hydrocarbons > chlorinated polycyclic compounds.

Parameters such as pH or temperature are influencing the performance due to the sensitivity of bacteria to such conditions. The humidity of the packing material inside the biofilter is ranging from 40% to 60% to get a high level of water required for bacteria and to minimize channeling due to dry zones. The packing material is an organic solid such as peat, wood or an inorganic compound (clay, calcium carbonate, ...). Mixtures of solid are also used to get a more efficient bacterial support as well as to buffer the liquid film and thus to control the pH range in the biofilter.

An example of the evolution of the pressure drop as a function of the gas flow velocity for a biofilter packed with wood bark (10/25 and 25/40) is given in Figures 15.5 and 15.6. The influence of the humidity or biomass presence on the packing solid is well shown. A biofilm generates a larger head loss due to a decrease of external porosity of the support.

The pressure drop in this system is difficult to evaluate using Ergun's equation (Coulson et al., 1994) due to difficulties to determine a real diameter of packing material. However, the Comitti and Renault (1989) relation can be successfully applied to biofilters packed with wood bark (Le Cloirec et al., 2001). The equation gives the evaluation of the pressure drop ($\Delta P/H$) as a function of the empty bed velocity (U_0):

$$\frac{\Delta P}{HU_0} = M^* U_0 + N^* \tag{15.1}$$

with

$$M^* = \left\{0.0413\left[1 - \left(1 - \frac{d_p}{c}\right)^2\right] + 0.0968\left(1 - \frac{d_p}{c}\right)^2\right\}\tau^3 \rho a_{vd} \frac{(1-\varepsilon)}{\varepsilon^3} \tag{15.2}$$

$$N^* = 2\eta\tau^2 a_{vd}^2 \left[1 + \frac{4}{a_{vd} c(1-\varepsilon)}\right]^2 \frac{(1-\varepsilon)^2}{\varepsilon^3} \tag{15.3}$$

with

ΔP: pressure drop (Pa)
H: bed depth (m)
U_0: empty bed velocity (m h^{-1})
d_p: packing material diameter (m)

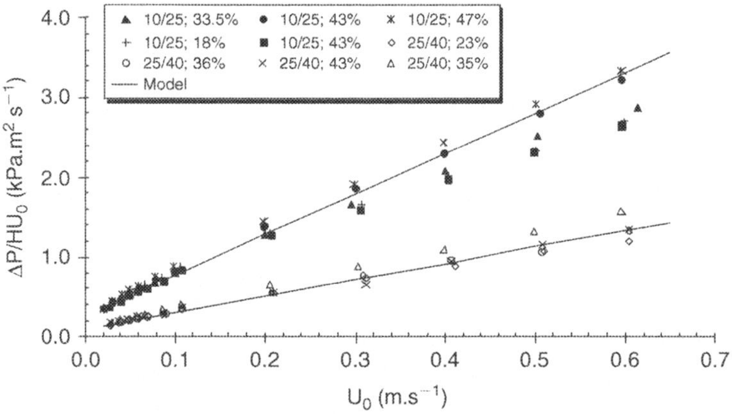

Figure 15.5 Linearized pressure drop ($\Delta P/HU_0$) in a wood bark biofilter as a function of the empty bed velocity, the size of the material and the humidity (closed symbol) in the process.

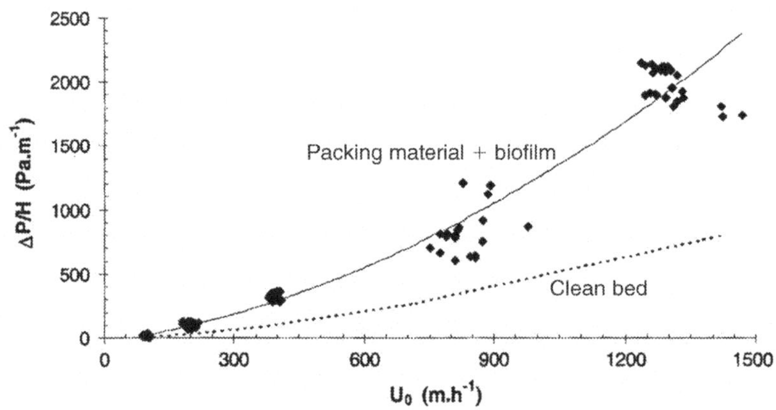

Figure 15.6 Pressure drop ($\Delta P/H$) as a function of gas velocity (U_0) and the presence of a biomass (close symbol) coating the wood bark.

c: bioreactor diameter (m)
τ: tortuosity (dimensionless)
ρ: fluid density (kg m^{-3})
a_{vd}: surface in contact with the fluid (m^2)
ε: external porosity (dimensionless)
μ: fluid viscosity (Pl or Pa · s)

Depending of the material characteristics, the parameters a_{vd} and τ are found:

$$1.8 < \tau < 2.9 \text{ and } 350 < a_{vd} < 950$$

It was found that the moisture content (40–60%) of the packing material has no strong effect on these two parameters.

15.3.3 Modeling a biofilter

15.3.3.1 Ottengraf's model

To describe mechanisms of transfer and biodegradation in a biofilter used in waste gas treatment, Ottengraf (1986) proposed a simple model. The organic compound in the gas phase diffuses in the biofilm coating a solid support particle where it is degraded. The hypothesis are as follows:

- Biodegradation occurs in the liquid phase-biofilm.
- Biofilm thickness is small compared to packing material diameter.
- Biomass concentration is supposed to be homogeneous in the reactor volume and constant as a function of time.
- Gas flow is a plug flow.
- There are no reactions between the chemical species.
- Mass transfer resistance in the gas phase is negligible.
- Regime is steady-state: there is no variation of operating conditions or in the biofilm (concentration of microorganisms, depth, ...).
- Equilibrium occurs at the gas–biofilm interface.

Assuming an elemental volume with a depth dz, the pollutant concentration evolution S_G (g m^{-3}) in the gas phase is given by the Fick's law:

$$U_G \frac{dS_G}{dz} = aD\left(\frac{dS_L}{dx}\right)_{x=0} \tag{15.4}$$

with the velocity in the empty biofilter U_G (m s^{-1}), the packing material specific surface area a (m^{2}m^{-3}), the diffusivity in the liquid phase D (m^{2}s^{-1}), the biofilm thickness x (m), the substrate concentration at the gas–biofilm interface $S_L^{x=0}$ (g m^{-3}). The initial condition is given assuming the pollutant concentration in the reactor inlet S_G^e to be constant:

$$S_G^{z=0} = S_G^e \tag{15.5}$$

On the depth dz, the mass balance and the substrate concentration evolution in the biofilm S_L(g m^{-3}), assuming a diffusional regime with a biodegradation rate R_s (g m^{-3}s^{-1}), are given by:

$$0 = D\left(\frac{d^2 S_L}{dz^2}\right) - R_s \tag{15.6}$$

For only one compound, Monod's equation gives the biodegradation reaction rate R_s:

$$R_s = \frac{-dS_L}{dt} = R_0 \frac{S_L}{(S_L + K_s)} \tag{15.7}$$

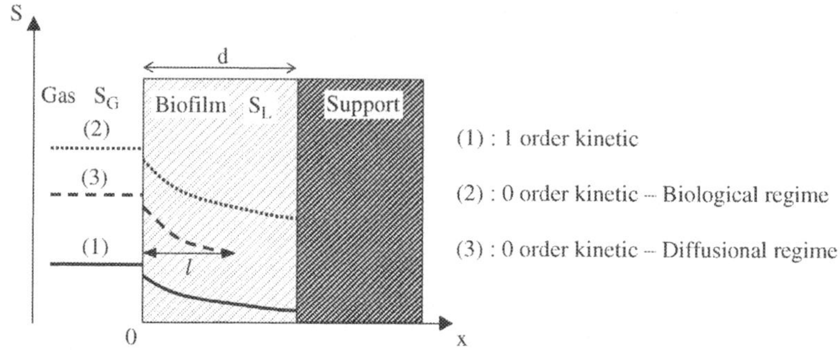

Figure 15.7 Substrate concentration profile in the biofilm as a function of the kinetic order (Ottengraf, 1986).

R_0 ($g\,m^{-3}s^{-1}$) is a function of the maximal biomass growth rate (μ_{max}) and depends on the concentration of microorganisms in the biofilm (X). K_s ($g\,m^{-3}$) is the Monod affinity constant. For carboneous substrates, K_s are low and range from 10^{-3} to $10^{-2}\,g\,m^{-3}$. Three cases can be considered for the biodegradation kinetics (Figure 15.7):

(1) First order kinetic: The substrate concentration is a limiting factor. The reaction rate is controlled by the pollutant diffusion inside the biofilm.
(2) Zero order kinetic-biological regime: the substrate concentration in the biofilm is high ($S_L \gg K_s$) but the diffusion into the biofilm is not a limiting factor.
(3) Zero order kinetic-diffusional regime: the substrate concentration in the biofilm is high ($S_L \gg K_s$) and the diffusion into the biofilm is a limiting factor.

Case I: First order kinetic 1 The substrate concentration is a limiting factor ($S_L \ll K_s$), then the biodegradation rate proceeds via a first order law:

$$R_S = \frac{-dS_L}{dt} = \frac{R_0}{K_s} S_L \tag{15.8}$$

Assuming a constant biofilm thickness δ, $x = \delta$ is achieved for a minimal substrate concentration:

$$\left(\frac{dS_L}{dx}\right)_{x=\delta} = 0 \tag{15.9}$$

Assuming an equilibrium at the gas–biofilm interface for $x = 0$:

$$S_L^{x=0} = \frac{S_G}{m} \tag{15.10}$$

with m ($\mathrm{g\,g^{-1}}$) the partition coefficient deduced from Henry's law. The analytical solution of the differential equation (15.6) with the limits (15.9) and (15.10):

$$\frac{S_L}{S_G/m} = \frac{\cosh[\phi_1(1-\sigma)]}{\cosh(\phi_1)} \tag{15.11}$$

$$\phi_1 = \delta\sqrt{\frac{R_0}{K_s D}} \tag{15.12}$$

with ϕ_1, the Thiele number and $\sigma = x/\delta$ the dimensionless length in the biofilm.

The Equation (15.9) gives the concentration gradient at the interface $(\mathrm{d}S_L/\mathrm{d}x)_{x=0}$ as a function of the substrate concentration in the gas phase S_G. The mass balance (15.4) is integrated along the biofilter length, Z, with the limit given by (15.5). Then the outlet pollutant concentration is:

$$\frac{S_G^s}{S_G^e} = \exp\left(-\frac{aZD}{mU_G\delta}\phi_1 \tanh\phi_1\right) \tag{15.13}$$

Case II: Zero order kinetic: biological regime If the substrate concentration is $S_L \gg K_s$, zero order kinetics apply (i.e. a biological regime):

$$R_S = \frac{-\mathrm{d}S_L}{\mathrm{d}t} = R_0 \tag{15.14}$$

The pollutant concentration in the biofilm has no effect on the biodegradation rate. In this case, the biodegradation rate is quick and the biofilm is considered active all along its thickness δ. The limits $x = 0$ and $x = \delta$ are used to integrate the Equation (15.6) with the Equations (15.9) and (15.10). Then, with a zero order kinetic, the analytical solution is:

$$\frac{S_L}{S_G/m} = 1 + \frac{1}{2}\frac{\left[\phi_0^2(\sigma^2 - 2\sigma)\right]}{S_G/S_G^e} \tag{15.15}$$

$$\phi_0 = \delta\sqrt{\frac{R_0 m}{DS_G^e}} \tag{15.16}$$

with ϕ_0 Thiele number for a zero order reaction.

The mass flux transferred at the biofilm interface is:

$$-D\left(\frac{\mathrm{d}S_L}{\mathrm{d}x}\right)_{x=0} = R_0\delta \tag{15.17}$$

The mass balance (15.4) is integrated along the biofilter length Z:

$$\frac{S_G^s}{S_G^e} = 1 - \frac{R_0 a \delta Z}{U_G S_G^e} \tag{15.18}$$

Case III: Zero order kinetic: diffusional regime Assuming a diffusional regime, that is the substrate diffusion velocity is low compared to the biological degradation rate, some biofilm zones are not active because substrate does not reach the bacteria. S_L is not degraded. Figure 15.7 gives the evolution of the substrate concentration in a biofilm. An active biofilm thickness λ is defined lower than the biofilm thickness δ.

A critical Thiele number is determined for a limiting biofilm diffusion assuming a pollutant concentration in the biofilm S_L close to zero at $\sigma = 1$. Then, Equation (15.15) gives a Thiele number as follows:

$$\phi_{cr} = \delta \sqrt{\frac{R_0 m}{D S_G}} = \sqrt{2} \tag{15.19}$$

To solve the mass balance in the biofilm, the limiting conditions are $x = 0$ (Equation 15.10) and $x = \lambda$:

$$\left(\frac{dS_L}{dx} \right)_{x=\lambda} = 0 \tag{15.20}$$

The solution of the differential equation (15.6) is:

$$\frac{S_L}{S_G/m} = 1 + \frac{1}{2} \frac{\phi_0^2}{S_G/S_G^e} \left(\sigma^2 - 2\sigma \frac{\lambda}{\delta} \right) \tag{15.21}$$

At $\sigma = \lambda/\delta$, $S_L = 0$, the penetration thickness λ is calculated with Equation (15.21):

$$\lambda = \sqrt{\frac{2DS_G}{mR_0}} \tag{15.22}$$

The mass flux transferred at the biofilm interface is:

$$-D \left(\frac{dS_L}{dx} \right)_{x=0} = R_0 \lambda \tag{15.23}$$

The outlet pollutant concentration in the gas phase is given by integrating Equation (15.4):

$$\frac{S_G^s}{S_G^e} = \left(1 - \frac{aZ}{U_G} \sqrt{\frac{R_0 D}{2 S_G^e m}} \right)^2 \tag{15.24}$$

Biofilter yield For the different cases, a biofilter yield (BY) can be determined as a function of the kinetic orders:

$$\mathrm{BY} = \frac{(S_G^e - S_G^s)}{S_G^e} \tag{15.25}$$

Table 15.3 overviews an abstract of the biofiltration treatment yields deduced from Equations (15.13), (15.18) and (15.24).

Evaluation of Ottengraf's model The Ottengraf's model is very simple and gives analytical solutions to differential equations. It has been validated in many studies with a good agreement between the calculated values and the experimental data (Diks & Ottengraf, 1991a, b). However, some hypotheses are discussed especially the interactions between the chemical species or the concentration levels in the gas or in the biofilm (Deshusses et al., 1995a, b; Devinny & Hodge, 1995; Shareefdeen et al., 1993; Veiga & Kennes, 2001). The model assumes a steady-state system. In fact generally in an industrial biofilter, the inlet pollutant concentration or the humidity level in the packing material fluctuate as a function of the amount of waste gas to be treated. The biomass concentration is assumed to be constant in the biofilm. It is obvious that a local growth of microbial populations occurs in the system due to support, operating conditions or substrate (Cox et al., 1999).

Table 15.3 Overview of kinetic constants and biodegradation yields as a function of the kinetic order (Ottengraf, 1986).

	Kinetic constant (K_i)	Biodegradation yield (BY)	Total biodegradation condition
First order kinetic	$K_1 = \frac{aD}{\delta}\phi_1 \tanh\phi_1$ (S^{-1})	$\mathrm{BY} = 1 - \exp\left(\frac{-K_1 Z}{mU_G}\right)$	$\frac{K_1 Z}{mU_G} \rightarrow \infty$
Zero order kinetic: biological regime ($\phi_{cr} < \sqrt{2}$)	$K_0 = R_0 a\delta$ $(\mathrm{g\,m^{-3}s^{-1}})$	$\mathrm{BY} = \frac{K_0 Z}{U_G S_G^e}$	$\frac{K_0 Z}{U_G S_G^e} \geq 1$
Zero order kinetic: diffusional regime ($\phi_{cr} > \sqrt{2}$)	$K_0 = R_0 a\delta$ $(\mathrm{g\,m^{-3}s^{-1}})$	$\mathrm{BY} = 1 - \left(1 - \frac{Z}{U_G}\sqrt{\frac{K_0 Da}{2S_G^e m\delta}}\right)^2$	$\frac{K_0 Z}{U_G S_G^e} \geq 2$

At this time, different improvements have been proposed (interactions between substrates and by-products, oxygen concentration, pollutant accumulation) with more complex equation systems and a larger number of adjustable parameters.

15.3.4 An example of biofiltration performances and modeling

In order to illustrate the utilization of the Ottengraf's model, an application of air treatment loaded with VOCs is presented in the next section (Le Cloirec et al., 2001; Ramirez-Lopez et al., 2000).

A biofilter was constructed with polycarbonate GE, 1.70 m in length and 0.45 m in width (Figure 15.8). A heterogeneous material as wood bark 10/25 (size in mesh) was used as packing material. The reactor was loaded with the support to a total depth of 1 m. A contaminated waste gas stream was produced by bubbling air in the liquid ethanol stock. Then, it was passed through the biofilter in an upflow air stream. The sampling ports were located at the inlet and outlet of the biofilter. The support humidity was controlled by intermittent injections of a nutrient solution based on oligoelements, microelements and vitamins (Le Cloirec et al., 2001; Ramirez-Lopez et al., 2000).

The inlet and outlet ethanol concentrations were measured by passing a polluted gas stream through a 1 L gas-sampling bulb. The ethanol concentration in the polluted gas stream was measured by a gas chromatograph (Perkin-Elmer Auto System; Perkin Elmer, USA). It was equipped with a capillary column and a flame ionization detector (FID). The performance of the biofilter was estimated as a function of the removal efficiency and the pollutant-loading rate.

Table 15.4 gives the operating parameters and the maximum biodegradation capacities for different periods. The lag time is determined by the time when a

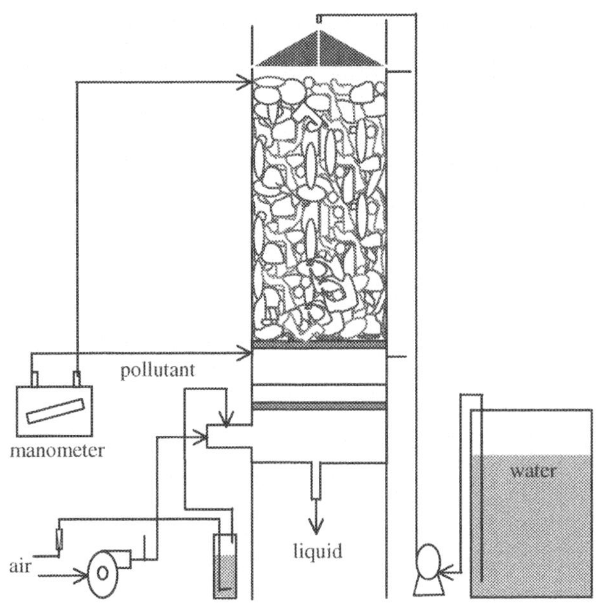

Figure 15.8 Experimental set-up of a biofilter packed with wood barks.

removal efficiency of 50% is achieved. The low pH is due to the production of acetic acid from ethanol oxidation. The critical concentration is the limit when the biofilter performance is maximal.

Figure 15.9 presents the biodegradation performances as a function of the volumetric load C_V ($g\,m^{-3}h^{-1}$). For $C_V = 220\,g\,m^{-3}h^{-1}$, the ethanol biodegradation is more than 95%. The performance decreases until 65% at $C_V = 380\,g\,m^{-3}h^{-1}$.

These experimental data were modeled by the Ottengraf model assuming a zero order kinetic and a diffusional regime. The performance following the different period are given by the equation:

$$\mathrm{BY} = 1 - \left(1 - \frac{\beta Z}{U_G\sqrt{S_G^e}}\right)^2 \tag{15.26}$$

with β a parameter which is a function of the molecular diffusivity, the pollutant partition coefficient, the kinetic constant, the specific surface area of the packing

Table 15.4 Operating conditions applied to a biofilter treating a biodegradable pollutant (ethanol) in air (Le Cloirec et al., 2001; Ramirez-Lopez et al., 2000).

Parameter	Unit	Period				
		I	II	III	IV	V
Time	day	51	74	29	21	21
Lag period	day	29	6	1	2–3	8
Gas velocity (U_G)	$m\,h^{-1}$	99	200	397	826	1288
Residence time	s	37	18	9	5	3
Inlet concentration (S_G^e)	$g\,m^{-3}$	0.52–1.47	0.2–1.88	0.1–0.67	0.04–0.57	0.04–0.17
Volumetric load (C_V)	$g\,m^{-3}h^{-1}$	51–125	35–376	40–266	39–484	56–210
Bed temperature	°C	18.1 ± 5.0	16.7 ± 2.0	18.6 ± 1.0	20.3 ± 2.0	24.2 ± 2.0
pH		6.3 ± 0.9	3.5 ± 1.0	3.5 ± 0.3	4.2 ± 0.3	4.5 ± 0.4
Removal rate	$g\,m^{-3}h^{-1}$	125	215	185	394	177
Critical concentration	$g\,m^{-3}$	–	1.46	0.56	0.57	0.17

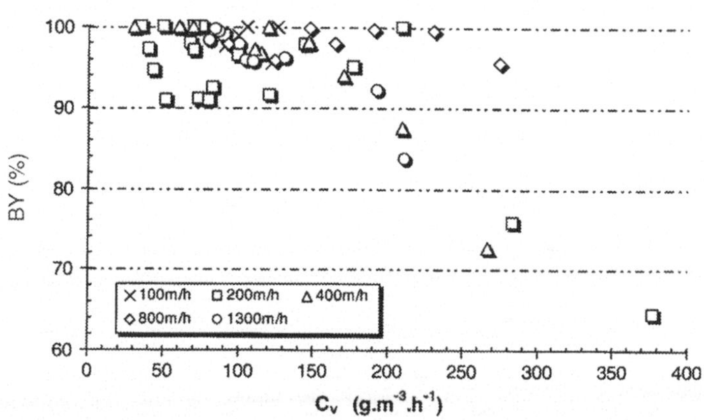

Figure 15.9 Biodegradation performance (BY) as a function of the volumetric load C_V.

material and depending also on temperature:

$$\beta = a\sqrt{\frac{R_0 D}{2m}} \tag{15.27}$$

Assuming β constant during the operation time, this parameter is evaluated from the experimental values for a yield more than 90% (diffusional regime).

The values (β) are reported in Table 15.5. An increase of this parameter for the periods I to III and a stabilization for the periods IV and V should be noted. These data could be analysed by an intensification of the microbial activity.

The Ottengraf's model (diffusional regime, zero order kinetic), adjusted by the β parameter was applied to the experimental data. Figure 15.10 shows the evolution of biodegradation rates of the biofilter as a function of the volumetric load and the limits of the model.

Table 15.5 Evolution of the biodegradation yields and β calculated by the Ottengraf model as a function of the operational periods.

Period	U_G (m h^{-1})	C_v (g m^{-3}h^{-1})	Biodegradation yield (BY) (%)	β (g$^{1/2}$m$^{-3/2}$h^{-1})
I	99	51–125	>99	89
II	200	35–176	>95	115
III	397	30–147	>95	168
IV	826	39–376	>95	331
V	1288	56–206	>90	298

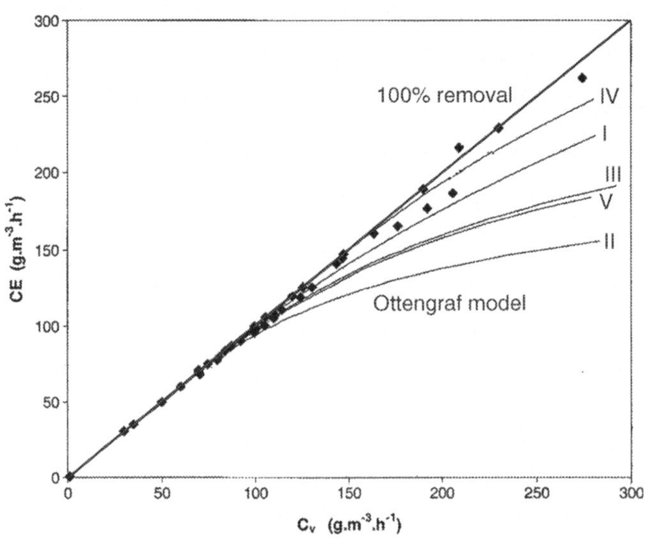

Figure 15.10 Evolution of biodegradation rates of the biofilter as a function of the volumetric load – experimental data and values calculated by the Ottengraf's model (diffusional regime, zero order kinetics).

15.4 BIOTRICKLING FILTERS

15.4.1 Principle, operating conditions and performance

In this process, the pollutant in the gaseous phase is transferred and biodegraded in a column packed with solid materials covered with biomass (Figure 15.11). The washing solution is recycled but due to evaporation in the treated air and to avoid an increase of by-product concentration, additional water is introduced in the buffer reactor at the bottom of the column. In this water, chemicals (base or acid solution) depending of the waste gas emission composition, can be added to get a good pH range for microorganisms.

Table 15.6 gives ranges of general operating conditions and obtained performance. The importance of gas and liquid velocities and residence times has to be noted. In order to get a significant gas–liquid interface, the specific surface area of the packing material is high. Due to a low degradation kinetic, column depths range from 2 to 15 m. Temperature and pH are in the classical range for a good microbial growth and pollutant degradation.

15.4.2 Modeling a biotrickling filter

In order to model the mass transfer and the biodegradation steps, three phases are assumed:

(1) Gaseous phase loaded with pollutant.
(2) Liquid phase absorbing the pollutant.
(3) The biofilm coating the packing material in the column.

Figure 15.12 presents the concentration profile of substrate in the gas–liquid–biofilm system. In the liquid, the global transfer coefficient K_L (m s^{-1}) is given by the following relation:

$$\frac{1}{K_L} = \frac{1}{mk_G} + \frac{1}{k_L} \tag{15.28}$$

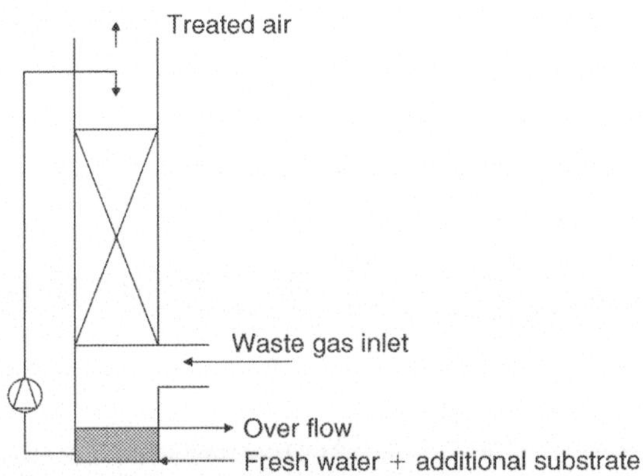

Figure 15.11 Schematic of a trickling filter used for waste gas emission treatment.

Table 15.6 General operating conditions applied in biotrickling filters and bioscrubbers.

Parameter	Value	Remarks
Liquid phase velocity (U_G) ($m\,h^{-1}$)	0.05–20	Compromise between mass transfer in water and the flooding point. High velocities increase the mass transfer and the biomass growth but give a biomass loss
Liquid hold up (%)	<5	
Gas phase velocity (U_L) ($m\,h^{-1}$)	100–1000	Limited by the head losses due to the clogging
Air residence time (s)	<60	Lower than for biofilters
Bed porosity (–)	0.5–0.95	High values avoid clogging
Specific surface area (S) ($m^2 m^{-3}$)	100–400	High values give a better mass transfer and a higher biomass concentration (biotrickling filter)
Filter depth (H) (m)	2–15	Light packing materials are required. High depth increases the head loss
Pressure drop (m H_2O)	0.005–0.05	As a function packing material and microorganism growth on the packing material
Water pH	4–9	Depending of the solubility of pollutants Easy to control
Temperature (T) (°C)	10–40	Mesophilic microorganisms Thermophilic microorganisms at higher temperature between 50°C and 70 °C
Lag time (day)	5–210	Function of the biodegradability of pollutants
Efficiencies (%)	90–99	Depending of biodegradability of molecules

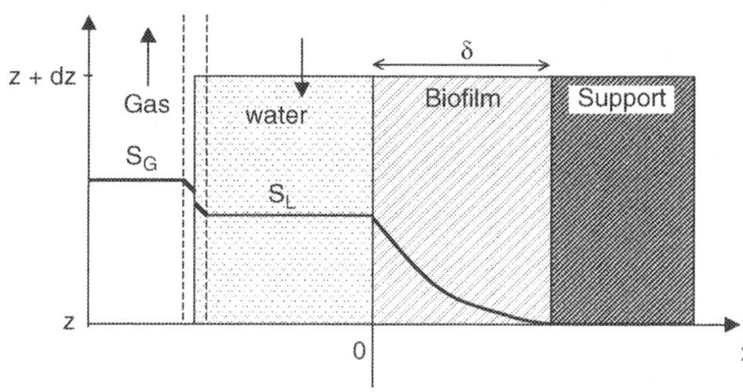

Figure 15.12 Substrate concentration profile in the gas–water–biofilm system.

where m ($g\,m^{-3}$ gas/$g\,m^{-3}$ water) is the partition coefficient between the gas and the liquid (Henry law). For a plug flow in the column, the mass balance in the gas phase is:

$$U_0 \frac{dS_G}{dz} = \pm K_L a \left(\frac{S_G}{m} - S_L \right) \tag{15.29}$$

with a is the packing material surface. For a counter-current system, the mass balance in the liquid phase is written:

$$U_L \frac{dS_L}{dz} = -K_L a \left(\frac{S_G}{m} - S_L \right) - Ja \tag{15.30}$$

with J (g s^{-1}) the substrate mass flow into the biofilm.

If the biomass concentration is a constant, and the biodegradation proceeds via Monod kinetics, J is given by:

$$J = \sqrt{2DK_s R_0 \left[\frac{S_L}{K_s} - \ln\left(1 + \frac{S_L}{K_s}\right) \right]} \tag{15.31}$$

In the previous equation, D is the diffusion coefficient of the substrate into the biofilm. S_L is the substrate concentration at the liquid–biofilm interface. If all the depth of the biofilm is used and the concentration is high compared to the Monod's constant K_s, the biodegradation kinetics are zero order. The maximal transfer flux is then:

$$J_{max} = R_0 \delta \tag{15.32}$$

The differential equation system ((15.25) and (15.26)) is solved using a numerical method (Runge–Kutta method). The flux is given by Equation (15.28) until a maximal value defined by (15.29). The initial conditions are the inlet pollutant concentration in the gas phase:

$$S_G^{z=0} = S_{Ge} \tag{15.33}$$

and the recirculation of the liquid phase which has the same substrate concentration in the inlet and outlet of the column.

$$S_L^{z=0} = S_L^{z=Z} \tag{15.34}$$

Some limits of the model can be mentioned:

- Transfer resistance at the liquid–biofilm interface is considered negligible. Therefore, the substrate consumption in the biofilm is over-estimated and thus the performance is also over-estimated.
- System is steady-state. The biofilm concentration is considered as a constant. In real biofilms, the biomass concentration changes as a function of the biofilm depth.
- A possible inhibition due to a high pollutant concentration or by-products is not taken into account.

15.5 BIOSCRUBBERS

15.5.1 Presentation of the process

The removal of gaseous pollutants with a bioscrubber is performed by an association of different stages of treatment: a pollutant-liquid transfer column, an activated sludge basin and possibly a settling tank (Figure 15.13). A packed column or an empty column

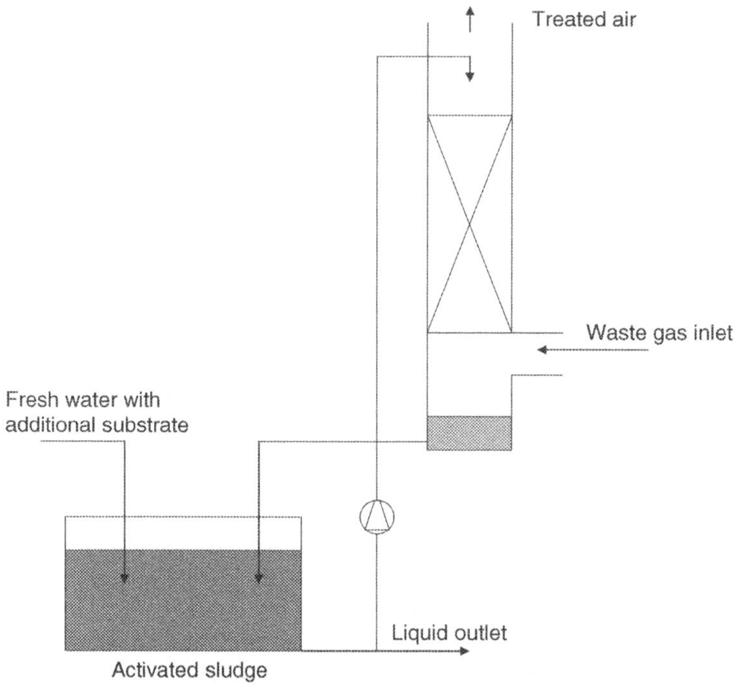

Figure 15.13 Schematic presentation of a bioscrubber.

with a spray injector are the principal types of gas absorption equipment. The operating conditions are close to those used in the biological trickling filters in terms of mass transfer in the column. Gas and liquid velocities are selected based on the solubility (Henry constant) of the pollutant. The absorbed VOC is degraded in the bioreactor and the solution, after sedimentation, is recycled in the column (Le Cloirec et al., 2005; Roustan, 2003).

Thus, a suspended-growth bioscrubber provides high elimination rates if the transfer step is well-controlled, and represents a low-cost solution for the deodorization of wastewater treatment facilities (Hansen & Rindel, 2000; Humeau et al., 2004).

15.5.2 Practical operating conditions and performance

Similar operating conditions presented in Table 15.6 are useful for a bioscrubbers. However, a significant residence time (several days) in the biological reactor is generally proposed in order to biodegrade transferred compounds. For recalcitrant organic molecules such as halogenated products or ethers, residence times of 40 or more days have been reported (Deshusses et al., 1995a, b; Kennes & Thalasso, 1998).

15.5.3 Modeling

15.5.3.1 Model description

The modeling approach is an arrangement of transfer equations with hydrodynamic parameters (Humeau et al., 2000, 2004; Le Cloirec et al., 2005). This theory allows to

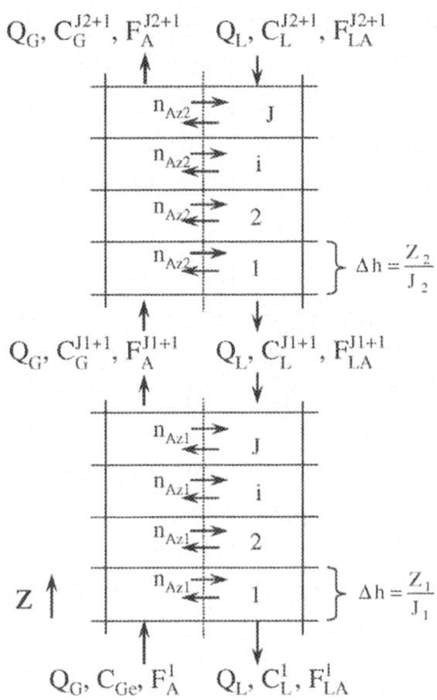

Figure 15.14 Schematic representation of the model of transfer characterization with two successive modes of contact.

determine the influence of operating conditions (gas and liquid loading rates, contact mode) and of the washing solution on the deodorization rate (Morris & Jackson, 1953; Roustan, 2003).

Several assumptions were retained to formulate the mass balance equations in the contactor of a bioscrubber, as the washing solution is generally an activated sludge suspension. First, the biodegradation of transferred pollutant is negligible in the gas–liquid tower. This assumption is often verified, as the residence times of the washing solution in the contactor are much shorter than in the activated sludge basin. Therefore, there is no concentration modification with time in the liquid phase, so the steady-state regime can be assumed. Secondly, a plug flow model, with and without axial dispersion, was used to represent the gas flow in the column. Thirdly, the tanks-in-series with mass exchange model was used to represent the washing solution flow. It was supposed that transfer operation occurred only in the dynamic phase of the flow, as no absorption can be carried out in the stagnant phase because its flow velocity equals zero.

From these hypotheses, the proposed model could be presented as in Figure 15.14. Each contact mode of the transfer column was divided into compartments with a height corresponding to its corresponding transfer height (Z) divided by the number J of a perfectly mixed reactor. Subsequently, subscript 1 and subscript 2 are used to describe the packing zone and the spraying zone, respectively (Levenspiel, 1999).

The set of equations to solve is based on the partial mass balances in the gas and liquid phases. Considering the axial dispersion of the gas phase, the partial mass balance in the gas phase on the height dz of the compartment i can be written following the expression:

$$-Q_G \frac{dC_G^i}{dz} + D_a S \frac{d^2C_G^i}{d^2z} - K_G(C_{Gz}^i - C_G^{iE})aSdz = 0 \qquad (15.35)$$

where Q_G is the gas flow rate ($m^3 s^{-1}$), C_G the volatile compound concentration in the gas phase ($mol\,m^{-3}$), D_a the axial dispersion of the gas flow ($m^2 s^{-1}$), K_G the overall transfer coefficient in the gas phase ($m\,s^{-1}$), C_G^E the gas concentration at the equilibrium with the liquid phase concentration ($mol\,m^{-3}$), a the specific surface area (m^{-1}) and S the surface of the column (m^2).

As for the partial mass balance in the liquid phase between the inlet and the outlet of the compartment i, it can be written as:

$$Q_L(C_L^{i+1} - C_L^i) + \int_0^{\Delta h} N_A^i aSdz = 0 \qquad (15.36)$$

where Q_L is the liquid flow rate ($m^3 s^{-1}$), C_L is the liquid concentration of the compound ($mol\,m^{-3}$) and N_A is the density of transfer flux ($mol\,m^{-2} s^{-1}$).

In a perfectly mixed reactor, the outlet concentration is the same as the concentration in the whole reactor. So it is possible to propose analytical solutions for the partial mass balance equations, for an elementary volume of compartment i, following the system of Equations:

$$\begin{cases} \text{Gas} & : C_G^i - C_G^{i-1} = (H'\,C_L^i - C_G^i)\left[1 - \exp\left(\alpha.\dfrac{Z}{J}\right)\right] & (15.37) \\ \text{Liquid} & : C_L^{i+1} - C_L^i = \gamma\,(C_G^i - C_G^{i-1}) & (15.38) \end{cases}$$

where H' is the Henry's law coefficient (dimensionless), Z is the height of the transfer column (m) and J is the number of tanks-in-series. α and γ are coefficients which expressions vary according to the considerations taken into account (Table 15.7).

In the specific case of H_2S (or similar acid or base compounds) absorption in water is influenced by the pH which is taken into account by dissociation reactions. This instantaneous reaction according to the pH (Astarita & Gioia, 1964; Littel et al., 1991), which enhances the transfer operation, is realized by the introduction of an acceleration factor E in the expression of the coefficient γ. This acceleration factor represents the fraction of transferred sulfur species with oxidation degree (-II) in the H_2S form at the pH of the solution considered, and is determined from the pKa values of H_2S/HS^- and HS^-/S^{2-}.

The values of the global transfer coefficients K_G are determined with the local transfer coefficients k_G and k_L. These coefficients, as well as the interfacial areas, are calculated from empirical relationships proposed by Onda et al. (1968) for the packed column and by Laddha and Degaleesan (1978) (cited by Perry & Green, 1984) for the spray column. These correlations allow the prediction of the experimental values of exchange coefficients to be reached with a precision of 20–30%. The effect of

Table 15.7 Expressions of α and γ according to operating parameters.

Plug flow of the gas phase	$\alpha = \dfrac{-K_G a S}{Q_G}$
Plug flow with axial dispersion of the gas phase	$\alpha = \dfrac{U_G}{2D_a}\left(-1 + \sqrt{1 + 4\dfrac{K_G a D_a}{U_G^2}}\right)$
Transfer without reaction in the liquid phase	$\gamma = \dfrac{-K_G a S}{\alpha Q_L}$
Transfer with reaction in the liquid phase	$\gamma = \dfrac{K_G a S}{\alpha Q_L}\underbrace{\left(1 + \dfrac{K_2}{[H_3O^+]} + \dfrac{K_1 K_2}{[H_3O^+]}\right)^{-1}}_{E}$

dynamic and static volumes V_1 and V_2 of the liquid phase in the column were taken into account using *a* corrective factor, applied following Equation (15.39) on the values of interfacial area *a* determined with the specific empirical relationships of each contact mode.

$$a_{\text{model}} = \left(1 - \frac{V_2}{V_1 + V_2}\right) a \tag{15.39}$$

The model should permit the resolution of the set of mass balance equations for each transfer zone. The calculation is realized for each transfer zone by the introduction of a partial transfer rate τ_1. The resolution of the mass balance equations is obtained using an iterative method, based on the estimation of the initial values of the global and partial transfer rates τ_0 and τ_{10}. For the iteration *k*, the values of the liquid concentrations at the outlet of the spray column $C_L^{J_1+1}$ and at the outlet of the column C_L^1 are deduced from the following Equations:

$$C_L^{J_1+1} = C_L^{J_2+1} + \gamma_2(\tau - \tau_1)\, C_{Ge} \tag{15.40}$$

$$C_L^1 = C_L^{J_1+1} + \gamma_1 \tau_1 C_{Ge} \tag{15.41}$$

The procedure of calculation stops when at the iteration *k*, the absolute value of the functions $|F(\tau_1)|$ and $|F(\tau)|$, given by Equations (15.42) and (15.43), are lower than $\varepsilon(\varepsilon = 10^{-6})$.

$$F(\tau_1) = \frac{\tau_1 C_{Ge}}{\beta_1} - \sum_{i=1}^{J_1}(C_G^{i-1} - H'C_L^i) \tag{15.42}$$

$$F(\tau) = \left(\frac{\tau - \tau_1}{\beta_2}\right) C_{\mathrm{Ge}} - \sum_{i=J_1+1}^{J_2} (C_{\mathrm{G}}^{i-1} - H'C_{\mathrm{L}}^{i}) \tag{15.43}$$

where:

$$\beta_1 = 1 - \exp\left(\alpha_1 \frac{Z_1}{J_1}\right) \tag{15.44}$$

$$\beta_2 = 1 - \exp\left(\alpha_2 \frac{Z_2}{J_2}\right) \tag{15.45}$$

The change of the value of τ at any iteration is affected by dichotomy between the maximal or minimal value of τ, according to the sign of the products $[F(\tau_k)\,F(\tau_{k-1})]$ and $[F(\tau_k)\,F(1)]$.

15.5.3.2 Comparison of experimental and computed results

The calculated results with the proposed model were compared to experimental results of the removal efficiencies of both ethanol and hydrogen sulfide, in order to validate the theoretical procedure (Figure 15.15). A Student's test, realized on the coefficient b of the regression line equation ($y_{\mathrm{calc}} = b\,y_{\mathrm{exp}}$), allows to conclude that the proposed model is appropriated for the prediction of the transfer efficiency with a 95% confidence interval, as it was shown that the difference between the two slopes were not significant. The low value of the R^2 coefficient ($R^2 = 0.9883$) could be explained by the experimental errors due to the measurement precisions (about 16% on transfer rates

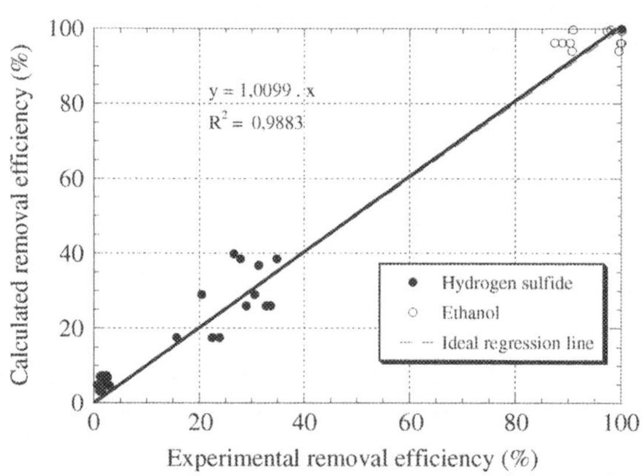

Figure 15.15 Comparison of calculated and experimental values of the compound removal efficiency of a bioscrubber treating ethanol or hydrogen sulfide present in air (Humeau et al., 2004; Le Cloirec et al., 2001).

of H_2S at pH = 8), and by the precision of the correlations used for the prediction of the exchange coefficients.

Thus, the elaborated model, from the residence time distribution characteristics, is able to reproduce the global trends concerning the influence of both the operating conditions (flow velocities of gas and liquid in the column, contact mode through hydrodynamic parameters) and the physico-chemical parameters of the washing solution (pH). It is then possible to define the influence of operating conditions on the transfer operation with a study of parametric sensitivity of the model.

15.5.3.3 Influence of the biodegradation step on the deodorization efficiency

The recycling of the washing solution after biological treatment in the activated sludge basin could affect the deodorization performance of a suspended-growth bioscrubber. Indeed, the biological oxidation of the transferred pollutants could lead to the formation of volatile and odorous by-products (Humeau et al., 2004; Le Cloirec et al., 2001). The concentrations of the by-products at the outlet of the biological reactor, represented by a mixed flow reactor, could not be negligible or zero.

It is possible to apply the previous model to estimate the impact of compounds present in the washing solution on the deodorization efficiency, by introduction of a by-product concentration value at the inlet of the liquid phase in the gas–liquid contactor (Figure 15.16).

To give an example, a simulation of gas–liquid transfer was conducted on the column with the two successive contact modes, with a gas velocity of $0.68\,m\,s^{-1}$, a liquid load of $0.577\,kg\,m^{-2}s^{-1}$ and an ethanol concentration of $50\,mg\,m^{-3}$ in the polluted

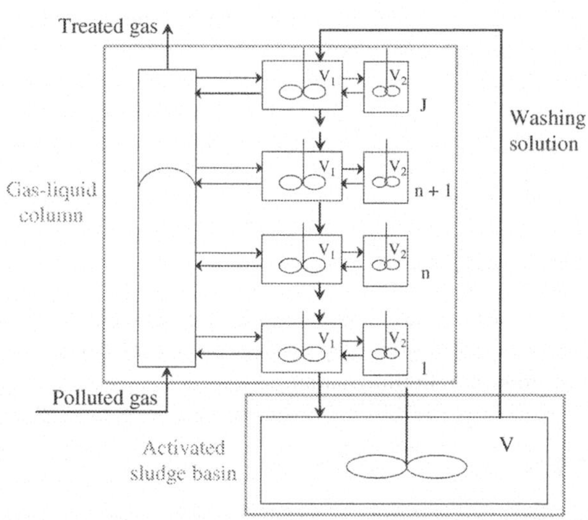

Figure 15.16 Model for the deodorization efficiency prediction of a suspended-growth bioscrubber.

gas. Then, two cases were considered: (1) where the complete oxidation of ethanol leads to the formation of CO_2 and water and (2) where a mole of transferred ethanol is converted in one mole of acetic acid (Henry's law constant = 2.455 atm (mole fraction)$^{-1}$). In this last case, the initial concentration of acetic acid in the washing solution at the inlet of the column is 77 mg L^{-1}. In both cases, the ethanol concentration in the treated gas is very low, as the calculated transfer efficiency is 99.9%. When no odorous compound is emitted from a complete oxidation of ethanol, the predicted concentration of acetic acid in the gas phase at the outlet of the column is 3 mg m^{-3} in the second case. This concentration is close to the detection limit of acetic acid (vinegar) odor generated by this molecule, which ranges between 0.025 and 6.5 mg m^{-3} (Martin & Laffort, 1991). These results show that the deodorization efficiency of a bioscrubber depends on the operating conditions, especially on the oxygen concentration in the biological reactor. If the operating conditions in the column are not able to oxygenate the basin properly, then undesired by-products as acetic acid could be generated and could reduce the overall deodorization efficiency of the process.

This theoretical approach emphasizes the great interactions between the transfer column and the activated sludge basin. A mismanagement of one of the two reactors could affect the deodorization efficiency of the suspended-growth bioscrubber, especially when the biodegradation by-products are volatile and odorous compounds.

15.6 CONCLUSIONS

In this chapter, different bioprocesses used to control waste gas have been presented: (1) biofilters, (2) biotrickling filters and (3) bioscrubbers. For each technology, principles, operating conditions, performances and modeling have been described. In these complex systems, however, the mechanisms of mass transfer and biodegradation are not well known. Some hypotheses of flow, mass transfer or biodegradation kinetics are used to simplify the modeling relations. A better understanding to approach the engineering of these bioprocesses is needed.

Strong collaborations and research programs between microbiologists, biochemical and chemical engineers are required to obtain realistic biodegradation mechanisms and performance sets of modeling equations. Then, a good design of processes and simulation of these systems will be obtained.

ACKNOWLEDGMENTS

The author gratefully acknowledges the French Ministry of Industry and ADEME (French Environmental and Energy Agency) for its partial financial support of the research programs presented in this chapter.

REFERENCES

Astarita, G. and Gioia, F. (1964) Hydrogen sulphide chemical absorption. *Chem. Eng. Sci.* **19**, 95–103.

Comiti, J. and Renaud, M. (1989) A new model for determining mean structure parameters for fixed beds from pressure drop measurements: application to beds packed with parallelepipedal particles. *Chem. Eng. Sci.* **44**(7), 1539–1545.

Coulson, J.M., Richardson, J.F. and Peacok, D.G. (1994) *Chemical Engineering*, 3rd edn. Pergamon, Elsevier Science, Oxford, UK.

Cox, H.H.J., Moerman, R.E., Van Baalen, S., Van Heiningen, W.N.M., Dodeema, H.J. and Harder, W. (1999) Performance of a styrene-degrading biofilter containing the yeast *Exophialia jeanselmei. Biotechnol. Bioeng*. **53**, 259–266.

Deshusses, M.A., Hammer, G. and Dunn, I.J. (1995a) Behavior of biofilters for waste air treatment. 1. Dynamic model development. *Environ. Sci. Technol*. **29**, 1048–1058.

Deshusses, M.A., Hammer, G. and Dunn, I.J. (1995b) Behavior of biofilters for waste air treatment. 2. Experimental evaluation of a dynamic model. *Environ. Sci. Technol*. **29**, 1059–1068.

Devinny, J.S. and Hodge, D.S. (1995) Formation of acidic and toxic intermediates in overloaded ethanol biofilters. *J. Air Waste Manage. Assoc*. February; **45**, 125–131.

Devinny, J.S., Deshusses, M.A. and Webster, T.S. (1999) *Biofiltration for Air Pollution Control*. Lewis Publishers, Boca Raton, FL.

Diks, R.M.M. and Ottengraf, S.P.P. (1991a) Verification studies of a simplified model for the removal of dichloromethane from waste gases using a biological trickling filter (Part I). *Bioprocess Eng*. **6**(3), 93–99.

Diks, R.M.M. and Ottengraf, S.P.P. (1991b) Verification studies of a simplified model for the removal of dichloromethane from waste gases using a biological trickling filter (Part II). *Bioprocess Eng*. **6**(4), 131–141.

Fass, S., Vaudrey, H., Vogel, T.M. and Block, J.C. (1999) Factors controlling the biodegradation of chemicals in soils. In *Bioavailability of Organic Xenobiotic in the Environment: Practical Consequences for the Environment* (eds. P. Bayes, J.C. Block, V.V. Goncharruk), Vol. 64, pp. 93–117. NATO ASI Series 2 Environment.

Hansen, N.G. and Rindel, K. (2000) Bioscrubbing, an effective and economic solution to odour control at wastewater treatment plants. *Wat. Sci. Technol*. **41**(6), 155–164.

Herrygers, V., Van Langenhove, H. and Smet, E. (2000) Biological treatment of gases polluted by volatile sulfur compounds. In *Environmental Technologies to Treat Sulfur Pollution* (ed. P. Lens and L.H. Pol), 1st edn, pp. 281–304. IWA Publishing, London.

Humeau, P., Baléo, J.N., Raynaud, F., Bourcier, J. and Le Cloirec, P. (2000) Flow characterization in a gas–liquid column: application to a bioscrubber for the deodorization of waste gases. *Water Sci. Tech*. **41**(12), 191–198.

Humeau, Ph., Pré, P. and Le Cloirec, P. (2004) Optimization of bioscrubber performances: experimental and modeling approaches. *J. Environ. Engn*. **130**(3), 314–321.

Kennes, C. and Thalasso, F. (1998) Waste gas biotreatment technology. *J. Chem. Technol. Biotechnol*. **72**, 303–319.

Kosteltz, A.M., Finkelstein, A. and Sears, G. (1996) What are the « Real opportunities » in biological gas cleaning for North America. *Proceedings of the 89th Annual Meeting and Exhibition*, Nashville, TN, 23–28 June, 15 pp.

Le Cloirec, P. (1998) *Les composés organiques volatils dans l'environnement*. Tec & Doc, Lavoisier, Paris, France, 734 pp.

Le Cloirec, P., Humeau, P. and Baléo, J.N. (1999) Processes for biological purification of air loaded with volatile organic compounds. *Biodépol*'99, Rennes, October 26 and 27, 25–41.

Le Cloirec, P., Humeau, Ph. and Ramirez-Lopez, E.M. (2001) Biotreatment of odours: control and performances of a biofilter and a bioscrubber. *Water Sci. Technol*. **44**(9), 219–226.

Le Cloirec, P., Andrès, Y., Gérente, C. and Pré, P. (2005) Biological treatment of waste gases containing volatile organic compounds. In *Biotechnology for Odour and Air Pollution Control* (eds. Z. Shareefdeen and A. Singh), pp. 281–302. Springer Verlag, Heidelberg, Germany, Chap. 13.

Levenspiel, O. (1999) *Chemical Reaction Engineering*, 3rd edn. John Wiley & Sons, New York.

Littel, R.J., Filmer, B., Versteeg, G.F. and Van Swaaij, W.P.M. (1991) Modelling of simultaneous absorption of H_2S and CO_2 in alkanolamine solutions: the influence of parallel and consecutive reversible reactions and the coupled diffusion of ionic species. *Chem. Eng. Sci*. **46**, 2303–2313.

Martin, G. and Lafford, P. (1991) *Odeur et Désodoristaiondans l'Environnement*. Tec & Doc, Lavoisier, Paris.

Morris, G.A. and Jackson, J. (1953) *Absorption Towers*. Butterworths Publications Ltd., London.

Onda, K., Takeuchi, H. and Okumoto, Y. (1968) Mass transfer coefficients between gas and liquid phases in packed columns. *J. Chem. Eng. Jpn.* **1**, 56–62.

Ottengraf, S.P.P. (1986) Exhaust gas purification. In *Biotechnology a Comprehensive Treatise* (eds. J.L. Rhem and G. Reed), Weinheim, Germany.

Ottengraf, S.P.P. (1987) Biological systems for waste gas elimination. *Trend. Biotechnol.* **5**, 132–136.

Perry, R.H. and Green, D.W. (1984) *Perry's Chemical Engineers' Handbook*. McGraw Hill, New York, USA.

Ramirez-Lopez, E.M., Montillet, A., Comiti, J. and Le Cloirec, P. (2000) Biofiltration of volatile organic compounds – application to air treatment. *Water Sci. Technol.* **41**(12), 183–190.

Roustan, M. (2003) *Transferts Gaz-Liquide dans les Procédés de traitement des Eaux et des Effluents Gazeux*. Tec & Doc, Lavoisier, Paris, France.

Shareefdeen, Z., Baltzis, B.C., Oh, Y.S. and Bartha, R. (1993) Biofiltration of methanol vapor. *Biotechnol. Bioeng.* **41**, 512–524.

Veiga, M.C. and Kennes, C. (2001) Parameters affecting performance and modeling of biofilters treating alkylbenzene-polluted air. *Appl. Microbiol. Biotechnol.* **55**, 254–258.

16

Fungal and dechlorinating biocatalysts in waste gas treatment

C. Kennes, Y. Jin and María C. Veiga

16.1 INTRODUCTION

Conventional biofilters using natural, mainly organic, filter beds as compost, peat or soil have been used for several decades for the treatment of polluted air. Such reactors were operated under empirical conditions using a rule-of-thumb strategy. Biofilters were originally exclusively used for the removal of odours, for example at wastewater treatment plants or composting facilities (Kennes & Veiga, 2001). In such systems, the indigenous microbial populations present in the filter beds act as biological catalysts able to degrade the odour-compounds. More recently, the long-term stability of conventional biofilters has been improved by using inert packing materials, often inoculated with undefined mixed cultures as activated sludge (Kennes & Thalasso, 1998; Kennes & Veiga, 2001). A lot of research was done over the past two decades to improve the performance of the conventional biofilters. Research has recently focused on the development of new reactor configurations and new biocatalysts (Kennes & Veiga, 2001). Consequently, the range of applications of gas-phase bioreactors for air pollution control has significantly increased. Nowadays, the technology can be applied to many different types of waste gases. Today, gas-phase bioreactors are not only used to treat odours at wastewater treatment plants or composting facilities; they are also useful tools for the treatment of industrial waste gases containing different organic or inorganic compounds. At the same time, higher loads have progressively been treated than in the original biofilters, often at relatively low residence times and with high removal efficiencies.

This chapter overviews the data available on a selection of new biocatalysts and biocatalytic activities used for air pollution control. It does not pretend to be an exhaustive overview and will focus on two major topics: fungal reactors and dechlorinating systems. The reader is referred to review papers for additional information of volatile organic compound (VOC) degradation in fungal reactors (Kennes & Veiga, 2004) or on some other recently studied catalytic activities in waste gas treatment as, for example NO-removal (Jin et al., 2005a).

16.2 FUNGAL ACTIVITIES USED IN GAS-PHASE BIOREACTORS

16.2.1 Introduction

A few authors have suggested the interest of using filamentous fungi in biofilters, mainly for the treatment of hydrophobic pollutants. The main reasons for this is that such organisms form a filamentous network (Figure 16.1) that would enhance the mass transfer of the pollutants from the gas phase to the biocatalyst, due to the large surface area of mycelia and the fact that the hyphae grow in the air, allowing direct gas–solid biocatalysis. Nevertheless, such increased mass transfer has not yet been demonstrated experimentally. Other advantages related to the use of fungi is their relatively high activity under extreme conditions, for example when the pH of the filter bed drops or when the water activity is low (Table 16.1). With a few exceptions (Veiga et al., 1999), most bacteria growing on aromatic VOCs are inhibited in acid environments. On the other hand, a potential problem with filamentous fungi is that, so far, only few such organisms have been identified and studied for their capability to degrade hydrophobic pollutants as alkylbenzenes, styrene, α-pinene, or other common aromatic air pollutants. Thus, very little information is available and much basic research is still needed. Besides, the few studied organisms seem to degrade a quite narrow range of substrates (Kennes & Veiga, 2004), which may be a problem when mixed pollutants are present in a specific waste gas. Nevertheless, it is also worth recalling that most of the recently published data concern pure cultures, while, as in the case of bacteria, mixed fungal populations are expected to develop in gas-phase bioreactors, exhibiting a wider metabolic diversity capable to degrade mixtures of pollutants.

16.2.2 VOC-degrading fungi

Table 16.2 presents data of pure filamentous fungal cultures able to use aromatic hydrophobic VOCs as sole source of carbon and energy. They were isolated from gas-phase biofilters. Table 16.2 shows that most studies have been done with either benzene, toluene, ethylbenzene, styrene or α-pinene.

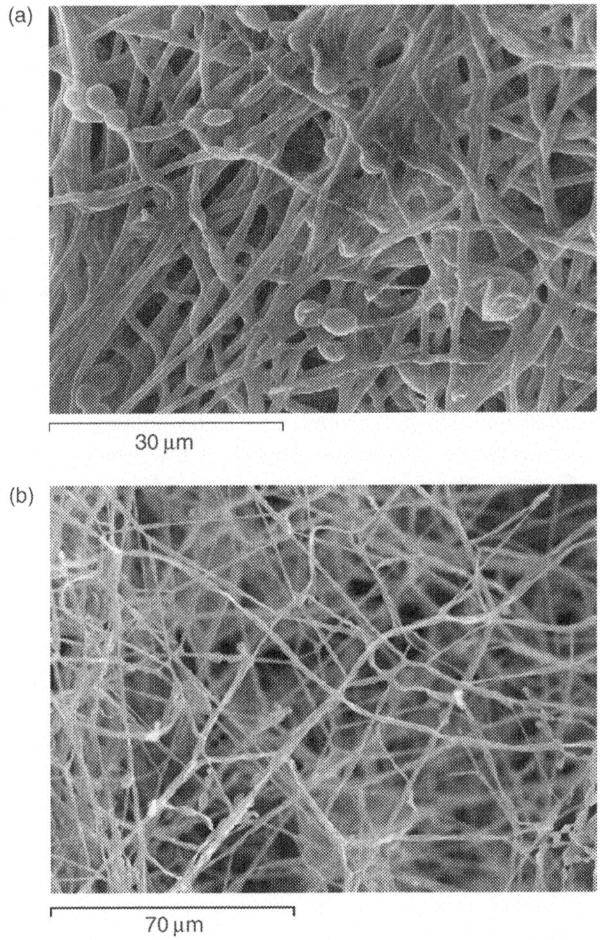

Figure 16.1 SEM pictures of filter-bed samples from gas-phase packed-bed bioreactors: (a) perlite colonized with the yeast-like fungus *Exophiala oligosperma* (Estévez et al., 2005b) and (b) lava rock colonized by a culture of the fungus *Ophiostoma* sp. (Jin et al., 2006).

Table 16.1 Advantages and drawbacks of filamentous fungi applied for the biofiltration of VOC containing waste gases.

Advantages	Drawbacks
Filamentous morphology favourable to mass transfer	Narrow substrate range
Active at relatively low pH	Faster pressure drop increase
Active at relatively low water activity	

Toluene is one of the most frequently encountered benzene compounds in waste gases. It is also the most widely studied alkylbenzene in biofiltration experiments performed at laboratory scale. This is the reason why most filamentous fungi growing on aromatic hydrophobic VOCs were isolated on toluene. Although only aromatic pollutants are listed in Table 16.2, some recent biofiltration studies have also been undertaken with *n*-hexane as air pollutant and *Aspergillus niger* (Spigno et al., 2003) or *Fusarium solani* (Arriaga et al., 2005) as fungal biocatalyst. The use of such organisms for the removal of *n*-hexane from waste gases is of practical interest. However, although the isolation of filamentous fungi degrading aromatic VOCs is rather recent, their growth on *n*-alkanes was shown already several decades ago (Kennes & Veiga, 2004).

Cladosporium sphaerospermum was the first organism that was unequivocally proven to grow on toluene and use that substrate as sole carbon and energy source (Weber et al., 1995). Around two-thirds of the pollutant is recovered as carbon dioxide and approximately one-third is converted to biomass. This is a very important characteristic when considering the biofiltration of polluted air, since the accumulation of intermediate metabolic products should be avoided. The accumulation of such intermediate metabolites was observed in ligninolytic fungi grown on BTEX (benzene, toluene, ethylbenzene and xylenes) mixtures, in which sometimes less than 5% of the original substrates was recovered as carbon dioxide in experiments with radiolabelled compounds (Yadav & Reddy, 1993). The accumulation of intermediate metabolites is quite usual in white-rot fungi as *Phanerochaete chrysosporium* (Kennes & Lema, 1994). Mass balance experiments are not always reported in manuscripts published in the scientific literature. Nevertheless, carbon dioxide is the major end product in most isolated organisms, as confirmed in batch assays for strains as *Cladophialophora* sp., *Exophiala oligosperma*, *Exophiala* sp. T4 and *Paecilomyces variotii* (Kennes & Veiga, 2004). For the other strains mentioned in Table 16.2, the products were most probably also mainly carbon dioxide and biomass. However, for those strains data of carbon dioxide concentrations were only reported during biofiltration studies which is not really a reliable information to do mass balance calculations, since other processes as endogenous respiration may contribute to carbon dioxide production.

Table 16.2 Fungal-biocatalysts isolated from biofilters and growing on VOCs as their sole source of carbon and energy.

Organism	VOC-substrate(s)	pH of activity (*)	Optimum temperature (°C)	References
Cladosporium sphaerospermum	Toluene	7 (optimal)	NR	Weber et al. (1995)
Exophiala jeanselmei	Styrene	3–5.7	22–33	Cox et al. (1996, 1997)
Exophiala lecanii-corni	Styrene Benzene Toluene Ethylbenzene	3.5–6.5	NR	Woertz et al. (2001)
Exophiala oligosperma	Toluene	3.9–6.9	30–35	Estévez et al. (2005a, b)
Paecilomyces variotii	Toluene	3.9–6.9	30–35	Estévez et al. (2005a, b)
Scedosporium apiospermum	Toluene	NR	30–37	García-Peña et al. (2001)
Trichosporon beigelii	Toluene	NR	30–35	Veiga et al. (1999)
Ophiostoma sp.	α-pinene	NR	30–35	Jin et al. (2005b)

(*) refers to the pH range studied in batch assays, and it is not excluded that a wider range could be suitable.
NR: not reported.

All the fungi listed in Table 16.2 grow under mesophilic conditions, at temperatures ranging between room temperature (around 20°C) and sometimes 35–37°C. One organism, namely *P. variotii*, was able to degrade toluene at 40°C, although with a lower toluene degrading activity than under optimal conditions of temperature (Estévez et al., 2005a). Higher temperatures were not checked. The fungal biodegradation of hydrophobic VOCs under thermophilic conditions has not been reported so far, although this might be an interesting characteristic presently evaluated in our group, since biofiltration at temperatures exceeding 30–35°C increases drying out problems and fungi are known to be more resistant than most bacteria to relatively dry environments.

16.2.3 Fungal gas-phase bioreactors

Filamentous fungi are interesting for the removal of hydrophobic compounds, among others in acidic environments and under conditions of low water content. Therefore, their main application will be in conventional biofilters. In other reactors as, for example, biotrickling filters, the water content is not limiting. Besides, pH fluctuations can be much more easily controlled in biotrickling filters by the continuous feed and recirculation of the aqueous phase. The configuration of biofilters and biotrickling filters is compared in Figure 16.2.

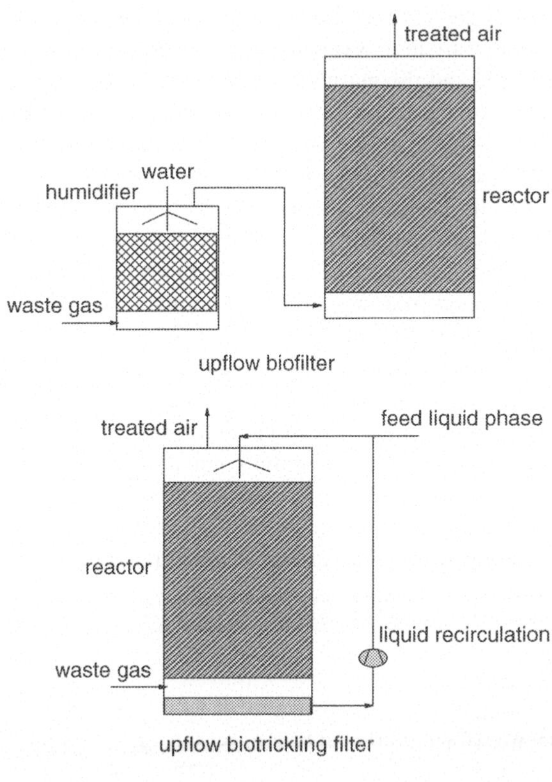

Figure 16.2 Typical configuration of a biofilter (upper) and a biotrickling filter (lower).

Performance data of fungal biofilters have been reported for reactors inoculated with either mixed cultures in which fungi became dominant or with pure fungal cultures. Data with pure cultures in which the inoculated strain remains dominant throughout the experimental study are scarce. This is, however, an important point since it allows to confirm that the collected performance data and observed phenomena (i.e. elimination capacity, pressure drop) are indeed exclusively the result of the presence of fungal biocatalysts. Otherwise, if heterogeneous bacterial–fungal communities develop, erroneous conclusions could be drawn on the effect of such filamentous fungi in biofiltration.

16.2.3.1 Water content

Only few studies have reported on the effect of the relative humidity of the waste gas and on the influence of the water content of the filter bed on the bioreactor performance. Although fungi are known to be more resistant to dry conditions than most bacteria, some experimental data show that they are still quite sensitive to drying out phenomena above all if such conditions are maintained for a long period of time of several weeks. In a recent study, fungal biofilters were fed toluene-polluted air with relative humidities (RH) ranging from 85% to 100% (Estévez et al., 2005b). The fungus *E. oligosperma* was used as inoculum and it was confirmed that it remained the single-dominant organism throughout the study which allows reliable comparison of results (Figure 16.1). In the biofilter fed water-saturated air (100% RH), a maximum elimination capacity of 95.6 $g/m^3/h$ was reached. Conversely, at an RH of 85%, the biofilter was more unstable and the maximum elimination capacity was only 48.7 $g/m^3/h$, under otherwise identical conditions. It is important to mention that a small volume of nutrient solution was added once a week to each reactor in order to avoid nutrient depletion. Such addition did not affect the RH of the waste gas, but it provided some water to the filter bed itself, which did also contribute to the results obtained at the lowest RH of 85%. Indeed, in another study it was shown that when feeding air with an RH of 80% to a biofilter packed with perlite, the water content of the packing material gradually dropped from about 65% to approximately 45% when no water was added to the filter bed (Cox et al., 1996). Conversely, at a RH of 100%, the water content remained basically constant over a 1-week period. A water content of 45% is usually still satisfactory for most fungal biofilters. In another study performed with a toluene-fed biofilter in which fungi were dominant, it appeared that the performance started to decrease somewhat when the water content of the filter bed dropped from 50–55% to 30% (Woertz et al., 2001). A significant drop of performance was also observed when the water content dropped below 30% for a few days. In all cases, the biofilters recovered their original performance when readjusting the water content to 50–65%.

16.2.3.2 pH

Although the optimal pH for fungal biodegradation of hydrophobic VOCs depends on the organism, it was observed that at least several *Exophiala* spp. exhibit their optimal activity around pH 5.5–6.0 (Estévez et al., 2005a). Some fungi may present a quite wide optimal pH. In that sense, the styrene degradation rate in *E. jeanselmei*

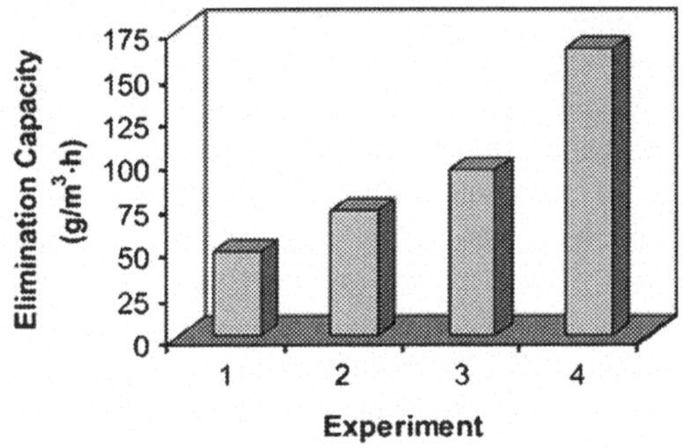

Figure 16.3 Maximum toluene elimination capacities under different conditions of RH and pH: (1) 85% RH, pH 4.5, (2) 85% RH, pH 5.9, (3) 85% RH, pH 7.5, (4) 100% RH, pH 7.5 (Estévez et al., 2005b).

hardly varies between pH 3 and 6 (Cox et al., 1997). The optimal pH is similar in batch and in biofiltration experiments. In another study, when feeding nutrient solutions with three different initial pHs of either 7.5, 5.9, or 4.5 to biofilters fed toluene-polluted air and inoculated with *E. oligosperma*, the best performance and the highest elimination capacity were reached when feeding the solution of pH 7.5. Under such condition, the pH of the drain water leaching from the reactor reached a value of 5.9 ± 0.5 which is known to be optimal for that organism (Estévez et al., 2005a), while its value dropped down to 2.5–3.5 in the other cases, that is at an initial pH of 5.9 or 4.5. The acidification results from the use of ammonium as nitrogen source with the release of protons in the medium during its consumption. The maximal removal efficiencies reached in a toluene-treating biofilter using *E. oligosperma* as biocatalyst are compared in Figure 16.3, for different RH and pH values (Estévez et al., 2005b).

Other authors report somewhat different results when using a similar medium with ammonium as nitrogen source (van Groenestijn et al., 2001). In their study, two similar solutions were fed, with their pH adjusted to either 4 or 8. In the first case, the pH gradually dropped to 2.3 over a period of 56 days, while it dropped down to 6.4 in the other case. The best performance was obtained with the pH of the nutrient solution adjusted to 4. Thus, it is possible that the optimal pH for the organisms present in the biofilters was closer to 2.3 than to 6.4. However, it is worth mentioning that the biofilters were originally inoculated with a complex mixed bacterial–fungal culture and that most probably different microbial populations developed in each reactor. Therefore, the pH was not the only variable parameter and the different elimination capacities achieved in each system may also have been the result of other factors than the pH value.

16.2.3.3 Pressure drop

It has been suggested, and sometimes observed, that the growth of filamentous organisms in packed bed bioreactors may result in a fast increase of the pressure drop (García-Peña et al., 2001; van Groenestijn et al., 2001). However, some authors have operated fungal biofilters for periods ranging between 5–6 months (Estévez et al., 2005b; Woertz et al., 2001) and almost 1 year (Cox et al., 1997) without significant variations in head losses nor performance. Cox et al. (1997) observed that the pressure drop remained basically negligible in a styrene-degrading biofilter even after 300 days operation. Woertz et al. (2001) reported pressure drops below 0.2 cm in a 30 cm packed-bed biofilter for a period of 5 months. After that period a rapid pressure drop increase was observed, which was concomitant with a stronger filamentous growth rather than a yeast growth. This could have resulted from a modification of both the gas- and liquid-phase feeding strategies. Other authors observed that the pressure drop remained below 6 mm water/m after 6 months operation of fungal biofilters treating high loads of either α-pinene (Jin et al., 2005b, 2006) or toluene (Estévez et al., 2005b).

If the pressure drop increases too much, several techniques are suitable for the removal of excess biomass from packed-bed gas-phase reactors, including backwashing and air sparging, using either water or aqueous solutions containing specific chemicals. The efficiencies of these different strategies have recently been compared (Mendoza et al., 2004). Other biomass cleaning methods suitable for fungal biofilters are based on the use of specific predators as mites (van Groenestijn et al., 2001), although little information is available on the performance of such techniques and on their possible implementation at industrial scale. It was observed that fungal biofilters in which mites were added exhibited a better performance with higher elimination capacities under long-term and stable conditions (van Groenestijn et al., 2001).

16.3 DECHLORINATING BIOCATALYSTS

16.3.1 Chlorinated compounds and air pollution

Research on the biological treatment of exhaust air contaminated with chlorinated compounds started less than two decades ago, during the 1980s. The literature published since then on that topic has been scarce and has focused almost exclusively on three types of pollutants: dichloromethane (DCM) (also called methylene chloride), chlorobenzenes and chlorinated ethylenes (mainly trichloroethylene, also known as TCE). Although incineration and carbon adsorption have traditionally been used to treat exhaust gases containing such chlorinated compounds, these technologies are rather costly and may even create new environmental problems such as the conversion of the original pollutants into new toxic compounds or the transfer of the contaminants from one phase, that is gas, to another one, that is activated carbon in adsorption processes. Therefore, bioprocesses have recently emerged as cost-effective alternatives to the conventional systems. It is worth noting that the annual production and use of these chlorinated chemicals follows a decreasing tendency as a result of their undesirable environmental impact. Chlorinated compounds are harder to be biodegraded than non-halogenated ones as alkanes, alcohols or alkylbenzenes. Therefore, residence times (largely) exceeding 1 min are the rule rather than the exception, while relatively low maximum elimination capacities are often reached.

DCM is used as a solvent in the chemical industry. As most halogenated compounds, it is considered a health hazard. However, it is more soluble in aqueous solutions than many other chlorinated pollutants and will thus not easily accumulate in body fat (Table 16.3). Its Henry coefficient is also rather low compared to other chlorinated volatile compounds.

Table 16.3 Water solubilities and Henry coefficients of some chlorinated air pollutants . It is worth noting that a quite wide range of values is reported in the literature. Depending on the source, fluctuations may sometimes reach more than 50%.

Pollutant	Formula	Solubility (mg/L, 25°C)	Henry's coefficient (−)
Perchloroethylene (PCE)	C_2Cl_4	133	0.723
Trichloroethylene (TCE)	C_2HCl_3	1104	0.392
Carbon tetrachloride (CT)	CCl_4	1155	1.244
Dichloromethane (DCM)	CH_2Cl_2	19366	0.089
1,2-Dichlorobenzene (1,2-DCB)	$C_6H_4Cl_2$	156	0.078
1,4-Dichlorobenzene (1,4-DCB)	$C_6H_4Cl_2$	74	0.100
Monochlorobenzene (MCB)	C_6H_5Cl	472	0.152

Gossett (1987); Mackay and Shiu (1981).

Chlorobenzenes are mainly used as solvents and as raw material for the production of other compounds. DDT is one such product, although its use has been banned. Phenol, aniline and poly (phenylene sulfide) are other examples of products that can be produced from chlorobenzene. Mono- and dichlorobenzenes (DCB) are the most widely used halogenated benzenes.

TCE is used as a solvent and chemical feedstock for the production of hydrofluorocarbon (HFC), used to substitute CFC-12 in refrigerant applications. It is also widely used in dry cleaning and as a cleaning and degreasing agent in metal working. Perchloroethylene (PCE) is suitable for similar purposes as TCE. TCE and PCE are still widely used today for their versatility and relatively low cost and since they are environmentally acceptable substitutes for other highly contaminating chlorinated chemicals, for example 1,1,1-trichloroethane.

16.3.2 Removal of chlorinated compounds from air in bioreactors

16.3.2.1 Introduction

The treatment of air contaminated with chlorinated compounds presents some common characteristics irrespective of the type of pollutant. In all cases, chloride is released in the medium and hydrogen chloride is formed as an end product, resulting in medium acidification. Although some authors have used conventional biofilters with organic filter beds for the treatment of chlorinated pollutants (Coutu et al., 2003; Delhoménie & Heitz, 2003; Ergas et al., 1994; Roberge et al., 2001), biotrickling filters are expected to be more stable and more performant and have been used by many researchers. One advantage of biotrickling filters in the treatment of air polluted with chlorinated compounds is the continuous feed and recirculation of an aqueous phase (Figure 16.2), which allows pH regulation and the removal of inhibitory compounds. The addition of a base as sodium hydroxide to neutralize the hydrochloric acid released from dechlorination represents one of the major operational costs of gas-phase reactors treating chlorinated pollutants (Kraakman, 2005).

16.3.2.2 Aerobic dechlorination of DCM

Biodegradation In the early 1980s, DCM was shown for the first time to be biodegradable under aerobic conditions (Brunner et al., 1980; Rittmann & McCarty, 1980). Nowadays, it is well known that this compound can be readily degraded by several different genera using it as their sole carbon and energy source. More recently, biodegradation was shown to be possible under anaerobic conditions as well (Braus-Stromeyer et al., 1993). The aerobic degradation of DCM yields 1 mol of carbon dioxide and 2 mol of hydrogen chloride, with medium acidification, according to the following reaction:

$$CH_2\text{—}Cl_2 + O_2 \rightarrow CO_2 + 2HCl$$

The overall reaction is quite similar under anaerobic conditions, except that 1 mol of methane would be formed instead of carbon dioxide.

For chlorinated methanes with a higher number of chlorines, as chloroform ($CH\text{—}Cl_3$) or carbon tetrachloride ($C\text{—}Cl_4$), anaerobic conditions are much more favourable for biodegradation, although co-metabolic aerobic degradation has occasionally been reported (Bartnicki & Castro, 1994; Strand & Shippert, 1986).

Biofilters Most studies on the treatment of DCM-polluted air in bioreactors have been undertaken with biotrickling filters. However, one study reports the removal of low concentrations of that compound, that is 3–50 ppmv, in a conventional compost-biofilter in the presence of toluene at empty bed residence times (EBRT) of 0.7– 1.0 min. The reason for choosing a conventional biofilter was that at low inlet concentrations, biofilters are more favourable for mass transfer than biotrickling filters due to higher specific surface areas and a smaller liquid film thickness (Ergas et al., 1994). The main problem linked to DCM degradation in such a biofilter was medium acidification, despite adding crushed oyster shells (130 $g/kg_{filter\ bed}$) to the compost:perlite (1:1) bed. Although acidification inhibited the activity of the dechlorinating organisms, increasing the pH again to near neutral values restored the original microbial activity. The solutions suggested to avoid a pH drop in conventional DCM-treating compost-biofilters were the formulation of media with an increased buffering capacity and periodical washing of the media with a buffer solution.

Biotrickling filters Most other researchers used biotrickling filters (Diks & Ottengraf, 1991a, b; Hartmans & Tramper, 1991; Okkerse et al., 1999a, b; Spigno et al., 2005). Biotrickling filters are more stable for DCM removal and can reach elimination capacities ranging from 50 $g/m^3/h$ with removal efficiencies of 100% to 200 $g/m^3/h$ with somewhat lower removals (Table 16.4). The corresponding inlet concentrations used in most studies ranged from 0.4 to about 10 g/m^3 (Diks & Ottengraf, 1991a, b; Hartmans & Tramper, 1991; Okkerse et al., 1999a). Fluctuations in the DCM supply does not affect the stability of the system under the above mentioned operation range (Hartmans & Tramper, 1991). In all the studies, the optimal pH was near neutral and the biocatalysts appear to be quite sensitive to medium acidification. Since biotrickling filters allow for pH control of the trickling phase, high elimination capacities can be maintained by means of pH regulators keeping the pH constant around 7 or 8.

Table 16.4 Comparison of the different biological DCM removal methods.

Bacteria	Reactor/packing	Co-substrate	O_2	Flow	pH	Inlet (ppm)	EBRT (min)	RE (%)	EC ($g/m^3/h$)	References
Hyphomicrobium sp. GJ21	BTF/ceramic Novalox saddles	None	A	D/U	8	10,000	0.48		157	Diks and Ottengraf (1991a, b)
Hyphomicrobium DM2, *Methylobacterium* DM4, and *Hyphomicrobium* DM20	BTF/polypropylene packing	Methanol	A	U	7 ± 0.5	209	0.269	59.3 87.2	103.51 12.7	Hartmans and Tramper (1991)
Unidentified DCM-degrading bacteria MC8b	BF/compost + perlite + crushed oyster shell	None	A	U	6.76	50	1	98	10.29	Ergas et al. (1994)
Activated sludge	BF/peat	11 VOCs*	A	D	3.5–4.5	14–57	0.6	32–100	6.4	Aizpuru et al. (2001)
Hyphomicrobium GJ21	BTF/polypropylene saddles	Methylmethacrylate	A	U	8	115	0.3	25–44	47–152	Okkerse et al. (1999a)
Hyphomicrobium GJ21	BTF/PVC	None	A	D	7.8–8	575	1	70	102	Okkerse et al. (1999b)

BF: biofilter; BTF: biotrickling filter; A: aerobic; AN: anaerobic; U: upflow; D: downflow; RE: removal efficiency; EC: elimination capacity.
*11 VOCs: methanol, acetone, methyl ethyl ketone, methyl isobutyl ketone, butyl and ethyl acetates, toluene, ethylbenzene, xylene, 1,2-dichloroethane.

It is worth mentioning that, since the pH of the trickling phase is neutralized just before it is recirculated, a pH gradient may appear along the height of the packed bed, generating a potential gradual decrease of the removal efficiency along the filter bed (Ottengraf & Diks, 1992). The continuous feed and recirculation of the trickling phase also allows adjusting the ionic strength as well as maintaining the NaCl concentration at values below 150 mM. Sodium chloride is formed during the neutralization procedure through sodium hydroxide addition. The *Hyphomicrobium* strains inoculated in many of the reported studies (Diks & Ottengraf, 1991a, b; Hartmans & Tramper, 1991; Okkerse et al., 1999a, b) are completely inhibited at NaCl concentrations reaching 300 mM. It is worth to note that the amount of NaOH required to neutralize the HCl formed can be used for mass balance calculations and to estimate the amount of DCM dechlorinated in a biotrickling filter. In all cases, the inoculation of a pure culture degrading DCM or a mixed culture adapted to that compound allows to shorten the start-up period from several weeks or even months to only a few days. This was observed both in biofilters and in biotrickling filters. Actually, it is well known that the inoculation of adapted cultures helps to shorten the start-up phase (Veiga & Kennes, 2001), but this is particularly true for halogenated pollutants.

It appeared that under standard conditions, the DCM conversion rate is limited by the diffusion of DCM, not by oxygen. The DCM penetration depth determines the active biofilm volume. At increasing pollutant concentrations in the gas phase, its flux and penetration depth increase. Oxygen transfer would become rate limiting at inlet DCM concentrations above about 7 g/m^3 (Okkerse et al., 1999b). Biotrickling filters can be operated either co-currently or counter-currently. However, no difference in performance was observed between these two modes of operation for DCM removal (Diks & Ottengraf, 1991a). One typical problem in biotrickling filters is clogging due to heavy biomass accumulation resulting from the continuous feed of a nutritive solution in such reactors. However, clogging of biotrickling filters fed DCM is a quite slow process. One reason for this is that DCM molecules have a low relative carbon content and thus a low molar carbon load for a given mass load compared to other VOCs as toluene, containing more than 90% carbon (Kennes & Thalasso, 1998). Another reason is the low growth rate and biomass yield on DCM (Okkerse et al., 1999b).

When the waste gas contains a mixture of volatile pollutants, it is quite common to observe a stratification of the biodegradation and of the microbial populations in the reactors (Aizpuru et al., 2001; Ergas et al., 1994; Okkerse et al., 1999a). Oxygenated and easily biodegradable compounds such as methanol or methylmethacrylate are usually biodegraded before the aromatic and halogenated ones (Aizpuru et al., 2001; Okkerse et al., 1999a). Therefore, dechlorinating populations are more abundant closer to the outlet of the bioreactors. However, it is also possible that one same organism degrades different compounds. In this case, the stratification along the filter bed develops for the metabolic activity rather than for specific organisms.

In one study, the feasibility of using a three-phase airlift bioreactor was evaluated and compared to a biotrickling filter (Zuber et al., 1999). However, although high removal efficiencies could be obtained at inlet DCM concentrations of 2 g/m^3 in the airlift bioreactor, biotrickling filters appeared to be more effective with calculated operating costs at full scale, which are half of those of the three-phase airlift reactor (Zuber et al., 1999).

16.3.2.3 Aerobic dechlorination of chlorobenzenes

Biodegradation As for most other chlorinated compounds, the first evidence of bacterial degradation of chlorinated benzenes was published only about two decades ago. Under aerobic conditions, chlorobenzene was proven to be used as sole carbon and energy source (Reineke & Knackmuss, 1984). Over the past two decades, several different strains have been isolated on these chlorinated substrates. In all cases, the aromatic ring is cleaved first, prior to the complete dechlorination. This results in the formation of hydrogen chloride and acidification of the medium as shown below:

$$C_6H_5Cl + 7O_2 \rightarrow 6CO_2 + HCl + 2H_2O$$

Complete mineralization takes place according to the following stoichiometric reaction when using ammonium nitrate as nitrogen source and considering biomass formation (Mpanias & Baltzis, 1998):

$$C_6H_5Cl + 3.69O_2 + 0.265NH_4NO_3 \rightarrow$$
$$2.65CH_{1.8}O_{0.5}N_{0.2} + 0.15H_2O + 3.35CO_2 + HCl$$

Biofilters Some reactor studies on the treatment of chlorobenzene-contaminated air have been performed with organic filter beds (Table 16.5). As already indicated for DCM, such conventional biofilters may be suitable when treating air polluted with low concentrations of chlorinated pollutants (Roberge et al., 2001) or when using media with a high buffering capacity (Coutu et al., 2003; Delhoménie & Heitz, 2003). The treatment of air polluted with low concentrations (i.e. <2 ppmv) of both 1,2- and 1,4-DCB was performed in a biofilter packed with a mixture of peat moss, animal manure, wood chips and DCB-contaminated soil, at an EBRT around 95 s. The pollutants proceeded from contaminated water of a refinery site. Stripping the pollutants from the water phase and treating the stripped, polluted air in a biofilter was claimed to be the most suitable alternative (Roberge et al., 2001). The system was designed for the removal of up to 1 g C/m^3/h. However, the removal efficiency appeared to be highly variable, resulting among others from load variations and from the sporadic presence of naphthalene. During the 4 months operation, the pH remained quite stable, around 7.5 and 8, because of the buffering capacity of the filter bed and the low pollutant concentration in the waste gas. A similar filter bed was also used for the treatment of 1,2-DCB as single pollutant (Coutu et al., 2003). The results of that study showed that with a filter bed composed of peat (56%, v/v), maple wood chips (20%, v/v), chicken manure (14%, v/v) and contaminated soil as inoculum (10%, v/v), the pH of the medium remained stable over the 2-month period during which 1,2-DCB was fed to the reactor. A maximum elimination capacity of 13 g/m^3/h was reached at an EBRT of 83 s. In another study, monochlorobenzene was fed to a conventional compost-biofilter (Delhoménie & Heitz, 2003). In this case, quite higher pollutant concentrations were fed, between 0.3 and 3.2 g/m^3, reaching removal efficiencies exceeding 90% at loads around 75 g/m^3/h. Again, the pH remained quite stable, although the authors do not indicate for how long. An aqueous solution containing urea (428.8 g urea/L), considered to act as a pH auto-buffering compound, was continuously added at a slow rate of 0.4 g N/h to the biofilter. The way the compost-biofilter was operated allows to classify this system half-way between a conventional biofilter and a biotrickling filter.

Table 16.5 Comparison of the different biological chlorobenzene removal methods.

Bacteria	Reactor/packing	Substrate	O_2	Flow	pH	Inlet (ppm)	EBRT (min)	RE (%)	EC ($g/m^3/h$)	References
Industrial sludge enrichments	BTF/perlite	MCB + 1,2-DCB	A	D	5–8.5	1.8	4.36		Up to 300	Oh and Bartha (1994)
Microbial consortium	BTF/ceramic saddles	MCB	A	U	6.8	100–956	3.8	75	56	Mpanias and Baltzis (1998)
Indigenous microorganism of the compost	BF/compost	MCB	A	U		65–695	1	90	70	Delhoménie and Heitz (2003)
Soil bacteria	BF/peat moss+ animal manure+ wood chip+soil	1,2-DCB+ 1,4-DCB	A	U	7.5	0.1–1.8	1.58	0–94	1	Roberge et al. (2001)
Sludge from wastewater treatment plant	BTF/structured packing	MCB+ 1,2-DCB	A	D	6.9	250–650	1.70	68	45.8 g C/m^3/h	Seignez et al. (2002)
Soil bacteria	BF/peat+maple wood chips + chicken manure + soil	1,2-DCB	A	U	6.71, 6.84	48–125	1.38	28–91.5	13	Coutu et al. (2003)
Sludge from wastewater treatment plant	BTF/structured packing	MCB+ 1,2-DCB	A	D	7	MCB, 1933 1,2-DCB, 466	1.9	95	75	Seignez et al. (2004)

Notations are as in Table 16.4.

The EBRT used in the above mentioned studies were between 60 and 95 s, while the residence times used in the biofiltration of more easily biodegradable organic compounds as alkanes, methanol or non-chlorinated benzenes is generally 1 min or less. Similar or even longer EBRT than 1 min were used in the removal of chlorobenzenes in biotrickling filters (see below).

Biotrickling filters Most biotrickling filters treating chlorobenzenes were operated with co-current flows of the air and water phases (Mpanias & Baltzis, 1998; Oh & Bartha, 1994; Seignez et al., 2004). Although the EBRT of the gas phase were between 1 and 5.7 min, the air contained quite high pollutant-concentrations resulting in relatively high loads. The maximum elimination capacities were typically 60–75 g/m^3/h, with near 100% removal efficiency (Mpanias & Baltzis, 1998; Seignez et al., 2004). Apparently values close to 200 g/m^3/h can be reached as well (Oh & Bartha, 1994). In biotrickling filters, the conversion of chlorobenzenes was found to increase when either the liquid flow rate was increased or the gas flow rate was decreased (Mpanias & Baltzis, 1998). The chlorobenzene-degrading microorganisms studied in gas-phase reactors are sensitive to pH fluctuations. Therefore, as for the other chlorinated pollutants, pH regulation is a prerequisite for successful reactor operation. In most studies, the optimal pH was around 7 (Mpanias & Baltzis, 1998; Seignez et al., 2004). Generally, the pH is maintained constant through the addition of sodium hydroxide. According to some batch studies undertaken with specific bacterial consortia, at least 80% of the optimal activity can be maintained at pH 5.0–8.5 for both monochlorobenzene and 1,2-DCB degradation (Oh & Bartha, 1994). The presence of relatively high concentrations of NaCl inhibits chlorobenzene degraders, as was similarly observed in DCM-degrading bacteria. Batch assays showed that in the presence of 200 mM NaCl, 80% and 40% of the maximum degrading activity could be maintained for monochlorobenzene and 1,2-DCB, respectively (Oh & Bartha, 1994). A similar trend was observed in biotrickling filters. In the absence of any inhibitory compound, monochlorobenzene is much easier to remove than 1,2-DCB in mixed waste gases (Seignez et al., 2004). A possible explanation could be the higher Henry constant of monochlorobenzene (Table 16.1). When operating biotrickling filters with shock loads, the accumulation of intermediate compounds was sometimes observed. It was suggested that these compounds could be mono- or dichlorocatechols (Seignez et al., 2002) or sometimes muconic acid (Seignez et al., 2004).

Once again, the start-up period could last up to several weeks or even months if non-adapted cultures were originally added to the filter bed (Delhoménie & Heitz, 2003). This period could significantly be reduced down to a few days when adapted cultures were inoculated (Mpanias & Baltzis, 1998; Seignez et al., 2004). Clogging problems have not been reported in such systems. The low growth yields on chlorobenzenes, reported in some early papers, can explain this. Typical values are 0.29 g biomass/g substrate for monochlorobenzene and 0.12 g/g for 1,2-DCB (Oh & Bartha, 1994). However, quite higher values were obtained later in other research groups, reaching 0.58 g/g (Mpanias & Baltzis, 1998) and 0.38–0.43 g/g (Seignez et al., 2001) for monochlorobenzene.

16.3.2.4 Aerobic dechlorination of TCE

Biodegradation All chlorinated ethylenes, except PCE, are biodegradable under aerobic conditions, but always through a co-metabolic process (Oldenhuis et al., 1991).

The higher the number of chlorines, the lower the removal rate under aerobic conditions. TCE can be completely mineralized, but the substrate itself and intermediate biodegradation products such as epoxides which are highly toxic may inhibit TCE removal. As was already mentioned for DCM and chlorobenzene, the complete dechlorination of TCE results in the release of three chlorines and the formation of hydrogen chloride with medium acidification. Many different co-substrates have been used for TCE removal, including acetone, ammonia, isoprene, isopropylbenzene, methane, phenol, propane, propene and toluene.

Biofilters and biotrickling filters In case of TCE, more different bioreactor configurations have been tested than for DCM or chlorobenzenes, including conventional biofilters, biotrickling filters, membrane bioreactors and bubble column bioreactors. For biofilters and biotrickling filters, a similar behaviour is found for TCE as for other chlorinated compounds, although its removal seems to be somewhat more difficult. In such reactors, primary carbon sources as phenol (Kim, 1997; Lee et al., 2003), toluene (Cox et al., 1998), propane (Lackley et al., 2002; Sukesan & Watwood, 1997) or methane (Speitel & McLay, 1993; Sukesan & Watwood, 1997) have been used (Table 16.6). Still in another laboratory-scale research, simulating the treatment of waste gases from microelectronic industries, TCE was mixed with both acetone and toluene (Den et al., 2003). Quite long residence times were used in most studies with EBRT as high as 5.6 (Sukesan & Watwood, 1997), 7 (Kim, 1997) or 9.9 min (Speitel & McLay, 1993). Besides, only partial degradation of the target pollutant is often reported. Complete TCE removal is achieved only if either low loads or relatively low inlet concentrations are fed to the reactors. Lee et al. (2003) report a 100% removal efficiency at loads of up to only 0.36 g/m^3/h in biotrickling filters. At higher loads, around 0.71 g/m^3/h, the removal efficiency quickly dropped. Kim (1997) mentions complete removal of TCE when operating a granular activated carbon (GAC) column at an EBRT of 7 min with a maximum concentration of 87 ppm TCE in the feed, while Speitel and McLay (1993) reached a maximum removal efficiency of 62.3% when feeding 74 ppm TCE to a reactor operated at an EBRT of 9.9 min. Sukesan and Watwood (1997) used a compost-biofilter fed 5–50 ppm (v/v) TCE at an EBRT of 5.6 min and reached maximum removal efficiencies of either 95% or less than 15% when using, respectively, a leaf debris compost or a wood chips-bark compost. Although some authors observe and mention that the pH drops as a result of the dechlorination process, no data are available on the effect of medium acidification on reactor performance in any of the publications. The addition of a non-ionic surfactant, Alfonic® 816–60 ethoxylate, at concentrations below its critical micelle concentration (5–50 mg/L) did not enhance TCE biodegradation in a GAC-column (Kim et al., 1998). The surfactant was partly biodegraded in the reactor.

Membrane bioreactors Some laboratory-scale studies have been published on the use of hollow fiber membrane bioreactors (HFMB) for waste gas treatment (Ergas, 2001). To the best of our knowledge such reactors have not been tested at industrial scale. One major problem with membrane reactors is the high cost of the membranes. Some of the studies done with HFMB focused on TCE as target compound. Generally, the polluted air is fed through the lumen of the system while an aqueous nutrient solution is fed through the shell, as shown in Figure 16.4 (Jin et al., 2005a).

Table 16.6 Comparison of the different biological TCE removal methods.

Bacteria	Reactor/packing	Co-substrate	O_2	Flow	pH	Inlet (ppm)	EBRT (min)	RE (%)	EC ($g/m^3/h$)	References
Secondary sludge	HFMB	Toluene	A/AN	TCE/bacteria, fixed (Lumen/Shell)	7.23–6.28	20	0.6	30		Parvatiyar et al. (1996)
Methylosinus trichosporium OB3b PP358	HFMB+ chemostat	Methanol	A	TCE/bacteria, suspended (Shell/Lumen)		26–35	1.6–5.0	92–96		Pressman et al. (2000)
Sludge from wastewater treatment plant	HFMB	Toluene	A	TCE/bacteria, suspended (Lumen/Shell)		26	0.06	36		Dolasa and Ergas (2000)
Pseudomonas cepacia G4	BS	Phenol	A	U	6.8	13–74	0.6–5.3	30–80	12	Hecht et al. (1995)
Phenol-oxidizing microorganisms	BF/GAC	Phenol	A	U	5.8–7.7	85	7	100	3.92	Kim (1997)
Heterotrophic microorganisms (1997)	BF/compost + GAC	Propane or methane	A			50	5.6	>97	2.8	Sukesan and Watwood (1997)
Burkholderia cepacia $PR1_{23}$ (TOM_{23C})	Fixed film bioreactor (sintered glass or activated carbon)		A	D		7.4–450	10		0.36–16.34	Sun and Wood (1997)
Pseudomonas putida F1	BF (Celite R635)	Toluene	A	D		11	7.25	30–40%	0.15	Cox et al. (1998)
	immobilized soil bioreactors + BF	Methane	A	U		255	1.62	63	32	Ramsay et al. (2001)

(Continued)

Table 16.6 (*Continued*).

Bacteria	Reactor/packing	Co-substrate	O_2	Flow	pH	Inlet (ppm)	EBRT (min)	RE (%)	EC ($g/m^3/h$)	References
Sludge from wastewater treatment plant	BF (GAC)		A	D		15–20	1.0–4.5	20–50		Den and Pirbazari (2002)
TCE-degrading consortia	BF (compost + pine bark + chopped kenaf + limestone)	Propane	A	U		50	20	98		Lackey et al. (2002)
Sludge from river	BF (compost)	Toluene + *m*-xylene + TCE + chloroform	A	D	6.3–7.2	1861	1.5	86.1	64	Yoon and Park (2002)
Burkholderia cepacia G4	2 BTFs (ceramic support)	Phenol	A	U		5–17	4.9	50	0.7	Lee et al. (2003)

BS: bioscrubber; ISBRs: immobilized soil bioreactors; other notations are as defined in Table 16.4.

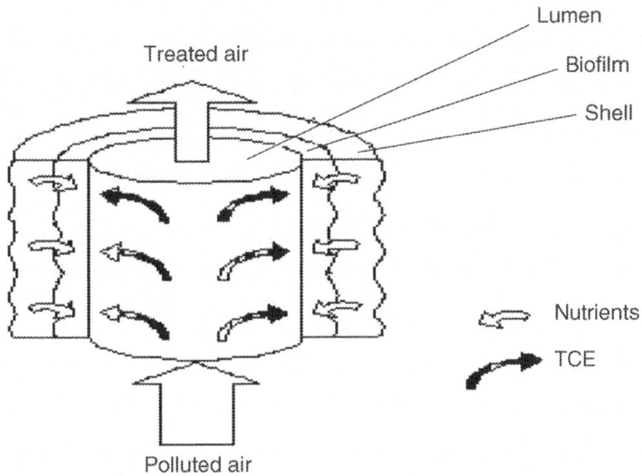

Figure 16.4 Schematic representation of an HFMB used for the treatment of TCE-contaminated air (Ergas, 2001).

In such a case, the biofilm grows on the shell-side of the membrane. The pollutant, that is TCE, diffuses through the membrane and is degraded on the shell-side. However, in one study the nutrient solution was fed through the lumen, while contaminated air flew through the shell of the system (Pressman et al., 2000). It was suggested that such operating strategy might reduce clogging problems although the data were not systematically compared to an HFMB fed polluted air through the lumen (Ergas, 2001).

Since only few papers are available on waste gas treatment in membrane reactors (Table 16.6), it is hard to draw conclusions on their performance. As with biofilters and biotrickling filters, quite low removal efficiencies are reached, even at low pollutant concentrations and sometimes high residence times. Parvatiyar et al. (1996) fed polluted air to the lumen of an HFMB. They worked at a relatively low residence time of 36 s, based on the internal lumen volume. However, when feeding 20 ppm TCE, only 20% removal efficiency was reached. It is worth mentioning that they initially fed toluene only and no TCE, in order to reach a good biofilm growth. Nevertheless, in a subsequent stage, when TCE was fed to the reactor, no toluene was added anymore. In another study, Pressman et al. (2000) fed polluted air through the shell of an HFMB. A strain of *Methylosinus trichosporium* OB3b was inoculated in the system. The organism was grown in a chemostat, whose effluent was subsequently sent to the fiber lumen of the HFMB. The reactor was operated for 20 days and partial TCE degradation was observed at shell residence times of 1.6–5 min.

Hardly any information is available on the influence of pH on the efficiency and stability of HFMB reactors. Only one paper reports that the pH gradually dropped from 7.23 to 6.68 over a 2-week period when feeding TCE at concentrations around 20 ppm (Parvatiyar et al., 1996). Nevertheless, although the pH is expected to drop further over time, no information was given on its value after that period.

Bubble column The efficiency of TCE removal has also been evaluated in a bubble column operated at a constant temperature of 25°C (Hecht et al., 1995). *Pseudomonas* *cepacia* G4 was inoculated in this system. A nutrient-solution containing 4.70 g phenol/L was fed to the bubble column, as well as air contaminated with TCE. The pH was maintained at 6.8 by controlled addition of a 2 M NaOH solution. The reactor was operated at a low dilution rate, minimizing the liquid flow rate, ensuring complete degradation of the soluble phenol, and preventing a liquid pollution problem. Studies were undertaken to evaluate the optimal dilution rate (D), as a too low dilution rate would reduce the steady-state biomass concentration in the reactor. At $D = 0.0073$/h, a superficial gas velocity of 4 cm/s, and inlet TCE gas phase concentrations of 70–400 mg_{TCE}/m^3, a maximum conversion rate of 12 $g/m^3/h$ was reached, with 40% removal efficiency. At lower superficial gas flow velocities, below 1 cm/s, removal efficiencies around 75–80% were reached but the TCE conversion rate decreased to 1–2 $g/m^3/h$.

16.3.2.5 Anaerobic dechlorination of highly chlorinated pollutants

Biodegradation Gas-phase biofilters used for waste gas treatment are usually fed polluted air and are thus aerobic in most cases, although it is well known that some anaerobic zones may occasionally form in such systems. On the other hand, research on the biodegradation of highly chlorinated solvents indicates that such compounds are recalcitrant to aerobic dechlorination. As a general rule, anaerobic dechlorination becomes easier with an increasing number of chlorines while the opposite is true for aerobic degradation. Compounds as PCE and carbon tetrachloride are only degradable by strict anaerobic organisms and several studies have been published on their anaerobic biodegradation (Kennes et al., 1998; Van Eeckert et al., 1998). Although many attempts have been made by microbiologists to obtain aerobic cultures degrading such compounds, none has been successful so far.

Intermediate products appearing during anaerobic PCE dechlorination are as mentioned hereafter:

$$\text{PCE} \longrightarrow \text{TCE} \longrightarrow \text{DCE isomers} \longrightarrow \text{Vinyl chloride} \longrightarrow \text{Ethene} \longrightarrow \text{Ethane}$$

Biofilters Although PCE is theoretically only biodegradable by aerobic microorganisms, the treatment of PCE-contaminated air was reported in a paper on a conventional biofilter packed with activated carbon as filter bed (Kim, 1997). However, the fate of PCE is not totally clear in that work. Mass balance calculations indicated a recovery of only 21% of the original substrate as chloride, and no information was given on the possible presence of chloride from other sources, as ammonium chloride which was added as nitrogen source to the medium. Other authors also observed the disappearance of PCE in aerobic reactors, with removal efficiencies ranging between 40% in a compost biofilter and 61% in an activated carbon biofilter (Devinny et al., 1995). Here again, the fate of these compounds is relatively unclear. It is believed that the removal took place in anaerobic zones within the particles of the support medium.

The anaerobic biofiltration of PCE-polluted air can be evaluated by feeding PCE to a biofilter with an inert carrier gas instead of air. This avoids the presence of oxygen and allows anaerobic dechlorination. Anaerobic waste gases containing highly chlorinated pollutants as PCE but hardly any oxygen may be found in landfill gas. The treatment of a simulated landfill gas was studied in a GAC-biofilter (Schwarz et al.,

1999). The gas was composed of carbon dioxide and methane as major compounds and 30–100 ppm PCE as pollutant. Sucrose was added as carbon source. Possible products from the anaerobic degradation of PCE, as TCE, dichloroethylene or vinyl chloride were not detected, but the accumulation of chloride in the medium was observed.

A two-step process could be considered as well, in which the highly chlorinated compounds are first transferred from the contaminated air stream to an aqueous phase in an absorption tower, to be treated subsequently in an anaerobic bioreactor. Besides, another two-stage system has recently been proposed, in which polluted air containing PCE and TCE was first submitted to ultraviolet oxidation to produce oxidation products more amenable to biodegradation in the second stage (Den & Pirbazari, 2000). This system allowed reaching quite high removal efficiencies although after photo-oxidation trace levels of compounds as phosgene, known to be more toxic than the parent compounds, were formed.

16.4 OTHER BIOCATALYTIC ACTIVITIES RECENTLY STUDIED IN GAS-PHASE BIOREACTORS

The development of some other biocatalytic activities has recently been studied in biofilters and biotrickling filters. One example is the removal of NO_x from waste gases in these gas-phase bioreactors (Jin et al., 2005a). Major anthropogenic sources of NO_x are motor vehicles, electric utilities and most activities that burn fuels (Jin et al., 2005a). NO_x can also be removed from waste gases by conventional biological (bioreactors) and absorption (scrubbers) processes, although such systems still need to be further studied in order to be cost effective at full scale (van der Maas et al., 2005). Another interesting and highly toxic pollutant is carbon monoxide (CO), which is present, among others, in industrial waste gases from synthetic-resin producing industries studied in our research group (Prado et al., 2005). CO is also a typical air contaminant of road tunnels. The possibility to treat air from road tunnels, containing both NO and CO has been demonstrated in pilot-scale biofilters reaching near complete removal of the pollutants (Robra et al., 1997).

16.5 CONCLUSIONS

Although little is known so far on the potential of filamentous fungi to degrade and assimilate non-oxygenated aromatic pollutants as carbon and energy source, their use in gas-phase biofilters has proven to lead to a very high reactor performance. Besides, they are more resistant than most bacteria to acid environments and to low water contents.

Dechlorinating organisms are also interesting biocatalysts for the treatment in bioreactors of waste gases containing chlorinated compounds. These systems need, however, further research and optimization. At present, they need to be operated at quite long residence times combined sometimes with low removal efficiencies of the chlorinated pollutants.

ACKNOWLEDGEMENT

Our on-going research on new biocatalysts for waste gas treatment is funded by the Spanish Ministry of Education and Culture through project CTM2004-00427 and the Xunta de Galicia through project PGIDIT05PXIC10304PN.

REFERENCES

Aizpuru, A., Malhautier, L., Roux, J.C. and Fanlo, J.L. (2001) Biofiltration of a mixture of volatile organic emissions. *J. Air Waste Manage. Assoc.* **51**, 1662–1670.

Arriaga, S., Muñoz, R., Hernández, S., Guieysse, B. and Revah, S. (2005) Treatment of *n*-hexane in fungal packed bed and stirred tank bioreactors containing two phases. In *Biotechniques for Air Pollution Control* (eds. C. Kennes and M.C. Veiga), pp. 337–342. University of La Coruña (Publisher), La Coruña, Spain.

Bartnicki, E.W. and Castro, C.E. (1994) Biodehalogenation, rapid oxidative metabolism of mono- and polyhalomethanes by *Methylosinus trichosporium* OB3b. *Environ. Toxicol. Chem.* **13**, 241–245.

Braus-Stromeyer, S.A., Hermann, R., Cook, A.M. and Leisinger, T. (1993) Dichloromethane as the sole carbon and energy source for an acetogenic mixed culture and isolation of a fermentative, dichloromethane-degrading bacterium. *Appl. Environ. Microbiol.* **59**, 3790–3797.

Brunner, W.B., Staub, D. and Leisinger, T. (1980) Bacterial degradation of dichloromethane. *Appl. Environ. Microbiol.* **40**, 950–958.

Coutu, C., Martineau, G., Guy, C. and Samson, R. (2003) Characterization of an organic filter medium for the biofiltration treatment of air contaminated with 1,2-dichlorobenzene. *J. Chem. Technol. Biotechnol.* **78**, 907–917.

Cox, H.H.J., Magielsen, F.J., Doddema, H.J. and Harder, W. (1996) Influence of the water content and water activity on styrene degradation by *Exophiala jeanselmei* in biofilters. *Appl. Microbiol. Biotechnol.* **45**, 851–856.

Cox, H.H.J., Moerman, R.E., van Baalen, S., van Heiningen, W.N.M., Doddema, H.J. and Harder, W. (1997) Performance of a styrene-degrading biofilter containing the yeast *Exophiala jeanselmei*. *Biotechnol. Bioeng.* **53**, 259–266.

Cox, C.D., Woo, H.J. and Robinson, K.G. (1998) Cometabolic biodegradation of trichloroethylene (TCE) in the gas phase. *Water Sci. Technol.* **37**, 97–104.

Delhoménie, M.-C. and Heitz, M. (2003) Elimination of chlorobenzene vapors from air in a compost-based biofilter. *J. Chem. Technol. Biotechnol.* **78**, 588–595.

Den, W. and Pirbazari, M. (2000) A novel ultraviolet oxidation and biotrickling filtration (UV-BTF) process to treat chlorinated VOCs in industrial air emissions. *AIChE-Annual Meeting*, 12–17 November.

Den, W. and Pirbazari, M. (2002) Modeling and design of vapor-phase biofiltration for chlorinated volatile organic compounds. *AIChE J.* **48**, 2084–2103.

Den, W., Huang, C. and Li, C.-H. (2003) Biotrickling filtration of volatile organic compounds from microelectronics industry. *J. Environ. Eng. ASCE.* **129**, 610–619.

Devinny, J.S., Webster, T.S., Torres, E.M. and Basrai, S.S. (1995) Biofiltration for removal of PCE and TCE vapors from contaminated air. *Hazard. Waste Hazard. Mater.* **12**, 283–293.

Diks, R.M.M. and Ottengraf, S.P.P. (1991a) Verification studies of a simplified model for the removal of dichloromethane from waste gases using a biological trickling filter (Part I). *Bioproc. Eng.* **6**, 93–99.

Diks, R.M.M. and Ottengraf, S.P.P. (1991b) Verification studies of a simplified model for the removal of dichloromethane from waste gases using a biological trickling filter (Part II). *Bioproc. Eng.* **6**, 131–140.

Dolasa, A.R. and Ergas, S.J. (2000) Membrane bioreactor for cometabolism of trichloroethene air emissions. *J. Environ. Eng-ASCE.* **126**, 969–973.

Ergas, S.J. (2001) Membrane bioreactors. In *Bioreactors for Waste Gas Treatment* (eds. C. Kennes and M.C. Veiga), pp. 163–177. Kluwer Academic Publishers, Dordrecht, The Netherlands.

Ergas, S.J., Kinney, K., Fuller, M.E. and Scow, K.M. (1994) Characterization of a compost biofiltration system degrading dichloromethane. *Biotechnol. Bioeng.* **44**, 1048–1054.

Estévez, E., Veiga, M.C. and Kennes, C. (2005a) Biodegradation of toluene by the new fungal isolates *Paecilomyces variotii* and *Exophiala oligosperma*. *J. Indust. Microbiol. Biotechnol.* **32**, 33–37.

Estévez, E., Veiga, M.C. and Kennes, C. (2005b) Biofiltration of waste gases with the fungi *Exophiala oligosperma* and *Paecilomyces variotii*. *Appl. Microbiol. Biotechnol.* **67**, 563–568.

García-Peña, E.I., Hernández, S., Favela-Torres, E., Auria, R. and Revah, S. (2001) Toluene biofiltration by the fungus *Scedosporium apiospermum* TB1. *Biotechnol. Bioeng.* **76**, 61–69.

Gossett, J.M. (1987) Measurement of Henry's law constants for C_1 and C_2 chlorinated hydrocarbons. *Environ. Sci. Technol.* **21**, 202–208.

Hartmans, S. and Tramper, J. (1991) Dichloromethane removal from waste gases with a trickle-bed bioreactor. *Bioproc. Eng.* **6**, 83–92.

Hecht, V., Brebbermann, D., Bremer, P. and Deckwer, W.D. (1995) Cometabolic degradation of trichloroethylene in a bubble column bioscrubber. *Biotechnol. Bioeng.* **47**, 461–469.

Jin, Y., Veiga, M.C. and Kennes, C. (2005a) Bioprocesses for the removal of nitrogen oxides from polluted air. *J. Chem. Technol. Biotechnol.* **80**, 483–494.

Jin, Y., Veiga, M.C. and Kennes, C. (2005b) Fungal biodegradation of α-pinene in gas-phase biofilter. In *Biotechniques for Air Pollution Control* (eds. C. Kennes and M.C. Veiga), pp. 321–327. University of La Coruña (Publisher), La Coruña, Spain.

Jin, Y., Veiga, M.C. and Kennes, C. (2006) Performance optimization of the fungal biodegradation of α-pinene in gas-phase biofilter. *Process Biochem.* **41**, 1722–1728.

Kennes, C. and Lema, J.M. (1994) Degradation of major compounds of creosotes (PAH and phenols) by *Phanerochaete chrysosporium*. *Biotechnol. Lett.* **16**, 750–764.

Kennes, C. and Thalasso, F. (1998) Waste gas biotreatment technology. *J. Chem. Technol. Biotechnol.* **72**, 303–319.

Kennes, C. and Veiga, M.C. (2001) *Bioreactors for Waste Gas Treatment*, Kluwer Academic Publishers, Dordrecht, The Netherlands.

Kennes, C. and Veiga, M.C. (2004) Fungal biocatalysts in the biofiltration of VOC-polluted air. *J. Biotechnol.* **113**, 305–319.

Kennes, C., Veiga, M.C. and Bhatnagar, L. (1998) Methanogenic and perchloroethylene-dechlorinating activity of anaerobic granular sludge. *Appl. Microbiol. Biotechnol.* **50**, 484–488.

Kim, O.J. (1997) Gaseous TCE and PCE removal by an activated carbon biofilter. *Bioproc. Eng.* **16**, 331–337.

Kim, O.J., Terkonda, P.K. and Lee, S.D. (1998) Gaseous CAH removal by biofiltration in presence and absence of a non-ionic surfactant. *Bioproc. Eng.* **19**, 253–259.

Kraakman, N.J.R. (2005) Biological waste gas purification of industrial methyl chloride emissions to protect the stratospheric ozone layer. In *Biotechniques for Air Pollution Control* (eds. C. Kennes and M.C. Veiga), pp. 353–360. University of La Coruña (Publisher), La Coruña, Spain.

Lackey, L.W., Gamble, J.R. and Boles, J.L. (2002) Bench-scale evaluation of a biofiltration system used to mitigate trichloroethylene contaminated air streams. *Adv. Environ. Res.* **7**, 97–104.

Lee, E.Y., Ye, B.D. and Park, S. (2003) Development and operation of a trickling biofilter system for continuous treatment of gas-phase trichloroethylene. *Biotechnol. Lett.* **25**, 1757–1761.

MacKay, D. and Shiu, W.Y. (1981) A critical review of Henry's law constants for chemicals of environmental interest. *J. Phys. Chem. Ref. Data.* **10**, 1175–1199.

Mendoza, J.A., Prado, O.J., Veiga, M.C. and Kennes, C. (2004) Hydrodynamic behaviour and comparison of technologies for the removal of excess biomass in gas-phase biofilters. *Water Res.* **38**, 404–413.

Mpanias, C.J. and Baltzis, B.C. (1998) An experimental and modelling study on the removal of mono-chlorobenzene vapour in a biotrickling filter. *Biotechnol. Bioeng.* **59**, 328–343.

Oh, Y.-S. and Bartha, R. (1994) Design and performance of a trickling air biofilter for chlorobenzene and *o*-dichlorobenzene vapors. *Appl. Environ. Microbiol.* **60**, 2717–2722.

Okkerse, W.J.H., Ottengraf, S.P.P., Diks, R.M.M., Osinga-Kuipers, B. and Jacobs, P. (1999a) Long term performance of biotrickling filters removing a mixture of volatile organic compounds from an artificial waste gas: dichloromethane and methylmethacrylate. *Bioproc. Eng.* **20**, 49–57.

Okkerse, W.J.H., Ottengraf, S.P.P., Osinga-Kuipers, B. and Jacobs, P. (1999b) Biomass accumulation and clogging in biotrickling filters for waste gas treatment. Evaluation of a dynamic model using dichloromethane as a model pollutant. *Biotechnol. Bioeng.* **63**, 418–430.

Oldenhuis, R., Oedzes, J.Y., van der Waarde, J.J. and Janssen, D.B. (1991) Kinetics of chlorinated hydrocarbon degradation by *Methylosinus trichosporium* OB3b and toxicity of trichloroethylene. *Appl. Environ. Microbiol.* **57**, 7–14.

Ottengraf, S.P.P. and Diks, R.M.M. (1992) Process technology of biotechniques. In *Biotechniques for Air Pollution Abatement and Odour Control Policies* (eds. A.J. Dragt and J. van Ham), pp. 17–31. Elsevier Science Publisher, Amsterdam, NL.

Parvatiyar, M.G., Govind, R. and Bishop, D.F. (1996) Treatment of trichloroethylene (TCE) in a membrane biofilter. *Biotechnol. Bioeng.* **50**, 57–64.

Prado, O.J., Veiga, M.C. and Kennes, C. (2005) Biological treatment of mixtures of toxic compounds emitted from formaldehyde resin-producing industries. In *Biotechniques for Air Pollution Control* (eds. C. Kennes and M.C. Veiga), pp. 391–396. University of La Coruña (Publisher), La Coruña, Spain.

Pressman, J.G., Georgiou, G. and Speitel, Jr, G.E. (2000) A hollow-fiber membrane bioreactor for the removal of trichloroethylene from the vapour phase. *Biotechnol. Bioeng.* **68**, 548–556.

Ramsay, B., Karamanev, D., Pierre, S., Lafontaine, C., Sakamoto, T. and Ramsay, J. (2001) Efficient TCE mineralization in a novel methanotrophic bioreactor system. *In Situ Aerat. Aerob. Remediat.* **6**, 171–178.

Reineke, W. and Knackmuss, H.J. (1984) Microbial metabolism of haloaromatics: isolation and properties of a chlorobenzene-degrading bacterium. *Appl. Environ. Microbiol.* **47**, 395–402.

Rittmann, B.E. and McCarty, P.L. (1980) Utilization of dichloromethane by suspended and fixed film bacteria. *Appl. Environ. Microbiol.* **39**, 1225–1226.

Roberge, F., Gravel, M.J., Deschênes, L., Guy, C. and Samson, R. (2001) Biofiltration of dichlorobenzenes. *Water Sci. Technol.* **44**, 287–293.

Robra, K.H., Wellacher, M., Kirchmeir, F., Leistentritt, R. and Pucher, K. (1997) Abluftreinigung für Straßentunnel und Tiefgaragen durch carboxidotrophe Mischpopulationen. In *Proceedings of the International Symposium Biological Waste Gas Cleaning*, VDI-Verlag, Maastricht, The Netherlands.

Schwarz, B.C.E., Devinny, J.S. and Tsotsis, T.T. (1999) Degradation of PCE in an anaerobic waste gas by biofiltration. *Chem. Eng. Sci.* **54**, 3187–3195.

Seignez, C., Vuillemin, A., Adler, N. and Péringer, P. (2001) A procedure for production of adapted bacteria to degrade chlorinated aromatics. *J. Hazard. Mater.* **B84**, 265–277.

Seignez, C., Atti, A., Adler, N. and Péringer, P. (2002) Effect of biotrickling filter operation parameters on dichlorobenzenes degradation. *J. Environ. Eng. ASCE* **128**, 360–366.

Seignez, C., Adler, N., Thoeni, C., Stettler, M., Péringer, P. and Holliger, C. (2004) Steady-state and transient-state performance of a biotrickling filter treating chlorobenzene-containing waste gas. *Appl. Microbiol. Biotechnol.* **65**, 33–37.

Speitel, Jr, G.E. and McLay, D.S. (1993) Biofilm reactors for the treatment of gas streams containing chlorinated solvents. *J. Environ. Eng. ASCE* **119**, 658–678.

Spigno, G., Pagella, C., Fumi, M.D., Molteni, R. and De Faveri, D.M. (2003) VOCs removal from waste gases: gas-phase bioreactor for the abatement of hexane by *Aspergillus niger*. *Chem. Eng. Sci.* **58**, 739–746.

Spigno, G., Tramelli, L., Galli, R., Pagella, C. and De Faveri, D.M. (2005) Biofiltration of dichloromethane vapours: isolation of microorganisms. In *Biotechniques for Air Pollution Control* (eds. C. Kennes and M.C. Veiga), pp. 345–351. University of La Coruña (Publisher), La Coruña, Spain.

Strand, S.E. and Shippert, L. (1986) Oxidation of chloroform in an aerobic soil exposed to natural gas. *Appl. Environ. Microbiol.* **52**, 203–205.

Sukesan, S. and Watwood, M.E. (1997) Continuous vapor-phase trichloroethylene biofiltration using hydrocarbon-enriched compost as filtration matrix. *Appl. Microbiol. Biotechnol.* **48**, 671–676.

Sun, A.K. and Wood, T.K. (1997) Trichloroethylene mineralization in a fixed-film bioreactor using a pure culture expressing constitutively toluene *ortho*-monooxygenase. *Biotechnol. Bioeng.* **55**, 674–685.

van der Maas, P., van der Brink, P., Utomo, S., Klapwijk, B. and Lens, P. (2005) Fe(II)$EDTA^{2-}$ regeneration, biomass growth and EDTA degradation in continuous BioDeNOx reactors. In *Biotechniques for Air Pollution Control* (eds. C. Kennes and M.C. Veiga), pp. 13–28. University of La Coruña (Publisher), La Coruña, Spain.

Van Eeckert, M.H.A., Schröder, T.J., Stams, A.J.M., Schraa, G. and Field, J.A. (1998) Degradation and fate of carbon tetrachloride in unadapted methanogenic granular sludge. *Appl. Environ. Microbiol.* **64**, 2350–2356.

van Groenestijn, J.W., Heiningen, W.N.M. and Kraakman, N.J.R. (2001) Biofilters based on the action of fungi. *Water Sci. Technol.* **44**, 227–232.

Veiga, M.C. and Kennes, C. (2001) Parameters affecting performance and modeling of biofilters treating alkylbenzene-polluted air. *Appl. Microbiol. Biotechnol.* **55**, 254–258.

Veiga, M.C., Fraga, M., Amor, L. and Kennes, C. (1999) Biofilter performance and characterization of a biocatalyst degrading alkylbenzene gases. *Biodegradation* **10**, 169–176.

Weber, F.J., Hage, K.C. and de Bont, J.A.M. (1995) Growth of the fungus *Cladosporium sphaerospermum* with toluene as the sole carbon and energy source. *Appl. Environ. Microbiol.* **61**, 3562–3566.

Woertz, J.R., Kinney, K.A., McIntosh, N.D.P. and Szaniszlo, P.J. (2001) Removal of toluene in a vapour-phase bioreactor containing a strain of the dimorphic black yeast *Exophiala lecanii-corni*. *Biotechnol. Bioeng.* **75**, 550–558.

Yadav, J.S. and Reddy, C.A. (1993) Degradation of benzene, toluene, ethylbenzene and xylenes (BTEX) by the lignin-degrading basidiomycete *Phanerochaete chrysosporium*. *Appl. Environ. Microbiol.* **59**, 756–762.

Yoon, I.K. and Park, C.H. (2002) Effects of gas flow rate, inlet concentration and temperature on biofiltration of volatile organic compounds in a peat-packed biofilter. *J. Biosci. Bioeng.* **93**, 165–169.

Zuber, L., Dunn, I.J. and Deshusses, M.A. (1997) Comparative scale-up and cost estimation of a biological trickling filter and a three-phase airlift bioreactor for the removal of methylene chloride. *J. Air Waste Manage. Assoc.* **47**, 969–975.

PART IV: Technologies for resource recovery

17

Wet scrubbing for the removal of NH_3 from waste gases

M. Roustan

17.1 INTRODUCTION

The removal of odorous substances from the atmosphere is one of the major issues concerning contemporary society. Odours are unpleasant and harmful to workers and, consequently, they cause public discontent. Some of these substances are hazardous to human health and environmentally dangerous. A major odorous compound found in emissions from agricultural and industrial activities is ammonia (NH_3). There are numerous sources of NH_3 including:

(1) Wastewater treatment plants. Ammonia is produced during water treatment processes and the treatment and management of biosolids (sludges).
(2) Animal husbandry and manure. NH_3 is the final product of the decay of nitrogenous organic compounds such as proteins, peptides and amino acids.
(3) Food industry.
(4) Chemical, petrochemical and drug industries.

The concentration of this component in air varies over a wide range, from some % (volume to volume) to less than some ppb, depending on the source and the dilution.

For example, the hydrolysis of animal waste produces air with an ammonia concentration ranging from 100 to 600 mg m^{-3}. In the cooked chicken industry, the concentration in NH_3 emissions coming from ovens varies from 13 to 50 mg m^{-3} depending on the production.

Ammonia is toxic and can be a significant localised pollutant in specific cases. Its most important effect in the atmosphere is its formation of the corrosive pollutants NH_4NO_3, NH_4HSO_4 and $(NH_4)_2SO_4$ (Manahan, 1997). In terms of odorous molecules, its human olfactory threshold is about 4 mg NH_3 m^{-3} of dry air (Devos et al., 1990).

In many cases, industrial emissions have to be treated. To remove ammonia, some classical treatments are available for industrial applications. Impregnated activated carbon packing filters are used to control low flow (some $m^3 h^{-1}$) emissions. For significant flow rates (5000–50,000 $m^3 h^{-1}$), the transfer of ammonia into an aqueous solution or a biological suspension is commonly carried out in scrubbers.

The objective of this chapter is to present the use of scrubbers to transfer NH_3 from a waste gas into water or acidified water. Some technologies using packed reactors or spray columns are described. The equations required to design such air purification processes are given and, for a specific example, a column is defined in terms of column diameter and height.

17.2 CHOICE OF SOLVENTS FOR ABSORPTION

Ammonia can be removed from a mixture of gases by absorption into a suitable liquid (solvent). The gas stream (essentially an ammonia–air mixture) passes through the liquid in which the ammonia dissolves, while the other gases pass out.

At equilibrium, gas solubility in a solvent corresponds to the concentration of the dissolved gas in the liquid, characterised by an equilibrium curve. At dilute concentrations, the distribution curve can be characterised by a straight line, expressed by a linear equation, known as Henry's law. Various units are used for the concentrations in the two phases, so the units of the constant of Henry's law vary (McCabe et al., 1993; Roustan, 2003; Sinnott, 1993). Some values of the constant for ammonia–water at 20°C are given in Table 17.1.

Table 17.1 Values of the Henry's law constants (*He*, *He'* or *He''* depending on the definition) for ammonia–water.

Expressions of Henry's law	Henry's law constant at 20°C
1 $y = \frac{He}{P_T} x = mx$ x, y: mole fraction of ammonia in liquid and in gas (at equilibrium) P_T: total pressure (Pa)	$He = 0.793 \times 10^5$ Pa m = 0.78 at atmospheric pressure
2 $p = He' C_L^*$ p: partial pressure (Pa) C_L^*: concentration of ammonia in water (mol m^{-3})	$He' = 1.43$ (Pa $m^3 mol^{-1}$)
3 $C_G = He'' C_L^*$ C_G, C_L^*: molar or mass concentration in gas and in water	$He'' = 0.00059$

Water is the cheapest, most used solvent, and indeed the absorption of ammonia from a gas stream can be carried out using water. Since ammonia is very soluble in water, these values are only valid for dilute concentrations in the two phases (for example, $x < 0.07$ at 10°C with x mole fraction of ammonia in liquid phase).

The Henry's law constant (*He*) varies with temperature. The constant m ($m = He/P_T$) increases when the temperature increases: $m = 0.48$ at 10°C; $m = 1.23$ at 30°C. The absorption of ammonia by water is normally accompanied by a heat of dissolution with, as a consequence, an increase in the temperature of the phases. With dilute concentrations ($x < 0.07$), the heat transfer between phases may be neglected.

Absorption followed by reaction in the liquid phase is often used in order to improve the removal of ammonia from the gas mixture. A dilute acid solution (sulfuric or hydrochloric acid) can be used to scrub ammonia from gas streams. In this case, a chemical reaction occurs in the liquid phase and the absorption of the gas by the liquid is influenced by this reaction, increasing both the absorption capacity of the liquid and the absorption rate.

17.3 FUNDAMENTALS OF MASS TRANSFER

17.3.1 Physical absorption

The well-known double film theory proposed by Witman (1923), presented in Figure 17.1, is usually used to describe the mass transfer during absorption. Let us consider the following hypothesis:

- On each side of the gas–liquid interface, there are two stagnant films of a certain thickness: a gas film and a liquid film.
- Transfer occurs in the direction of the concentration gradient.
  The mechanism of transport of the component into the two films is purely diffusional.
- The absorption is a steady-state process.
- There is no resistance at the interface itself.
- At the interface, the concentrations are at equilibrium and can be described by Henry's law, $C_{Gi} = He''C_{Li}$.

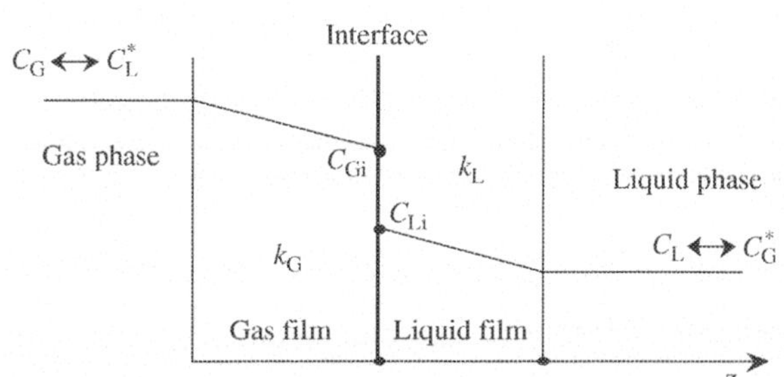

Figure 17.1 Double film model.

In a steady-state process, the rate of mass transfer of ammonia through the gas film is the same as that through the liquid film. The flux relationships (mass transfer per time through a given area) can be expressed in four different ways, using film coefficients and film driving forces (the difference between the bulk concentration and the interface concentration) or overall coefficients and driving forces in the gas and liquid phases. They are written as:

$$N_A = k_G(C_G - C_{Gi}) = K_G(C_G - C_G^*) = k_L(C_{Li} - C_L) = K_L(C_L^* - C_L) \quad (17.1)$$

where:
N_A: transferred flux (kmol m^{-2}s^{-1})
k_L: liquid-film mass transfer coefficient (m s^{-1})
k_G: gas-film mass transfer coefficient (m s^{-1})
K_L: overall liquid mass transfer coefficient (m s^{-1})
K_G: overall gas mass transfer coefficient (m s^{-1})
C_G: gas-phase concentration of component A (kmol m^{-3})
C_L: liquid-phase concentration of component A (kmol m^{-3})
C_G^*: gas-phase concentration of component A that would exist if the gas were in equilibrium with the liquid concentration C_L (kmol m^{-3})
C_L^*: liquid-phase concentration of component A that would exist if the liquid were in equilibrium with the gas concentration C_G (kmol m^{-3})
C_{Gi}: interface concentration of component A, gas phase (kmol m^{-3})
C_{Li}: interface concentration of component A, liquid phase (kmol m^{-3}).

At the interface, the concentrations C_{Gi} and C_{Li} are in equilibrium, given by Henry' law, $C_{Gi} = He'' C_{Li}$. The overall mass transfer coefficients can be expressed as:

$$\frac{1}{K_L} = \frac{1}{k_L} + \frac{1}{He'' k_G} \quad (17.2)$$

$$\frac{1}{K_G} = \frac{1}{k_G} + \frac{He''}{k_L} \quad (17.3)$$

Ammonia is very soluble in water and the value of the Henry's law constant He'' is small (0.00059). The values of k_G and k_L are in the range of 1×10^{-2} and 1×10^{-4}, respectively. Therefore, Equation (17.3) shows that the overall mass transfer coefficient K_G is approximately equal to k_G. Nevertheless, this result must be verified for each set of operating conditions in different absorbers.

The mass transfer resistance for ammonia absorption is thus essentially in the gas film. Since the gas-film controls the mass transfer, it is necessary to choose a gas–liquid contactor that provides a very large effective interfacial area and great turbulence in the gas phase. Therefore, packed columns and spray columns are the most suitable devices for ammonia absorption.

17.3.2 Absorption associated with an irreversible chemical reaction

In case the absorption is performed with an irreversible reaction, the dissolved gas reacts with a dissolved reactant B in the liquid according to a second-order irreversible reaction: $A + zB = P$. The conditions in the gas phase are not changed and are similar to those described for physical absorption (see Section 17.3.1). However, in the liquid film, a chemical reaction occurs that may have an effect on the rate of absorption. Several types of behaviour may be identified (Danckwerts, 1970), depending on the comparison of the rate of diffusion of component A through the liquid film and the reaction kinetics of A with B.

The effect of the chemical reaction is that it accelerates the removal of A from the interface. The enhancement factor, E, is the factor by which the reaction increases the amount absorbed compared to physical absorption. Factor E can be evaluated from relationships between dimensionless numbers: the Hatta number (Ha) and the limiting enhancement factor (E_L):

$$Ha = \frac{\sqrt{k_2 C_{BL} D_{AL}}}{k_L} \tag{17.4}$$

$$E_L = 1 + \frac{D_{BL} C_{BL}}{z D_{AL} C_{Ai}} \tag{17.5}$$

Table 17.2 summarises the different regimes depending on the values of the Hatta number.

An analytical solution giving $E = f(Ha, E_L)$ is not available. Van Krevelen and Hoftijer proposed an approximate equation (Danckwerts, 1970) that allows E to be calculated with little error (about 10%):

$$E = \frac{Ha\sqrt{\dfrac{E_L - E}{E_L - 1}}}{\tanh\left[Ha\sqrt{\dfrac{E_L - E}{E_L - 1}}\right]} \tag{17.6}$$

Table 17.2 Different regimes according to Hatta values.

	Ha	Main characteristics
1	$Ha < 0.3$	Very slow or slow chemical reaction, which takes place in the bulk of the liquid. The reaction has negligible effect, $E = 1$.
2	$0.3 < Ha < 3$	The reaction is moderately fast, one part in the film, another part in the bulk of the liquid ($C_{AL} = 0$), $E > 1$.
3	$Ha > 3$	Fast reaction. The component A reacts completely in the liquid film, $E > 1$. If $Ha > 10\ E_L$, the reaction can be considered as a pseudo-first order and then $E = Ha$.
4	Ha very high	– Very fast reaction: the dissolved gas reacts instantaneously with B on a reaction plane inside the film, $E = E_L$ – Instantaneous reaction on the interface, mass transfer resistance in the gas-film controls the rate of absorption.

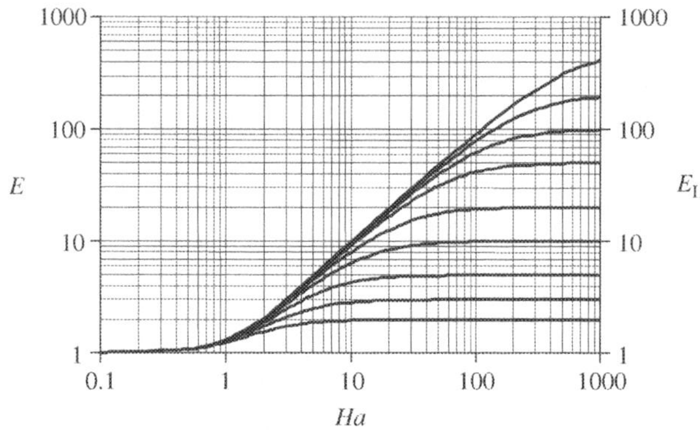

Figure 17.2 Evolution of $E = f(Ha, E_L)$ as the second-order reaction proposed by Van Krevelen and Hoptijer (1948).

Figure 17.2 shows the factor E as a function of Ha and E_L, calculated according to Equation (17.6). This curve indicates clearly the different types of behaviour that can be found when a dissolved gas reacts with a reactant in the liquid phase.

Taking into account the enhancement factor, the Equations (17.2) and (17.3) can be written as follows:

$$\frac{1}{K_L} = \frac{1}{Ek_L} + \frac{1}{He''k_G} \tag{17.7}$$

$$\frac{1}{K_G} = \frac{1}{k_G} + \frac{He''}{Ek_L} \tag{17.8}$$

For ammonia absorption in dilute solutions of HCl or H_2SO_4, the reaction may be considered as instantaneously taking place at the gas–liquid interface. In these conditions, $C_{Gi} = 0$ and then $C_{Li} = 0$. The rate of absorption is entirely controlled by the gas-film resistance (transport of A across the gas film) and the value of the flux is given after arranging Equation (17.1):

$$N_A = k_G C_G \tag{17.9}$$

17.4 WET SCRUBBING TECHNOLOGIES

Since the mass transfer of ammonia is always controlled by the gas-film resistance, the types of equipment for wet scrubbing are (i) packed columns or (ii) spray columns. They generally operate in counter-current mode.

17.4.1 Packed tower

17.4.1.1 Operating principle

In an absorption packed column (or tower), the gas–liquid contact is continuous and the phases operate in a counter-current mode. The liquid flows down the column and

Table 17.3 Main characteristics of different packing materials used in packed tower scrubbers.

Types of packing	Diameter D (mm)	Interfacial area a^* ($m^2 m^{-3}$)	Porosity ε (%)	Packing factor F ($m^2 m^{-3}$)
Raschig rings (steel)	25	184	86	449
	38	128	90	272
	51	95	92	187
Pall rings (plastic)	25	171	90	207
	38	131	91	128
	51	102	92	102
Berl saddles (ceramic)	25	250	68	360
	38	151	71	215
	51	105	72	148
Structured packing	na*			
Mellapak®		125–500	90–98	
Rombopak®		150–450	unknown	

*na: not applicable.

is distributed in the form of films over the packing surface. The upflow gas forms the continuous phase. The liquid holdup is relatively low (<20%). A packed tower contains packing in order to provide a large surface area, a high interfacial area between the gas and the liquid. The packing is designed to have an open structure, to create low resistance to the gas flow (a low pressure drop for the gas) and to promote uniform liquid distribution over the packing surface.

17.4.1.2 Packing material

Many types and shapes of packing are available to satisfy these requirements and they fall into two main groups: random packing and structured packing.

The principal kinds of random packing (rings, saddles) are shown in Table 17.3. They are available in a variety of materials: metal, ceramic, plastic (polypropylene). They are characterised by geometrical and physical properties: size (d), porosity (ε), bulk density (ρ), surface area (a^*) and packing factor (F). In general, the recommended size of the packing depends on the size of the column (diameter D) and must be in the range $10 < D/d < 80$. The pressure drop is in the range of 200–400 Pa m^{-1}.

Structured packing consists of a material that is folded and arranged with a regular geometry. This provides a very high surface area with a high void fraction. The size of the packing is equal to the size of the column and the height of one element is in the range of 0.15–0.30 m. The main characteristics of structured packing are given in Table 17.3.

With random packing, the interfacial area decreases when the size of the packing increases. In the case of structured packing, the interfacial area is independent of the size of the packing. The cost of structured packing per cubic metre is significantly higher than that of random packing, thus a thorough economic comparison must be made for the final choice of the packing.

17.4.2 Spray column

In a spray column, the liquid phase is distributed in the form of droplets. The gas upflow forms the continuous phase. As in packed columns, the liquid holdup is low (<10%). There is a large variety of mechanical devices to disperse a liquid in a gas phase. They can be classified into three categories (for more technology details, see Coulson et al., 1991):

(1) Pressure nozzles (hollow and solid cone).
(2) Two-fluid nozzles (venture atomiser).
(3) Rotary devices (flat disks, disks with vanes, bowl-shaped cups).

The most important characteristic of these devices is the size distribution of the droplets produced. Fine droplets are favoured because the exchange interfacial area between gas and liquid is very large. However, if they are too fine, droplets may escape from the absorber.

Some equations are available for the prediction of the mean diameter of the droplets. For example, in the case of a solid cone spray, the droplet size is given by:

$$\frac{d_g}{D} = k_d \, Re^x \, We^y \tag{17.10}$$

where:

$$Re = \frac{U_L D \rho_L}{\mu_L}; \; We = \frac{U_L \mu_L}{\sigma}$$

U_L: liquid velocity through the nozzle
k_d, x and y: constants.

Murty (1982) proposed the following values: $k_d = 57$, $x = -0.48$ and $y = -0.18$. The physico-chemical properties of the liquid (surface tension, viscosity, density) have a great effect on the droplet size. Since few relationships are available for the calculation of the droplet mean diameter, it is recommended to ask the manufacturers for the available data for each type of spray device.

17.5 DESIGN OF PACKED AND SPRAY COLUMNS

The design of packed and spray columns is based on the writing of the mass balance equations and on the introduction of the terms "height of transfer unit" (HTU) and "number of transfer unit" (NTU). The design method is developed further by McCabe et al. (1993), Roustan (2003) and Sinnott (1993).

17.5.1 Physical absorption

17.5.1.1 Design equation

For the design of an absorption column based on physical absorption, the following assumptions are made:

- The concentration of the solute (ammonia) in the gas phase is small, the equations can be used with little error for mixtures with up to 10% solute, say less than 10%.

- Gas and liquid flow rates can be considered constant throughout the column (change in molar flow rate is neglected) so the equilibrium line and the operating line are straight lines.

Consider a packed or a spray column as absorber. The absorber can be designed using the four different ways (Equation (17.1)) and writing the mass balances in a height section dz. The differential volume is Ωdz and the total effective exchange area is $a^0\Omega dz$. The amount absorbed in section dz is $-Q_G dC_G$ or $Q_L dC_L$. The equations can be written as follows:

$$\begin{aligned} -Q_G dC_G &= K_G a^0 \Omega \, dz \, (C_G - C_G^*) = k_G a^0 \Omega \, dz \, (C_G - C_{Gi}) \\ Q_L dC_L &= K_L a^0 \Omega \, dz \, (C_L^* - C_L) = k_L a^0 \Omega \, dz \, (C_{Li} - C_L) \end{aligned} \tag{17.11}$$

where:
Q_G, Q_L: gas and liquid volumetric flow rates ($m^3 s^{-1}$)
a^0: exchange area of the wetted packing or the droplets, by unit volume of column ($m^2 m^{-3}$)
Ω: cross-section of the absorber (m^2)
C_{Gi} and C_{Go}: molar concentration of ammonia in the gas at the inlet and outlet of the column, respectively.

Choosing the overall gas coefficient, Equation (17.11) can be rearranged for integration as follows:

$$\int_0^Z dz = \frac{Q_G}{K_G a^0 \Omega} \int_{C_{GS}}^{C_{GE}} \frac{dC_G}{(C_G - C_G^*)} \tag{17.12}$$

The column height is given by:

$$Z = \frac{Q_G}{K_G a^0 \Omega} \int_{C_{GS}}^{C_{GE}} \frac{dC_G}{(C_G - C_G^*)} \tag{17.13}$$

In Equation (17.13), the term $Q_G/K_G a^0 \Omega$ has units of length (m) and is called "the height of an overall gas-phase transfer unit" (HTU_{OG}). The term $\int_{C_{GS}}^{C_{GE}} dC_G/(C_G - C_G^*)$ is "the number of overall gas-phase transfer units" (NTU_{OG}).

Thus, the height of the absorber is $z = HTU_{OG} \times NTU_{OG}$.

Where the operating and equilibrium curves are straight lines (which corresponds to the conditions of dilute systems), Equation (17.13) can be integrated after introducing the absorption factor $A = \frac{1}{m} \times \frac{1}{G}\big|_{molar}$ and the removal efficiency E_G of the component from the gas phase $E_G = (C_{Gi} - C_{GO})/C_{Gi}$.

The number of transfer units (NTU_{OG}) is given by:

$$\mathrm{NTU}_{\mathrm{OG}} = \frac{A}{A-1}\,\mathrm{Ln}\,\frac{A-E_{\mathrm{G}}}{A(1-E_{\mathrm{G}})} \tag{17.14}$$

and the height of the absorber (z) becomes:

$$Z = \frac{Q_{\mathrm{G}}}{K_{\mathrm{G}}a^0\Omega} \times \frac{A}{A-1}\,\mathrm{Ln}\,\frac{A-E_{\mathrm{G}}}{A(1-E_{\mathrm{G}})} \tag{17.15}$$

From Equation (17.15), the efficiency of the removal of the component is given by:

$$E_{\mathrm{G}} = \frac{A(1-\mathrm{e}^{-\alpha})}{A-\mathrm{e}^{-\alpha}} \tag{17.16}$$

with

$$\alpha = \frac{K_{\mathrm{G}}a^0\Omega\, Z(A-1)}{A\,Q_{\mathrm{G}}} = \frac{A-1}{A} \times \mathrm{NTU}_{\mathrm{OG}}$$

If the absorption factor A is equal to 1, then $\mathrm{NTU}_{\mathrm{OG}} = \dfrac{E_{\mathrm{G}}}{1-E_{\mathrm{G}}}$.

If the absorption factor A is <1, E_{G} is always less than or equal to A.

17.5.1.2 Packed column

For the design of the packed column, it is necessary to calculate first the diameter of the column from flooding conditions (Coulson et al., 1991), then the value of the wetted area of the packing and finally the values of the mass transfer coefficients k_{L} and k_{G} in order to determine the overall gas mass transfer coefficient K_{G} (Roustan, 2003; Sinnott, 1993).

Many correlations have been published for predicting the main parameters of packed columns (Coulson et al., 1991; Perry et al., 1997). The majority are shown in Table 17.4.

The height of the column is deduced from correlation equation (17.15) if the efficiency of the column is imposed. From a practical point of view, the recommended range for the absorption factor A is 1.2–2. Values around 1.4–1.5 are frequently used.

17.5.1.3 Spray column

The droplet size is in the range of 400–1600 μm. They settle in a gas phase flowing in counter-current flow. If U_{GS} designates the cross-sectional velocity and U_{g} the terminal velocity of the droplet in a quiet gas phase, the real velocity U_{R} of the droplet is $U_{\mathrm{R}} = U_{\mathrm{g}} - U_{\mathrm{GS}}$. The absolute velocity of the droplet towards the gas velocity is

Table 17.4 Determination of the packed column design parameters for operating on physical absorption.

Column diameter D_C from flooding conditions (LOBO method) (m)	$X = \frac{L_M}{G_M}\sqrt{\frac{\rho_G}{\rho_L}}; Y = \frac{F}{g}\left(\frac{\rho_{H2O}}{\rho_L}\right)^{0.2}\frac{\rho_G}{\rho_L}\left(\frac{\mu_L}{\mu_{H2O}}\right)^{0.2} U^2_{Flood}$ $Y = \exp(-4X^{0.25}) \quad 0.02 \leqslant X \leqslant 4$ Percentage flooding 60%: $U_{Greal} = 0.6\, U_{flood}$ Required diameter: $D_c = \sqrt{\frac{4Q_G}{pU_{Greal}}}$
Wetted packing area a^0 ($m^2 m^{-3}$)	$\frac{a^0}{a^*} = 1 - \exp\left(-1.45\left(\frac{\sigma_C}{\sigma_L}\right)^{0.75}(a * d)^{-0.35}\, Ga_L^{0.05} We_L^{0.2}\right)$
Mass transfer coefficient k_G ($m\, s^{-1}$)	$Sh_G = 5.23(a * d)^{-1.70}\, Re_G^{0.70}\, Sc_G^{0.33}$ $Sh_G = \frac{k_G d}{D_{AG}}, Re_G = \frac{U_{GS} d\rho_G}{\mu_G}, Sc_G = \frac{\mu_G}{\rho_G D_{AG}}$
Mass transfer coefficient k_L ($m\, s^{-1}$)	$k_L = 0.0051\left(\frac{\mu_L g}{\rho_L}\right)^{0.33}(a * d)^{-0.27}\left(\frac{a^0}{a^*}\right)^{-0.66} Re_L^{0.66} Sc_L^{-0.50}$ $Re_L = \frac{U_{LS} d\rho_L}{\mu L}, Sc_L = \frac{\mu_L}{\rho_L D_{AL}}$
Overall mass transfer	$\frac{1}{K_G} = \frac{1}{k_G} + \frac{He''}{k_L}$

$U_g + U_{GS}$, a value that enables the calculation of the Reynolds number of the droplet. The recommended velocity U_{GS} is about 50% of the droplet terminal velocity U_g. The height z of the spray tower and the efficiency E_G are deduced from Equation (17.15) and the parameters given in Table 17.5.

17.5.2 Absorption with an irreversible chemical reaction

17.5.2.1 Design equations

Where an instantaneous reaction takes place at interface, Equations (17.6) and (17.8) give:

$$-Q_G dC_G = k_G a^0 \Omega \, dz \, C_G \tag{17.17}$$

After integration it becomes:

$$Z = \mathrm{Ln}\frac{1}{1 - E_G} \times \frac{Q_G}{k_G a^0 \Omega} \tag{17.18}$$

Table 17.5 Determination of the spray column design parameters for operating on physical absorption.

Residence time of droplets (s)	$\tau = \dfrac{Z}{U_R} = \dfrac{Z}{U_g - U_{GS}}$
Interfacial area ($m^2 m^{-3}$)	$a^0 = \dfrac{6Q_L \tau}{d_g \Omega Z}$
Mass transfer coefficient k_G ($m s^{-1}$)	$Sh_G = 2 + 0.6 Re_g^{0.50} Sc_G^{0.33}$ $Sh_G = \dfrac{k_G d_g}{D_{AG}}, \; Re_g = \dfrac{U_{abs} d_g \rho_G}{\mu_G}$ $Sc_G = \dfrac{\mu_G}{\rho_G D_{AG}}$
Mass transfer coefficient k_L ($m s^{-1}$) (inside the droplet)	$Sh = \dfrac{k_L d_g}{D_{AL}}$ #17
Overall mass transfer coefficient K_G ($m s^{-1}$)	$\dfrac{1}{K_G} = \dfrac{1}{k_G} + \dfrac{He''}{k_L}$

where:

$$\text{NTU}_G = \text{Ln}\frac{1}{1 - E_G}\,; \; \text{HTU}_G = \frac{Q_G}{k_G a^0 \Omega}$$

The removal efficiency (E_G) is given by:

$$E_G = 1 - \exp\left(-\frac{k_G a^0 \Omega Z}{Q_G}\right) = 1 - e^{-\text{NTU}_G} \tag{17.19}$$

17.5.2.2 Packed column

For the design of the packed column, it is necessary to calculate first the diameter of the column from flooding conditions, then the value of the wetted area of the packing and finally the value of the mass transfer coefficient k_G. Then, application of Equations (17.18) and (17.19) gives z or E_G.

17.5.2.3 Spray column

From Equation (17.19) and the data given in Table 17.5, the removal efficiency E_G or the height z can be deduced from the following equation:

$$E_G = 1 - \exp\left[-\frac{6k_G}{d_g} \times \frac{Q_L}{Q_G} \times \frac{Z}{(U_g - U_{GS})}\right] = 1 - \exp^{-NTU_G} \quad (17.20)$$

17.6 EXAMPLES OF DESIGN

17.6.1 Absorption of NH_3 by water in a packed tower

A gas stream, containing 3% (volume) NH_3 and 97% air, is to be scrubbed with water in a packed column equipped with ceramic Raschig rings (diameter 51 mm). The gas flow rate is 1000 $m^3 h^{-1}$ and the expected removal efficiency is 96.7%. The column is operating at atmospheric pressure with an inlet temperature of 20°C for the two phases. The selected absorption factor $A = 1.50$. What is the height of the column?

To simplify the analysis, the temperature of the two phases is assumed to be constant and equal to 20°C. The physical properties of the gas are taken as those for air, as the NH_3 concentration is low. The results are shown in Table 17.6.

Table 17.6 Results of the design of a packed tower for NH_3 absorption in water.

Parameters	Data
Packing diameter	50.8 mm
Gas flow rate	0.33 $kg\,s^{-1}$
Absorption factor	1.5
Water flow rate	0.25 $kg\,s^{-1}$
Column diameter	0.44 m
Wetted interfacial area a^0	39 $m^2 m^{-3}$
k_G	0.066 $m\,s^{-1}$
k_L	$1.1 \times 10^{-4} m\,s^{-1}$
K_G	0.049 $m\,s^{-1}$
HTU_{OG}	0.94 m
NTU_{OG}	7.1
Height z	6.7 m

The overall mass transfer coefficient K_G is a little different from k_G but the resistance is essentially located in the gas phase. The application of the different equations (17.13)–(17.16) and Table 17.4 gives a HTU_{OG} value of about 1 m. For random packing, typical HTU_{OG} values are 0.3–1 m, depending on the operating conditions, types and sizes of packing. The NTUs depend on the removal efficiency of ammonia. For 96.7%, NTU_{OG} is 7.1, but if the E_G becomes 99.5%, NTU is 12.6.

17.6.2 Absorption of NH_3 by a dilute acid solution in a packed tower

Table 17.7 Results of the design of the packed tower.

Parameters	Calculated values
Gas flow rate	$0.33\,kg\,s^{-1}$
Liquid flow rate	$0.25\,kg\,s^{-1}$
Absorption factor	1.5
Column diameter	0.44 m
k_G	$0.063\,m\,s^{-1}$
$HTU_{OG} = HTU_G$	0.43 m
NTU_{OG}	3.4
Height Z	1.5 m

The data are similar to these presented in Section 17.6.1. The liquid phase is now a dilute acid solution (HCl or H_2SO_4). The results of the design calculations for NH_3 absorption in dilute acid solution are shown in Table 17.7.

This example shows very clearly the effect of absorption followed by reaction in the liquid phase. The height of the column, 1.5 m, much is reduced compared with the height (6.7 m) using water as absorbent. Another advantage of absorption plus reaction is the increase in the mass transfer coefficient and consequently the decrease in HTU_{OG} (reduced by about 50% compared to absorption with water).

17.6.3 Absorption of NH_3 by a dilute acid solution in a spray column

A spray column is to be designed to absorb NH_3 from a gas mixture by scrubbing the gas with a dilute acid solution. The gas flow rate is $30\,m^3\,s^{-1}$ (temperature 20°C and atmospheric pressure). The height of the absorber is 5 m and the droplet size is 1 mm (settling velocity $4.2\,m\,s^{-1}$). What is the liquid flow rate when the removal efficiency is 99.5%?

Equations (17.13)–(17.16) and Table 17.5 will be used to calculate the solution to the previous question.

Inside the absorber, the gas velocity U_{GS} is 50% U_t, so $U_{GS} = 2\,m\,s^{-1}$. The cross-section of the absorber is equal to $\Omega = 30/2 = 15\,m^2$ and then $D_C = 4.4$ m. The velocity of the droplet, $2 + 4.2 = 6.2\,m\,s^{-1}$, leads to a Reynolds number for the droplet of 413. The Frössling equation allows the mass transfer coefficient to be calculated as $k_G = 0.2\,m\,s^{-1}$. The ratio Q_L/Q_G is 2×10^{-3}. The flow rate of the liquid solution would be $216\,m^3\,h^{-1}$, to be sprayed by several nozzles at the top of the absorber.

17.7 CONCLUSIONS

Ammonia can be removed efficiently from gas mixtures by using absorption into several aqueous solvents. Water can be used due to the high solubility of ammonia in water. However, the best results are obtained with dilute acid solutions due to the presence of an instantaneous reaction in the liquid phase. This chemical reaction increases the exchange potential and the mass transfer coefficient. Also, the height of the absorption column is greatly reduced when dilute acid is used as solvent rather than water alone.

Two main technologies are available for the scrubbing of ammonia from waste gases: packed columns or spray towers. The final choice depends on the type of gas mixture. In the case dust is also present in the gas phase, spray columns are recommended.

REFERENCES

Coulson, J.M., Richardson, J.F., Backhurst, J.R. and Harker, J.H. (1991) *Chemical Engineering*, Vol. 2, 4th edn. Butterworth Heinemann, Oxford.

Danckwerts, P.V. (1970) *Gas–Liquid Reactions*, McGraw-Hill Book Company, New York.

Devos, M., Patte, F., Rouault, J., Laffort, P. and Van Genert, L.J. (1990) *Standardized Human Olfactory Threshold*, Oxford University Press, Oxford.

Manahan, S.E. (1997) *Environmental Science and Technology*, Lewis Publishers, Boca Raton, FL.

McCabe, W.L., Smith, J.C. and Hariott, P. (1993) *Unit Operations of Chemical Engineering*, 5th edn. McGraw-Hill, New York.

Murty, K.N. (1982) Predict spray-nozzle performance, *Chem. Eng.* **8**, 118.

Perry, R.H., Green, D.W. and Maloney, J.O. (1997) *Chemical Engineers' Handbook*, 7th edn. McGraw Hill, New York.

18

CO_2 capture in energy production

D. Aaron and C. Tsouris

18.1 INTRODUCTION

Approximately 7 Gt of carbon are emitted to the earth's atmosphere each year as a result of human activity. A large portion of this carbon is in the form of gaseous CO_2, and approximately 30% of this CO_2 comes from fossil fuel power plants. In parallel to rising levels of atmospheric CO_2, the earth's temperature is increasing, and it is believed that the two phenomena are related. Therefore, reduction of CO_2 emissions is currently an important area of research (Aaron & Tsouris, 2005). Separation and sequestration of CO_2 are near-term goals for emission reduction and they are critical and challenging topics for researchers. Many traditional separation processes are being considered for CO_2 capture, including absorption using solvents or solid sorbents, pressure-swing adsorption (PSA), and temperature-swing adsorption (TSA) using various solid sorbents, cryogenic distillation, and membrane separation. Several novel technologies are emerging as well. Both traditional and emerging processes are reviewed in this chapter.

18.2 ABSORPTION

18.2.1 Working principle

The process of CO_2 absorption by a liquid solvent or solid matrix is currently being investigated for scrubbing CO_2 from flue gas streams. Absorption is a process that relies on a solvent's chemical affinity with a solute to preferably dissolve one species over another. In CO_2 absorption processes, a liquid solvent is used that preferentially dissolves CO_2, but not N_2, H_2O, NO_x, SO_x, CO, O_2, or any particulates (Li et al., 2003). The CO_2-rich solution is typically pumped to a stripping column where the CO_2 is stripped from the solution and the solvent recycled for a new batch of flue gas. Optimal conditions for absorption are low temperature and high pressure. In addition, most solvents are easily degraded or deactivated by compounds such as fly ash, other particulates, SO_x, and NO_x, so the absorption step must take place after electrostatic precipitation (ESP) and flue gas desulfurization (FGD). In a typical absorption process, the CO_2-lean exhaust gas is either emitted to the atmosphere or possibly used in other applications (e.g. chemical production). Absorption is often considered a base case for CO_2 separation schemes because it is well known and understood in industrial applications.

18.2.2 Use of monoethanolamine as a solvent

An effective, economical, and traditional solvent that can be used for CO_2 absorption is monoethanolamine (MEA). The flue gas from a fossil fuel power plant is passed through a column in which MEA selectively absorbs CO_2 (Figure 18.1). This CO_2-rich solution is then pumped to a tower in which thermodynamic conditions are manipulated to release the CO_2. High-pressure and low-temperature favor absorption while low-pressure and high-temperature result in regeneration of the solvent. Temperature manipulation (specifically, to release the CO_2 and regenerate the MEA) makes up to

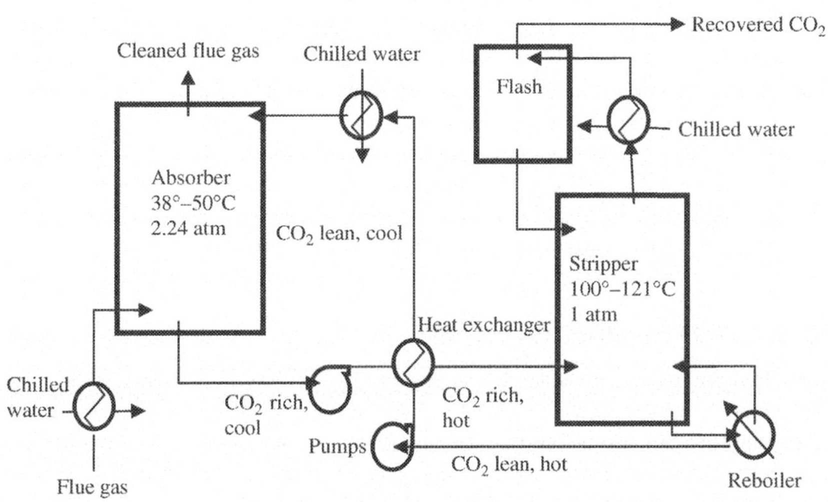

Figure 18.1 Flow sheet of chemical CO_2 absorption system with MEA solvent (Yeh & Pennline, 2001).

70–80% of the operating cost. The main area of improvement for this process is finding new solvents or optimizing the existing method to minimize regeneration costs (Veawab et al., 2001).

In Figure 18.1, the flue gas is cooled prior to entering the absorption chamber. The minimum temperature of operation should be above the condensation point of the gas and the freezing point of the solvent. After absorption, the CO_2-rich solution is passed through a heat exchanger to recover much of the heat from a hot stream coming from the regenerator. It then goes into the stripper where it is heated to release the CO_2 in solution (if pressure control is involved, the pressure is lowered). The CO_2-lean solution is then pumped back into the absorber for reuse. There are many points where a chilled water stream exchanges heat with the solvent stream, which constitutes a cost-effective method to conserve energy (Yeh & Pennline, 2001).

The absorption process has many design issues that must be overcome. The most important issue is the regenerability of the solvent (the regeneration process is the same regardless of the solvent employed). Since it is desirable to carry out the process with as little energy penalty as possible, the solvent activity must have an optimum balance between absorption and desorption rates. Absorption of CO_2 at low temperatures (below 50°C) and relatively low CO_2 partial pressure (normal flue gas fraction of CO_2 is approximately 0.15) can occur if the solvent is highly attractive to the solute, but this high attraction between the solvent and the CO_2 results in significant regeneration energy. If the solvent does not have a high affinity with CO_2, regeneration is simple, but little loading is possible.

Another concern is the oxygen content of the flue gas. A high oxygen concentration can corrode carbon steel facilities and cause excessive amine loss. Hindered amines (organic amine compounds with large attached groups that hinder some solvent–oxygen interactions) offer some resistance against degradation of the solvent. Acidic gases, such as SO_x and NO_x, are also a major concern. It is preferable to keep levels of these acidic gases below 0.001% because they form stable salts with the amines used for absorption. Since typical SO_x scrubbers often remove 90% of the SO_x in the flue gas, solvent degradation can be a major issue if MEA is used. Fly ash and NO_x compounds create the same problems as those presented by SO_x compounds; any amount of these components will cause at least some solvent degradation.

Finally, the high-temperature associated with flue gases can degrade the solvents and lower the solubility of CO_2. The maximum ideal temperature for MEA is 45°C. Thus, the temperature needs to be decreased to approximately 45°C in the SO_x scrubber, upstream from the separation column (Chapel et al., 1999).

18.2.3 FGD product recycling

One method that has been studied by Taulbee et al. (1997) uses the separated SO_x from FGD to form a solvent that absorbs CO_2. Limestone-based compounds are currently used to separate SO_x from flue gas streams prior to absorption. The limestone-SO_x compounds can then be slurried, becoming the liquid solvent for absorption. Laboratory experiments show that up to 0.0257 $m^3\,CO_2$/kg of FGD product can be absorbed. An attractive feature of this design is that many coal power plants are either already equipped with FGD units or are upgrading to include these units in order to comply with tightening environmental standards. Additional transport equipment for the slurry to be moved to the absorber is necessary. The overall cost for an absorption unit using FGD products as the solvent is lower than a conventional absorption process.

18.2.4 Advantages and disadvantages for absorption

CO_2 absorption is a well-established method of capturing CO_2 and the mechanisms and involved thermodynamics are well known. The fact that CO_2 absorbents can be regenerated is possibly the most attractive feature that motivates renewed research efforts focused upon improvement of the absorption process. The main advantage of absorption is that there are solvents that can be repeatedly regenerated. MEA has been used for over 60 years in the chemical industry, while new solvents operating on the same principles are currently being investigated (Herzog, 1999). CO_2 absorption is a well-established method of separating CO_2 and the mechanisms and involved thermodynamics are well known.

When the CO_2-rich solution is sent into the stripper, the solvent can be regenerated, thus reducing the cost of materials. However, different conditions demand different solvents. For low partial pressures of CO_2 (<15 vol.%), liquid solvents like MEA are preferable. For high partial pressures of CO_2 (>15 vol.%), solid absorbents, such as lithium hydroxide and lithium zirconate, are better because they can absorb more CO_2 and are more easily regenerated (Audus, 1997). Also, the separated CO_2 stream from the stripping column is of high purity, often up to 95%. This is pure enough for most sequestration methods, including aquifer injection, coal bed methane production, and enhanced oil recovery (Holloway, 1997).

While absorption does have strong advantages, the total cost (including addition of new solvent and other operating and maintenance (O&M) costs) is relatively high, about \$40–70/ton CO_2 separated, as reported by Chakma and Tontiwachwuthikul (1999). The exact life of a batch of solvent was not reported for MEA or any other solvent, but solvent generally degrades before it is replaced. The energy penalty of the entire absorption process (no O&M costs included) is approximately 0.0764 kWh/kg CO_2 (\$15.14/ton CO_2), as reported by Gottlicher and Pruschek (1997). Note that the cost reported by Gottlicher and Pruschek is only the energy penalty necessary to operate the process. The cost reported by Chakma includes the cost of purchasing new solvent and other operational costs.

In addition to high regeneration costs, sulfur compounds (SO_x gases) present in flue gases during absorption degrade the solvents currently considered for absorption. Approximately 8.75 kg of solvent are lost for each ton of CO_2 separated (Chapel et al., 1999). The loss of solvent occurs in the form of salts that do not dissociate in solution; they precipitate out. This often requires absorption with MEA to take place after the FGD step.

18.2.5 Process improvements

18.2.5.1 New solvents

Improvement is necessary for absorption to be viable for CO_2 separation in the future. New solvents, such as the new PSR™ solvents reported by Veawab et al. (2001), are being investigated. The PSR solvents are a family of solvents, currently under development, that require less heating for regeneration and are able to absorb more CO_2 than MEA. PSR is a trade name for this solvent whose chemical formula is currently unreleased. These solvents absorb 20–80% more CO_2 than MEA per unit volume of solvent, which allows them to be circulated more slowly through the system; this decreases pump costs. PSR solvents are also more resistant to degradation from flue gases and cause less corrosion to equipment.

Possibly the most promising absorption process employs the KS-1, KS-2, and KS-3 solvents being developed by Mitsubishi Heavy Industries (MHI). This family of solvents shows higher CO_2 loading per unit mass of solvent, milder regeneration conditions, and little corrosion, degradation, or amine loss. A novel packing material, KP-1, has also been developed that will further improve this process. Commercial-scale use of the solvent and packing has been in place at a fertilizer plant in Malaysia since 1999 (MHI CO_2 web site). The separated CO_2 is used in urea production, mitigating storage costs.

In addition to the Mitsubishi process, the Econamine FG™ process also shows lower overall costs. This process improves upon regular MEA absorption by using hindered amines in solution. Hindered amines are similar to regular amine solvents, but specific functional groups are added to these solvents to diminish or prevent degradation. This also lowers steam costs for regeneration, thereby lowering the overall cost. Audus (1997) reports that the Econamine FG process costs $18.70/ton CO_2 in O&M costs when CO_2 is 13% of the flue gas by volume. This process is currently being modified to further decrease the cost.

18.2.5.2 New equipment configuration

In addition to new solvents and packing configurations, alterations of the equipment used in the absorption process are being investigated. Fisher et al. (2005) have performed rigorous simulations of a MEA absorption system with various equipment configurations. The base case setup is illustrated in Figure 18.2, but includes compression after CO_2 exits the stripping column. This compression is included in all of the process configurations explored as a necessary prerequisite for transport of the separated CO_2. In addition to the base case, heat recovery from the exiting CO_2 and multistage compression in the stripping column were considered. Four scenarios were actually simulated: base case, heat recovery, multistage compression, and multistage compression combined with heat recovery. Figure 18.2 is an illustration of multistage compression combined with heat recovery used in the simulations.

In these simulations, the goal of 90% CO_2 recovery was met, and the associated cost for each configuration was calculated using costing software and simulations. It was determined that multipressure stripping with heat recovery reduced the cost of MEA absorption by 9.8% compared to simple stripping and compression. The implementation of this process would cause the cost of electricity to be $0.0599/kW-h, while the base cost without any CO_2 capture was $0.025/kW-h. Despite this increase, this modification shows great promise, especially if the configuration changes are coupled with better solvents.

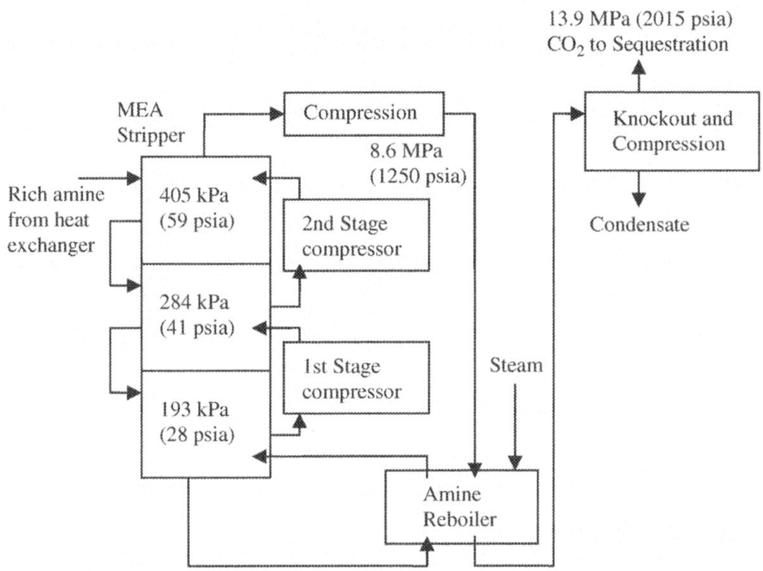

Figure 18.2 Multistage compression and heat recovery in the simulations by Fisher et al. (2005).

Table 18.1 Selection of materials and conditions in CO_2 absorption.

Absorbent	Absorption temperature (°C)	Absorption pressure (atm)	Desorption temperature (°C)	Desorption pressure (atm)
MEA	50	2.24	120	1
PSR	50	2.24	110	1

18.2.5.3 Other possible improvements to absorption

Another approach to improve absorption is to increase contact between the solvent and the flue gases in the absorption column. The addition of packing is used to maximize flue gas-solvent contact as well as decrease residence time. In a recent experiment performed by Yeh and Pennline (2001), a column with gases and conditions typical of flue gases entering an absorption chamber was packed with different materials such as BX gauze, Flexipac, and intalox saddle (made by Norton Chemical Process Products). The saddle packing was tested with both random and structured configurations. Absorption took place at approximately 50°C for the MEA, regardless of which packing was used. Regeneration took place at 120°C. This process is often used in current applications at most plants with MEA scrubbers. The structured packing showed the greatest solvent utilization at $43.9 \pm 0.5\%$ (fraction of solvent that absorbed CO_2), while the random packing achieved only 28.6% solvent utilization.

In summary, different packing materials and organic solvents have various effects on the absorption rate and regeneration energy for the system. The right combination of solvent, packing, and operating conditions could yield a process that cleans flue gases for a fraction of the current costs. A list of materials and conditions reported thus far for the separation of CO_2 by absorption is given in Table 18.1.

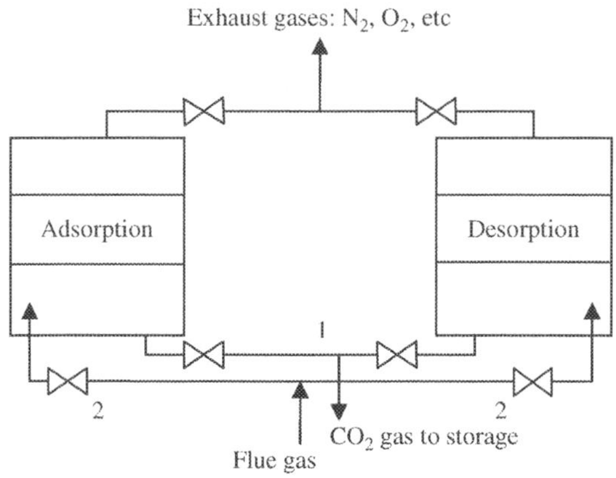

Figure 18.3 Flow chart of a PSA system.

18.3 ADSORPTION

18.3.1 Working principle

Adsorption is another method that can be used to separate CO_2 from flue gases generated by fossil fuel power plants. Due to the interaction energy between adsorbent and guest molecules, CO_2 molecules are attracted to and trapped by surface groups of the adsorbent. Conditions can be manipulated to facilitate adsorption or desorption. Many solids have the capability to selectively adsorb CO_2 into small cracks, pores, or onto their external surfaces under specific temperature and pressure conditions. Adsorbents such as activated carbon, molecular sieves, and zeolites have all been considered for adsorption processes.

The two traditional methods for adsorption are PSA and TSA. In either case, adsorption rates depend on temperature, partial pressures of CO_2, surface forces (interaction energy between adsorbent and CO_2), and pore size or available surface area of the adsorbent (Meisen & Shuai, 1997). Riemer et al. (1994) report that PSA is superior to TSA due to its lower-energy demand and faster regeneration rate.

Figure 18.3 shows a PSA where two chambers pass the pressure back and forth in a continuous cycle. Flue gases are sent into one of the two chambers and pressure is applied, allowing adsorption of CO_2. The pressure is then transferred to the other chamber where a new volume of waste gas has been introduced. As that chamber is pressurized, the first chamber is depressurized and the CO_2 is collected. The cycle then continues in a switching mode as waste gases are sequentially injected into the two chambers. The lines and valves shown beneath the tanks in Figure 18.3 control the flow of the CO_2 out of one tank (line 1) and the flow of untreated flue gases to the other (line 2).

18.3.2 Process improvements

18.3.2.1 Electrical desorption

A new class of adsorbents, based on a carbon fiber composite molecular sieve (CFCMS), can adsorb CO_2 and then release it when an electric current is applied across the adsorbent. Similar to PSA or TSA, electricity is the means to liberate the CO_2 trapped in the adsorbent. The adsorbent used is electrically conductive; electric current passed through the CFCMS raises the temperature, allowing for CO_2 desorption. Desorption was achieved on a CFCMS (2.5 cm diameter, 7.5 cm length cylinder) using 1 V direct-current potential and 5 amps. The carbon CFCMS proposed for this process is carbon based because of the high electrical conductivity and high specific surface area of carbon (Judkins & Burchell, 1997).

18.3.2.2 Hydrotalcite and water–gas shift

Perhaps the most promising use of adsorption is in conjunction with the water–gas shift process. Water–gas shift is a catalytic process that yields decarbonized H_2 fuel gas to a boiler for combustion (Allam et al., 2004). In this process, CO and steam are fed to a reactor where CO_2 and hydrogen gas form according to the following reaction:

$$CO + H_2O \leftrightarrow CO_2 + H_2$$

Hydrotalcite (HTC) adsorbs CO_2, pushing equilibrium to favor H_2 and CO_2. The advantage is twofold: less CO and more H_2 are present in the boiler feed.

This sorbent-enhanced water–gas shift (SEWGS) process shows exceptional promise for future power generation. The adsorption step proceeds as the PSA described above, but is operated simultaneously to the water–gas shift process. Allam et al. (2004) have performed ASPEN simulations to determine the efficiency and net power output of a plant operating with the SEWGS process, using HTC as the adsorbent. Experiments were performed to determine the properties of HTC under high-temperature conditions (>400°C). Based on these experiments and simulations, it was concluded that a plant employing air for combustion could achieve 48% efficiency, while the use of pure O_2 in combustion could achieve a plant efficiency of 47%. The air combustion is more efficient, but also uses more fuel for duct heating of the incoming air and has larger equipment sizes to accommodate N_2 in the system. Additionally, air combustion removes more CO_2 per year than O_2 combustion, but this is at least partially attributed to capturing CO_2 from the increased fuel usage. Further design modifications and simulations are being performed to characterize design parameters of this process at an industrial scale.

18.3.3 Advantages and disadvantages for adsorption

Unlike liquid absorption where the solvent can form corrosive solutions with flue gases, physical adsorption only requires vessels capable of withstanding small pressure changes. PSA, like chemical absorption, is based on regenerability of the adsorbent. As shown in Figure 18.3, the adsorbent can be reused many times for CO_2 separation. Gottlicher and Pruschek (1997) report that the energy cost for adsorption is projected to be \$6.94/ton CO_2 removed at CO_2 concentrations of 28–34 mol%. The energy cost associated with the treatment of gas with CO_2 concentrations of 10–11.5 mol% (the most common CO_2 concentration in flue gas) is expected to be four times greater than that at 28–34 mol%.

There are three significant drawbacks that make adsorption currently unfavorable as a stand-alone process. The first is that the system cannot easily process high concentrations of CO_2; good separation is achieved at low percentages of CO_2 (1–2 mol%). Most power plants have much higher concentrations of CO_2 in flue gases, approximately 12%. The second is that available adsorbents are not selective enough for CO_2 separation from flue gases. The adsorbents' ability is usually based on pore size. When CO_2 is the target to be selectively adsorbed, gases smaller than CO_2 can also penetrate the pores. N_2 is the gas that most commonly fills up pore space in adsorbents. This makes the process less efficient as a lower degree of CO_2 separation can be achieved in each cycle. The purity of the CO_2 stream is consequently also affected by the adsorbent used, since various adsorbents will adsorb varying amounts of N_2 (the most common impurity). A third drawback is that adsorption is slow. For typical materials, the residence time for maximum adsorption depends on the adsorbent, but 20 min is a reasonable estimate (Satyapal et al., 2001). When dealing with large volumes of flue gas, as in a power plant, this is just too slow to be practical.

Despite these disadvantages, physical adsorption can play a satisfactory role in a hybrid system. Since it requires a low concentration of CO_2 for optimum performance, it could be placed after another separation process. Research is currently being conducted to find more selective adsorbents, adsorbents with higher capacities, better operating conditions, and more efficient packing structures. If these goals can be met, adsorption can become a viable method for separating CO_2 in the future.

18.4 CRYOGENIC DISTILLATION

18.4.1 Working principle

Another separation process that has received some attention is cryogenic distillation. In this process, flue gas is sent to equipment designed to cool and compress the stream to conditions that will cause CO_2 to liquefy and leave the other components in the gaseous state (the triple point (point where solid, liquid, and gas all coexist) for CO_2 is 5.1 atm and −57.6°C). Pressure and temperature can be varied, but cryogenic separation operates at low temperature and high pressure.

After leaving the boiler, flue gas enters a chamber cooled by a refrigerant stream. Water is condensed to avoid chemical corrosion and particulate matter is removed by appropriate traps. CO_2, O_2, and N_2 are then directed to the column for cryogenic fractionation/distillation. Some of the CO_2-lean gases (mostly O_2) are recycled to the boiler while the high-purity liquid CO_2 is stored (Meratla, 1997).

18.4.2 Advantages and disadvantages for cryogenic distillation

Cryogenic distillation has one advantage over most other separation processes: its product is liquid CO_2, which is ready for transport via pipeline or tanker for sequestration. In addition to being ready for transport, the CO_2 recovery is very high (CO_2 purity after distillation can exceed 95%). The cryogenic process, however, is extremely energy intensive. The energy required to keep the system cool makes the current process cost ineffective. If H_2O is in the stream when the gases enter the chamber, it can freeze and clog the pipes (the water could also corrode the equipment). The cost for cryogenic

Table 18.2 Materials and conditions for membrane diffusion.

Material	Type	Temperature (°C)	Pressure (atm)
Organic	Solid polymer	25	1.01
Glycerine carbonate	Gas absorption	23	Transmembrane pressure = 0.4 atm
Ideal metallic or ceramic	Solid inorganic	350	1.00

distillation is predicted by lab-scale experiments performed by Gottlicher and Pruschek (1997) to be approximately \$32.7/ton CO_2 separated. This can be compared to the absorption cost of \$15.14/ton CO_2 removed and the adsorption cost of \$27.8/ton CO_2.

Refrigeration under pressure offers similar benefits to cryogenic separation, with similar drawbacks. While the cost for cooling under pressure is lower than that for cryogenic separation, more energy is required to pressurize the gases. To reduce the cost, a heat exchanger may be introduced to assist in cooling entering gases and heating exiting gases. The exiting gases may also power a turbine to generate energy, offering some repayment for the energy required in the rest of the process. The energy penalty for this process depends on the degree of separation. If 90% of the CO_2 is captured, the overall efficiency of the plant is decreased from 38% (before any CO_2 removal) to 26%; the efficiency is 31% when 50% of the CO_2 is removed. The base-case efficiency of 38% was an arbitrary decision by Schussler and Kummel (1989).

18.5 MEMBRANE SEPARATIONS

18.5.1 Working principle

The use of membranes has been established as a means of separating CO_2 from light hydrocarbons in the fuel and chemical industries. Membranes are also traditionally used to separate hydrogen gas from various other gases. Ceramic, metallic, polymeric, and solid–liquid are all forms of membranes used today (Audus, 1997). A list of membranes and their operating conditions is shown in Table 18.2.

The selectivity of polymeric membranes is based on their ability to interact with the target molecule. Whatever molecule is to be separated interacts with the membrane and is allowed to diffuse across, either by solution-diffusion or absorption-diffusion mechanisms. The principle of ceramic and metallic membranes, which are porous, is that only gases of a certain size are able to pass through the pores of the membrane. So, these membranes act as a sieve to separate CO_2 from larger gas molecules. The flue gases from a fossil fuel power plant are sent at approximately atmospheric pressure into a vessel that is divided by a membrane; the CO_2 passes through the membrane into another part of the chamber where it is collected at a lower pressure (typically 10% of the feed pressure).

In Figure 18.4, the flue gases simply enter the separation tank and the CO_2 diffuses across the membrane. The pressure on the permeate side is approximately 10% of the feed side; this pressure difference, established by a vacuum pump, will continually pull CO_2 across the membrane. Sometimes SO_x, NO_x, H_2O, O_2, CO, and

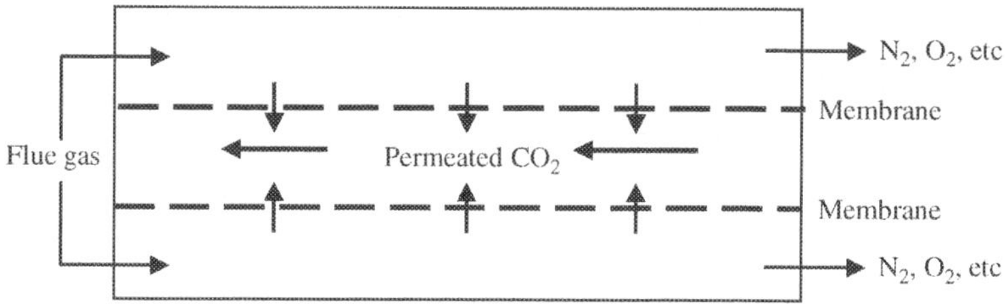

Figure 18.4 Simple membrane separation tube for selective permeation of CO_2.

particulates are separated prior to entering the membrane unit because they can harm the membrane. If this is the case, the membrane primarily has to separate CO_2 from N_2.

With a two-stage membrane system, the same process takes place, but the gases that permeate the membrane in the first chamber are again separated in a second separation vessel. This approach provides greater separation and a gas stream that is more suitable for carbon sequestration. The unpermeated gases from both units are sent to the same receiver for further treatment. Tokuda et al. (1997) report that double separation yields a much higher-purity stream of CO_2: 89.1% instead of 46.4% at 25°C with a single stage. However, it generally costs twice as much as the traditional amine separation processes (Tam et al., 2001).

In addition to the membranes mentioned above, gas absorption membranes are being investigated. These consist of microporous solid membranes impregnated with a liquid absorbent. Thus, gas absorption membranes are a hybrid of a solid membrane and liquid absorbent. In this arrangement, CO_2 selectively diffuses across the membrane and is captured and removed by the liquid absorbent. This allows for individual control of gas and liquid flows and minimizes complications such as flooding, channeling, and foaming. The gas absorption membrane system is a hybrid of solid membranes and liquid absorption. Research performed by Kovvali and Sirkar (2002) shows that the thickness of the gas absorption membrane plays an integral role in the effective permeance of the system. A membrane that is 10 μm thick would be 20 times more permeable than a membrane that is 200 μm thick. However, Kovvali and Sirkar have found that membrane thickness does not significantly influence selectivity. These experiments used glycerine carbonate as the liquid and either Celgard or poly(vinylidene fluoride) as the polymeric solid.

18.5.2 Advantages and disadvantages for membrane diffusion

The greatest asset to membrane separation is simplicity. While PSA requires special equipment for swinging the pressure and cryogenic distillation must endure extreme temperatures, the only equipment necessary for membrane separation is the membrane and fans. The flue gases must be compressed somewhat before separation (the ideal pressure is approximately 1.01 atm) but this compression is much smaller than that necessary for PSA. Membrane separators do suffer from a trade-off between selectivity and permeability to CO_2. Also, many organic membranes do not perform well at the high temperatures typical of flue gases exiting the boiler. Boiler exit temperature is approximately 370°C while the temperature following ESP (particle removal) and FGD is approximately 150°C. The difference between the pressure of the flue gases and the permeated gases (a factor of 10) also causes problems for membranes that are not structurally stable.

Gas absorption membranes suffer the same drawbacks that absorption does: the incoming gases may contain chemicals that foul or otherwise damage the liquid component of the membrane (Meisen & Shuai, 1997). Gas absorption membranes are also less structurally stable than what metal or ceramic membranes are predicted to be.

Research is being done to find new metallic, ceramic, and alumina membranes. These membranes would withstand better the high temperature of the entering gases and the pressure changes associated with the gas flow through the chamber. Mowbray (1997) considers 350°C a reasonable temperature for testing these membranes. Since gas absorption membranes have higher permeability and selectivity for CO_2, methods of making the solvent stage of the membrane more chemically stable are being investigated. The presence of water vapor in the pores of membranes has also been studied by Park and Lee (1995). It was found that, when water vapor clogs micropores, permeability decreases for both N_2 and CO_2. However, the decrease of N_2 permeability was so great that its permeability was immeasurable while CO_2 permeability decreased by only a factor of 10 (small decrease compared to the decrease in N_2). Additionally, the membranes in development must be able to withstand the presence of SO_x and other compounds that typically pose problems for CO_2 separation units, since they are located before the FGD unit.

Development of more stable solid and gas absorption membranes will determine whether membrane separation can stand alone as a CO_2 capture process or whether it may have to be part of a hybrid separation system.

18.6 CO_2 PRECIPITATION

18.6.1 Hydrate formation and dissociation

18.6.1.1 Working principle

One method for separating CO_2 from flue gases is to use hydrate formation. Hydrates are ice-like structures in which water forms a cage with cavities where small gases such as CO_2 can be trapped. While other gases can get trapped in these cavities, CO_2 is often the most likely occupant (a better hydrate former than other gases, under induced conditions). The maximum weight fraction of CO_2 in a unit of hydrate is 0.31 g CO_2/g H_2O (Pruschek et al., 1997).

Since water readily forms hydrates with CO_2, this process can capture CO_2 and allow the waste gases to escape. Hydrates do not form above 10°C or much below 4 atm. Conditions on the ocean seafloor where CO_2 hydrates are being researched for sequestration are approximately 2–4°C and pressure varying with depth (Brewer et al., 1999). In the case of syngas separation (Figure 18.5), the CO_2 hydrate slurry goes to the slurry separator where the H_2 is removed. This is a highly pure stream and can be used for hydrogen fuel cells or chemical reactions. This process is most likely to be used in an integrated gasification combined cycle (IGCC) application, but a similar process could be adapted for a standalone CO_2 hydrate separator at a pulverized coal plant. Using heat from the ammonia cooling exchangers, the hydrates are melted and the CO_2-rich stream is collected for sequestration or utilization. This process is also known as the SIMTECHE process (Deppe, 2004).

18.6.1.2 Addition of tetrahydrofuran for hydrate formation

Investigations performed by Kang and Lee (2000) sought to maximize CO_2 hydrate formation. One barrier to the use of hydrate formation is the high pressure involved. It was found that the addition of 1 mol% tetrahydrofuran (THF) lowered the pressure necessary for hydrate formation from 82.4 to 4.69 atm. In addition, it was found that lowering the temperature in the presence of THF by only 6 K increased the mol% of CO_2 in hydrate form from 39% to 61%. THF is used only in small quantities (1–3 mol% solution in water) and it can be recycled easily with the water after CO_2 has been liberated. The optimum conditions determined in this experiment were approximately 2°C and 3.95 atm, 70 mol% CO_2 + 30 mol% N_2 gas mixture, and 3 mol% THF (3% when THF is added to water). Further research is being conducted to determine the effect of varying concentrations of THF and other hydrate promoters.

18.6.1.3 Advantages and disadvantages of hydrate formation

The advantage of using hydrate formation for separating CO_2 from flue gas or syngas in an IGCC reactor is that it may not be more energy intensive than traditional processes (chemical absorption, PSA, etc.) and the product can be easier transported than CO_2 gas. Conventional amine scrubbers, PSA, and other methods impose energy penalties of up to 35% for coal-fired plants. Hydrate formation, while not actually performed in a plant, is believed to impose an energy penalty of only 4.4% in an IGCC system (Ubis et al., 2000).

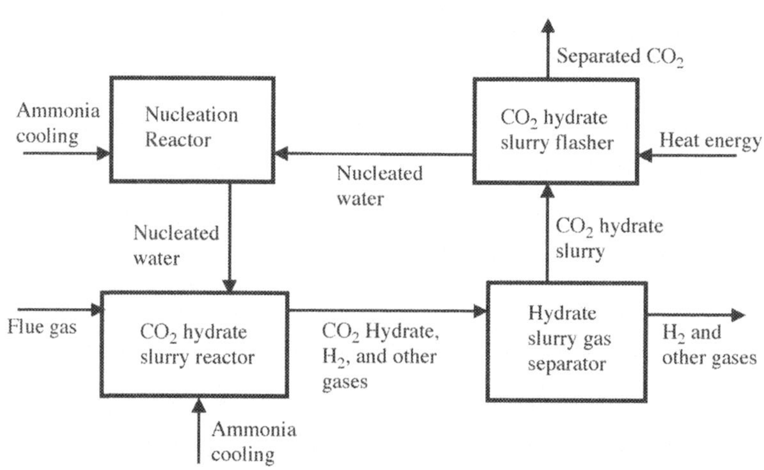

Figure 18.5 Flow diagram of a CO_2 hydrate separation process.

Hydrate formation is limited for current separation applications because it requires high pressures (~88 atm). While PSA requires high pressure, it is not as high as that for hydrate formation. In addition, the temperature for hydrate formation is quite low, ~4°C, especially when compared to the temperature of flue gases from coal plants. A significant amount of energy would be required to cool the flue gases to a suitable temperature, and then more energy would be required for pressurization.

While some conversion rates were as high as 35% (i.e., 35% of the CO_2 that entered the reactor was captured in the hydrate form), better phase contacting could increase the efficiency of hydrate separation. Other major areas of improvement include the removal of the heat liberated during hydrate formation and the achievement of steady flow despite multiple phases (liquid CO_2, hydrate, liquid water, and ice) present in the process.

18.6.2 Ammonium carbonation

18.6.2.1 Working principle

Separation of CO_2 from flue gas streams can be achieved by reacting the CO_2 with ammonia gas and water vapor in a gas-phase reaction (Figure 18.6). The reaction proceeds according to the following equation:

$$CO_2 + NH_3 + H_2O \rightarrow NH_4HCO_3\ (s)$$

This reaction can be achieved by bubbling raw flue gas through an aqueous ammonia solution. The NH_4HCO_3 forms a stable solid while the N_2 and other gases present in the flue gas stream continue through for release or treatment. It should be noted here that the solid product is to be used as a soil fertilizer, instead of being regenerated to recover the CO_2. In experiments done by Li et al. (2003), the flue gas was bubbled through a water bath kept at 40°C and then sent to the reactor to be bubbled through the ammonia solution. In bench-scale testing, the reaction proceeded at room temperature and ambient pressure, indicating that the energy penalty for this method would be lower than most current methods (such as amine scrubbing, adsorption, etc.). Maximum CO_2 removal percentages were reached after approximately 400 min of residence time. The flow rate ratio (NH_3: flue gas) that yielded the highest CO_2 conversion percentage (48%) was 0.12; lower flow rate ratios yielded lower conversion rates. While a higher

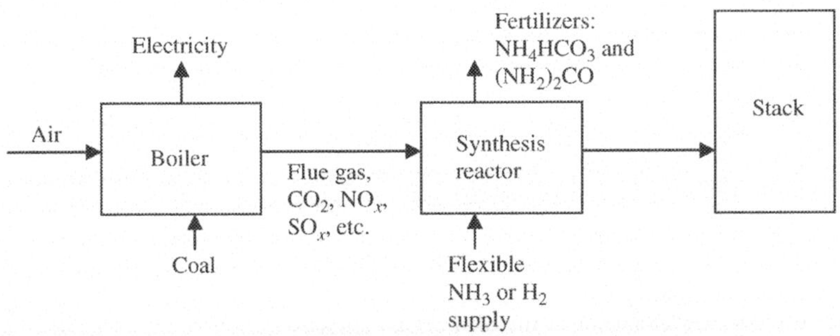

Figure 18.6 Flow diagram of ammonium carbonation to produce fertilizer.

NH_3 to flue gas ratio yields a higher conversion rate, computer modeling suggests that 20–60% conversion would be the best balance between conversion and cost. Another promising result of this study is that the residence time in the reactor can be cut from 188 to 47 s if the flow rate of the gas mixture is quadrupled (these times correspond to the 20–60% conversion, which takes less time than the 100% conversion). Decreasing residence time had no measurable effect on the conversion rate. This shows that, if a large-scale version of this process is used, large volumes of flue gas can be efficiently cleaned.

18.7 SUMMARY OF DEVELOPMENT STATUS

Absorption using chemical solvents is already in commercial application. The solvents currently used are not as efficient as the next generation being developed. These new solvents are in the pilot plant testing stage. Physical adsorption is used in some chemical commercial applications, but not for CO_2 separation. Development of an adsorbent that can function despite the high CO_2 concentration and is selective enough to be economically viable is still in lab testing. Cryogenic distillation is also used in industry to produce pure CO_2, but not on the scale involved in flue gas separations. The energy requirements and other costs associated with this method inhibit progress. Membrane diffusion is also used in some chemical applications, but not for CO_2 separation yet. Development of membranes capable of surviving and functioning well at pre-FGD conditions is necessary before testing can get out of the lab test phase. In all of the processes mentioned above, trace constituents of the flue gas are emitted to the atmosphere or treated further in some other process. Hydrate formation, electrical desorption and ammonium carbonation separations are promising novel concepts still in laboratory testing.

REFERENCES

Aaron, D. and Tsouris, C. (2005) Separation of CO_2 from flue gas: a review. *Separ. Sci. Technol.* **40**(1–3), 321–348.

Allam, R., Chiang, R., Hufton, J., Weist, E., White, V. and Middleton, P. (2004) Power generation with reduced CO_2 emissions via the sorption enhance water gas shift process. Presented at the *Third Annual Conference on Carbon Capture and Sequestration*, Alexandria, VA, May 3–6.

Audus, H. (1997) Greenhouse gas mitigation technology: an overview of the CO_2 capture and sequestration studies and further activities of the IEA greenhouse gas R&D programme. *Energy* **22**(2/3), 217–221.

Brewer, P.G., Friederich, G., Peltzer, E. and Orr, Jr. F.M. (1999) Direct experiments on the ocean disposal of fossil fuel CO_2. *Science* **284**, 943–945.

Chakma, A. and Tontiwachwuthikul, P. (1999) *CO_2 Separation from Combustion Gas Streams by Chemical Reactive Solvents*. Online Library: Combustion Canada.

Chapel, D., Ernest, J. and Mariz, C. (1999) Recovery of CO_2 from flue gases: commercial trends. Presented at the *Canadian Society of Chemical Engineers Annual Meeting*, Saskatchewan, Canada, 4–6 October.

Deppe, G. (2004) Development in the SIMTECHE process – CO_2 capture by formation of hydrates. Presented to the *3rd Annual Conference on Carbon Capture and Sequestration*, Washington, DC, May.

Fisher, K.S., Beitler, C., Curtis, R., Searcy, K., Rochelle, G., Jassim, M. and Figueroa, J.D. (2005) Integrating MEA regeneration with CO_2 compression to reduce CO_2 capture costs. Presented at the *Fourth Annual Conference on Carbon Capture and Sequestration*, Alexandria, VA, 2–5 May.

Gottlicher, G. and Pruschek, R. (1997) Comparison of CO_2 removal systems for fossil-fuel power plant processes. *Energ. Convers. Manage.* **38**(Suppl.), S173–S178.

Herzog, H. (1999) *An Introduction to CO_2 Separation and Capture Technologies.* MIT Energy Laboratory, August. http://web.mit.edu/energylab/www/hjherzog/publications.html.

Holloway, S. (1997). An overview of the underground disposal of carbon dioxide. *Energ. Convers. Manage.* **38**(Suppl.), S193–S198.

Judkins, R.R. and Burchell, T.D. (1997) A novel carbon fiber based material and separation technology. *Energ. Convers. Manage.* **38**(Suppl.), S99–S104.

Kang, S.P. and Lee, H. (2000) Recovery of CO_2 from flue gas using gas hydrate: thermodynamic verification through phase equilibrium measurements. *Environ. Sci. Technol.* **34**, 4397–4400.

Kovvali, A.S. and Sirkar, K.K. (2002) Carbon dioxide separation with novel solvents as liquid membranes. *Ind. Eng. Chem. Res.* **41**, 2287–2295.

Li, X., Hagaman, E., Tsouris, C. and Lee, J.W. (2003) Removal of carbon dioxide from flue gas by ammonia carbonation in the gas phase. *Energ. Fuel.* **17**(1), 69–74.

Meisen, A. and Shuai, X. (1997) Research and development issues in CO_2 capture. *Energ. Convers. Manage.* **38**(Suppl.), S37–S42.

Meratla, Z. (1997) Combining cryogenic flue gas emission remediation with a CO_2/O_2 combustion cycle. *Energ. Convers. Manage.* **38**(Suppl.), S147–S152.

Mitsubishi Heavy Industries Research. http://www.mhi.co.jp/machine/recov_co2/index. htm

Mowbray, J. (1997) Ceramic membrane to combat global warming. *Membr. Technol.* **92**, 11–12.

Park, Y.I. and Lee, K.H. (1995) The permeation of CO_2 and N_2 through asymmetric polyetherimide membrane. *Energ. Convers. Manage.* **36**(6–9), 423–426.

Pruschek, R., Oeljeklaus, G., Haupt, G., Zimmermann, G., Jansen, D. and Ribberink, J.S. (1997) The role of IGCC in CO_2 abatement. *Energ. Convers. Manage.* **38**(Suppl.), S153–S158.

Riemer, P.W.F., Webster, I.C., Omerod, W.G. and Audus, H. (1994) Results and full fuel cycle study plans from the IEA greenhouse gas research and development programme. *Fuel* **73**(7), 1151–1158.

Satyapal, S., Filburn, T., Trela, J. and Strange, J. (2001) Performance and properties of a solid amine sorbent for carbon dioxide removal in space life support applications, *Energ. Fuel.* **15**, 250–255.

Schussler, U. and Kummel, R. (1989) Carbon dioxide removal from fossil fuel power plants by refrigeration under pressure. *Proceedings of the 24th Intersociety* **4**(6–11), 1789–1794.

Tam, S.S., Stanton, M.E., Ghose, S., Deppe, G., Spencer, D.F., Currier, R.P., Young, J.S., Anderson, G.K., Le, L.A. and Devlin, D.J. (2001) A high pressure carbon dioxide separation process for IGCC plants. Presented at the *First National Conference on Carbon Sequestration*, 14–17 May.

Taulbee, D.N., Graham, U., Rathbone, R.F. and Robl, T.L. (1997) Removal of CO_2 from multicomponent gas streams using dry-FGD wastes. *Fuel* **76**(8), 781–786.

Tokuda, Y., Fujisawa, E., Okabayashi, N., Matsumiya, M., Takagi, K., Mano, H., Haraya, K. and Sato, M. (1997) Development of hollow fiber membranes for CO_2 separation. *Energ. Convers. Manage.* **38**(Suppl.), S111–S116.

Ubis, T., Bressan, L. and O'Keefe, L. (2000) The 800MW PIEMSA IGCC project. Presented at *2000 Gasification Technologies Conference*, 8–11 October.

Veawab, A., Aroonwilas, A., Chakma, A. and Tontiwachwuthikul, P. (2001) Solvent formulation for CO_2 separation from flue gas streams. Presented at the *First National Conference on Carbon Sequestration*, 14–17 May.

Yeh, J.T. and Pennline, H.W. (2001) Study of CO_2 absorption and desorption in a packed column. *Energ. Fuel.* **15**(2), 274–278.

19

Use of hydrates in gas treatment

V. Yakushev

19.1 GAS HYDRATES AND THEIR PROPERTIES

Gas hydrates are solid crystalline compounds of gas and water, looking like snow (porous hydrate) or ice (monolithic hydrate), sometimes coloured depending on the gas composition (Figure 19.1). Molecules of gas are encapsulated in the cavities created by water molecules (Figure 19.2). The general chemical formula of a gas hydrate is $M \times nH_2O$, where M is gas molecule and n is hydrate number showing the water/gas molecular ratio. There are two molecular structures of gas hydrates spread in nature: cubic structure 1 (CS-1, usually for small gas molecules, Figure 19.3) and cubic structure 2 (CS-2, for larger gas molecules). They are different in cavity size able to accept gas molecules. There are some other structures of gas hydrates formed in laboratories, but they are practically not spread in nature. The hydrate number of the majority of individual natural gases varies from 5.67 (CS-2 with full filling by gas molecules of small and large structure cavities) to 17 (CS-2 with only large cavities filling by gas molecules). The ideal formula for CS-1 is $M \times 5.75\ H_2O$, for CS-2 $M \times 5.67\ H_2O$. For methane (CH_4), the hydrate number usually varies from 5.9 to 6.2 (CS-1), whereas for CO_2 (also CS-1) the hydrate number varies in the range 7.0–7.6 depending on hydrate-formation conditions (Istomin & Yakushev, 1992).

Figure 19.1 "Burning ice" – burning methane hydrate on a hand.

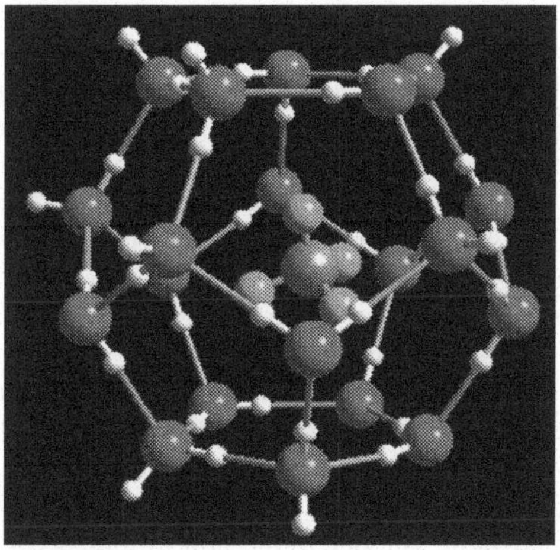

Figure 19.2 A model of a gas molecule (CH_4) entrapped in the cavity formed by water molecules in a hydrate structure.

Gas hydrates can be formed when water and gas are in contact in a certain range of pressure and temperature. For different gases these conditions are different. Figure 19.4 shows P/T hydrate-formation conditions of CH_4, CO_2 and some other gases. Methane, for example, can form hydrates at temperature 273 K (0°C) only if the pressure is below

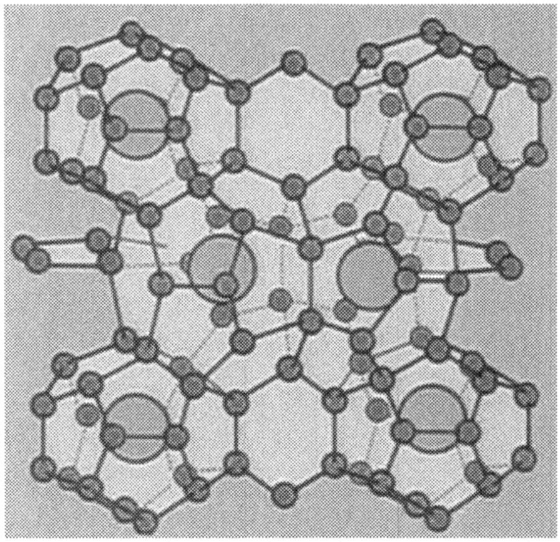

Figure 19.3 Methane and carbon dioxide hydrate structure (CS-1): methane (or carbon dioxide) molecule inside water lattice in hydrate structure.

2.6 MPa (Figure 19.4, bold line for CH_4). The higher the temperature, the higher pressure is required for hydrate formation.

Being encapsulated in the water lattice, gas molecules are in concentration corresponding to a pressure 160–180 atm (in free phase). The pressure required for hydrates formation is much less. For methane hydrate, it is 30–50 atm in the temperature range +2 to + 4°C (Figure 19.4). This property to concentrate gas at relatively low pressure and separate gas and water mixtures has attracted attention for a while (see e.g. Barduhn, 1968). Also when water forms the hydrate lattice, it loses salts and other admixes (only deionized water forms the hydrate lattice) and if hydrates are forming from gas mixture, different gas molecules come into the hydrate structure with a different intensity. This creates the base for water purification and gas separation technologies. There are data on isotopic separation of gas and water molecules in the course of hydrate formation. Another property of hydrate formation important for technological applications is a large heat of hydrate formation/decomposition (Handa, 1986). This property on the one hand slows the process of hydrate formation/decomposition and on the other hand makes the process of fuel gas storage and transport in the hydrate state more safe due to the slow hydrate decomposition in case of accidents.

The main properties of gas hydrates, important for technology development are:

(1) One volume of gas hydrate contains up to 180 volumes of gas.
(2) Specific heat of hydrate formation is 30–50% higher than for ice formation.
(3) The majority of natural gases form hydrates only at the contact surface with water (ice), not from solution.

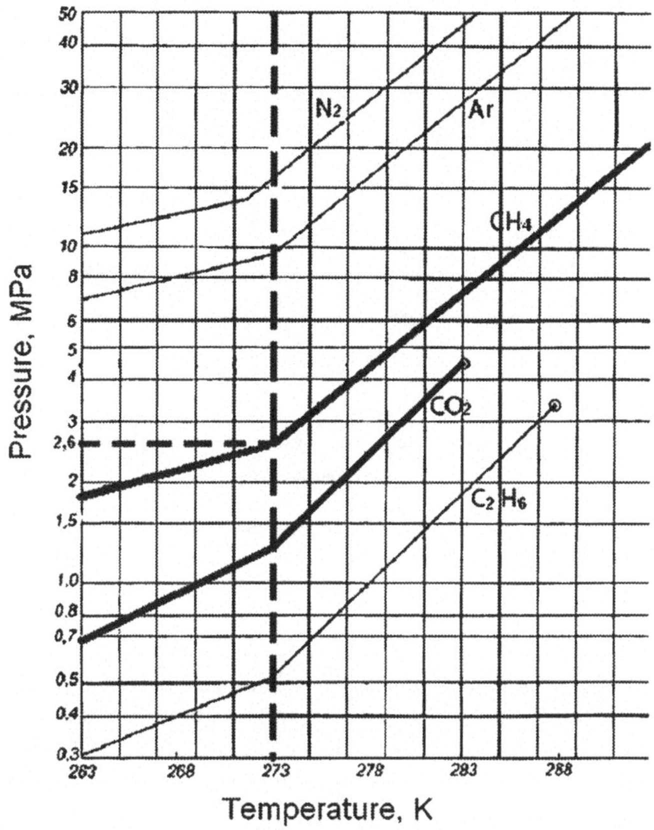

Figure 19.4 P/T hydrate-formation conditions for methane (thicker line) and carbon dioxide (thinner line) and some other gases (Istomin & Yakushev, 1992).

(4) The CH_4 hydrate specific density is lower (910 kg/m^3), but the CO_2 hydrate density (1100 kg/m^3) is higher than that of marine water density.
(5) Gas hydrate-formation results in gas mixture separation and water desalination.

According to kinetic properties, considerable volumes of hydrates can be formed only when there is a wide area of gas–water contact, a good heat exchange with the environment and there is enough gas to form hydrate. These requirements, plus the necessity of high-pressure vessels to store and transport gas in the hydrate state prevented the development and wide application of hydrate technologies in the natural gas industry. Nevertheless, some attempts to use the hydrate-formation process for gases separation, storage and water desalination have been made since the forties of 20th century.

19.2 SELF-PRESERVATION PHENOMENON

In general, the curve of hydrate-formation conditions looks like the one shown in Figure 19.5 for methane hydrate (Sloan, 1990). High-pressure vessels are needed for

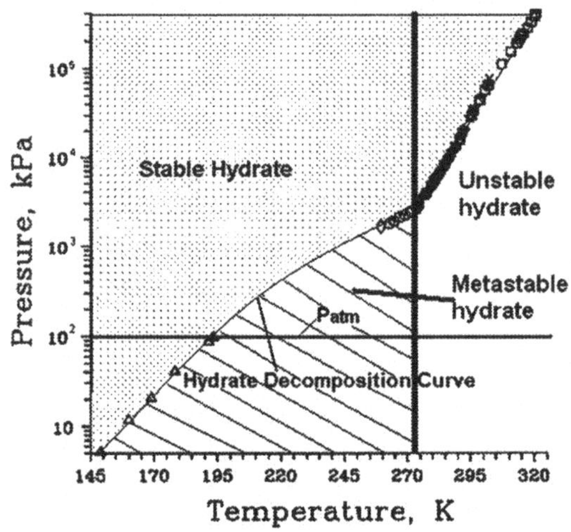

Figure 19.5 Methane hydrate-formation/decomposition conditions (modified from Sloan, 1990).

hydrate formation and storage according to this curve. This is the main challenge for wide application of hydrate technologies for gas storage and transport. But recently a new property of gas hydrates – the self-preservation phenomenon – was discovered and studied. This property opens perspectives to use hydrate technologies.

Figure 19.5 shows, that if there is a need to avoid high-pressure vessels and to store hydrates at atmospheric pressure, the temperature of the storage medium must be less than −83°C. So special low-temperature materials are needed to store the methane hydrate and strong freezing machines also are needed. But special experiments with relatively large pieces of gas hydrates at subzero temperature have shown one more thermodynamic area of gas hydrates existence: metastability. This area of metastability is shown in Figure 19.5 below the equilibrium curve and in the subzero temperatures range (lined area).

It was established in the course of experiments, that the self-preservation phenomenon is characteristic only for monolithic species of gas hydrates and hydrate-containing sediments (Yakushev, 1991). If pressure above the hydrate piece is reduced below the equilibrium point, but the surrounding temperature is subzero, the initial surface decomposition of the hydrate quickly results in isolating ice film formation and the decomposition stops (Figure 19.6). The gas content of the hydrate is slightly reduced, but is sometimes negligible. The formed self-preserved methane hydrates can be stored at atmospheric pressure and reasonable temperature (−5°C to −20°C) for a few weeks without considerable loss of its gas content (Yakushev, 2000). In nature self-preserved hydrates form sediments saturated by relic hydrates in hard-frozen permafrost above the top level of the current hydrate stability zone.

Self-preserved hydrates can be stored at temperatures below 0°C for a short (few minutes) or an indefinitely long (years and more) time depending on the hydrate sample structure, possibility for water sublimation from the sample surface, mechanical and

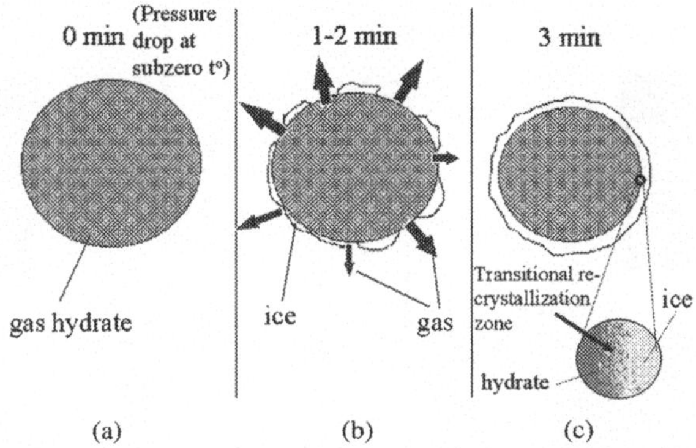

Figure 19.6 Formation of self-preserved hydrate particles (temperature −5°C to −15°C). (a) Hydrate particle at the moment of gas pressure reduction down to atmospheric value. (b) Partial surface decomposition of a hydrate particle (residual water freezing on the hydrate surface) and (c) isolating ice film formation on hydrate particle surface (film thickness 0.1–0.2 mm), hydrate metastability.

electro-magnetic impact, storage pressure and temperature (Yakushev, 1991). It was established experimentally that the minimum methane hydrate particle size (diameter) needed for self-preservation realization is 0.2 mm (Yakushev, 2000). But much depends on the procedure of hydrate sample formation and storage. Analysis of different self-preserved methane hydrate samples with different inner structure has shown that the particle size of the hydrate can be even less than 0.2 mm if these particles are surrounded by ice formed at hydrate-formation conditions (Yakushev, 2000). Three types of sample structure have been defined and studied for their stability with time at the same storage conditions: monolithic hydrate (mono- or poly-crystalline), hydrate sealed by ice and porous (dispersed) hydrate (Figure 19.7).

Dispersed or porous methane hydrate samples have the lowest stability: they had completely converted to ice in 2 weeks. Monolithic hydrates had more stability: according to our observations, methane hydrates with such a structure can exist more than 5 months (observation period) when stored at temperatures between −5°C to −10°C in closed boxes filled by neutral liquid (e.g. mineral oil). Hydrates sealed by ice at hydrate-formation conditions had the greatest stability: agglomerated ice-methane hydrate still contained hydrates after 2 years of storage at temperature −6°C (Yershov et al., 1990).

The gas hydrate self-preservation phenomenon has initiated studies of a new kind of phase equilibrium of water crystal structures: equilibrium with imposed kinetic barriers (e.g. Stern et al., 1996). Further development of this scientific direction should result in the production of hydrate samples, where hydrates of different gases compose different envelopes within one sample. For example, methane hydrate samples (high pressure of dissociation) could be surrounded by a carbon dioxide, propane or even tetrahydrofuran hydrate envelope (low pressure of dissociation) and be stored for a long time at atmospheric pressure and temperatures of about 273 K without notable decomposition.

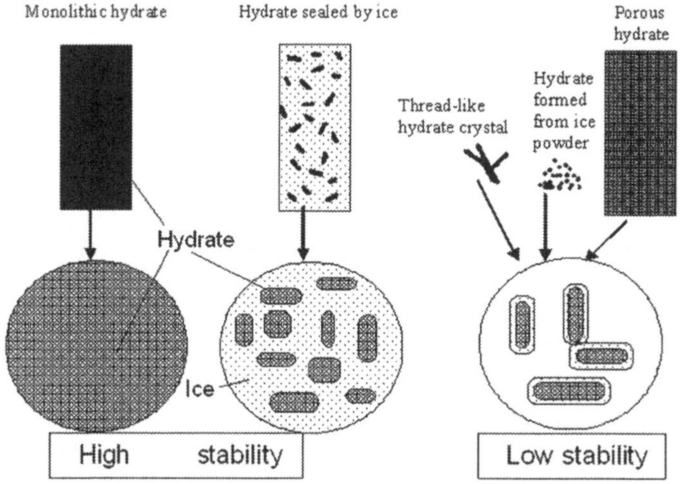

Figure 19.7 Scheme of different hydrate sample structures and their relative stability in self-preserved state.

The self-preservation phenomenon thus expanded thermodynamic conditions of gas hydrate existence in nature and created a theoretical base for the development of principally new technologies of gas storage and transportation.

19.3 HYDRATE TECHNOLOGIES FOR BIOGAS SEPARATION AND TRANSPORTATION

The ability of gas hydrates to separate gas mixtures when formed and stored at atmospheric pressure, keeping gas inside the hydrate in a concentrated state could be used for biogas processing, transport, storage and use. Mitsui Engineering and Shipbuilding Co, Ltd (Mitsui Co. presentation, 2005) has developed a pilot plant for biogas purification and transportation to the local market (Figure 19.8). Utilization of biogas, which is produced from organic waste by fermentation, is increasingly substituting fossil fuels. Biogas is generally composed of about 60% methane gas and 40% carbon dioxide. A pilot plant is currently being used for the study of methane purification into higher concentration and its transformation to the hydrate state for transportation to the market in a preserved form ($t = -20$°C) (Figure 19.9).

According to the technology, gathered waste gas is fed to a compressor, elevating gas pressure enough for hydrate formation. Then gas is coming to the first reactor for hydrate formation, where primary hydrate formation takes place. CO_2 accumulates presumably in the hydrate phase, the free phase is enriched by methane. Then the free phase is fed to a secondary hydrate-formation reactor and the process is repeated. The outcoming gas is enriched enough with methane for feeding it to a third reactor, where hydrate formation takes place, hydrates are freezing (self-preservation) and after pelletizing could be transported to the market by usual freezing lorries. To liberate the

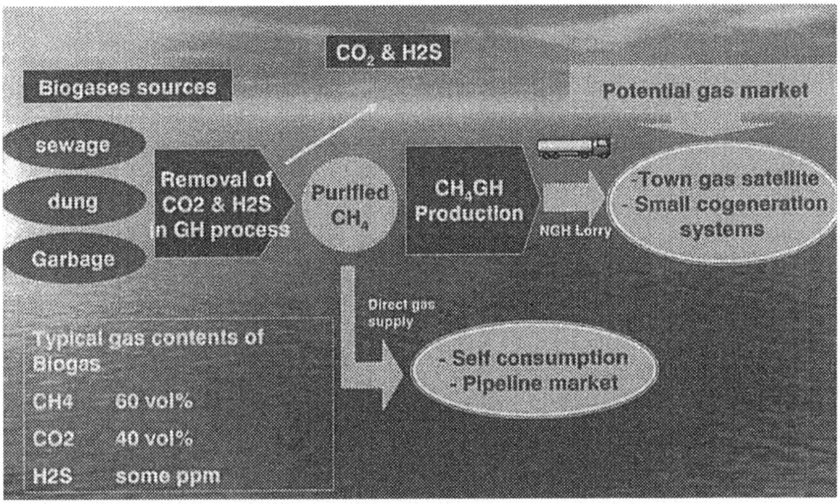

Figure 19.8 Mitsui Co. technologic scheme of the GH (gas hydrate) process of biogas purification, concentration and transportation to the market.

Figure 19.9 Experimental plant of Mitsui Co. (Kitami Sewage Center) for biogas purification using the hydrate-formation process.

fuel gas from hydrates only heating above 0°C is needed. The liberated gas could be fed to a local pipeline network or directly to a single consumer.

CO_2-enriched gas from hydrates in the primary and secondary reactors could be directed to separation units or mixed with new incoming waste gas volumes. One more advantage of such technology is water desalination during hydrate formation. If seawater is used in the process, fresh water could be received after hydrate decomposition. It is very important for areas with limited energy and fresh water supply.

The hydrate technology for waste gases includes two main elements: gas separation and hydrate preservation/storage. The gas separation process is patented in USA and

Japan (USA Patent WO0174472 (2001), Japan Patents 2002-249790 (2002), 2003-138281 (2003), 2003-181243 (2003)). Preservation and storage methods are patented mainly in Japan (e.g. Japan Patent 2005-076640 (2005)). There are also some suggestions on CO_2 conversion to organic substances using the hydrate-formation process at the surface of activated ZnO (US5482599 (1996)).

19.4 HYDRATE TECHNOLOGIES OF CO_2 SEQUESTRATION IN THE COURSE OF OFFSHORE NATURAL GAS PRODUCTION

There are two general challenges for the world energy business attributed to fossil fuels utilization:

(1) More production of the most environmentally friendly fuel: natural gas.
(2) Sequestration of CO_2 produced by large fuel consumers (exhausting gases).

Usually these two problems were tried to be solved separately. But in the last decade there are more and more industrial attempts to inject CO_2 to oil and natural gas reservoirs to enhance production. Hydrate technologies open new opportunities to combine a solution of these two problems. The natural gas industry now goes to more and more deep waters (now gas is produced from deposits under more than 1000 m water depth). In the same time, production from such a great water depth makes production platforms very expensive. Also the conventional ways (pipeline, liquefied natural gas) of produced gas transportation to the seashore are expensive. Hydrate technologies could reduce the cost of production and transportation in deep waters considerably, because hydrates can be formed from raw gas without conventional separation (hydrate-formation processes can separate the C_1–C_4 fraction from more heavy ones) and this process could be made on the hydrate carrier (tanker). So production platforms could be of a smaller size and reduced cost. Marine transportation of natural gas by hydrate carriers is also considerably more effective than by liquid natural gas (LNG) or gas-to-liquid (GTL) carriers (Gudmundsson et al., 1995).

But one of the main obstacles for such technology development is the large heat of hydrate formation. So much energy should be generated offshore to chill hydrate-formation tanks and to accelerate the hydrate-formation process. To avoid this, the heat requires for another hydrate decomposition could be used. If heat is recuperated, these processes could compensate each other. The other hydrate to be used can be CO_2 hydrate formed in the hydrate carrier docked to marine terminal. CO_2 should be separated from exhausting gases and collected before hydrate formation. Then CO_2 hydrates should be transported to offshore natural gas production platforms where CO_2 hydrates should be decomposed simultaneously with natural gas hydrate formation in neighbouring tanks. The released CO_2 can be injected to the sea floor providing hydrate formation (according to the curve on Figure 19.4) or to production reservoirs for enhanced production, or just in permeable reservoirs for burial.

The concept of this technology is shown in Figure 19.10 (Russian Patent 2198285 (2001); Yakushev, 1998). Natural gas produced offshore is converted to the hydrate in

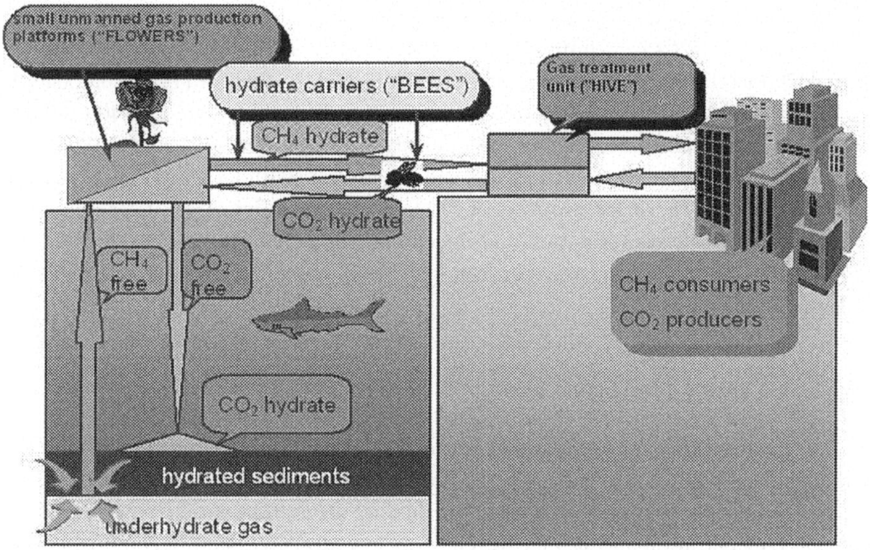

Figure 19.10 The concept of "Flowers and bees" hydrate technology of natural gas production and CO_2 sequestration offshore (Yakushev, 1998).

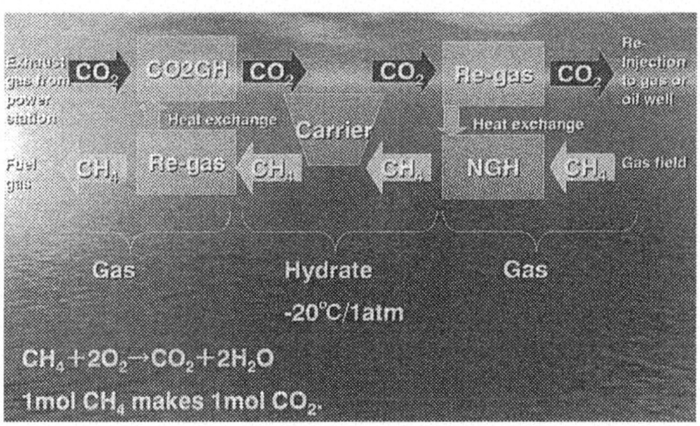

Figure 19.11 The concept of hydrate technology for offshore natural gas production and CO_2 sequestration using hydrate as carriers of CH_4 and CO_2 (Mitsui Co., 2005). NGH: natural gas hydrate; gas: free gas state; CO_2GH: CO_2 gas hydrate.

the carrier, then transported to shore, where hydrate decomposition, treatment and supply to a consumer (e.g. power station) takes place. The CO_2 produced by the power station should be separated from exhausted gas, collected and fed to the shore terminal and converted into a CO_2 hydrate in the hydrate carrier. It is then transported by carrier to a marine production platform, where it is decomposed and CO_2 is injected to water depth, providing hydrate formation or to a submarine reservoir. Thus there is a carbon recycling process. A similar concept has been suggested by Mitsui Co. in 2005 (Figure 19.11). Such technologies could solve not only the problem of CO_2 sequestration, but also considerably reduce the costs of natural gas production offshore.

One more gas hydrate technology now developing is simple CO_2 sequestration on the sea floor as CO_2 hydrate (Japan Patents 03-128792 (1990), 03-164419 (1991), 03-164420 (1991), 04-011920 (1992), 04-083528 (1992), 04-292775 (1992), 05-004041 (1993), 05-004042 (1993), 05-004043 (1993), 05-038429 (1993), 2003-284940 (2003)). The carbon dioxide hydrate has a greater density than seawater, so it can be deposited on the sea floor and be dissolved by this water with time without a serious impact on the environment. There are two ways of CO_2 sequestration: injection of gaseous CO_2 to the depth of hydrate formation and injection of liquid CO_2 to the depth of liquid CO_2 stability. In the second case, a hydrate film isolating liquid CO_2 from seawater will be formed.

19.5 CONCLUSION

Hydrate technologies for natural and waste gases just began to develop. Only pilot plants and technologic concepts are suggested now. But the properties of gas hydrates like the self-preservation phenomenon, gas mixture separation during hydrate formation, gas concentration and storage at relatively low pressures, slow decomposition in bulk state due to large heat of phase transition make these concepts very promising, especially for biogas purification, safe transportation of gas to the consumer and industrial CO_2 sequestration offshore. Hydrate technological concepts are at their initial stage of development now and have a great potential for their cost reduction in the course of further development.

REFERENCES

Barduhn, A.I. (1968) The state of the crystallization processes for desalting saline water. *Desalination*, **5**(2), 173–184.

Gudmundsson, J.S., Hveding, F. and Borrehaug, A. (1995) Transport of natural gas as frozen hydrate. *Proceedings of the 5th International Offshore and Polar Engineering Conference*, pp. 282–288. The Hague, The Netherlands, 11–16 June.

Handa, Y.P. (1986) Calorimetric determinations of the composition, enthalpies of dissociation and heat capacities in the range 85 to 270 K for clathrate hydrates of xenon and krypton. *J. Chem. Thermodyn.* **18**, 891–902.

Istomin, V.A. and Yakushev, V.S. (1992) *Gas Hydrates at Natural Conditions*, 236 pp. Nedra Publisher, Moscow, (in Russian).

Japan Patent 03-128792 (1990) Method and device for carbon dioxide gas in spot deep in water.

Japan Patent 03-164419 (1991) Treatment of gaseous carbon dioxide.

Japan Patent 03-164420 (1991) Device for treating gaseous carbon dioxide.

Japan Patent 04-011920 (1992) Gaseous carbon dioxide treatment apparatus for submarine boat.

Japan Patent 04-083528 (1992) Method and apparatus for immodilizing carbon dioxide in ocean.

Japan Patent 04-292775 (1992) Abandoning method for carbon dioxide gas into submarine bottom ground.

Japan Patent 05-004041 (1993) Device for immobilizing carbon dioxide to deep sea.

Japan Patent 05-004042 (1993) Device for immobilizing carbon dioxide to deep sea.

Japan Patent 05-004043 (1993) Treating device for carbon dioxide.

Japan Patent 05-038429 (1993) Treatment of carbon dioxide.

Japan Patent 2002-249790 (2002) Method for composition control of mixed gas including gas hydrate formation process and method for concentration of the mixed gas.

Japan Patent 2003-138281 (2003) Separating agent for gas and method and apparatus for separating and concentrating gas.

Japan Patent 2003-181243 (2003) Method for treating waste gas, soot and dust.

Japan Patent 2003-284940 (2003) Method for generating gas hydrate in sea, gas hydrate generator and underwater storage system of carbon dioxide.

Japan Patent 2005-076640 (2005) Storing method and device for natural gas hydrate.

Mitsui Co. presentation (2005) At *The SOJITZ/Gazprom Meeting.* Osaka, Japan, 21 April.

Russian patent 2198285 (2001) Method of natural gas production from offshore gas and gas hydrate fields – "Flowers and Bees".

Sloan, E.D. (1990) *Clathrate Hydrates of Natural Gases*, 641 pp. Marcel Dekker Inc.

Stern, L., Kirby, S.H. and Durham, W.B. (1996) Peculiarities of methane clathrate hydrate formation and solid-state deformation, including possible superheating of water ice. *Science* **273**, 1843–1848.

USA Patent US5482599 (1996) Photochemical method for converting carbon dioxide gas to organic substances.

USA Patent WO0174472 (2001) Methods of selectively separating CO_2 from a multicomponent gaseous stream using CO_2 hydrate promoters.

Yakushev, V.S. (1991) Composition, structure and properties of frozen hydrate-containing soils. Ph.D. Thesis, Moscow State University, 122 pp. (in Russian).

Yakushev, V.S. (1998) Offshore system of natural gas production and carbon dioxide disposal ("Flowers and Bees" hydrate technology). *Proceedings of the International Symposium on Methane Hydrates: Resources in the Near Future?* pp. 205–217. JNOC-TRC, Chiba-City, Japan, 20–22 October.

Yakushev, V.S. (2000) A review of research on self-preservation of gas hydrates. In *Abstracts of Forum on Feasibility Studies on Gas Hydrate for Development of Energy Resources*, pp. 112–117. NEDO, Mielparque Tokyo, Minato-ku, Tokyo, Japan, 15–16 March.

Yershov, E.D., Lebedenko, Y.P., Chuvilin, E.M. and Yakushev, V.S. (1990) Experimental studies of agglomerate ice-methane hydrate microstructure. *Inzhenernaya Geol.* [*Eng. Geol.*] **3**, 38–44 (in Russian).

20

Application of zeolites in environmental catalysis

C. Henriques, M.F. Ribeiro and F.R. Ribeiro

20.1 INTRODUCTION

> Rarely in our technological society does the discovery of a new class of inorganic materials result in such a wide scientific interest and kaleidoscopic development of applications as has happened with the zeolite molecular sieves (Breck, 1974)

Chemical industry and related activities become, in the last years, one of the main targets concerning environmental protection. Market and raw-materials availability are no longer the only main aspect: environmental issues become also of crucial importance.

Chemical industry and academic researchers have focused to overcome the drawbacks of chemical processes. At the base of this effort are chemical technologies, including catalytic processes. For this, two main aspects have been extensively studied and developed: (i) replacement of existing more polluting processes, by new environmentally friendly technologies that use less harmful raw-materials and catalysts, producing less undesirable by-products – *Green Chemistry* and (ii) abatement of gaseous pollutants (e.g. NO_x, VOCs, SO_x and CO). Both aspects are based on the development of higher active, selective, stable and non-aggressive catalysts that permit the operation of chemical processes at softer conditions and are able to be re-used after adequate regeneration (Bellussi & Perego, 2000).

Zeolites and other zeotype materials (in general, metal-oxide solids which can form crystalline material with inner pores and cavities in the range of kinetic diameters of reacting molecules – from ca. 4 to 18 Å – or in the range 20 to 500 Å – mesoporous materials), that present simultaneously strong acid sites (Brönsted and Lewis type) and metal active sites (cations, metal clusters, metal oxides) both located inside a regular framework with molecular-scale dimensions, are particularly well adapted to be used with the objective of solving environmental problems, since conventional catalysts are, in most cases, unable to achieve such performances (Marcus & Cornier, 1999).

20.2 ZEOLITES IN CATALYSIS

Zeolites are aluminosilicate minerals that occur in nature and are synthesized for catalysis and adsorption purposes. They were first discovered in 1756 by the Swedish scientist Cronstedt, when an unidentified silicate mineral was heated and seemed to boil (in Greek *zeo* = boil and *lithos* = stone). The term *molecular sieve* was introduced by McBain (1932) when he found that the mineral chabazite selectively adsorbed molecules smaller than 5 Å in diameter. Today, more than 130 different structures are known, though most of them are not found in nature. Zeolites are microporous high-internal surface area crystalline materials with open, three dimensional frameworks consisting of the combination of TO_4 tetrahedra, where T is normally assigned to a Si or Al atom. Adjacent tetrahedrons are linked by their corners via a common oxygen atom, resulting in structures where a negative charge is associated with each aluminium tetrahedron, balanced by cations or protons (Guth & Kessleer, 1999) (see Figure 20.1).

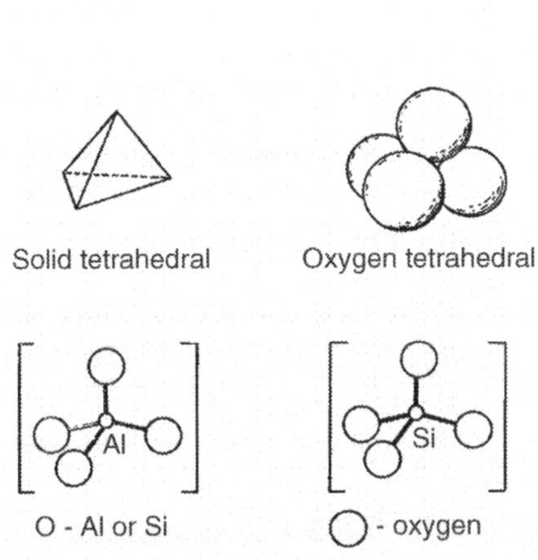

Figure 20.1 Zeolite structure – primary building units.

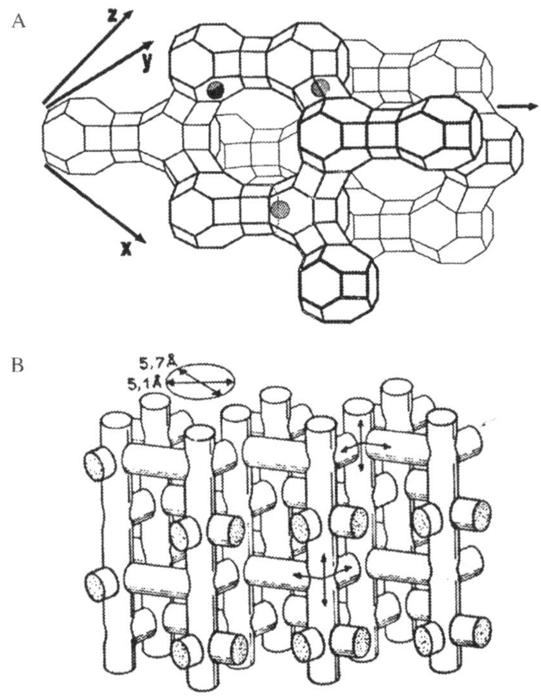

Figure 20.2 Examples of zeolite structures – A: Y Zeolite (FAU structure) and B: ZSM-5 (MFI structure).

Since the early 1960s, zeolites have played a key and increasing role in the refining and petrochemical industries. Zeolite micropores offer unique reaction fields for catalytic reactions based on narrow uniform pore diameters (in the order of many industrial relevant molecules – *confinement effects*) and a wide variety of channel geometries, ion-exchange properties and high density of acid sites. Most zeolite structures are stable up to 500–700°C (Sie, 1994) (see Figure 20.2). Zeolites offer considerable scope for surface modifications. Ion-exchange properties and framework crystallinity enable to stabilize functional metal ions in a variety of oxidation states and coordination geometries, within a well-defined microporous structure.

20.3 ZEOLITE CATALYSTS AND GREEN CHEMISTRY

Green chemistry is used to describe the design of chemical products and processes that reduce or eliminate the use and generation of hazardous substances (Anastas & Warner, 1998). From raw materials to solvents, synthesis and processing, green chemistry seeks ways to produce materials in a more human health and environment friendly way.

Metal-exchanged zeolites (and even mesoporous materials, like MCM-41) are finding wide applications in environmentally safe processes for example, replacement of hazardous catalysts in existing processes, namely traditional homogeneous catalysts because of their ready regenerability, shape-selectivity and easy separation and recovery.

The design and use of new catalysts are simultaneously achieving the dual goals of environmental protection and economic benefits.

The unique properties of zeolites can provide effective environmental solutions (Bellussi & Perego, 2000; Panov, 2000), replacing many hazardous acidic catalysts and eliminating some intermediate steps in certain processes to reduce the overall waste output and energy use. Some examples are:

- *Phenol* production based on the direct hydroxylation of benzene by N_2O over Fe/ZSM-5 (Monsanto; BIC and Solutia Inc.), eliminates cumene as intermediate, preventing by-production of acetone.
- *Hydroquinone and catechol* production by direct liquid phase phenol oxidation with H_2O_2 on TS-1 (Enichem).
- Liquid phase *cumene* production over Beta zeolite (Enichem), avoiding the use of mineral acids.
- Complete elimination of by-production of inorganic sulfate alongside with *caprolactam* (intermediate for nylon-6) production, by direct ammoxidation of cyclohexanone with H_2O_2 and ammonia on TS-1 (Enichem), dramatically reducing the number of processing steps.
- Methanol to gasoline process over ZSM-5 catalysts (Mobil), increasing biomass exploitation.

Catalysis will certainly continue to offer even greater improvements in efficiency, selectivity, energy reduction and rate enhancement.

20.4 ABATEMENT OF NO_x IN OXYGEN-RICH STREAMS WITH ZEOLITE CATALYSTS

Nitrogen oxides (NO and NO_2) are mainly produced by the combustion of fossil fuels. NO_x reduction is a key target of clean air strategies, as it constitutes a major contributor to acid rains and, in the presence of organic compounds, gives rise to photochemical oxidants and PAN (peroxyacylnitrates). Therefore, legislation requires a strict limit for NO_x emissions.

20.4.1 Catalytic NO_x removal

Catalytic methods for removing NO_x from exhaust gases are usually classified into (i) decomposition ($2NO \rightarrow N_2 + O_2$), that constitutes the easiest and cheapest process, but strongly inhibited by the presence of O_2, (ii) non-selective catalytic reduction (NSCR) that uses non-selective reducing agents (like CO and H_2) and is applied under rich-burn conditions and (iii) selective catalytic reduction (SCR), applied under lean-burn condition, in the presence of selective reducing agents (like NH_3, hydrocarbons and oxygenated hydrocarbons).

20.4.1.1 Decomposition

Decomposition processes are generally related with the elimination of N_2O produced in nitric acid plants and during manufacture of adipic acid (used in the manufacture

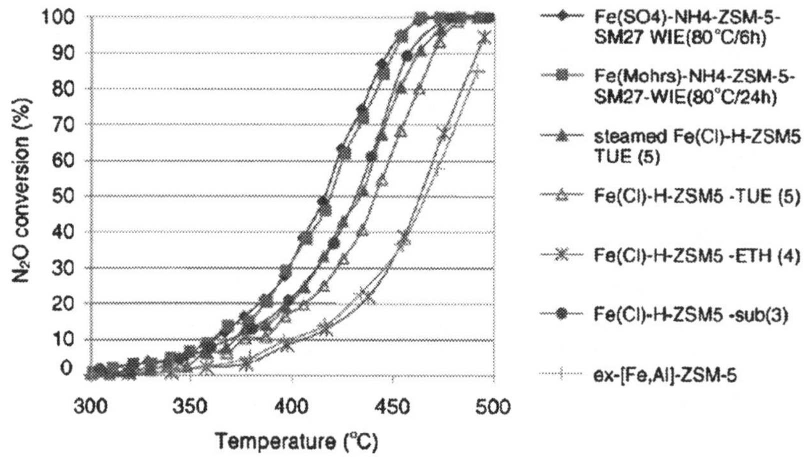

Figure 20.3 N_2O decomposition, over iron-ZSM5 prepared by different methods (30 000 h^{-1}) (from Pieterse et al., 2004).

of nylon). Current technologies include Cu or Co supported in Y and ZSM-5 zeolites. Promising results, obtained with highly active Ru-exchanged HNaUSY, have been also reported for N_2O decomposition at low temperatures (120–200°C) (Chang et al., 1994). Recently, iron-based MFI zeolites, prepared by wet-ion exchange, show good activity for N_2O decomposition together with good (hydrothermal) stability (Pieterse et al., 2004) (see Figure 20.3).

20.4.1.2 NSCR technologies

Three-way catalytic converters (TWC) used in gasoline-powered ignition engines are an example of non-selective reduction. Nitrogen oxides are "non-selectively" reduced by hydrocarbons and by carbon monoxide that are present in the exhaust stream which contains no excess oxygen (Heck & Farrauto, 2002).

In stationary sources (nitric acid plants, for instance) NSCR technology is used to reduce NO_2 to N_2. In this case, it is necessary to first catalytically reduce excess O_2 by injecting a fuel, like natural gas. Once O_2 is removed, NO_2 is then catalytically reduced to N_2 by the remaining injected fuel. Non-selective technologies, applied to stationary sources, have been almost completely replaced by SCR, using NH_3 as reductant (Farrauto & Bartholomew, 1997).

20.4.1.3 SCR technologies

SCR technologies, in stationary sources, are currently based on the addition of NH_3 or, in some cases, urea that selectively reacts to reduce NO_x. The most broadly used catalyst is based in V_2O_5/TiO_2: this is a medium-temperature working catalyst with high resistance to sulfur compounds (Farrauto & Bartholomew, 1997). Zeolite-based catalysts have been also applied, although in a higher temperature range, with very high NO_x conversions and NH_3 total destruction (Heck, 1999). The wide use of

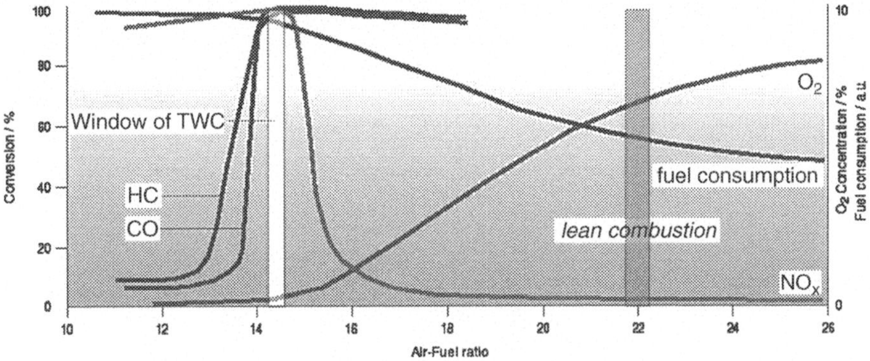

Figure 20.4 Work-window of TWC systems (from Matsumoto, 2000).

natural gas-based power plants originates a new approach to this problem, where methane can be used as a selective reductant (CH_4-SCR) (Li & Armor, 1992; 1994).

20.4.1.4 SCR under lean-burn conditions

The use of lean-burn (air/fuel ≈22) engines in transportation (both diesel and gasoline direct injection) constitutes a new challenge as three-way catalysts are no longer active in oxygen-rich conditions (see Figure 20.4, from Matsumoto, 2000).

For the past 15 years, this subject corresponded to one of the most promising efforts of both academic and industrial researchers and engineers, in order to solve this very specific problem.

In general, the reduction of the NO_x content under oxygen-rich conditions can be accomplished by using hydrocarbons as "selective reductants" that preferentially react with NO_x than with oxygen (HC-SCR).

Iwamoto and co-workers (1981, 1986) first claimed that the Cu/ZSM-5 catalyst was active for NO decomposition. Some years later, two research groups (Held et al., 1990; Iwamoto et al., 1990), independently showed the possibility of the catalytic reduction of NO with hydrocarbons (C_2–C_3), in the presence of an oxidizing atmosphere, using the same catalysts.

The study and development of catalysts for the Selective Catalytic Reduction with hydrocarbons (HC-SCR) had begun. Since then, this catalyst and other zeolite-based catalysts have been studied intensively.

Cu/zeolite-based catalysts (namely ZSM-5 and Mordenite) have shown to be the most active with large spectra of hydrocarbons (C_2–C_{10}, paraffins and olefins or oxygenated compounds) as reductants. Different zeolite structures (ZSM-5 and mordenite) were compared (see Figure 20.5).

Although an enormous amount of papers and patents exists, copper-based catalysts never reached a total development and practical application. In fact, Cu-zeolite systems have several critical problems: (i) poisoning with water and sulfur compounds, (ii) low thermal stability, (iii) moderate activity, (iv) N_2 selectivity (at higher temperatures HC reacts preferentially with oxygen than with NO) and (v) narrow working temperature window.

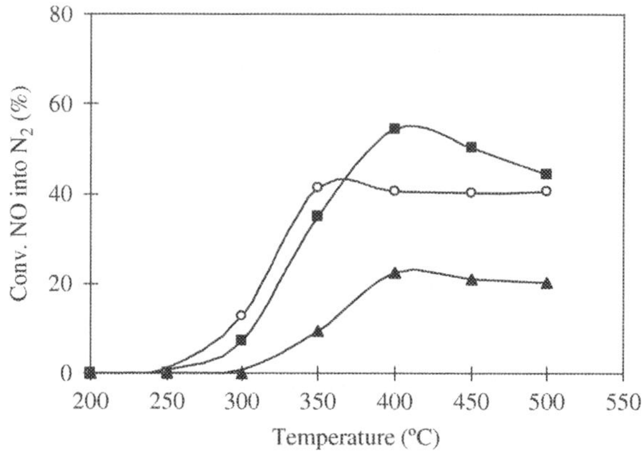

Figure 20.5 Influence of zeolite structure on NO SCR with propene: ■ – Cu/ZSM-5, ○ – Cu/MOR, ▲ – Cu/Y: Cu = 1.4 wt%, Si/Al ≈23 (from Torre-Abreu et al., 1999).

Different modifications of catalysts were tested: variation of zeolites Si/Al ratio (by dealumination techniques), different Cu loading, introduction of co-cations (alkaline and alkaline-earth elements or noble metals) and introduction of Cu by several methods.

The presence of water vapour in exhaust gases showed to have not only an inhibition effect on HC-SCR reaction but also leads to an increase of irreversible deactivation.

Active Pt-containing zeolites operate at significantly lower temperatures than copper, which seems to constitute an important advantage. However, these catalysts present a high selectivity to N_2O, which constitutes a major problem.

20.4.2 New developments

Nevertheless, there is a considerable interest in finding a direct HC-SCR (*DeNO$_x$*) catalyst with appropriate performances (high activity and selectivity, good thermal and hydrothermal stability, resistance to sulfur, wide temperature working window). Recently, Ag-supported catalysts have received much attention, especially in the presence of small amounts of hydrogen. The addition of H_2 has a particular effect at lower temperatures, by increasing the rate of NO_x reduction by more than an order of magnitude (Burch, 2004).

With the increased use of natural gas, the development of new catalytic systems became an important goal. Li and Armor (1992, 1993) first claimed that Co-exchanged zeolites present a promising performance for SCR with methane. As observed with Cu-zeolite catalysts, this system is also sensitive to water and sulfur compounds, and its activity is too low. Bimetallic catalysts, namely exchanged Pd–Co zeolites, were found to be less water sensitive than Co samples (see Figure 20.6). The introduction of co-cations could change adsorption properties of these catalysts, improving water resistance.

Further developments must be attained until practical use can be foreseen. Despite the important drawbacks to be surmounted, the enormous evolution in catalytic emissions control obtained in the last years give an optimistic perspective.

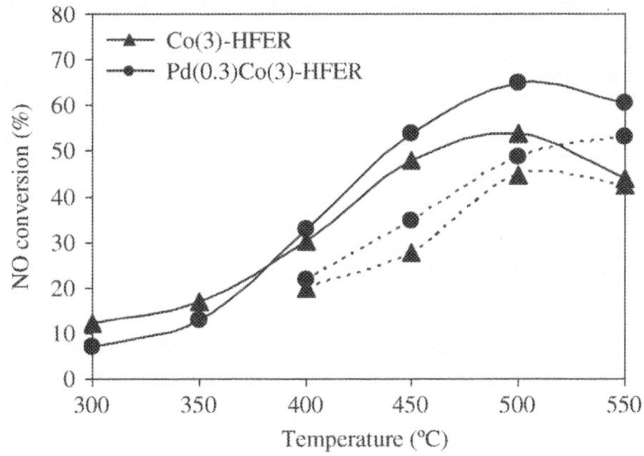

Figure 20.6 Influence of Pd loading in Co-FER catalyst on SCR-NO with methane: —dry tests; ---- 2% H_2O tests (from Ferreira, et al., 2006).

20.5 VOCs ABATEMENT USING ZEOLITES

20.5.1 Sources of VOCs

Volatile organic compounds (VOCs) can be considered as a major source of air pollution. They can be produced from a variety of industrial and commercial processes. VOCs include produced hydrocarbons (HC) such as methane obtained from biological decompositions. Nevertheless, most are produced from anthropogenic (man-made) sources, mainly released from motor vehicles and from chemical sources (mainly solvents, paints, inks and glues). Globally, they include hydrocarbons, halogenated hydrocarbons and oxygenated compounds.

Individual components of VOCs are known for harmful effects on humans, such as toxic effects (some are carcinogenic and mutagenic substances). However, the main environmental effect of VOCs is their role in the formation of smog. In the presence of NO_x, VOCs are precursors to the formation of ground level (tropospheric) ozone which is the primary constituent of smog. These compounds can cause respiratory damage, eye irritation and destruction of vegetation. Some VOCs also contribute to the global warming, as they are powerful infrared absorbers.

As far as motor vehicles are concerned, the main problem is related with the very first minutes after motor ignition (cold start), before three-way catalysts reach their appropriate temperature (250–300°C). During this period, VOCs are emitted, despite the significant amount of effort expended over the past 30 years, to develop a method to treat automobile exhaust and to fulfil the increasingly stringent vehicle emission regulations. Even in automobiles equipped with modern three-way catalysts, the majority of HC emissions (up to 80%) (Czaplewski et al., 2003), that occur during that period, are simply released to the atmosphere. A possible solution to the cold-start problem consists on trapping HC emissions during this period; these products will be released when catalysts reach its light-off temperature and consequently burned. Zeolites were found to be preferred adsorbents, due to their stability under these conditions.

Despite the great number of proposed systems, the trapping of light HC remains an important negative aspect of the system, as they normally are released before catalysts reach their optimum temperature. Possible solutions involve the choice of the appropriate zeolite structure (pore size) and the study of the effect of parameters like hydrophobicity, Si/Al ratio and hydrothermal aging on the adsorption capacity and desorption temperature of the emission traps (Burke et al., 2003; Heck & Farrauto, 2002).

20.5.2 VOCs removal and reuse

Several methods are nowadays considered for VOCs removal, both destructive and non-destructive: (i) by changing industrial processes and increase leakage control; (ii) by concentration and recovery (with techniques like condensation, adsorption, absorption and membranes; (iii) by destruction (transformation into CO_2 and H_2O or even into an inorganic acid, as in halogenated hydrocarbons) by combustion (incineration), catalytic oxidation or biofiltration. In Figure 20.7, a selection map (lowest cost based) of VOCs abatement technologies is presented (Moretti, 2002).

20.5.2.1 Adsorption

Among recovery techniques, adsorption is a good technology when VOCs concentrations are low. Zeolites are particularly well adapted to be used as adsorbents: hydrophobic forms are less sensitive to moisture (when compared with activated carbon, for instance). Furthermore, zeolites are non-combustible, which is important when streams containing high flammable VOCs are treated. Zeolites can be regenerated and VOCs recovered.

20.5.2.2 Decomposition

When destructive techniques must be applied (Moretti, 2002), catalytic oxidation offers a high destructive efficiency, low operating temperature (300–450°C instead

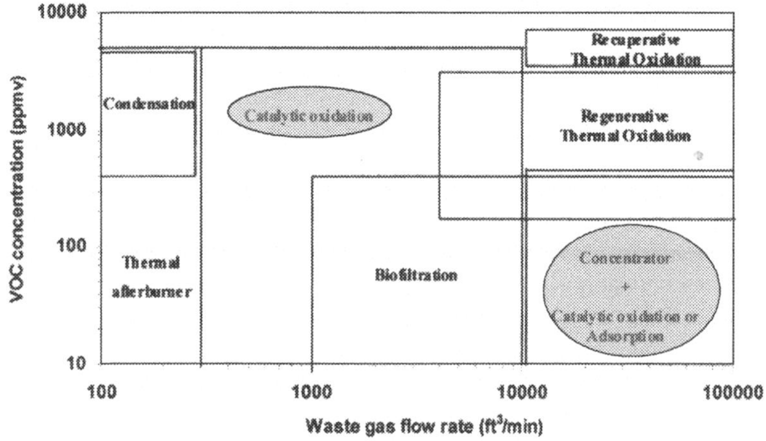

Figure 20.7 Selection map (lowest cost based) of VOCs abatement technologies (from Moretti, 2002).

of 650–1100°C in incineration processes) and lower capital cost, which allow smaller units (Heck & Farrauto, 2002). The main disadvantage of this process is that catalysts can be sensitive to poisoning by compounds present in VOC streams (as sulfur compounds) or even by reaction products (as inorganic acids, with halogenated VOCs).

Metal oxides (in particular mixed oxides) and metal-supported catalysts (namely noble metals) are well established as efficient catalysts for complete oxidation of VOCs. The choice between the use of transition metal oxides or noble metals depends on numerous aspects (nature of VOC to be transformed, contaminants present in the stream, among others). Nevertheless, it is usually recognized that noble metals are more active than metal oxides. They also have a higher resistance to poisons (Everaert & Baeyens, 2004; Spivey, 1987), although they are more affected by the presence of water and by the presence of Cl_2 and HCl (produced when oxidizing chlorinated VOCs).

Among supported metal catalysts, not only noble metals (Pt and Pd) but also transition metals like Cr (Chatterjee & Greene, 1991) and Cu (Antunes et al., 2001), supported in zeolites (Faujasite, Beta, ZSM-5) are stated as suitable for the total oxidation of both non-chlorinated and chlorinated VOCs (Guisnet et al., 1999; Pinard et al., 2004a, 2004b).

The positive role of zeolite has been related to the higher reducibility of noble metals when supported on this type of materials (Li & Armor, 1994; Neyestanaki et al., 1995). Parameters as zeolite dealumination and acidity seem to play an important role in catalyst activity and stability, depending on the HC molecule. On the other hand, an enhancement of the hydrophobic character of the used zeolite results on an increasing catalyst resistance to moisture (Baek et al., 2004; Burch et al., 1999).

20.6 CONCLUSION

The utilization of catalysts in order to reach the goals of green chemistry represents an enormous success, in attaining both environmental and economic benefits simultaneously. Furthermore, the wide use of catalytic systems to minimize gaseous harmful effluents constitutes an enormous achievement since the last 40 years.

Zeolites have played a key role in the construction of new catalytic systems that give rise to new chemical processes as well as in the resolution of major pollutions problems.

Further design and development of new catalysts will activate otherwise inert raw materials and will enhance reaction rates and desired products selectivity.

A considerable work is also foreseen to surpass the actual limitations in NO_x and VOCs abatement. Metallo-zeolites are expected to play a key role in these tasks (Wichterlová et al., 2003).

In the near future, areas such as power generation, selective oxidation of light alkanes and natural gas conversion will require highly structured catalysts. Zeolites are foreseen to constitute the base of such systems. Deep insight in understanding the structure, function, nature and stabilization process of active sites and their relation with activity–selectivity must be accomplished.

REFERENCES

Anastas, P.T. and Warner, J.C. (1998) *Green Chemistry: Theory and Practice*, Oxford University Press, New York.

Anastas, P.T., Kirchhoff, M.M. and Williamson, T.C. (2001) Catalysis as a foundational pillar of green chemistry. *Appl. Catal. A* **221**, 3–13.

Antunes, A.P., Ribeiro, M.F., Silva, J.M., Ribeiro, F.R., Magnoux, P. and Guisnet, M. (2001) Catalytic oxidation of toluene over CuNaHY zeolites: coke formation and removal. *Appl. Catal. B* **33**, 149–164.

Baek, S.-W., Kim, J.-R. and Ihm, S.-K. (2004) Design of dual functional adsorbent/catalyst system for the control of VOCs by using metal-loaded hydrophobic Y-zeolites. *Catal. Today* **93–95**, 575–581.

Bellussi, G. and Perego, C. (2000) Industrial catalytic aspects of the synthesis of monomers for nylon production. *CATTECH* **4**, 4–16.

Breck, D.W. (1974) *Zeolite Molecular Sieves: Structure, Chemistry and Use*, Wiley, New York.

Burch, R. (2004) Knowledge and know-how in emission control for mobile applications. *Catal. Reviews* **46**, 271–333.

Burch, R., Crittle, D.J. and Hayes, M.J. (1999) C-H bond activation in hydrocarbon oxidation on heterogeneous catalysts. *Catal. Today* **47**, 229–254.

Burke, N.R., Trimm, D.L. and Howe, R.F. (2003) The effect of silica:alumina ratio and hydrothermal ageing on the adsorption characteristics of BEA zeolites for cold start emission control. *Appl. Catal.* **46**, 97–104.

Chang, Y.-F., McCarty, J.G., Wachsman, E.D. and Wong, V.L. (1994) Catalytic decomposition of nitrous oxide over Ru-exchanged zeolites. *Appl. Catal. B* **4**, 283–299.

Chatterjee, S. and Greene, H.L. (1991) Oxidative catalysis of chlorinated hydrocarbonsby metal-loaded acid catalysts. *J. Catal.* **130**, 76–85.

Czaplewski, K.F., Reitz, T.L., Kim, Y.J. and Snurr, R.Q. (2003) One-dimensional zeolites as hydrocarbon traps. *Micropor. Mesopor. Mater.* **56**, 55–64.

Everaert, K. and Baeyens, J. (2004) *J. Hazard. Mater. B* **109**, 113–139.

Farrauto, R.J. and Bartholomew, C.H. (1997) *Fundamentals of Industrial Catalytic Processes, Blackie Academic & Professional*, Chapman & Hall, London.

Ferreira, A.P., Henriques, C., Ribeiro M.F. and Ramôa Ribeiro, F. (2006) SCR of NO with methane on Pd-Co ferrierite: effect of alkali and alkali-earth cations on catalytic activity and stability (to be submitted).

Guisnet, M., Dégé, P. and Magnoux, P. (1999) Catalytic oxidation of volatile organic compounds 1. Oxidation of xylene over a 0.2wt% Pd/HFAU(17) catalyst. *Appl. Catal. B* **20**, 1–13.

Guth, J.-L. and Kessleer, H. (1999) Synthesis of aluminosilicate zeolites and related silica-based materials in catalysis and zeolites. In *Fundamentals and Applications*. (eds. J. Weitkamp and L. Puppe), Springer-Verlag, Berlin.

Heck, R.M. (1999) Catalytic abatement of nitrogen oxides-stationary applications. *Catal. Today* **53**, 519–523.

Heck, R.M. and Farrauto, R.J. (2002) *Catalytic Air Pollution Control*, 2nd edn, Wiley-Interscience, John Wiley & Sons, New York.

Held, W., König, A., Richter, T. and Puppe, L. (1990) Catalytic NO_x reduction in net exhaust gas, SAE 900496.

Iwamoto, M., Yokoo, S., Sakai, K. and Kagawa, S. (1981) Catalytic decomposition of nitric oxide over copper(II)-exchanged, Y-type zeolites. *J. Chem. Soc. Faraday Trans.* **1**, 1629–1638.

Iwamoto, M., Furukawa, H., Mine, Y., Uemura, F., Mikuriya, S. and Kagawa, S. (1986) Copper(II) ion-exchanged ZSM-5 zeolites as highly active catalysts for direct and continuous decomposition of nitrogen monoxide. *J. Chem. Soc. Chem. Commun.* **16**, 1272–1273.

Iwamoto, M., Yahiro, H., Yu-u, Y., Shundo, S. and Mizuno, N. (1990) *Shokubai* **32**, 430–439.

Li, Y. and Armor, J.N. (1992) Catalytic reduction of nitrogen oxides with methane in the presence of excess oxygen. *Appl. Catal. B* **4**, L31–L40.

Li, Y. and Armor, J.N. (1993) Selective catalytic reduction of NO_x with methane over metal exchange zeolites. *Appl. Catal. B* **2**, 239–256.

Li, Y. and Armor, J.N. (1994) Catalytic combustion of methane over palladium exchanged zeolites. *Appl Catal. B* **3**, 275–282.

Marcus, B.K. and Cornier, W.E. (1999) Going green with zeolites. *CEP* **95**, 47–53.

Matsumoto, S. (2000) *Cattech* **4**, 102–109.

McBain, J.W. (1932) *The Sorption of Gases and Vapours by Solids*, Routledge and Sons, London.

Moretti, E.C. (2002) Reduce VOC and HAP emissions. *CEP* **98**, 30–40.

Neyestanaki, A.K., Kumar, N. and Lindfors, L.-E. (1995) Catalytic combustion of propane and natural gas over Cu and Pd modified ZSM zeolite catalyst. *Appl Catal. B* **7**, 95–111.

Panov, G.I. (2000) Advances in oxidation catalysts. *Cattech* **4**, 18–32.

Pieterse, J.A.Z., Booneveld, S. and Van den Brink, R.W. (2004) Evaluation of Fe-zeolite catalysts prepared by different methods for the decomposition of N_2O. *Appl. Catal. B* **51**, 215–228.

Pinard, L., Mijoin, J., Ayrault, P., Canaff, C. and Magnoux, P. (2004a) On the mechanism of the catalytic destruction of dichloromethane over Pt zeolite catalysts. *Appl. Catal. B* **51**, 1–8.

Pinard, L., Magnoux, P., Ayrault, P. and Guisnet, M. (2004b) Oxidation of chlorinated hydrocarbons over zeolite catalysts 2. Comparative study of dichloromethane transformation over NaX and NaY zeolites. *J. Catal.* **221**, 662–665.

Sie, S.T. (1994) Past, present and future role of microporous catalysts in the petroleum industry In *Advanced Zeolite Science and Applications* (eds. J.C. Jansen, H.G. Stöker Karge and J. Weitkamp) *Stud. Surf. Sci. Catal.* **85**, 587–631.

Spivey, J.J. (1987) Complete catalytic oxidation of volatile organics. *Ind. Eng. Chem. Res.* **26**, 2165–2180.

Torre-Abreu, C., Henriques, C., Ribeiro, F.R., Delahay, G. and Ribeiro, M.F. (1999) Selective catalytic reduction of NO on copper-exchanged zeolites: the role of the structure of the zeolite in the nature of copper-active sites. *Catal. Today* **54**, 407–418.

Tsou, J., Pinard, L., Magnoux, P., Figueiredo, J.L. and Guisnet, M. (2003) Catalytic oxidation of volatile organic compounds (VOCs). Oxidation of o-xylene over Pt/HBEA catalysts. *Appl. Catal. B* **48**, 371–379.

Wichterlová, B., Sobalík, Z. and Dědeček, J. (2003) Redox catalysis over metallo-zeolites. Contribution to environmental catalysis. *Appl. Catal. B* **41**, 97–114.

21

Operation of fuel cells with biomass resources (hydrogen and alcohols)

C. Lamy

21.1 INTRODUCTION

Discovered in England in 1839 by Sir William Grove (Grove, 1839a, b) a fuel cell (FC) is an electrochemical device, which transforms directly the heat of combustion of a fuel (hydrogen, natural gas, methanol, ethanol, hydrocarbons, etc.) into electricity. The fuel is electrochemically oxidized at the anode, without producing any pollutants (only water and/or carbon dioxide are released in the atmosphere), whereas the oxidant (oxygen from the air) is reduced at the cathode. This process does not follow the Carnot's theorem, so that higher-energy efficiencies are expected: 40–50% in electrical energy, 80–85% in total energy (electricity + heat production).

There is now a great interest in developing different kinds of fuel cells with several applications (besides the first and until now unique application in space programs) depending on their nominal power: stationary electric power plants (100 kW to 10 MW), power train sources (20–200 kW) for the electrical vehicle (bus, truck and

 Waste Gas Treatment for Resource Recovery edited by Piet N.L. Lens, Christian Kennes, Pierre Le Cloirec and Marc Deshusses. ISBN: 1843391279. Published by IWA Publishing, London, UK.

individual car), electricity and heat production for buildings and houses (5–20 kW), auxiliary power units (1–100 kW) for different uses (automobiles, aircrafts, space launchers, space stations, etc.) and portable electronic devices (1–100 W); for example, cell phones, computers and cam recorders (Gosselink, 2002; Ströbel et al., 2002).

For most applications, hydrogen is the convenient fuel, but it is not a primary fuel, so that it has to be produced from different sources: water, fossil fuels (natural gas, hydrocarbons, etc.), biomass resources or different wastes (industrial wastes, municipal wastes, agricultural wastes). Moreover the production, storage and distribution of hydrogen are still strong limitations for the development of such techniques (Gosselink, 2002; Ströbel et al., 2002), so that other fuels, particularly those that are liquid under ambient temperature and pressure, are more convenient due to their ease of handling and distribution.

Therefore, alcohols begin to be considered as valuable alternative fuels, because they have a high energy density (6–8 kWh kg^{-1}, compared to 33 kWh kg^{-1} for pure hydrogen without storage and to about 11 kWh kg^{-1} for gasoline) and they can be produced from renewable sources (e.g. bioethanol from biomass feedstock). Ethanol is an attractive fuel for electric vehicles, since it can be easily produced in great quantity by the fermentation of sugar-containing raw materials from agriculture (corn, sugar beet, sugar cane, etc.) or from different wastes (e.g. agriculture wastes containing lignocellulose residues). In addition, in some countries like Brazil, ethanol is already distributed through the car fuel station network for use in conventional cars with internal combustion engines.

With a hydrogen-rich gas, obtained from fossil or biomass fuels and containing only a small amount (from a few % to a few ppm, depending on the type of fuel cell) of carbon monoxide (which acts as a poisoning species for platinum catalysts), or with alcohols (methanol, ethanol) which can react directly at the fuel cell anode, low-temperature fuel cells are more convenient for mobile and portable applications. Proton exchange membrane fuel cell (PEMFC), which can operate from ambient temperature to 70–80°C, and the direct ethanol fuel cell (DEFC), which works at higher temperatures (up to 120–150°C), will be particularly discussed in this chapter. After recalling the working principles and the different kinds of fuel cells, the processes involved in the transformation of biomass resources and in the treatment of waste organic compounds to get hydrogen sufficiently pure (content of CO lower than a few tens of ppm) to feed a PEMFC will be discussed. Finally, the DEFC in which ethanol reacts directly at the fuel cell anode (without the necessity of a fuel processor) will be presented in more details.

21.2 PRINCIPLES AND DIFFERENT TYPES OF FUEL CELLS

21.2.1 Working principles of a fuel cell

The principles of the fuel cell are illustrated in Figure 21.1. The electrochemical cell consists of two electrodes, an anode and a cathode, which are electronic conductors, separated by an electrolyte (e.g. a proton exchange membrane in a PEMFC or in a DEFC), which is an ionic conductor (as the result of proton migration and diffusion).

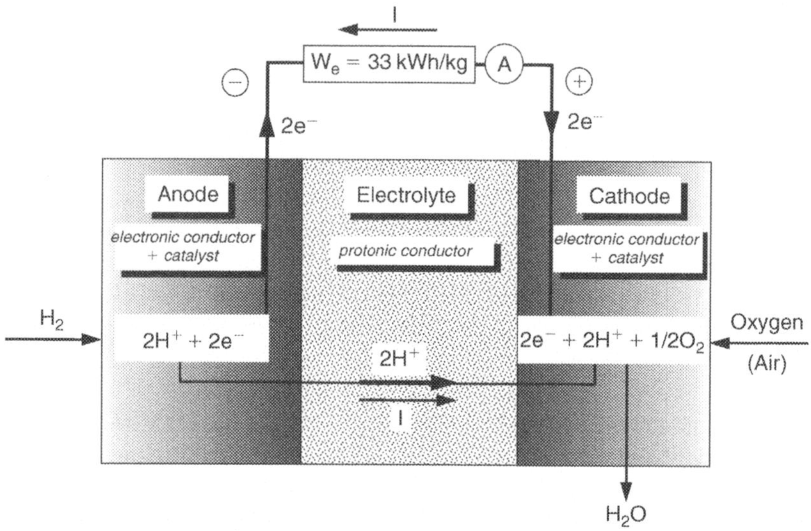

Figure 21.1 Schematic diagram of a hydrogen/oxygen fuel cell.

An elementary electrochemical cell converts directly the chemical energy of combustion in oxygen (i.e. ΔG, the Gibbs free energy change) of a given fuel (hydrogen, natural gas, hydrocarbons, kerosene, alcohols, biomass resources and wastes) into electricity (Appleby & Foulkes, 1989; Kordesch & Simader, 1996; Lamy & Léger, 1994; Tilak et al., 1981; Vielstich et al., 2003).

21.2.1.1 *FC thermodynamics*

At the anode (negative pole of the cell) the electro-oxidation of hydrogen takes place as follows:

$$H_2 \rightarrow 2H^+ + 2e^- \quad E_1^o = 0.000 \text{ V/SHE} \tag{21.1}$$

whereas the cathode (positive pole) undergoes the electro-reduction of oxygen, that is:

$$O_2 + 4H^+ + 4e^- \rightarrow 2H_2O \quad E_2^o = 1.229 \text{ V/SHE} \tag{21.2}$$

where E_i^o are the electrode potentials versus the standard hydrogen (reference) electrode (SHE).

This corresponds to the overall combustion reaction of hydrogen in oxygen:

$$H_2 + \tfrac{1}{2}O_2 \rightarrow H_2O \tag{21.3}$$

with the thermodynamic data, under standard conditions:

$$\Delta G^o = -237 \text{ kJ mol}^{-1}; \quad \Delta H^o = -286 \text{ kJ mol}^{-1} \text{ of } H_2$$

The protons produced at the anode crossover the membrane, ensuring the electrical conductivity inside the electrolyte, whereas the electrons liberated at the anode reach the cathode (where they reduce oxygen) through the external circuit. This process produces an electric energy, $W_e = nFE^o_{eq} = -\Delta G^o$, corresponding to an energy mass density of the fuel of $W_{el} = -\Delta G^o/(3600M) = 32.9\,\text{kWh kg}^{-1}$, where $M = 0.002$ kg is the molecular weight of hydrogen.

The standard electromotive force (emf), E^o_{eq}, at equilibrium (no current flowing) is then calculated as follows:

$$E^o_{eq} = -\frac{\Delta G^o}{nF} = \frac{237 \times 10^3}{2 \times 96485} = E^o_2 - E^o_1 = 1.229\,\text{V} \approx 1.23\,\text{V} \qquad (21.4)$$

with $F = 96485$ C, the Faraday constant (i.e. the absolute value of the electric charge of 1 mol of electrons) and $n = 2$, the number of electrons exchanged per hydrogen molecule.

The working of the cell under reversible thermodynamic conditions does not follow Carnot's theorem, so that the theoretical energy efficiency, defined as the ratio between the produced electric energy ($-\Delta G^o$) and the heat of combustion ($-\Delta H^o$) at constant pressure, is:

$$\varepsilon^{rev}_{eq} = \frac{W_e}{(-\Delta H^o)} = \frac{nFE^o_{eq}}{(-\Delta H^o)} = \frac{\Delta G^o}{\Delta H^o} = 1 - \frac{T\Delta S^o}{\Delta H^o} = \frac{237}{286} = 0.83 \qquad (21.5)$$

for the hydrogen/oxygen fuel cell.

This theoretical efficiency is much greater (by a factor of 2) than that of a thermal combustion engine, producing the reversible work, W_r, according to the Carnot's theorem:

$$\varepsilon^{rev}_{therm} = \frac{W_r}{(-\Delta H^\circ)} = 1 - \frac{Q_2}{Q_1} = 1 - \frac{T_2}{T_1} = 0.43 \qquad (21.6)$$

with, for example, $T_1 = 350°$C and $T_2 = 80°$C.

21.2.1.2 FC kinetics

However under working conditions, with a current density j, the cell voltage $E(j)$ decreases greatly as the result of three limiting factors: (i) the over-voltages η_a and η_c at both electrodes due to a rather low reaction rate of the electrochemical processes involved (η is defined as the difference between the working electrode potential E_i and the equilibrium potential E^o_i), (ii) the ohmic drop $R_e j$ both in the electrolyte and interface resistances R_e and (iii) mass transfer limitations for reactants and products (Figure 21.2).

The cell voltage, $E(j)$, defined as the difference between the cathode potential E_2 and the anode potential E_1, can thus be expressed as:

$$\begin{aligned} E(|j|) &= E_2(|j|) - E_1(|j|) = E^o_2 + \eta_c - (E^o_1 + \eta_a) - R_e\,|j| \\ &= E^o_{eq} - (|\eta_a| + |\eta_c| + R_e\,|j|) \end{aligned} \qquad (21.7)$$

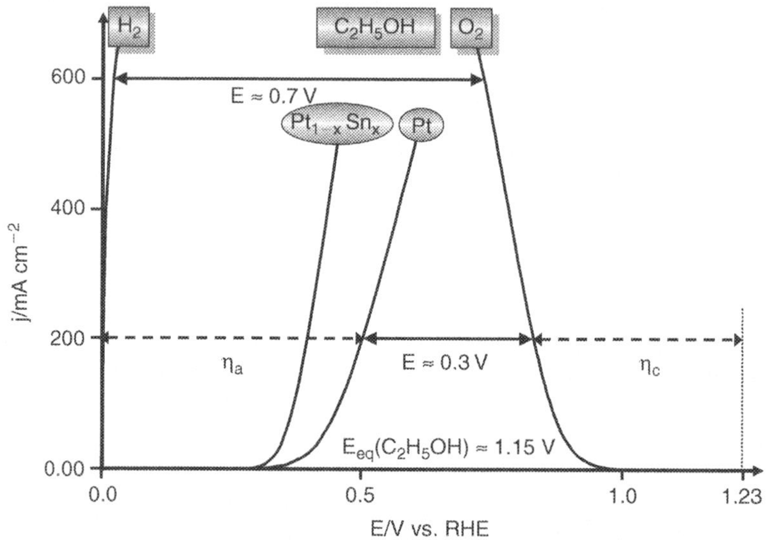

Figure 21.2 Comparative current density (j) versus potential (E) curves for H_2 and EtOH electro-oxidation at different Pt-based catalytic anodes and oxygen electro-reduction at a Pt cathode.

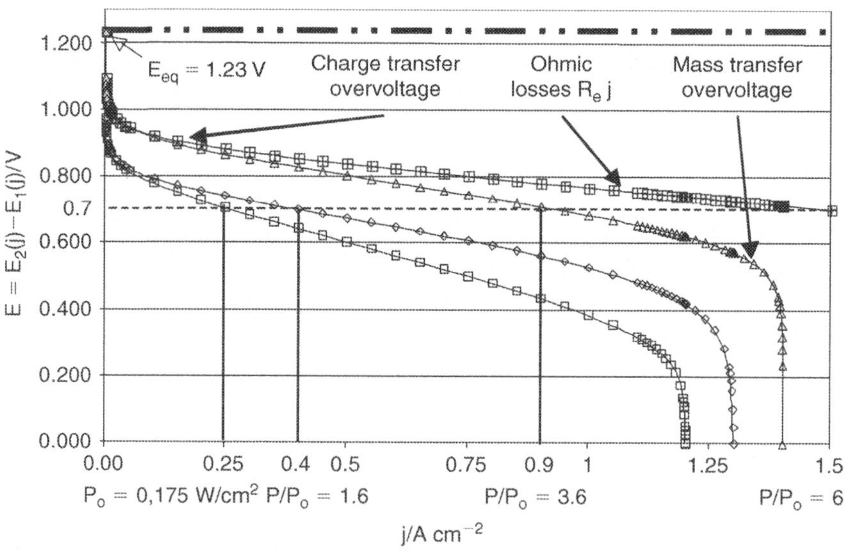

Figure 21.3 Theoretical $E(j)$ electric characteristics of a H_2/O_2 fuel cell: (□) $j_o = 10^{-8}$, $R_e = 0.30, j_l = 1.2$; (◇) $j_o = 10^{-8}, R_e = 0.15, j_l = 1.3$; (Δ) $j_o = 10^{-6}, R_e = 0.15, j_l = 1.4$;

where the over-voltages η_a (anodic reaction, i.e. oxidation of the fuel with $\eta_a > 0$) and η_c (cathodic reaction, i.e. reduction of the oxidant with $\eta_c < 0$) take into account both the low kinetics of the electrochemical reactions involved (activation polarization) and the limiting rate of mass transfer (concentration polarization) – see Figure 21.3.

For a PEMFC, fed with reformate hydrogen and air, the working cell voltage is typically 0.8 V at 500 mA cm^{-2}, which leads to a voltage efficiency ε_E of:

$$\varepsilon_E = \frac{E(j)}{E^o_{eq}} = 1 - \frac{|\eta_a(j)| + |\eta_c(j)| + R_e|j|}{E^o_{eq}} = \frac{0.8}{1.23} = 0.65 \quad (21.8)$$

The overall energy efficiency (ε_{cell}) thus becomes:

$$\varepsilon_{cell} = \frac{W_e}{(-\Delta H^o)} = \frac{n_{exp} F\, E(j)}{(-\Delta H^o)} = \frac{n_{exp}}{n_{th}} \frac{E(j)}{E^o_{eq}} \frac{n_{th} F\, E^o_{eq}}{(-\Delta H^o)} = \varepsilon_F \varepsilon_E \varepsilon^{rev}_{eq} \quad (21.9)$$

where the Faradaic efficiency $\varepsilon_F = n_{exp}/n_{th}$ is the ratio between the number of electrons n_{exp} effectively exchanged in the overall reaction and the theoretical numbers of electrons n_{th} for complete oxidation of the fuel (to H_2O and CO_2). $\varepsilon_F = 1$ for hydrogen oxidation, but $\varepsilon_F = 4/6 = 0.67$ for methanol oxidation stopping at the formic acid stage (4 electrons exchanged instead of 6 electrons for complete oxidation to CO_2).

As an example the overall energy efficiency of an H_2/O_2 fuel cell, working at 0.8 V under 500 mA cm^{-2}, is:

$$\varepsilon^{H_2/O_2}_{cell} = 1 \times 0.65 \times 0.83 = 0.54 \quad (21.10)$$

whereas that of a DEFC working at 0.5 V under 100 mA cm^{-2} (assuming complete oxidation to CO_2) would be:

$$\varepsilon^{C_2H_5OH/O_2}_{cell} = 1 \times 0.44 \times 0.97 = 0.43 \quad (21.11)$$

$$\text{since } \varepsilon_E = \frac{0.5}{1.14} = 0.44 \text{ and } \varepsilon^{rev}_{eq} = \frac{1325}{1366} = 0.97$$

with $\Delta G^o = -1325$ kJ mol^{-1}; $\Delta H^o = -1366$ kJ mol^{-1} of ethanol (see Section 21.4.2.1).

These energy efficiencies are better than those of the best thermal engines (diesel engines), which have energy efficiency of the order of 0.40.

Therefore, it appears that the only way to increase significantly the overall energy efficiency is to increase ε_E and ε_F; that is, to decrease the over-voltages η and the ohmic drop $R_e j$, since ε^{rev}_{eq} is given by the thermodynamics (one can increase it slightly by changing the pressure and temperature operating conditions). For the hydrogen/oxygen fuel cell $\varepsilon_F \sim 1$ usually, but it can be much lower in the case of incomplete combustion of the fuel (see e.g. the case of the DEFC in Section 21.4.2.1). The decrease of $|\eta|$ is directly related to the increase of the rate of the electrochemical reactions occurring at both electrodes. This is typically the field of electrocatalysis, where both the action of the electrode potential and of the catalytic electrode material synergistically increase the reaction rate v (Appleby, 1983; Bockris & Reddy, 1972; Sakellaropoulos, 1981).

Indeed, the current intensity, I, is proportional to the rate of reaction v, that is, $I = nFv$. For heterogeneous reactions, v is proportional to the surface area A of the interface, so that the kinetics of electrochemical reactions are better defined by the

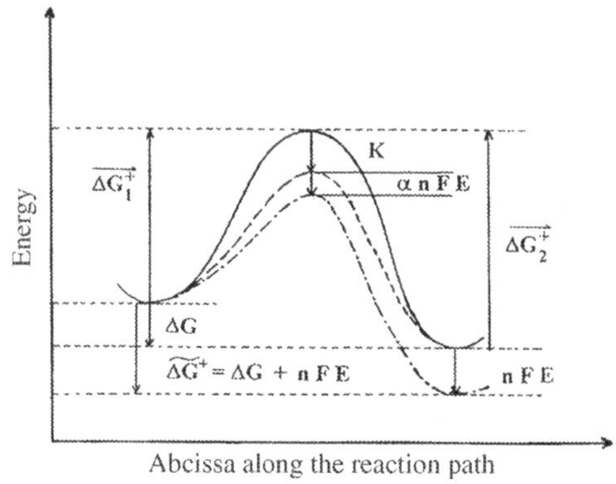

Figure 21.4 Activation barrier for an electrochemical reaction. K is the decrease in activation energy due to the electrode catalyst and αnFE that due to the electrode potential E.

intrinsic rate $v_i = v/A$ and the current density $j = I/A = nFv_i$. The electrical characteristics $j(E)$ can then be obtained by introducing the exponential behavior of the rate constant with the electrochemical activation energy, that is $\widetilde{\Delta G}^+ = \Delta G_0^+ - \alpha nFE$ which comprises two terms: the first one (ΔG_0^+) is the chemical activation energy, and the second one (αnFE) is the electric part of the activation energy. The latter is a fraction α ($0 \leqslant \alpha \leqslant 1$) of the total electric energy coming from the applied electrode potential E, where α is called the charge transfer coefficient (Figure 21.4). In the theory of absolute reaction rate, one obtains, for a first-order electrochemical reaction (the rate of which is proportional to the reactant concentration c_i):

$$j = nFv_i = nFk(T,E)c_i = nFk_0 c_i \, e^{-(\Delta G_0^+/RT)} e^{(\alpha nFE/RT)} = j_0 e^{(\alpha nFE/RT)} \quad (21.12)$$

This last equation contains the two essential activation terms met in electrocatalysis: (i) an exponential function of the electrode potential E and (ii) an exponential function of the chemical activation energy ΔG_0^+. By modifying the nature and structure of the electrode material, one may decrease ΔG_0^+, thus increasing j_0, as the result of the catalytic properties of the electrode. This leads to an increase of the reaction rate j.

On the other hand, this equation allows us to understand the effect of impurities on the $E(j)$ characteristics of the fuel cell. For example, carbon monoxide, which acts as a strong poison of platinum catalysts at both electrodes of low-temperature fuel cells (Figure 21.5), is strongly adsorbed on the electrode surface, and blocks the adsorption of hydrogen (or oxygen) thus increasing ΔG_0^+. Figure 21.5 illustrates the effect of CO on the $E(j)$ curves of a PEMFC fed with a reformate gas containing 5–100 ppm of CO.

The way to overcome this difficulty is to increase the working temperature, for example a phosphoric acid fuel cell (PAFC) (see Section 21.2.2.2 and Table 21.1) working at 180°C can tolerate 1% of CO in the reformate gas, whereas for a PEMFC the CO content must be <10 ppm). Alternatively, the platinum catalyst can be modified by addition of other metals to decrease the adsorption energy of CO.

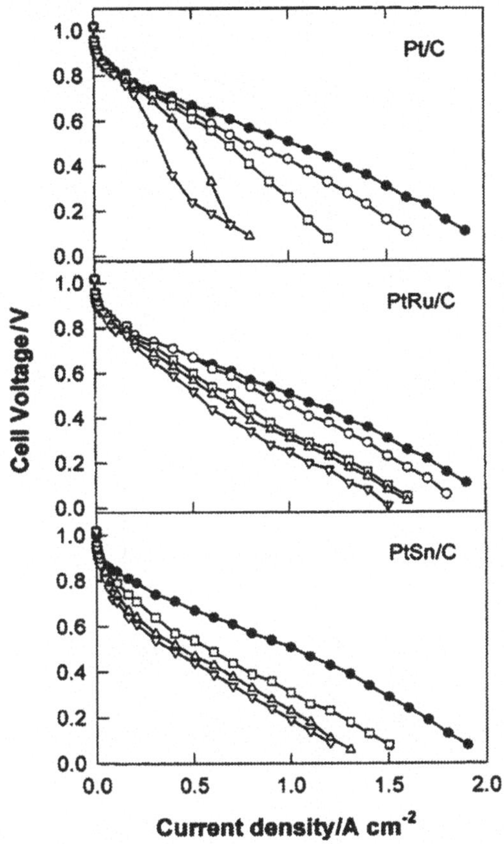

Figure 21.5 Single cell performance plots of a H_2/O_2 fuel cell with Pt/C, Pt–Ru/C and Pt–Sn/C anodes at 85°C for several CO concentrations in the reformate gas: (•) 0 ppm; (○) 5 ppm; (□) 20 ppm; (Δ) 50 ppm and (∇) 100 ppm (Lee et al. 1999).

21.2.2 Different types of a fuel cell

Different types of fuel cells have been developed (Lamy et al., 2000) and are classified mainly according to (i) the type of fuel, (ii) the operating temperature range and/or electrolyte or (iii) the direct or indirect utilization of fuel (see Table 21.1). The state of the art and the different applications of fuel cells are given in Table 21.1.

21.2.2.1 Fuels for fuel cells

For fuel cells operating at low (<100°C) and intermediate temperatures (up to 200°C), H_2 and H_2/CO_2 (with minimal amounts of CO) are the ideal fuels. The H_2/CO_2 gas mixture is produced by steam reforming/shift conversion, or partial oxidation/shift conversion of the primary or secondary organic fuels (see Section 21.3.4). Hydrogen is a secondary fuel, and like electricity, is an energy carrier. On a large scale, hydrogen is produced from the primary fuels, that is natural gas, coal or oil. For high-temperature fuel cell systems (>650°C), a mixture of H_2, CO, CO_2, produced by steam reforming can be used in fuel cells quite efficiently (power plant efficiency of >50%).

Table 21.1 Status of fuel cell technologies.

Type of fuel cell (operating fuel; operating temperature)	Power rating[a] (kW)	Fuel efficiency[b] (%)	Power density[a] ($mW\,cm^{-2}$)	Lifetime[b] (h)	Capital cost[b] ($/kW)	Applications
PAFC (CH_4, CH_3OH, oil; 200°C)	200–10,000	40–45	200–300	30,000–40,000	200–3000	On-site integrated energy systems, transportation (fleet vehicles), load leveling
MCFC (CH_4, coal ?; 650°C)	100–5000	50–55	150–300	10,000–40,000	1250	Base load and intermediate load, power generation – cogeneration
SOFC (CH_4, coal ?; 800–1000°C)	25–5000	50–60	200–400	8,000–40,000	1500	Base load and intermediate load, power generation – cogeneration
PEMFC (H_2, CH_3OH; 25–100°C)	0.01–250	40–50	500–1000	10,000–100,000	50–2000	Transportation, stand-by power, portable power, space stations
DMFC (CH_3OH; 25–150°C)	0.001–10	30–45	50–200	1,000–10,000	1000	Portable power, stand-by power, transportation (?), APU
AFC (H_2; 80°C)	20–100	65	250–400	3,000–10,000	1000	Space flights, space stations, transportation, APU

Notes:
AFC: Alkaline fuel cell.
DMFC: Direct methanol fuel cell.
MCFC: Molten carbonate fuel cell.
PAFC: Phosphoric acid fuel cell.
PEMFC: Proton exchange membrane fuel cell.
SOFC: Solid oxide fuel cell.
[a]Higher values are projected.
[b]Lower values are projected.

These fuels (pure H_2, H_2/CO_2, $H_2/CO/CO_2$) can also be produced from renewable energy sources – biomass, solar, windmills, hydroelectric power. Pure H_2 can also be generated by electrolysis using nuclear power plants. Hydrogen is the most electroactive fuel for fuel cells operating at low and intermediate temperatures. Methanol and ethanol are the most electroactive alcohol fuels, and, when they are electro-oxidized directly at the fuel cell anode (instead of being transformed in a hydrogen-rich gas), the fuel cell is called direct alcohol fuel cell (DAFC), either a DMFC (with methanol) or a DEFC (with ethanol).

21.2.2.2 Hydrogen fed fuel cells

Hydrogen/oxygen (air) fuel cells are classified according to the type of electrolyte used and the working temperature:

- Solid oxide fuel cell (SOFC) working between 700°C and 1000°C with a solid oxide electrolyte such as yttria stabilized zirconia (ZrO_2–8% Y_2O_3).
- Molten carbonate fuel cell (MCFC) working at about 650°C with a mixture of molten carbonates (Li_2CO_3/K_2CO_3) as electrolyte, both of them being high-temperature fuel cells.
- PAFC working at 180–200°C with a porous matrix of PTFE-bonded silicon carbide impregnated of phosphoric acid as electrolyte. This medium temperature fuel cell is now commercialized by ONSI (USA) mainly for stationary applications.
- Alkaline fuel cell (AFC) working at 80°C with concentrated potassium hydroxide as electrolyte. This kind of fuel cell, developed by IFC (USA), is now used in the space shuttle.
- PEMFC working at around 70°C with a polymer membrane electrolyte, such as Nafion®, which is a solid protonic conductor.

21.2.2.3 Methanol and ethanol fed fuel cells

Besides hydrogen as a fuel, methanol or ethanol can be directly converted into electricity in a DAFC, a great progress which resulted from the use of a proton exchange membrane acting both as an electrolyte (instead of aqueous electrolytes previously used) and as a separator preventing the mixing of fuel and oxidant. DAFC can work at moderate temperatures (30–50°C) for portable applications, but now the tendency is to look for new membranes less permeable to alcohol and working at higher temperatures (120–150°C) to increase the rate of the electrochemical reactions involved (oxidation of alcohol and reduction of oxygen) and to manage better the heat produced, either to use it in a cogeneration system or to evacuate it in transportation applications.

The three latter fuel cells (AFC, PEMFC and DAFC) are low-temperature fuel cells. In this chapter, the discussion will be focused on these fuel cells, particularly the PEMFC and the DAFC, since they can accommodate biomass fuels, either after fuel processing to obtain reformate hydrogen or directly with bioethanol.

For these low-temperature fuel cells the development of catalytic materials is essential to activate the electrochemical reactions involved. This concerns the electro-oxidation of the fuel (reformate hydrogen containing some traces of CO, which acts as a poisoning species for the anode catalyst; ethanol which has a relatively low reactivity at low temperatures) and the electro-reduction of the oxidant (oxygen) which is still the source of high energy losses (up to 30–40%) due to the low reactivity of oxygen at the best platinum-based electrocatalysts.

21.3 WASTE GASES AND BIOMASS TREATMENTS FOR LOW-TEMPERATURE FUEL CELLS

This part will present the different physico-chemical processes involved in the transformation of waste gases and biomass resources into a hydrogen-rich gas (Lens et al., 2005). Then it will discuss the different ways to upgrade this gas mixture to the hydrogen quality suitable for its use in a PEMFC (Trogisch & Baaske, 2004).

21.3.1 Thermo-chemical transformation of solid biomass

Starting from solid biomass resources (lignocellulosic feedstock, agricultural and industrial wastes, etc.), the first step is to transform them into a convenient slurry to feed a gasification or a fermentation reactor (Riedel et al., 2003; Rohde et al., 2004; Schulz, 2003).

Gasification and pyrolysis involve the conversion of solid or liquid organic materials into permanent gases (CO, CO_2, CH_4, H_2). Pyrolysis is the decomposition of biomass by heating it in the absence of any reactant, whereas gasification involves the reaction of carbonaceous compounds with air, oxygen, steam, carbon dioxide, etc. Pyrolysis taking place at medium temperature (400–600°C) leads mostly to liquid products (called "pyrolysis oil" or "green oil"), whereas gasification taking place at higher temperatures (650–900°C) yields mainly gaseous products.

In all cases the different gasification processes give a mixture of hydrogen and carbon monoxide, called the "syngas", from which different fuels can be synthesized: methanol, hydrocarbons (mainly paraffins using the Fisher–Tropsch synthesis), hydrogen, etc.

The basic chemical reactions involved are described below, where biomass compounds $C_xH_yO_z$ (with an average composition $C_6H_9O_4$) can be any feedstock from biomass, including methane-containing biogas. The enthalpy (ΔH) and Gibbs free energy (ΔG) of the reactions are calculated under standard conditions ($T = 298\,K$) and at elevated temperatures ($T = 1200\,K$) using the thermodynamic data given in the *Handbook of Chemistry and Physics* (Lide, 1992).

21.3.1.1 Steam reforming

Water steam reforming is the most interesting process involved, because it provides extra hydrogen from water:

$$C_xH_yO_z + H_2O \longrightarrow CO + CO_2 + H_2 \tag{21.13}$$

for example in the case of methane (natural gas or biogas) taken as a typical example:

$$CH_4 + 2H_2O \longrightarrow 4H_2 + CO_2 \tag{21.14}$$

with $\Delta H_{298\,K} = +165\,kJ\,mol^{-1}$ and $\Delta H_{1200\,K} = +193\,kJ\,mol^{-1}$ (endothermic reaction)

and $\Delta G_{298\,K} = +113\,kJ\,mol^{-1}$ and $\Delta G_{1200\,K} = -74.5\,kJ\,mol^{-1}$, or

$$CH_4 + H_2O \longrightarrow 3H_2 + CO \tag{21.15}$$

with $\Delta H_{298\,K} = +206\,kJ\,mol^{-1}$ and $\Delta H_{1200\,K} = +226\,kJ\,mol^{-1}$ (endothermic reaction)
and $\Delta G_{298\,K} = +142\,kJ\,mol^{-1}$ and $\Delta G_{1200\,K} = -77.7\,kJ\,mol^{-1}$.

These reactions are endothermic, needing external heating and a nickel catalyst to work at reasonable temperatures (650–950°C). The overall process, under given conditions of temperature and pressure ($T = 900°C$; $P = 8\,bar$), leads to a mixture of gases (e.g. 55% H_2, 11% CO, 5% CO_2 and 28% H_2O).

21.3.1.2 Water gas shift reaction

In addition, the water gas shift reaction (WGSR) produces extra hydrogen, by reaction of CO with water, as follows:

$$CO + H_2O \longrightarrow CO_2 + H_2 \tag{21.16}$$

with $\Delta H_{298\,K} = -41.2\,kJ\,mol^{-1}$ and $\Delta H_{1200\,K} = -32.9\,kJ\,mol^{-1}$ (exothermic reaction)
and $\Delta G_{298\,K} = -28.7\,kJ\,mol^{-1}$ and $\Delta G_{1200\,K} = +3.20\,kJ\,mol^{-1}$.

This process is favoured at low temperatures ($T < 200°C$).

21.3.1.3 Partial oxidation

Another process, which is exothermic all over the temperature range, is partial combustion of biomass with air (or pure oxygen) written as follows in the case of methane:

$$CH_4 + \tfrac{1}{2}O_2 \longrightarrow 2H_2 + CO \tag{21.17}$$

with $\Delta H_{298\,K} = -35.9\,kJ\,mol^{-1}$ and $\Delta H_{1200\,K} = -23.1\,kJ\,mol^{-1}$ (exothermic reaction)
and $\Delta G_{298\,K} = -86.6\,kJ\,mol^{-1}$ and $\Delta G_{1200\,K} = -259\,kJ\,mol^{-1}$.

This is a non-catalytic process, which leads to a mixture of gases; for example, 40% H_2, 20% CO and 40% N_2 (when using air instead of pure oxygen).

Steam reforming and partial oxidation can be combined together in the so-called autothermal reforming (ATR) process with no net exchange of heat, the stoichiometric amount of water and air being adjusted so that $\Delta H \approx 0$ (e.g. in the case of methane):

$$CH_4 + O_2 + H_2O \longrightarrow H_2 + CO + CO_2 \quad \text{with } \Delta H \approx 0 \tag{21.18}$$

21.3.1.4 Cracking

At elevated temperatures ($T > 700°C$) and low pressure, methane cracking does occur:

$$CH_4 \longrightarrow 2H_2 + C \tag{21.19}$$

with $\Delta H_{298\,K} = +74.6\,kJ\,mol^{-1}$ and $\Delta H_{1200\,K} = +90.1\,kJ\,mol^{-1}$ (endothermic reaction)
and $\Delta G_{298\,K} = +50.5\,kJ\,mol^{-1}$ and $\Delta G_{1200\,K} = -41.3\,kJ\,mol^{-1}$,

leading to carbon particulates which may deactivate the nickel catalyst. Similarly CO can decompose at lower temperatures ($T < 700°C$) producing again coke (Boudouard reaction), as follows:

$$2CO \rightarrow CO_2 + C \tag{21.20}$$

with $\Delta H_{298\,K} = -172\,kJ\,mol^{-1}$ and $\Delta H_{1200\,K} = -169\,kJ\,mol^{-1}$ (exothermic reaction)
and $\Delta G_{298\,K} = -120\,kJ\,mol^{-1}$ and $\Delta G_{1200\,K} = +39.6\,kJ\,mol^{-1}$.

21.3.1.5 Impurities

When using biomass feedstock some solid residues are also formed, such as tar, ashes, etc., together with some impurities such as sulfur, halogens (e.g. chlorides) and alkali, which must be eliminated to sustain the process.

Flash pyrolysis working at higher temperatures (800–1500°C) also produces a mixture of hydrogen and carbon monoxide, together with some impurities (particulates, ashes, etc.).

21.3.2 Transformation of liquid biomass

Biomass containing liquid compounds can be obtained, either by pyrolysis at moderate temperatures (400–600°C) of solid biomass leading to pyrolysis oil (green oil) or by fermentation of sugar-containing feedstock leading to bioethanol.

Starting from lignocellulosic biomass, coming either from agricultural residues or energy crops, first of all thermo-chemical pre-treatments give a mixture of cellulose and lignin, followed by enzymatic hydrolysis of cellulose providing sugar intermediates (mainly hexoses, but also some pentoses). Then bioethanol is obtained by glucose fermentation and separation of the ethanol (Haagensen et al., 2005).

Ethanol can then be either used in a DEFC, which converts its combustion energy into electricity (see Section 21.4.2), or transformed by steam reforming into a hydrogen-rich gas:

$$C_2H_5OH + 3H_2O \rightarrow 6H_2 + 2CO_2 \tag{21.21}$$

with $\Delta H_{298\,K} = +173\,kJ\,mol^{-1}$ and $\Delta H_{1200\,K} = +213\,kJ\,mol^{-1}$ (endothermic reaction)
and $\Delta G_{298\,K} = +64.9\,kJ\,mol^{-1}$ and $\Delta G_{1200\,K} = -310\,kJ\,mol^{-1}$.

This last process is a catalytic one (with a nickel catalyst), working between 600°C and 750°C, which provides a mixture of gases; for example, 52% H_2, 14% CO, 8% CO_2 and 25% H_2O under given experimental conditions ($T = 700°C$ and $P = 1$ bar).

Again the gas mixture can be enriched into hydrogen through the catalyzed WGSR at high temperatures (HTWGSR) around 350–500°C, followed by reaction at lower temperatures (LTWGSR) around 180–250°C. This finally gives, as a typical example, a mixture of 66% H_2, 0.9% CO, 21% CO_2 and 12% H_2O.

21.3.3 Utilization of biogas

Biogas, a mixture of methane (CH_4) and carbon dioxide (CO_2), is produced from wastewater, solid wastes and biomass feedstock (including energy crops), by anaerobic digestion of organic matter with microorganisms. With liquid wastes (waste slurries, waste water) from industrial or domestic origin, biogas is easily produced by anaerobic digestion (wet fermentation process).

Starting from solid wastes, containing cellulose, hemicellulose, lignocellulose, lipids, proteins, etc., the first step is to solubilize these organic compounds by enzymatic hydrolysis with bacteria. Then acetogenesis occurs due to the combined action of diverse groups of microorganisms leading to acetate and hydrogen, besides carbon dioxide. In the final step acetate and hydrogen are converted to methane by acetotrophic and hydrogenotrophic methanogenesis (Meulepas et al., 2005).

The composition of the biogas greatly depends on the biomass sources (i.e. industrial wastes, agricultural wastes, sewage gas or landfill gas). It contains mainly CH_4 (between 50% and 75%) and CO_2 (between 50% and 25%), together with small amounts of contaminants and impurities like hydrogen sulfide (100 ppm to a few %), halogenated hydrocarbons (a few 100 ppm), ammonia, siloxanes, etc., and some particulates. The concentration of sulfur compounds and halides (particularly chlorides) must be reduced to a few ppm, because they are strong poisons for the reformer (nickel-based) and PEMFC (platinum-based) catalysts. Particulates can be removed by using gas filters or ceramic membranes.

One way to use upgraded biogas in a PEMFC is to remove CO_2 by gas scrubbing (e.g. with water, caustic soda solution or monoethanolamine), adsorption on a large surface area material (e.g. activated carbon), CO_2 liquefaction or membrane separation (Schulte-Schulze Berndt, 2005). Then methane has to be further cleaned up and transformed in hydrogen by steam reforming or autothermal processes (see Section 21.3.1). Hydrogen is further purified using pressure swing absorption (PSA), which eliminates most of the residual impurities. PSA operates on two working pressures: high pressure (8–16 bar) to adsorb the undesired compounds and low pressure (1–2 bar) to regenerate the adsorber (Schulte-Schulze Berndt, 2005). The remaining traces of CO can be further reduced by preferential oxidation or methanation (see Section 21.3.4) to a level below 10 ppm acceptable for PEMFC.

Another way to produce hydrogen is to make methane to react with carbon dioxide contained in the biogas in the so-called "dry reforming process" (Olsbye et al., 1997):

$$CH_4 + CO_2 \longrightarrow 2H_2 + 2CO \tag{21.22}$$

with $\Delta H_{298\,\mathrm{K}} = +247\,\mathrm{kJ\,mol^{-1}}$ and $\Delta H_{1200\,\mathrm{K}} = +259\,\mathrm{kJ\,mol^{-1}}$ (endothermic reaction)
and $\Delta G_{298\,\mathrm{K}} = +171\,\mathrm{kJ\,mol^{-1}}$ and $\Delta G_{1200\,\mathrm{K}} = -80.9\,\mathrm{kJ\,mol^{-1}}$.

This process is favored at elevated temperatures (800–1000°C) for a CH_4/CO_2 ratio close to 1.

21.3.4 "Syngas" upgrading to high-quality hydrogen

The "syngas" may be further transformed into hydrogen by the WGSR at high temperatures (HTWGSR) around 350–500 °C (with a FeCr catalyst), followed by reaction at lower temperatures (LTWGSR) around 180–250°C (with a Cu Zn catalyst):

$$CO + H_2O \longrightarrow H_2 + CO_2 \tag{21.16}$$

This leads to a gas mixture with a CO content between 0.5% and 2% at the exit of the WGS reactors (Ballerini et al., 2006a). This is still too high to feed a low-temperature fuel cell (PEMFC, AFC) with Pt-based catalysts. To lower the CO content below a few tens of ppm, further treatments are thus required (Ballerini et al., 2006b), such as:

- preferential oxidation (PrOx) of CO to CO_2 in the presence of hydrogen in the reformate gas (selective oxidation):

$$CO + \tfrac{1}{2}O_2 \rightarrow CO_2 \tag{21.23}$$

with $\Delta H_{298\,K} = -283\,kJ\,mol^{-1}$ and $\Delta H_{1200\,K} = -282\,kJ\,mol^{-1}$ (exothermic reaction)
and $\Delta G_{298\,K} = -257\,kJ\,mol^{-1}$ and $\Delta G_{1200\,K} = -178\,kJ\,mol^{-1}$.

This reaction can be done at around 100–200°C using a noble metal catalyst (Pt, Rh, etc.).

- methanation which is the reaction of CO with hydrogen leading to methane, an inert gas for PEMFC:

$$CO + 3H_2 \rightarrow CH_4 + H_2O \tag{21.24}$$

with $\Delta H_{298\,K} = -206\,kJ\,mol^{-1}$ and $\Delta H_{1200\,K} = -226\,kJ\,mol^{-1}$ (exothermic reaction)
and $\Delta G_{298\,K} = -142\,kJ\,mol^{-1}$ and $\Delta G_{1200\,K} = +77.7\,kJ\,mol^{-1}$.

This is the reverse reaction of the methane steam reforming process and is favored at low temperatures (<200°C) on noble metal catalysts.

- membrane separation using thin layers of supported Pd alloys (Pd/Ag, Pd/Cu) which preferentially absorb hydrogen (Ballerini et al., 2006a).

All these treatments, particularly membrane separation, allow to lower the CO content of the hydrogen reformate to a few ppm, which is suitable to feed a PEMFC.

21.4 LOW-TEMPERATURE FUEL CELLS (PEMFC AND DAFC)

21.4.1 PEMFC

PEMFC is nowadays the most advanced low-temperature fuel cell technology (Lamy et al., 2003; Stevens et al., 2000), because it can be used in several applications (space, electric vehicle, stationary power plants, portable electronics). The progress made in one application is greatly beneficial to the others.

21.4.1.1 Principle of a PEMFC

An elementary PEMFC consists of a thin (10–200 μm) solid polymer electrolyte (a protonic membrane, such as Nafion®), on which the electrode structures (fuel anode and oxygen cathode) are pasted on both sides, giving a membrane–electrode assembly (MEA; Figure 21.6). A single cell delivers a cell voltage of 0.9–0.5 V (instead of the theoretical emf of 1.23 V under standard equilibrium conditions), depending on

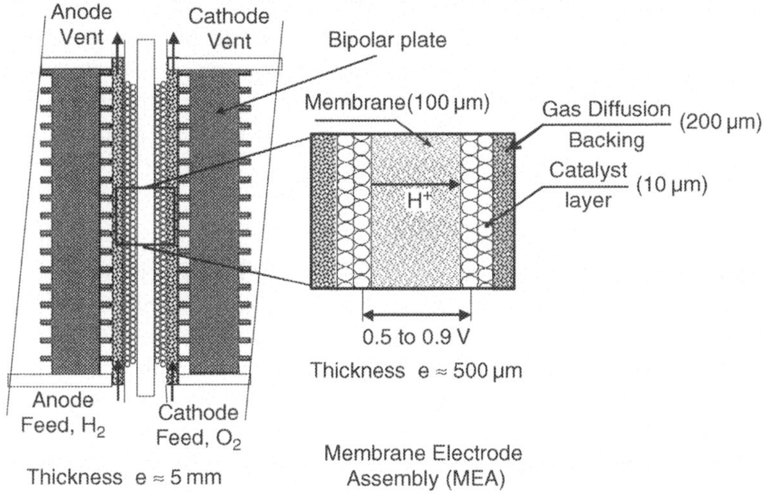

Figure 21.6 Schematic representation of a PEMFC elementary cell (and its MEA).

the working current density. Many elementary cells, electrically connected by bipolar plates, are assembled together (in series and/or in parallel) to reach the nominal voltage (such as 48 V for electric vehicles) of the fuel cell stack.

In PEMFC working at low temperature (20–90°C) several problems need to be solved before technological development of fuel cell stacks for different applications. This concerns the components of the elementary cell; that is, the proton exchange membrane, the electrode (anode and cathode) catalysts, the MEA and the bipolar plates (Lamy et al., 2003; Stevens et al., 2000).

21.4.1.2 The proton exchange membrane

The proton exchange membrane, which is a solid polymer electrolyte, plays a key role in the PEMFC. It allows the electrical current to pass through it thanks to its H^+ ionic conductivity and prevents any electronic current through it in order that electrons are obliged to circulate in the external electric circuit to produce the electric energy corresponding to the combustion reaction of the fuel. Moreover, it must avoid any gas leakage between the anodic and cathodic compartments, so that no chemical combination between hydrogen and oxygen is directly allowed. It must also be mechanically, thermally (up to 150°C in order to increase the working temperature of the cell) and chemically stable. Finally, its lifetime must be sufficient for practical applications (4000–40,000 h).

Many PEMFC have a Nafion® membrane, which partially fulfills these requirements, since its thermal stability is limited to 100°C, and its protonic conductivity strongly decreases at higher temperatures because of its dehydration. On the other hand, it is not completely impermeable to liquid fuels (such as alcohols). This becomes more important as the membrane is thin (a few 10 μm). Furthermore its actual cost is high (>500 €/m^2), so that its use in a PEMFC for an electric car is not yet cost competitive.

Therefore, new membranes are investigated with improved stability and conductivity at higher temperatures (up to 150°C). For power FC, the increase of temperature will increase the rate of the electrochemical reactions occurring at both electrodes (i.e. the current density at a given cell voltage) and the specific power. Furthermore, thermal management and heat utilization will be improved, particularly for residential applications with heat cogeneration, and for mobile applications to exhaust excess heat.

21.4.1.3 The electrode catalysts

One of the main problems of low-temperature (20–80°C) PEMFCs is the relatively low kinetics of the electrochemical reactions involved, such as oxygen reduction at the cathode and fuel (hydrogen from a reformate gas or alcohols) oxidation at the anode. The reaction rates can only be increased by the simultaneous action of the electrode potential and electrode material. Moreover, increasing the working temperature from 80°C to 150°C would strongly increase (by a factor of at least 100–1000) the rates of the electrochemical reactions (thermal activation). All these combined effects would increase the cell voltage by ca. 0.1–0.2 V, since at room temperature the anode and cathode overvoltages are close to 0.2–0.4 V, which decreases the cell voltage by 0.4–0.7 V leading to values close to 0.5–0.7 V, instead of the 1.23 V theoretical cell voltage.

The investigation of new electrocatalysts, particularly Pt-based catalysts, more active for oxygen reduction and fuel oxidation (hydrogen from reformate gas or alcohols) is thus an important point for the development of proton exchange membrane cells (PEMCs) (Lamy et al., 2001, 2002; Ralph & Hogarth, 2002; Tarasevich et al., 1983).

21.4.1.4 The MEA

The realization of the MEA is a crucial point to build a good fuel cell stack. The method currently used consists in hot-pressing (at 130°C and 35 kg/cm^2) the electrode structures on the polymer membrane (Nafion®). This gives non-reproducible results (in terms of interface resistance) and this is difficult to industrialize. New concepts must be elaborated, such as the continuous assembly of the three elements in a rolling tape process (like in the magnetic tape industry), or successive deposition of the component layers (microelectronic process), etc.

21.4.1.5 The bipolar plates

The bipolar plates, which separate both electrodes of neighboring cells (one anode of a cell and one cathode of the other), have a triple role:

- to ensure the electronic conductivity between two neighboring cells.
- to allow the distribution of reactants (gases and liquids in the case of alcohols) to the electrode catalytic sites, and to evacuate the reaction products (H_2O and CO_2 in the case of alcohols).
- thermal management inside the elementary cell by evacuating the excess heat.

The bipolar plates are usually fabricated with non-porous machined graphite or corrosion-resistant metal plates. Distribution channels are engraved in these plates. Metallic foams can also be used for distributing the reactants. One key point is to ensure a low ohmic resistance inside the bipolar plate and at the contact with the MEA.

Another point is to use materials with high corrosion resistance in the oxidative environment of the oxygen cathode.

21.4.1.6 Auxiliary and control equipment

A more detailed picture of a PEMFC system, including the auxiliary and control equipments, is given in Figure 21.7.

Fuel supply is usually from liquid hydrogen or pressurized gaseous hydrogen. For other fuels a fuel processor is needed, which includes a reformer, gas shift reactors and purification reactors, in order to decrease the amount of CO to an acceptable level (a few tens of ppm), which would otherwise poison the platinum-based catalysts. This equipment is still heavy and bulky, and limits the dynamic response of the fuel cell stack, particularly for the electric vehicle in some urban driving cycle.

On the other hand, the other auxiliary equipments depend greatly on the stack characteristics: air compressor, whose characteristics are related to the pressures supported by the proton exchange membrane, humidifiers for the reacting gases with controlled humidification conditions, preheating of gases to avoid condensation phenomena, hydrogen recirculation and purging systems of the anode compartment, cooling system for the MEAs, control of pressure valves and/or of gas flows, DC/DC or DC/AC electric converters.

The system control must ensure correct working of the system, not only under steady state conditions, but also during power transients. All the elementary cells must be controlled (cell voltage of each elementary cell, if possible) and the purging system must be activated in case of technical hitch.

According to different applications (stationary power plants, power sources for electric vehicle, auxiliary power units, etc.) different specifications and system design are required.

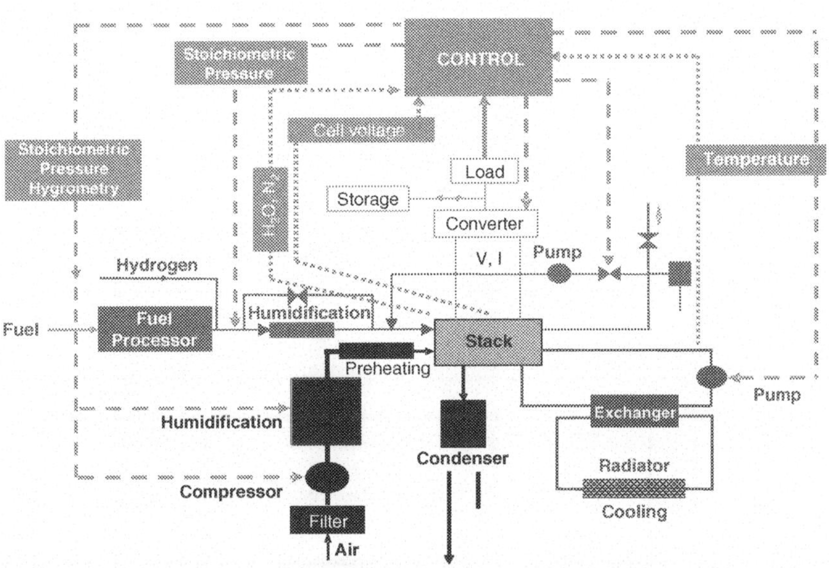

Figure 21.7 Detailed scheme of a PEMFC system with its auxiliary and control equipments.

21.4.2 DEFC

21.4.2.1 Principle of the DEFC

The schematic diagram of a DEFC is shown in Figure 21.8. The DEFC transforms directly the Gibbs energy of combustion of ethanol into electricity, without a fuel processor (Figure 21.8). This greatly simplifies the system, reducing its volume and cost (Lamy et al., 2001, 2002). Recent development of DEFC is due to the use of a proton exchange membrane as electrolyte, instead of a liquid acid electrolyte, as previously done.

At the anode (negative pole of the cell), the electro-oxidation of ethanol takes place as follows:

$$CH_3CH_2OH + 3H_2O \longrightarrow 2CO_2 + 12H^+ + 12e^- \quad E_1^o = 0.085\,\text{V versus SHE} \tag{21.25}$$

whereas the electro-reduction of oxygen does occur at the cathode (positive pole):

$$O_2 + 4H^+ + 4e^- \longrightarrow 2H_2O \quad E_2^o = 1.229\,\text{V versus SHE} \tag{21.26}$$

where E_1^o are the electrode potentials versus SHE. This corresponds to the overall combustion reaction of ethanol into oxygen:

$$CH_3CH_2OH + 3O_2 \longrightarrow 2CO_2 + 3H_2O \tag{21.27}$$

with the thermodynamic data, under standard conditions:

$$\Delta G^o = -1325\,\text{kJ mol}^{-1} \quad \Delta H^o = -1366\,\text{kJ mol}^{-1} \tag{21.28}$$

This gives a standard electromotive force (emf) at equilibrium:

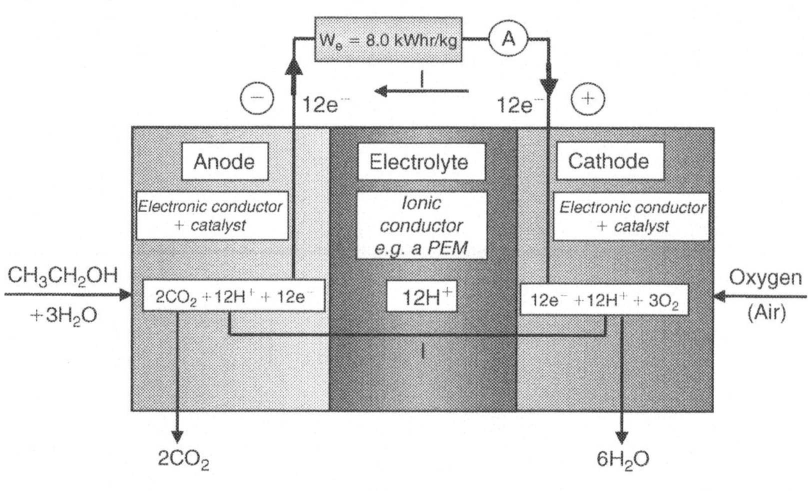

Figure 21.8 Schematic diagram of a DEFC.

$$E^{\circ}_{eq} = -\frac{\Delta G^{\circ}}{nF} = \frac{1325 \times 10^3}{12 \times 96485} = E^{\circ}_2 - E^{\circ}_1 = 1.144\ \text{V} \qquad (21.29)$$

with $n = 12$, the number of electrons exchanged per molecule for complete oxidation to CO_2. The corresponding electrical energy, $W_e = nFE^{\circ}_{eq} = -\Delta G^{\circ}$, leads to a mass energy density $W_{el} = -\Delta G^{\circ}/(3600M) = 8\,\text{kWh kg}^{-1}$, where $M = 0.046$ kg is the molecular weight of ethanol.

The theoretical energy efficiency, under reversible standard conditions, defined as the ratio between the electrical energy ($-\Delta G^{\circ}$) and the heat of combustion ($-\Delta H^{\circ}$) at constant pressure, is:

$$\varepsilon^{rev}_{eq} = \frac{W_e}{(-\Delta H^{\circ})} = \frac{nFE^{\circ}_{eq}}{(-\Delta H^{\circ})} = \frac{\Delta G^{\circ}}{\Delta H^{\circ}} = 1 - \frac{T\,\Delta S^{\circ}}{\Delta H^{\circ}} = \frac{1325}{1366} = 0.97 \qquad (21.30)$$

But, under working conditions, with a current density j, the cell voltage $E(j)$ is lower than E°_{eq} (see Equation (21.7)), so that the practical energy efficiency, for a DEFC working at 0.5 V and 50 mA cm^{-2} with complete oxidation to CO_2, would be:

$$\begin{aligned}\varepsilon^{C_2H_5OH/O_2}_{cell} &= \frac{W_e}{(-\Delta H^{\circ})} = \frac{n_{exp} F\, E(j)}{(-\Delta H^{\circ})} \\ &= \varepsilon_F \times \varepsilon_E \times \varepsilon^{rev}_{eq} = 1 \times 0.437 \times 0.97 = 0.424\end{aligned} \qquad (21.31)$$

since the potential efficiency is $\varepsilon_E = E(j)/E^{\circ}_{eq} = 0.5/1.14 = 0.437$ and the faradaic efficiency is $\varepsilon_F = n_{exp}/n_{th} = 1$, for complete oxidation to CO_2. This is quite similar to that of the best thermal engine (diesel engine). But, if the reaction process stops at the acetic acid stage, which involves the transfer of 4 electrons (instead of 12 for complete oxidation), the efficiency will be reduced by one-third, reaching only 0.15.

21.4.2.2 DEFC fuel cell tests

The preparation of electrodes and MEA were described elsewhere (Dubau et al., 2003). Fuel cell tests (determination of the cell voltage E and power density P versus the current density j) were carried out in a single DEFC with 5 or 25 cm^2 geometric surface area electrodes using a Globe Tech test bench, purchased from ElectroChem, Inc., USA (Figure 21.9).

The cell voltage E and current density j were recorded using a high-power potentiostat (Wenking model HP 88) interfaced with a variable resistance in order to fix the current applied to the cell and with a PC to apply constant current sequences and to store the data.

The reaction products at the outlet of the anode side of the DEFC were analyzed quantitatively by HPLC, as described by Rousseau et al. (2005). Large surface area electrodes (25 cm^2) were used in order to have a sufficient amount of products. The current density was kept constant and the voltage of the cell was simultaneously measured as a function of time. The ethanol flow rate was chosen close to 2 mL min^{-1} in

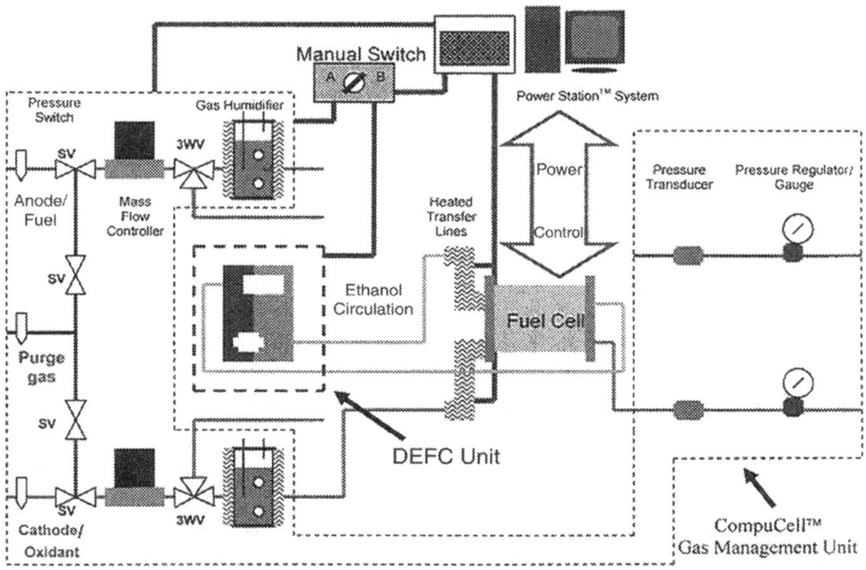

Figure 21.9 Typical configuration of a DEFC Globe Tech test station.

Table 21.2 Chemical yields in AAL, AA and CO_2 for Pt/C, Pt–Sn (90:10)/C and Pt–Sn–Ru (86:10:4)/C catalysts during ethanol electro-oxidation under DEFC operating conditions at 80°C for 4 h.

	AAL/%	AA/%	CO_2/%
Pt/C XC72 60 wt.%	47.5	32.5	20
Pt–Sn (9:1)/C XC72 60 wt.%	15.4	76.9	7.7
Pt–Sn–Ru (86:10:4)/C XC72 60 wt.%	15.2	75	9.8

order to perform long-term experiments (for at least 4 h) to get enough reaction products.

The experiments were carried out using Pt/C, Pt–Sn/C and Pt–Sn–Ru/C catalysts, and in each case no other reaction products than acetaldehyde (AAL), acetic acid (AA) and CO_2 were detected. The addition of tin to platinum not only increases the activity of the catalyst toward the oxidation of ethanol and therefore the electrical performance of the DEFC, but also changes greatly the product distribution: the formation of CO_2 and AAL is lowered, whereas that of AA is greatly increased (Table 21.2).

Typical electrical performances obtained with several Pt-based electrocatalysts are shown in Figure 21.10. The use of platinum alone as anode catalyst leads to poor electrical performance, the open circuit voltage (OCV) being lower than 0.5 V and the maximum current density reaches only 100 mA cm^{-2}, leading to a maximum power density lower than 7 mW cm^{-2} at 110°C. The addition of Ru, and above all of Sn to platinum in the anode catalyst greatly enhances the electrical performance of the DEFC by increas-

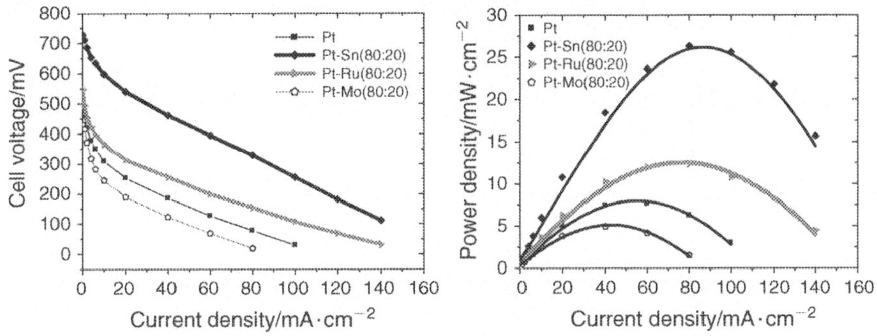

Figure 21.10 Fuel cell characteristics of a 5 cm^2 DEFC recorded at 110°C. Influence of the nature of the bimetallic catalysts (80:20 atomic ratio with 30% loading). Anode catalyst: 1.5 mg cm^{-2}; cathode catalyst: 2 mg cm^{-2} (40% Pt/XC72 from E-TEK); membrane: Nafion® 117; ethanol concentration: 1 M.

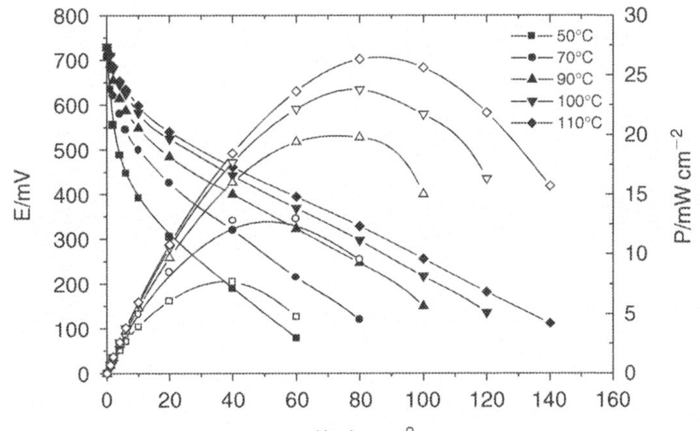

Figure 21.11 Fuel cell characteristics of a 25 cm^2 DEFC recorded with a 30% Pt–Sn (90:10) catalyst. Influence of the working temperature. Anode catalyst: 1.5 mg cm^{-2} (30% Pt–Sn (90:10)/XC72); cathode catalyst: 2 mg cm^{-2} (40% Pt/XC72 from E-TEK); membrane: Nafion® 117; ethanol concentration: 1 M.

ing the OCV up to 0.75 V, which indicates that the modified catalysts are less poisoned by adsorbed species coming from ethanol than the Pt/C catalyst. For Pt–Sn (80:20) current densities up to 150 mA cm^{-2} are reached at 110°C, giving a maximum power density greater than 25 mW cm^{-2}, which means electrical performances four times higher than those obtained with Pt/C. The increase in the electrical performance indicates that the bimetallic catalyst is more active for ethanol electro-oxidation than the Pt/C catalyst.

With the best electrocatalyst, that is Pt–Sn (90:10)/XC72, the effect of temperature on the cell voltage E and power density P versus current density j characteristics is shown in Figure 21.11. It appears clearly that increasing the temperature greatly increases the

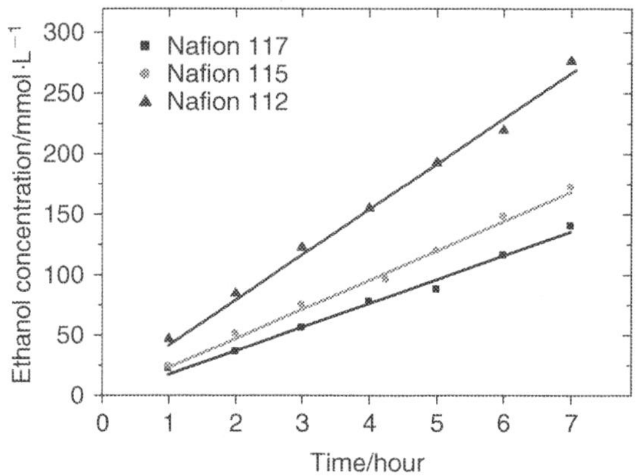

Figure 21.12 Behavior of different Nafion® membranes. Ethanol permeability measurements of different Nafion® membranes (T = 25°C).

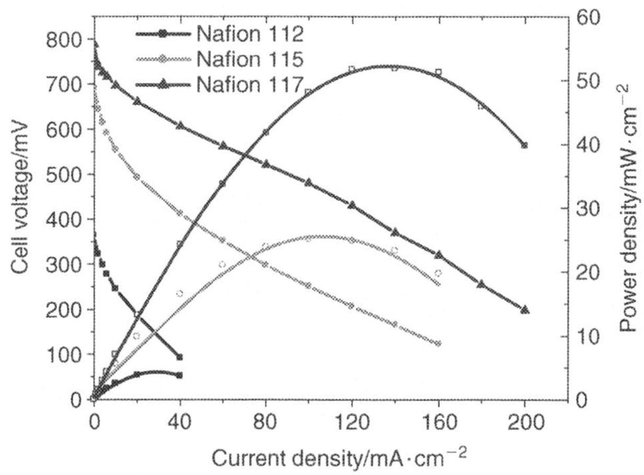

Figure 21.13 Fuel cell characteristics of a DEFC recorded at 90°C with a 60% Pt–Sn (90:10)/XC72 catalyst for different Nafion® membranes.

performance of the cell, from a maximum power density close to 5 mW cm^{-2} at 50°C to 25 mW cm^{-2} at 110°C, that is almost five times higher.

This confirms the difficulty of oxidizing ethanol at low temperatures and the necessity to work at temperatures higher than 100°C to enhance the electrode kinetics and, thus, the performance of the DEFC.

Finally the performance of a DEFC with an anode containing a higher amount of platinum (i.e. 60 wt.% instead of 30 wt.% for the previous experiments) was determined with three Nafion® membranes, N117, N115, N112 of different thickness (180, 125, 50 μm, respectively) to investigate the effect of ethanol crossover through the membrane. Figure 21.12 shows the ethanol crossover as measured by following the ethanol concentration in a second cell compartment, containing initially no ethanol, separated by the Nafion® membrane from the first one containing 1 M ethanol. The crossover rate through N117 is about twice lower than that through N112. The better behavior of N117 is confirmed in Figure 21.13 showing the comparative electrical characteristics of a DEFC having as electrolyte one of the three Nafion® membranes and an anode with a higher platinum loading (60% Pt–Sn (90:10)/XC72): the DEFC with N117 displays the highest OCV (0.8 V) and leads to a power density of 52 mW cm^{-2} at 90°C.

21.5 CONCLUSIONS

Low-temperature fuel cells (PEMFC and DEFC) can operate with biofuels from biomass resources (energy crops, lignocellulosic feedstock, agricultural and industrial wastes, etc.). For PEMFC, biomass feedstocks or fermentation products (bioethanol, biogas) have to be processed to obtain hydrogen of sufficient purity (CO content lower than a few ppm). This involves several physico-chemical processes, including gasification and pyrolysis, water steam reforming, partial oxidation or ATR. These processes lead to "syngas", a mixture of H_2 and CO, in a ratio depending on the origin of the biomass resources and on the process involved. Further treatments are necessary to lower the CO content; for example WGSR followed by hydrogen upgrading using PSA, preferential oxidation, methanation or membrane separation. Bioethanol can feed directly a DEFC, which is a very promising power source for portable electronics, particularly for cell phones.

REFERENCES

Appleby, A.J. (1983) Electrocatalysis. In *Comprehensive Treatise of Electrochemistry* (eds. B.E. Conway, J.O'M. Bockris, E. Yeager, S.U.M. Khan and R.E. White), Vol. 7, Chap. 4, pp. 173–239, Plenum Press, New York.

Appleby, A.J and Foulkes, F.R. (1989) *Fuel Cell Handbook*, pp. 203–500, Van Nostrand Reinhold, New York.

Ballerini, D., Casanave, D., Forestiere, A., Lacombe, S. and Montagne, X. (2006a) L'éthanol et l'ETBE. In: *Les Biocarburants* (eds. D. Ballerini and N. Alazard-Toux), Chap. 2, pp. 124–125, Editions Technip, Paris.

Ballerini, D., Girard, Ph., His, S., Marty, E., Montagne, X., Napoli, A. and Seiler, J.-M. (2006b) La transformation de la biomasse lignocellulosique par voie thermochimique. In: *Les Biocarburants* (eds. D. Ballerini and N. Alazard-Toux), Chap. 5, pp. 253–255, Editions Technip, Paris.

Bockris, J.O'M. and Reddy, A.K.N. (1972) *Modern Electrochemistry*, Vol. 2, p. 1141, Plenum Press, New York.

Dubau, L., Coutanceau, C., Garnier, E., Léger, J.-M. and Lamy, C. (2003) Electrooxidation of methanol at platinum–ruthenium catalysts prepared from colloidal precursors: atomic composition and temperature effects. *J. Appl. Electrochem.* **33**, 419–429.

Gosselink, J.W. (2002) Pathways to a more sustainable production of energy: sustainable hydrogen – a research objective for Shell. *Int. J. Hydrogen Energy* **27**, 1125–1129.

Grove, W.R. (1839a) *Phil. Mag.* **14**, 127.

Grove, W.R. (1839b) *Phil. Mag.* **15**, 287.

Haagensen, F., Torry-Smith, M. and Ahring, B.K. (2005) Fermentation of biomass to alcohols. In: *Biofuels for Fuel Cells* (eds. P. Lens, P. Westermann, M. Haberbauer and A. Moreno), Chap. 10, pp. 169–193, IWA Publishing, London.

Kordesch, K. and Simader, G. (1996) *Fuel Cells and their Applications*, VCH, Weinheim.

Lamy, C. and Léger, J.-M. (1994) Les piles à combustible – Application au véhicule électrique. *J. Phys.* **IV 4**, C1-253–281.

Lamy, C., Léger, J.-M. and Srinivasan, S. (2000) Direct methanol fuel cells – from a 20th century electrochemists' dream to a 21st century emerging technology. In: *Modern Aspects of Electrochemistry* (eds. J.O'M. Bockris and B.E. Conway), Vol. 34, Chap. 3, pp. 53–117, Plenum Press, New York.

Lamy, C., Belgsir, E.M. and Léger, J.-M. (2001) Electrocatalytic oxidation of aliphatic alcohols – application to the direct alcohol fuel cells (DAFC). *J. Appl. Electrochem.* **31**, 799–809.

Lamy, C., Lima, A., Le Rhun, V., Delime, F., Coutanceau, C. and Léger, J.-M. (2002) Recent advances in the development of direct alcohol fuel cells (DAFC). *J. Power Sources* **105**, 283–296.

Lamy, C., Kauffmann, J.M. and Grenier, J.C. (2003) Les piles à Combustible et leur gestion. *Livre Blanc "Programme ENERGIE du CNRS"*, pp. 21–41, Ministère Délégué à la Recherche et aux Nouvelles Technologies, Paris.

Lee, S.J., Mukerjee, S., Ticianelli, E.A. and McBreen, J. (1999) Electrocatalysis of CO tolerance in hydrogen oxidation reaction in PEM fuel cells *Electrochim. Acta* **44**, 3283–3293.

Lens, P., Westermann, P., Haberbauer, M. and Moreno, A. (2005) *Biofuels for Fuel Cells*, IWA Publishing, London.

Lide, D.R. (1992) *Handbook of Chemistry and Physics*, 73rd edn, CRC Press, Boca Raton, CA.

Meulepas, R.J.W., Nordberg, Å., Mata-Alvarez, J. and Lens, P.N.L. (2005) Methane production from wastewater, solid waste and biomass. In: *Biofuels for Fuel Cells* (eds. P. Lens, P. Westermann, M. Haberbauer and A. Moreno), Chap. 7, pp. 121–138, IWA Publishing, London.

Olsbye, U., Wurzel, T. and Mleczko, L. (1997) *Industr. Engineer. Chem. Research* **36**, 510.

Ralph, T.R. and Hogarth, M.P. (2002) Catalysis for low temperature fuel cells. Part I: The cathode challenges. *Plat. Met. Rev.* **46**, 3–14.

Riedel, T., Schulz, H., Schaub, G., Jun, K.-W. and Lee, K.-W. (2003) Fischer–Tropsch on iron with H_2/CO and H_2/CO_2 as synthesis gases: the episodes of formation of the Fischer–Tropsch regime and construction of the catalyst. *Top. Catal.* **26**, 41–54.

Rohde, M., Unruh, D., Pias, P., Lee, K.-W. and Schaub, G. (2004) Fischer–Tropsch synthesis with CO_2-containing syngas from biomass: kinetic analysis of fixed bed reactor model experiments. *Studies Surf. Sci. Catal.* **153**, 97–102.

Rousseau, S., Coutanceau, C., Lamy C. and Léger J.-M. (2006) Direct ethanol fuel cell: electrical performances and reaction products distribution under operating conditions with different platinum-based anodes. *J. Power Sources* **158**, 18–24.

Sakellaropoulos, G.P. (1981). In: *Advances in Catalysis* (eds. D.D. Eley, H. Pines and P.B. Weisz), p. 218, Academic Press, New York.

Schulte-Schulze Berndt, A. (2005) Biogas upgrading with pressure swing adsorption versus biogas reforming. In *Biofuels for Fuel Cells* (eds. P. Lens, P. Westermann, M. Haberbauer and A. Moreno), Chap. 22, pp. 414–429, IWA Publishing, London.

Schulz, H. (2003) Major and minor reactions in Fischer–Tropsch synthesis on cobalt catalysts. *Top. Catal.* **26**, 73–85.

Stevens, P., Novel-Catin, F., Hammou, A., Lamy C. and Cassir, M. (2000) Piles à combustible. In *Traité Génie Electrique* (ed. R. Bonnefille), Vol. D3 340, pp. 1–28, Techniques de l'Ingénieur, Paris.

Ströbel, R., Oszcipok, M., Fasil, M., Rohland, B., Jörissen, L. and Garche, J. (2002) The compression of hydrogen in an electrochemical cell based on a PE fuel cell design. *J. Power Sources* **105**, 208–215.

Tarasevich, M.R., Sadkowski, A. and Yeager, E. (1983) Oxygen electrochemistry. In *Comprehensive Treatise of Electrochemistry* (eds. B.E. Conway, J.O'M. Bockris, E. Yeager, S.U.M. Khan and R.E. White), Vol. 7, Chap. 6, pp. 301–398, Plenum Press, New York.

Tilak, B.V., Yeo, R.S. and Srinivasan, S. (1981) Electrochemical energy conversion-principles. In *Comprehensive Treatise of Electrochemistry* (eds. J.O'M. Bockris, B.E. Conway, E. Yeager and R.E. White), Vol. 3, pp. 39–122, Plenum Press, New York.

Trogisch, S. and Baaske, W.E. (2004) *Biogas Powered Fuel Cells*. Trauner Verlag, Linz.

Vielstich, W., Lamm, A. and Geisteiger, H. (2003) *Fuel Cells: Fundamental, Technology and Applications*. Wiley, Chichester.

22

Resource recovery from synthesis gas

J. Sipma, A.J.M. Stams and P.N.L. Lens

22.1 INTRODUCTION

22.1.1 Synthesis gas

Synthesis gas, a mixture of H_2, carbon monoxide (CO) and carbon dioxide (CO_2), is produced by thermal gasification of a wide range of organic matter, including fossil fuels and biomass. Synthesis gas is the main source for bulk H_2 production. Large-scale cost effective H_2 production is desired not merely for its use as a future energy carrier, but also because it is an important chemical reactant in reductive processes in both the chemical industry as well as in specific wastewater treatment processes. For the latter purpose, especially the use of H_2 as electron donor in biological sulfate reduction of sulfate/sulfite-rich wastewaters which are low in organic matter content has been shown successful both at moderate (30°C) and elevated (55°C) temperatures (Van Houten et al., 1994, 1997). In contrast to chemical production processes, wastewater treatment generally does not result in the production of valuable compounds. Therefore, electron donors should be cheap. Synthesis gas as electron donor has been proposed as a cheap alternative for pure H_2 gas in biotechnological sulfate reduction processes (Du Preez & Maree, 1994; Du Preez et al., 1992; Van Houten & Lettinga, 1996). However, the presence of CO limits its direct application, because of its toxic and poisoning effects, and extensive purification of the H_2 may be required prior to its utilization.

22.1.2 The chemical water–gas shift reaction

Synthesis gas is nowadays purified in so-called water–gas shift reactors, in which the CO present is reacted with water over a metal catalyst yielding CO_2 and H_2. This reaction is known as the water–gas shift reaction:

$$CO + H_2O \longrightarrow H_2 + CO_2 \qquad \Delta G^{o} = -20\,\text{kJ}\,\text{mol}^{-1}$$

The water–gas shift reaction is predominantly carried out in a two-stage catalytic process. In the first stage, also known as high-temperature shift conversion, which proceeds between 315°C and 510°C using a reduced iron catalyst, exit CO concentrations ranging from 2% to 5% can be obtained. Water to CO ratios of 2–4 are used to force the equilibrium toward hydrogen production (Armor, 1999; Bartish & Drissel, 1978). The second stage requires a more active zinc oxide–copper oxide catalyst, which enables operation at lower, thermodynamically more favorable, temperatures. The resulting enriched H_2 gas can have a CO content as low as 0.2%. The residual CO generally is hydrogenated to methane using a reduced nickel catalyst after removal of bulk CO_2 (Bartish & Drissel, 1978).

Catalytic processes are generally easily poisoned by hydrogen sulfide (Bredwell et al., 1999). Therefore, a biological alternative, tolerating higher hydrogen sulfide concentrations and operating under more moderate conditions (pressure, temperature) will become undoubtedly of interest for all industries, which produce or use synthesis gas.

22.1.3 The biological water–gas shift reaction

The discovery of fast-growing anaerobic thermophilic bacteria capable of performing the same CO oxidation with liberation of H_2 (as reviewed in Sipma et al., 2006a), presents an interesting alternative for the above-mentioned chemical catalytic process. These bacteria have been designated as *hydrogenogens* due to their capacity to metabolically convert CO coinciding with the formation of H_2 (Svetlichnyi et al., 2001), analog to the use of the term methanogens for methane producing microorganisms. Furthermore, the terms *hydrogenogenic* and *hydrogenogenesis* were proposed to refer to this type of metabolism (Svetlichnyi et al., 2001).

This chapter overviews the studies concerning the occurrence of hydrogenogenic CO conversion, its microbiological aspects and practical aspects of biotechnological H_2 purification, as well as synthesis gas utilization in biodesulfurization.

22.2 HYDROGENOGENIC CO CONVERSION IN ANAEROBIC GRANULAR SLUDGE

22.2.1 Occurrence of hydrogenogenic CO-converting bacteria in sludges from anaerobic wastewater treatment plants

Hydrogen production from CO has only been described so far for a few mesophilic bacteria, that is some phototrophic organisms (Dashekvicz & Uffen, 1979; Jung et al., 1999a; Kerby et al., 1995; Uffen, 1976) and a facultative anaerobic bacterium (Jung et al., 1999b). All these organisms have in common that their growth under anaerobic conditions in the dark is rather slow. Sipma et al. (2003) screened the sludges from several full-scale anaerobic bioreactors, operated at 30–35°C and treating different types of wastewater for their CO conversion capacity. All sludge types converted CO to methane and/or acetate when incubated at 30°C. Inhibition experiments, using 2-bromoethanesulfonate (BES) and vancomycine, showed that CO conversion to methane at 30°C occurred via acetate, not via H_2. The occurrence of acetogenic CO conversion at 30°C is not surprising regarding the reported number of mesophilic homoacetogens capable of CO metabolism (Sipma et al., 2006a).

The incubations at 55°C revealed a fast evolution of H_2 upon CO conversion with five out of seven sludge samples from anaerobic bioreactors. In case methane was generated in CO supplemented incubations, inhibition experiments revealed that CO was converted to H_2, but that direct methane formation from CO did not occur in any incubation conducted at an initial partial CO pressure (P_{CO}) exceeding 160 kPa (Sipma et al., 2003). Four of the H_2 producing sludges were growing at 30–35°C, but the lag phase for hydrogenogenic CO conversion at 55°C was very short (<4 days). This suggests short generation times of the responsible microorganisms. Hydrogenogenic CO conversion at high P_{CO} appears thus to be a rather common phenomenon, despite all anaerobic thermophilic hydrogenogenic microorganisms isolated so far originated from locations with volcanic activity (Sipma et al., 2006a). According to Symonds et al. (1994) such locations contain up to 1.5% CO. On the other hand, the enzymes required for hydrogenogenic CO conversion, that is carbon monoxide dehydrogenase (CODH) and hydrogenases, are widespread in anaerobic microorganisms (Sipma et al., 2006a). The key enzyme for CO metabolism, CODH, not only functions in the CO oxidation, but it is also one of the key enzymes in the synthesis or cleavage of acetyl-CoA (Ferry, 1995).

22.2.2 Effect of P_{CO} and H_2 on CO conversion routes at 55°C

The P_{CO} may influence the product formation and CO conversion due to toxic effects of CO. CO acts as an inhibitor of some types of hydrogenases (Daniels et al., 1977; Davidova et al., 1994; Kim et al., 1984; Lupton et al., 1984) and some sulfate-reducing bacteria growing on organic substrates as well (Davidova et al., 1994). Sipma et al. (2004) investigated the effect of the initial P_{CO} and the presence of H_2 on the CO

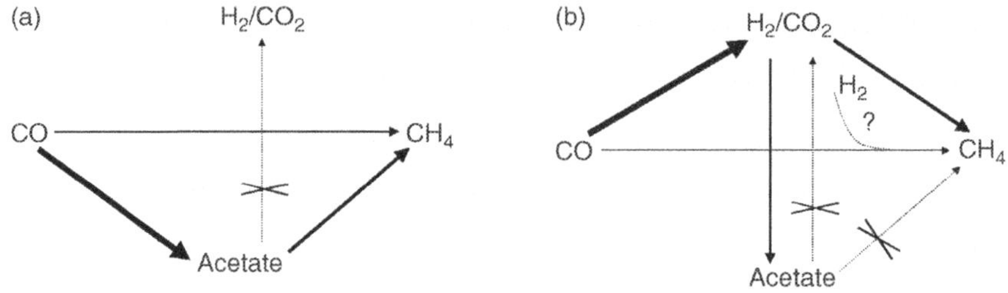

Figure 22.1 Schematic presentation of the main CO conversion routes by anaerobic full-scale wastewater treating sludge samples at (a) 30°C and (b) 55°C. The dotted lines with cross represent conversion routes that were absent.

Table 22.1 Stoichiometry and standard values of Gibbs free energy changes of reactions involved in the anaerobic conversion of CO. Gibbs free energy changes were calculated using the data set compiled by Amend and Shock (2001).

Reactions	$\Delta G^{o\prime}$ (25°C) (kJ mol^{-1})	$\Delta G'_{55}$ (55°C) (kJ mol^{-1})
(1) $CO + H_2O \rightarrow CO_2 + H_2$	−20.0	−22.3
(2) $4CO + 2H_2O \rightarrow 3CO_2 + CH_4$	−210.8	−207.5
(3) $4CO + 4H_2O \rightarrow CH_3COO^- + 2HCO_3^- + 3H^+$	−165.7	−145.5
(4) $2HCO_3^- + 4H_2 + H^+ \rightarrow CH_3COO^- + 4H_2O$	−104.2	−92.7
(5) $HCO_3^- + 4H_2 + H^+ \rightarrow CH_4 + 3H_2O$	−135.4	−127.5
(6) $CH_3COO^- + H_2O \rightarrow CH_4 + HCO_3^-$	−31.2	−34.9

conversion routes at 55°C. From the hydrogenogenic CO-converting sludge samples, two were selected for a more detailed study, Eerbeek (paper mill) and Nedalco (distillery) sludge. The main difference between these sludge samples was the absence of methane formation with Nedalco sludge under an atmosphere of 150 kPa CO. CO conversion routes were not dependent on the supplied amount of CO, although with Eerbeek sludge at P_{CO} values up to 50 kPa some methane formation occurred directly from CO in the presence of vancomycine. Nevertheless, the rate of this methanogenesis from CO was so low that its contribution to the overall CO conversion pathway at 55°C (Figure 22.1) was negligible. The absence of direct CO conversion at high P_{CO} (>150 kPa) is attributed to a toxic effect of CO for these methanogens, which most likely also prevented the production of methane in Nedalco sludge. Nedalco sludge rapidly converted H_2/CO_2 into methane, but traces of CO, left after CO conversion was completed, were apparently sufficient to inhibit methanogenesis completely. The possible route of direct methane production from the reduction of CO with H_2 (question mark in Figure 22.1), could not be distinguished from a direct methanogenic CO conversion route from CO alone, but in view of the very similar rates, the presence of H_2 does not stimulate methane production. The main reactions involved in CO conversion are shown in Figure 22.1 and reactions are tabulated in Table 22.1.

The energy yield of hydrogenogenic CO conversion is relatively low and as a result of H_2 and CO_2 production, the Gibbs free energy change declines along with the

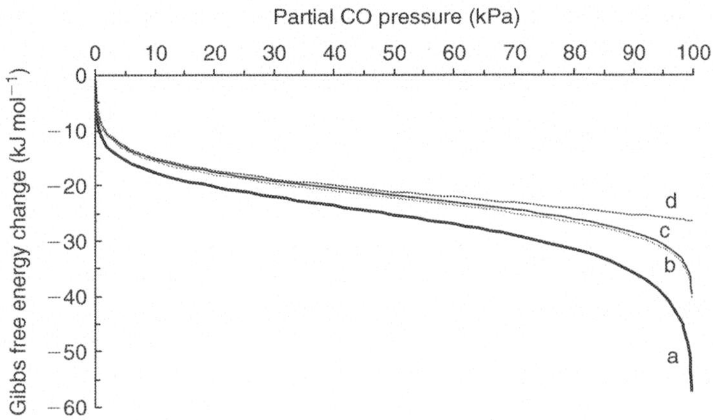

Figure 22.2 Theoretical evolution of the Gibbs free energy change at 55°C during the conversion of 100 kPa CO to H_2 and CO_2, and the effect of the initial presence of H_2 and/or CO_2. The different hypothetical gas compositions are: (a) only CO, (b) CO + 100 kPa H_2, (c) CO + 100 kPa CO_2 and (d) CO + 50 kPa H_2 and 50 kPa CO_2. Calculations performed for batch incubations (gas volume of 70 mL; liquid volume of 50 mL).

conversion of CO. Furthermore, the initial presence of large amounts of H_2 as is typical for synthesis gas decreases the energy yield of CO conversion further. This is illustrated in Figure 22.2, which presents theoretical Gibbs free energy changes of hydrogenogenic CO conversion (reaction (1) in Table 22.1) and the effect of the initial presence of H_2 and/or CO_2.

The slightly higher decrease of ΔG in the presence of CO_2 is due to the higher solubility of CO_2. Results of experiments with Eerbeek sludge revealed that the initial presence of H_2 decreased the CO conversion rates (Sipma et al., 2004), but it did not prevent the depletion of CO to undetectable levels (about 400 ppm for the gas chromatograph used). The observed differences in CO conversion rates can be attributed to the initial differences in ΔG values, which likely resulted in higher biomass yields and consequently higher conversion rates. The thermodynamic estimations suggest that the apparent minimum free energy requirement would be even smaller than $-7\,\mathrm{kJ\,mol\,CO^{-1}}$, as this value was estimated assuming a final CO concentration of 400 ppm (Sipma et al., 2004).

22.3 APPLICATION OF SYNTHESIS GAS IN THERMOPHILIC BIODESULFURIZATION

22.3.1 Microbiology

In general, sulfate-reducing bacteria are found to be highly sensitive for CO toxicity (Davidova et al., 1994; Mörsdorf et al., 1992). Despite the capability of various sulfate-reducing bacteria to convert CO at concentrations up to 20 kPa, higher concentrations were found to inhibit growth completely (Davidova et al., 1994; Klemps et al., 1985; Lupton et al., 1984; Mörsdorf et al., 1992). Recently was found that the organisms *Desulfotomaculum kuznetsovii* and *Desulfotomaculum thermobenzoicum* spp.

thermosyntrophicum can use CO as sole carbon and energy source for sulfate reduction up to 50 kPa CO in the gas phase (Parshina et al., 2005a). Nevertheless, incubations with Eerbeek sludge were capable of sulfate reduction when incubated at $P_{CO} > 160$ kPa (Sipma et al., 2004), suggesting that CO-rich synthesis gas could be used as electron donor for biological sulfate reduction.

A moderately thermophilic sulfate reducer ($T_{opt} = 55°C$), strain CO-1-SRB, could be isolated from Eerbeek sludge (Parshina et al., 2005b). The unique property of this isolate is that CO serves as the sole energy and carbon source, both in the presence and absence of sulfate. CO was converted to H_2 and CO_2, and in the presence of sulfate, the H_2 formed was used for sulfate reduction. Phylogenetic analysis of the 16S rRNA gene sequence placed CO-1-SRB to the genus *Desulfotomaculum*, closely resembling *Desulfotomaculum nigrificans* DSM 574^T and *Desulfotomaculum* sp. RHT-3 (Mori et al., 2000), that is 99% and 100% similarity respectively. Nevertheless, the latter strains were completely inhibited at high levels of CO and only metabolized CO in the presence of sulfate, whereas strain CO-1-SRB grew well at 200 kPa CO in the headspace. Based on phylogenetic and physiological features strain CO-1-SRB was described as a new species, that is *Desulfotomaculum carboxydivorans* (Parshina et al., 2005b).

As a result of its unique properties to convert CO to H_2 and in the presence of sulfate to reduce sulfate hydrogenotrophically, *D. carboxydivorans* may be applicable in both synthesis gas purification and biodesulfurization at elevated temperatures (e.g. flue gas desulfurization). *D. carboxydivorans* grows rapidly at 200 kPa CO, pH 7.0 and 55°C, with a generation time of 100 min, producing nearly equimolar amounts of H_2 and CO_2 from CO (Sipma et al., 2006d). At the start of batch incubations extremely high specific CO conversion rates, exceeding 0.8 mol CO $(\text{g protein})^{-1}\text{h}^{-1}$, were observed, which rapidly decreased as a result of increased biomass concentrations. This indicates that CO mass transfer from the gas to the liquid phase limits the CO conversion rate and not the activity of the biomass. The biomass yield of *D. carboxydivorans* amounted to 0.42 mg protein produced per mmol CO converted (Sipma et al., 2006d). As *D. carboxydivorans* is capable to grow on various substrates, the fast growth rates with high biomass yield on glucose may be interesting for use as co-substrate to stimulate biomass production (Sipma et al., 2006d). In the presence of sulfate, the biomass yield is about four times that of growth on CO alone (Sipma et al., 2006d).

22.3.2 H_2S inhibition

In the presence of sulfate, sulfide production occurs at a P_{CO} exceeding 100 kPa with a maximal specific sulfate reduction rate of 32 mmol $(\text{g protein})^{-1}\text{h}^{-1}$. Such a sulfate reduction rate is promising for biodesulfurization processes at elevated temperatures, although the culture's tolerance for sulfide is rather low and pH dependent, that is a maximum of 9 and 5 mM sulfide were produced at pH 7.2 and 6.5, respectively (Sipma et al., 2006d). The growth of *D. carboxydivorans* with sulfate, thiosulfate and sulfite as electron acceptor (Parshina et al., 2005b), is interesting in view of the treatment of flue gas scrubbing water as the major part of SO_x forms sulfite upon dissolving in an alkaline solution (Janssen et al., 2000).

Employing *D. carboxydivorans* for both hydrogenogenic CO conversion and sulfate reduction in a bioreactor might require special measures to maintain sufficient low sulfide concentrations due to its sensitivity to sulfide (Sipma et al., 2006d). These measures may include capture of H_2S from the gas recycle or application of a low HRT. A sulfide concentration of 10 mM, which inhibits growth at a pH of about 7.4, would require a HRT as low as 2.7 h considering a sulfate reduction rate of 90 mmol L^{-1} day^{-1}. Reactor operation at very low HRT (<3 h) is feasible and promotes stable sulfate reduction (Sipma et al., 2006c). Despite the high CO conversion efficiencies (>95%) obtained at this HRT and most of the CO-derived H_2 being used for sulfate reduction (>87%), methanogens remained present and upon increasing the HRT, methanogenic activity rapidly increased at the expense of sulfate reduction. The maximal sulfate reduction rate achieved was about 17 mmol L^{-1} day^{-1}, and the rate was limited by the amount of CO supplied and the CO conversion efficiency (about 85%) at higher CO loads (106 mmol L^{-1} day^{-1}), probably as a result of limited biomass retention in the reactor.

22.3.3 Competition between sulfate reduction and methanogens

Thermophilic sulfate reduction with CO as electron donor has been well established in a gas-lift reactor inoculated with Eerbeek granular sludge (Sipma et al., 2006b). Despite the high CO conversion capacity of the biomass present, the sulfate reduction capacity was rather limited due to strong competition between methanogens and sulfate reducers for the CO-derived H_2. Although, the methanogens appeared to be more sensitive to pH and temperature shocks imposed to the reactor, they could not be completely eliminated. As a result of the high growth rates of the methanogens present (generation time 4.5 h), their recovery from shocks proceeded fast and they readily consumed over 90% of the CO-derived H_2. Thus, suppressing methanogens is of utmost importance when a sulfate-reducing system is desired (Sipma et al., 2006b).

Although CO is a direct methanogenic substrate, methanogens present in granular sludge are quite sensitive to CO, as indicated by the absence of methane production by crushed granular sludge (Sipma et al., 2003). However, this suppression of methanogenesis is presumably only a short-term effect of CO toxicity as reactor sludges produced methane from CO despite that the sludge was finely dispersed (Sipma et al., 2006b). Adaptation of methanogens to increasing levels of CO has indeed been reported previously (O'Brien et al., 1984). Furthermore, investigations in the utilization of CO-derived H_2 revealed that the methanogens consume H_2 at higher rates than the sulfate reducers and are further capable of growth at more extreme conditions of pH and salinity than the sulfate reducers present, presumably *D. carboxydivorans* (Sipma et al., 2006b). Complete elimination of methanogenesis from sludge could be obtained by treating the sludge for a short term at high temperatures (90 min at 95°C; Sipma et al., 2004) as found in batch experiments. With this procedure living vegetative cells are killed, but spores remain almost unaffected. Elimination of methanogenesis has been achieved successfully by heat treatment (Oh et al., 2003) or acid–base treatment (Chen et al., 2002). However, heat pretreatment of the sludge alone was not sufficient to prevent methanogenesis in gas-lift reactors (Sipma et al., 2006b), due to the presence of a small number of methanogens in the reactor and their fast generation times. Nevertheless, we consider the application of such a heat treatment as quite promising for practice, especially, since such a heat treatment does not pose any environmental risk, that is potentially harmful chemicals are not released. Moreover, a heat treatment in principle needs to be performed only once. In contrast, chemical inhibitors, such as BES, acetylene (Oremland & Capone, 1988) and chloroform (Chidthaisong & Conrad, 2000; Scholten et al., 2000) represent a potential risk when they are released into the environment. Furthermore, adaptation might result in increased operational costs due to an increasing demand of chemical inhibitors to maintain the same level of inhibition. Weijma et al. (2002) applied a continuous dosing of $2\,g\,L^{-1}$ BES to a bioreactor. This was found to be ineffective for suppressing methanogenesis as complete inhibition of methanogenesis was obtained for 2 days only. In contrast, Sparling et al. (1997) successfully inhibited methane formation in a H_2 producing solid substrate anaerobic digester by the addition of 1% v/v acetylene in the headspace.

22.3.4 Biomass retention

CO is a good electron donor for sulfate reduction at elevated temperatures (55°C), although special attention has to be paid to biomass retention as well as elimination of methanogens. Biomass retention using carrier material as, for example, employed by Van Houten et al. (1997) ensures the need for a complete elimination or suppression of methanogens. In view of the need for biomass retention, the use of a submerged anaerobic membrane bioreactor (SAMBaR) represents an interesting option as shown by Vallero et al. (2005) for sulfate reduction at high salinity using a pure culture of *Desulfobacter halotolerans*. The short generation time of *D. carboxydivorans* on CO at 55°C ($t_d = 1.7\,h$; Sipma et al., 2006d) compared to that of *Desulfobacter halotolerans* ($t_d = 40\,h$) suggests that the start-up of a thermophilic CO fed sulfate-reducing submerged membrane bioreactor may be fast. In order to increase the biomass concentration of *D. carboxydivorans*, the supply of some glucose during the start-up phase could be considered (Sipma et al., 2006d). Assuming that *D. carboxydivorans* is the most dominant sulfate reducer in the biomass of the reactor presented by Sipma et al. (2006c), the estimated specific sulfate reduction rates would be between 0.7 and 1.0 mol $(g\,protein)^{-1}\,day^{-1}$. This corresponds well to the specific sulfate reduction rates of *D. carboxydivorans* (0.8 mol $(g\,protein)^{-1}\,day^{-1}$; Sipma et al., 2006d). Therefore, the application of biomass retention membranes in a CO or synthesis gas fed sulfate-reducing bioreactor has potential using either a pure culture of *D. carboxydivorans* or a heat-treated sample of Eerbeek sludge.

22.4 DESIGN OF A BIOLOGICAL SYNTHESIS GAS PURIFICATION PROCESS

22.4.1 Product specification

One of the major points in the design of a biological synthesis gas purification process is the required product specification, that is especially the permitted levels of CO. The biological achievable minimal CO concentrations with a pure culture of *D. carboxydivorans* were found to be determined by the thermodynamics rather than by effects of pH or CO affinity of the bacterium (Sipma et al., 2006d). Thermodynamic calculations performed with the results obtained in Sipma et al. (2004) already indicate that the minimal ΔG for CO conversion is $<-7\,\text{kJ}\,(\text{mol CO})^{-1}$, as at this value the CO concentration is below the detection limit of 400 ppm for the gas chromatograph used in that experiment. This is lower than the values obtained from chemostat and field studies, in which catabolic activity was observed when the associated free energy change was at least $-10\,\text{kJ}\,\text{mol}^{-1}$ catabolic reaction (Seitz et al., 1990; Westermann, 1994). Nevertheless, CO conversion by *D. carboxydivorans* proceeds as long as $\Delta G < 0$, which suggests a very efficient coupling of energy conservation to CO conversion. The results presented in Sipma et al. (2006d) show that the assessed minimal CO concentration could be even as low as 4–5 ppm, depending on the initial gas composition and P_{CO}. Thermodynamic calculations, assuming that CO conversion stops at $\Delta G = 0$, were compared to actual measurements of CO concentration, showing a good correlation between them (Sipma et al., 2006d). The initial composition, that is the presence of H_2 and CO_2 as is typical of synthesis gas, affects the minimal achievable CO concentrations (Sipma et al., 2006d). This is illustrated by the minimal CO concentration for synthesis gas derived from coal, that is 120 ppm at a total gas pressure of 101 kPa, whereas at 1 MPa the minimal CO concentration cannot be decreased to <1% (Sipma et al., 2006d).

Low exit CO concentrations imply that the CO concentrations in the gas phase of the reactor are low as well, which negatively affects the mass transfer of CO from the gas to the liquid phase. This would result in the need to use voluminous reactors. Operation of a synthesis gas purification reactor at high P_{CO} would greatly enhance the mass transfer, but results in higher exit CO concentrations. From the growth curve of *D. carboxydivorans* (Figure 22.3), it is evident that the CO conversion activity of the biomass is much higher than the CO flux to the liquid phase as the maximal CO conversion rates of $0.8\,\text{mol CO}\,(\text{g protein})^{-1}\,\text{h}^{-1}$ were obtained at the start of the incubations when the biomass concentration was low.

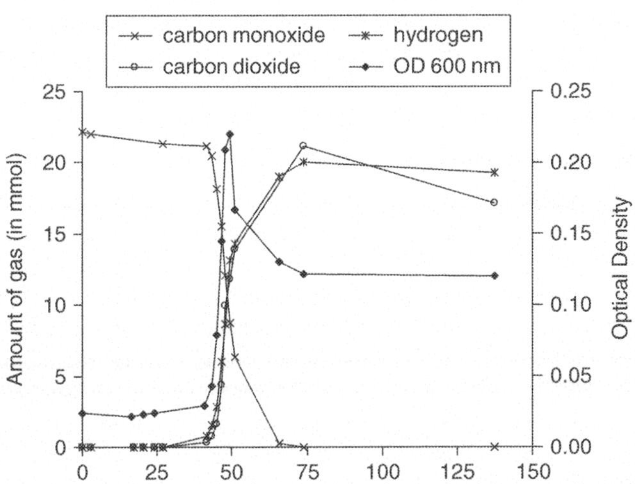

Figure 22.3 Growth curve of *D. carboxydivorans* grown on 160 kPa CO in the absence of sulfate at 55°C.

22.4.2 Product gas purification

Synthesis gas purification at high P_{CO} requires a gas separation technology to provide a high-purity H_2 gas or the selective removal of one of the products to maintain $\Delta G < 0$. The latter could be achieved by physical absorption of CO_2, for example by hot potassium carbonate or other processes using amine, selexol, sulfinol or rectisol solutions (Czuppon et al., 1995; Chapter 18). This offers the possibility for selective CO_2 sequestration and geological storage (Gale, 2004), thus preventing its emission to the atmosphere. Liberation of high-purity CO_2 could be even useful for reuse in chemical synthesis of, for example, methanol (Pruschek et al., 1997) or in greenhouse horticulture (Chapter 24).

The combination of separation processes with a biological CO conversion process would enable high-rate CO conversions due to high mass transfer of CO from the gas to the liquid phase at high P_{CO}. Complete CO conversions could be obtained by recycling the unconverted CO back to the reactor after removal of both CO_2 and H_2. The hydrogen present in the product gas coming from the bioreactor can be separated from the other gasses by, for example, a pressure swing adsorption (PSA) process or by selective gas separation membranes (Koros & Mahajan, 2000; Perry et al., 1997). Although, PSA is usually preferred when high-purity gases are required (Koros & Mahajan, 2000), the use of gas separating membranes seem particularly attractive to be used in a synthesis gas purification process. The reason is its simplicity and continuous operation in-line with the bioreactor. The H_2 purity of the product gas with the use of membranes is maximally 99.9% H_2 at recoveries up to 98% (Air Liquide technical specifications). PSA systems operate between 500 and 5000 kPa, and for selective gas membranes pressure up to 8 MPa can be handled (Air Liquide technical specifications). This indicates a further interest for operation of the bioreactor at elevated pressures as synthesis gas is also produced at high pressures (>2 MPa; Czuppon et al., 1995), and alternating depressurizing–pressurizing represents additional operation costs. However, the upper limit of P_{CO} needs to be determined to demonstrate the feasibility of operating a bioreactor for synthesis gas at relatively high pressures. A proposed synthesis gas purification process employing both H_2 separation and selective CO_2 recovery is schematically depicted in Figure 22.4.

22.4.3 Bioreactor design

For the hydrogenogenic CO-converting bioreactor two concepts are particularly interesting, that is biotrickling filters (Plaggemeier & Lämmerzahl, 2000) and monolith biofilm reactors (Ariga et al., 1986). Synthesis gas purification in principle requires only a small liquid stream, which supplies the biomass with nutrients and prevents too high salinity. Furthermore, thin liquid films are beneficial for the mass transfer of gaseous compounds toward the biomass. In both reactor types, the biomass may be growing as a biofilm or the use of trickling cell suspensions over the surface of the carrier material (Lens et al., 2003) could be considered. Bioreactors using cell suspensions, consisting of individual cells or small flocs have lower mass transfer resistances compared to biofilm reactors, as the sole mass transfer barrier in cell suspension bioreactors is transfer of the compounds from the gas to the liquid phase (Nicolella et al., 2000). The theoretical volumetric CO conversion rate in a biotrickling filter, assuming

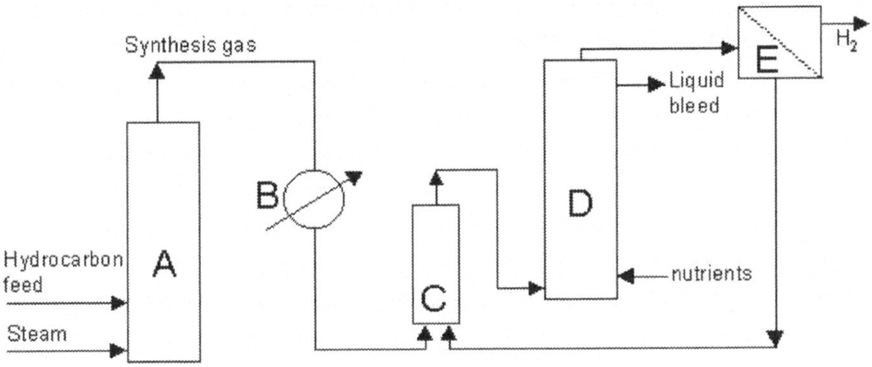

Figure 22.4 Proposed synthesis gas purification process including (A) steam reforming reactor, (B) heat recovery, (C) CO_2 absorption, (D) the biological CO conversion reactor and (E) selective recovery of a high-purity H_2 gas using a gas-selective membrane.

a k_La of $0.015\,s^{-1}$ (Klasson et al., 1992), was estimated at $40\,m^3 m^{-3}$ reactor volume under standard conditions (25°C and 101.325 kPa) with synthesis gas derived from coal gasification at P_{CO} of 200 kPa. Since the CO content of coal-derived synthesis gas is about 50%, it represents a treatment of $80\,m^3 m^{-3}$ reactor volume of synthesis gas.

As biological reactions occur in the water phase, the use of poorly soluble gaseous substrates usually suffers from mass transfer limitations. This is also the case for CO conversion by *D. carboxydivorans* evidenced from the extreme high maximal specific conversion rates (Sipma et al., 2006d). Research into new bioreactor concepts with considerably higher mass transfer rates could facilitate the use of bioreactors as alternative for currently employed chemical catalytic transformations such as CO conversion.

22.5 CONCLUSION

The occurrence of thermophilic microorganisms capable of high-rate hydrogenogenic CO conversion as well as sulfate reduction in the presence of high levels of CO enables the use of CO containing synthesis gas in biotechnological applications, such as synthesis gas purification or the use of CO-rich synthesis gas as electron donor in the treatment of sulfate-rich wastewaters.

REFERENCES

Air Liquide at http://www.airliquide.com/en/business/industry/oil_gas/equipment/

Amend, J.P. and Shock, E.L. (2001) Energetics of overall metabolic reactions of thermophilic and hyperthermophilic archaea and bacteria. *FEMS Microbiol. Rev.* **25**, 175–243.

Ariga, O., Kimura, M., Taya, M. and Kobayashi, T. (1986) Kinetic evaluation and characterization of ceramic honeycomb-monolith bioreactor. *J. Ferment. Technol.* **64**(4), 327–334.

Armor, J.N. (1999) The multiple roles for catalysis in the production of H_2. *Appl. Catal. A* **176**(2), 159–176.

Bartish, C.M. and Drissel, G.M. (1978) Carbon monoxide. In: *Encyclopedia of Chemical Technology* (ed. Kirk-Othmer), 3rd edn, pp. 772–793. John Wiley & Sons, New York.

Chen, C.-C., Lin, C.-Y. and Lin, M.-C. (2002) Acid–base enrichment enhances anaerobic hydrogen production process. *Appl. Microbiol. Biotechnol.* **58**, 224–228.

Chidthaisong, A. and Conrad, R. (2000) Specificity of chloroform, 2-bromoethanesulfonate and fluoroacetate to inhibit methanogenesis and other anaerobic processes in anoxic rice field soil. *Soil Biol. Biochem.* **32**, 977–988.

Czuppon, T.A., Knez, S.A. and Newsome, D.S. (1995) Hydrogen. In: *Encyclopedia of Energy, Technology and the Environment* (eds. A. Bisio and S. Boots), Vol. 3, pp. 1752–1782. John Wiley & Sons, Inc., New York.

Daniels, L., Fuchs, G., Thauer, R.K. and Zeikus, J.G. (1977) Carbon monoxide oxidation by methanogenic bacteria. *J. Bacteriol.* **132**, 118–126.

Dashekvicz, M.P. and Uffen, R.L. (1979) Identification of a carbon monoxide-metabolizing bacterium as a strain of *Rhodopseudomonas gelatinosa* (Molisch) van Niel. *Int. J. Syst. Bacteriol.* **29**, 145–148.

Davidova, M.N., Tarasova, N.B., Mukhitova, F.K. and Karpilova, I.U. (1994) Carbon monoxide in metabolism of anaerobic bacteria. *Can. J. Microbiol.* **40**, 417–425.

Du Preez, L.A. and Maree, J.P. (1994) Pilot-scale biological sulphate and nitrate removal utilizing producer gas as energy source. *Water Sci. Technol.* **30**, 275–285.

Du Preez, L.A., Odendaal, J.P., Maree, J.P. and Ponsonby, M. (1992) Biological removal of sulphate from industrial effluents using producer gas as energy source. *Environ. Technol.* **13**, 875–882.

Ferry, J.G. (1995) CO dehydrogenase. *Annu. Rev. Microbiol.* **49**, 305–333.

Gale, J. (2004) Geological storage of CO_2: What do we know, where are the gaps and what more needs to be done? *Energy* **29**, 1329–1338.

Janssen, A.J.H., Dijkman, H. and Janssen, G. (2000) Novel biological processes for the removal of H_2S and SO_2 from gas streams. In *Environmental Technologies to Treat Sulfur Pollutions; Principles and Engineering* (eds. P.N.L. Lens and L.W. Hulshoff Pol), pp. 265–280. IWA Publishing, London.

Jung, G.Y., Jung, H.O., Kim, J.R., Ahn, Y. and Park, S. (1999a) Isolation and characterization of *Rhodopseudomonas palustris* P4 which utilizes CO with the production of H_2. *Biotechnol. Lett.* **21**, 525–529.

Jung, G.Y., Kim, J.R., Jung, H.O., Park, J.Y. and Park, S. (1999b) A new chemoheterotrophic bacterium catalyzing water–gas shift reaction. *Biotechnol. Lett.* **21**, 869–873.

Kerby, R.L., Ludden, P.W. and Roberts, G.P. (1995) Carbon monoxide-dependent growth of *Rhodospirillum rubrum*. *J. Bacteriol.* **177**, 2241–2244.

Kim, B.H., Bellows, P., Datta, R. and Zeikus, J.G. (1984) Control of carbon and electron flow in *Clostridium acetobutylicum* fermentations: utilization of carbon monoxide to inhibit hydrogen production and to enhance butanol yields. *Appl. Environ. Microbiol.* **48**, 764–770.

Klasson, K.T., Ackerson, M.D., Clausen, E.C. and Gaddy, J.L. (1992) Bioconversion of synthesis gas into liquid or gaseous fuels. *Enzyme Microb. Technol.* **14**, 602–608.

Klemps, R., Cypionka, H., Widdel, F. and Pfennig, N. (1985) Growth with hydrogen, and further physiological characteristics of *Desulfotomaculum* species. *Arch. Microbiol.* **143**, 203–208.

Koros, W.J. and Mahajan, R. (2000) Pushing the limits on possibilities for large scale gas separation: Which strategies? *J. Membrane Sci.* **175**, 181–196.

Lens, P., Gastesi, R. and Lettinga, G. (2003) Use of a cell suspension bioreactor for biological flue gas desulfurisation. *Biodegradation* **14**(3), 229–240.

Lupton, F.S., Conrad, R. and Zeikus, J.G. (1984) CO metabolism of *Desulfovibrio vulgaris* strain Madison: physiological function in the absence or presence of exogenous substrates. *FEMS Microbiol. Lett.* **23**, 263–268.

Mori, K., Hatsu, M., Kimura, R. and Takamizawa, K. (2000) Effect of heavy metals on the growth of a methanogen in pure culture and coculture with a sulfate-reducing bacterium. *J. Biosci. Bioeng.* **90**, 260–265.

Mörsdorf, G., Frunzke, K., Gadkari, D. and Meyer, O. (1992) Microbial growth on carbon monoxide. *Biodegradation* **3**, 61–82.

Nicolella, C., van Loosdrecht, M.C.M. and Heijnen, S.J. (2000) Particle based biofilm reactor technology. *Trend. Biotechnol.* **18**, 312–320.

O'Brien, J.M., Wolkin, R.H., Moench, T.T., Morgan, J.B. and Zeikus, J.G. (1984) Association of hydrogen metabolism with unitrophic or mixotrophic growth of *Methanosarcina barkeri* on carbon monoxide. *J. Bacteriol.* **158**, 373–375.

Oh, S.-E., Ginkel, S. van and Logan, B.E. (2003) The relative effectiveness of pH control and heat treatment for enhancing biohydrogen gas production. *Environ. Sci. Technol.* **37**, 5186–5190.

Oremland, R.S. and Capone, D.G. (1988) Use of "specific" inhibitors in biogeochemistry and microbial ecology. *Adv. Microb. Ecol.* **10**, 285–383.

Parshina, S.N., Kijlstra, S., Henstra, A.M., Sipma, J., Plugge, C.M. and Stams, A.J.M. (2005a) Carbon monoxide conversion by thermophilic sulfate-reducing bacteria in pure culture and in co-culture with *Carboxydothermus hydrogenformans*. *Appl. Microbiol. Biotechnol.* **68**, 390–396.

Parshina, S.N., Sipma, J., Nakashimada, Y., Henstra, A.M., Smidt, H., Lysenko, A.M., Lens, P.N.L., Lettinga, G. and Stams, A.J.M. (2005b) *Desulfotomaculum carboxydovorans* sp.nov., a novel sulfate-reducing bacterium capable of growth at 100% CO. *Int. J. Syst. Evol. Microbiol.* **55**, 2159–2165.

Perry, R.H., Green, D.W. and Maloney, J.O. (1997) *Perry's Chemical Engineers' Handbook*, 7th edn. McGraw-Hill, New York, USA.

Plaggemeier, T. and Lämmerzahl, O. (2000) Treatment of waste gas pollutants in trickling filters. In *Biotechnology. Vol. 11C. Environmental Processes III; Solid Waste and Waste Gas Treatment, Preparation of Drinking Water* (eds. J. Klein and J. Winter), pp. 333–344. Wiley-VCH, Weinheim, Germany.

Pruschek, R., Oeljeklaus, G., Haupt, G., Zimmerman, G., Jansen, D. and Ribberink, J.S. (1997) *Energ. Convers. Manage.* **38**, s153–s158.

Scholten, J.C.M., Conrad, R. and Stams, A.J.M. (2000) Effect of 2-bromo-ethane sulfonate, molybdate and chloroform on acetate consumption by methanogenic and sulfate-reducing populations in freshwater sediment. *FEMS Microbiol. Ecol.* **32**, 35–42.

Seitz, H.J., Schink, B., Pfennig, N. and Conrad, R. (1990) Energetics of syntrophic ethanol oxidation in defined chemostat cocultures. I. Energy requirement for H_2 production and H_2 oxidation. *Arch. Microbiol.* **155**, 82–88.

Sipma, J., Lens, P.N.L., Stams, A.J.M. and Lettinga, G. (2003) Carbon monoxide conversion by anaerobic bioreactor sludges. *FEMS Microbiol. Ecol.* **44**, 271–277.

Sipma, J., Meulepas, R.J.W., Stams, A.J.M., Lettinga, G. and Lens, P.N.L. (2004) Effect of CO, H_2 and sulfate concentration on thermophilic (55°C) carbon monoxide conversion routes in two anaerobic bioreactor sludges. *Appl. Microbiol. Biotechnol.* **64**, 421–428.

Sipma, J., Henstra, A.M., Parshina, S.N., Lens, P.N.L., Lettinga, G. and Stams, A.J.M. (2006a) Microbial hydrogenogenic CO conversions with applications in synthesis gas purification and bio-desulfurization. *Crit. Rev. Biotechnol.* (in press).

Sipma, J., Lettinga, G., Stams, A.J.M. and Lens, P.N.L. (2006b) Moderate thermophilic hydrogenogenic CO conversion in a sulfate fed gas lift reactor: competition between methanogens and sulfate reducers for CO-derived H_2. *Biotechnol. Progr.* (submitted).

Sipma, J., Osuna, M.B., Parshina, S.N., Lettinga, G., Stams, A.J.M. and Lens, P.N.L. (2006c) Effect of hydraulic retention time on sulfate reduction in a carbon monoxide fed thermophilic gas lift reactor. *Appl. Microbiol. Biotechnol.* (submitted).

Sipma, J., Osuna, M.B., Parshina, S.N., Henstra, A.M., Lettinga, G., Lens, P.N.L. and Stams, A.J.M. (2006d) Physiological study of *Desulfotomaculum carboxydivorans* for use in biotechnological synthesis gas conversions. *Environ. Sci. Technol.* (submitted).

Sparling, R., Risbey, D. and Poggi-Varaldo, H.M. (1997) Hydrogen production from inhibited anaerobic composters. *Int. J. Hydrogen Energ.* **22**, 563–566.

Svetlitchnyi, V., Peschel, C., Acker, G. and Meyer, O. (2001) Two membrane-associated NiFeS-carbon monoxide dehydrogenases from the anaerobic carbon-monoxide-utilizing eubacterium *Carboxydothermus hydrogenoformans*. *J. Bacteriol.* **183**, 5134–5144.

Symonds, R.B., Rose, W.I., Bluth, G. and Gerlach, T.M. (1994) Volcanic gas studies: methods, results, and applications. In *Volatiles in Magmas: Mineralogical Society of America Reviews in Mineralogy* (eds. M.R. Carroll and J.R. Holloway), Vol. 30, pp. 1–66.

Uffen, R.L. (1976) Anaerobic growth of a *Rhodopseudomonas* species in the dark with carbon monoxide as sole carbon and energy substrate. *Proc. Natl. Acad. Sci. USA* **73**(9), 3298–3302.

Vallero, M.V.G., Lettinga, G. and Lens, P.N.L. (2005) High rate sulfate reduction in a submerged anaerobic membrane bioreactor (SAMBaR) at high salinity. *J. Membrane Sci.* **253**, 217–232.

Van Houten, R.T. and Lettinga, G. (1996) Biological sulphate reduction with synthesis gas: microbiology and technology. In Progress in Biotechnology (eds. R.H. Wijffels, R.M. Buitelaar, C. Bucke and J. Tramper), Vol. 11, pp. 793–799. Elsevier, Amsterdam.

Van Houten, R.T., Hulshoff Pol, L.W. and Lettinga, G. (1994) Biological sulphate reduction using gas-lift reactors fed with hydrogen and carbon dioxide as energy and carbon source. *Biotechnol. Bioeng.* **44**, 586–594.

Van Houten, R.T., Yu Yun, S. and Lettinga, G. (1997) Thermophilic sulphate and sulphite reduction in lab-scale gas-lift reactors using H_2/CO_2 as energy and carbon source. *Biotechnol. Bioeng.* **55**, 807–814.

Weijma, J., Bots, E.A.A., Tandlinger, G., Stams, A.J.M., Hulshoff Pol, L.W. and Lettinga, G. (2002) Optimisation of sulphate reduction in a methanol-fed thermophilic bioreactor. *Water Res.* **36**, 1825–1833.

Westermann, P. (1994) The effect of incubation temperature on steady-state concentrations of hydrogen and volatile fatty acids during anaerobic degradation in slurries from wetland sediments. *FEMS Microbiol. Ecol.* **13**, 295–302.

23

Product recovery from H_2S-containing gases

A. González-Sánchez and S. Revah

23.1 INTRODUCTION

The world's energy requirements have strongly increased due to the growing industrialization, especially from emerging economies. Fossil fuels still represent the main energy source around the world. Among the various energy sources, gaseous fuels have been increasingly used because of their easier exploitation and their somewhat more environmentally friendly impact. However gaseous fuels, including non-fossil ones such as biogas, commonly contain significant concentrations of hydrogen sulfide (H_2S), and other sulfur species such as carbon disulfide (CS_2), carbonyl sulfide (COS) and organic sulfides in lower concentrations. Table 23.1 shows the typical composition of several gaseous fuels.

After combustion of sulfur-containing fuels, huge amounts of sulfur are released into the atmosphere as SO_2, which is further oxidized to SO_3 and resulting in acid rain. Thus, desulfurization processes with minimum impact to the environment are needed. Desulfurization treatments can be performed either before or after combustion. Only a few post-combustion processes are currently economically viable, as any treatment will need to deal with large exhaust air streams and issues related to the

 Waste Gas Treatment for Resource Recovery edited by Piet N.L. Lens, Christian Kennes, Pierre Le Cloirec and Marc Deshusses. ISBN: 1843391279. Published by IWA Publishing, London, UK.

Table 23.1 Typical composition of various gaseous fuels.

	Composition % mol. (v/v)							
Gaseous fuels	Other S[a]	H_2S	H_2	CO_2	CH_4	N_2	CO	HC[b]
Natural gas	≈0.25	3.8–8.6	≈0.3	2–18	36–71	54–58	–	≈3.0
Distillery biogas	≈0.003	1.2–2.5	≈0.04	22–28	58–64	1.5–5	–	–
Refinery fuel gas	≈0.03	4.5–7.5	≈20	11–15	≈1.5	50–59	≈18	≈2
Coke oven gases	≈0.03	0.3–4.8	≈15.0	24–28	≈50	≈0.15	≈10	≈0.02
Coal gasification	≈0.017	0.6–2.0	≈21	15–26	5–26	≈27	9–22	≈0.6

Adapted from Pandey and Malhotra (1999).
[a] CS_2, Mercaptans and thiophene.
[b] HC: Hydrocarbons.

management of by-products (Fenouil et al., 1994; Ma et al., 2000). In the pre-combustion processes, fuels can be desulfurized and sulfur can be disposed in an environmentally adequate manner or even transformed to useful products (Janssen et al., 2000a). The preferred alternative is to convert the sulfur species to elemental sulfur, as it can easily be recovered by physical methods. Elemental sulfur can be produced by physicochemical or biological technologies and be further sold for different applications such as fertilizer, sulfuric acid and H_2S production (Amelin et al., 1966; Kijlstra et al., 2001; Tichý et al., 1994; Wen et al., 2003).

Emerging biological technologies operating at ambient conditions and with reduced energy and chemical inputs are, under certain conditions, a feasible and economic alternative to the physicochemical processes that have relatively high energy, chemical, catalyst and disposal costs. Several biological processes to treat gaseous fuels have been investigated, but only a few have been deployed at a large scale. In this chapter, the main biological processes for H_2S removal with sulfur recovery from gas fuel streams are described.

23.2 BIOLOGICAL DESULFURIZATION OF GASEOUS FUELS WITH RECOVERY OF ELEMENTAL SULFUR

23.2.1 Pre-combustion desulfurization

Biological H_2S removal processes from fuel gases require an initial absorption into an aqueous phase where specialized microorganisms catalyse its oxidation. Some of these, called sulfide-oxidizing bacteria, can obtain energy for cell functioning from the oxidation of dissolved sulfide species to sulfate. Under certain conditions which are reviewed below, elemental sulfur is formed.

Table 23.2 shows some sulfide-oxidizing bacteria that have been applied for the desulfurization of gaseous fuels. Depending on their energy metabolism, sulfide-oxidizing bacteria can either be phototrophic or chemotrophic. Phototrophic (purple or green) bacteria use light as energy source to reduce CO_2 to carbohydrates, while the chemotrophic (colourless) sulfur bacteria obtain energy from the chemical aerobic

Table 23.2 Sulfide-oxidizing bacteria for desulfurization of gaseous fuels containing H_2S.

Organism	Energy	pH	References
Thiobacillus thioparus	Chemotroph	6–8	Schlegel (1995)
Thiobacillus denitrificans	Chemotroph	6–8	Schlegel (1995)
Thiobacillus sp. W5	Chemotroph	7–9	Visser et al. (1997a)
Thiobacillus thiooxidans	Chemotroph	2–5	Schlegel (1995)
Thiobacillus ferrooxidans	Chemotroph	2–6	Schlegel (1995)
Thiobacillus sp. A1	Chemotroph	5–7	Alcántara et al. (2004)
Thioalkalivibrio denitrificans	Chemotroph	7.5–10.5	Sorokin et al. (2001)
Thioalkalimicrobium cyclicum	Chemotroph	7.5–10.5	Sorokin et al. (2002)
Oscillatoria limnetica	Phototroph	6.8	Cohen et al. (1975)
Chlorobium	Phototroph	6.8–7.0	Hurse and Keller (2004)
Chlorobium	Phototroph	6.8–7.0	Syed and Henshaw (2003)

oxidation of reduced sulfur compounds. These bacteria grow at a wide pH range (Table 23.2).

23.2.1.1 *Aqueous iron processes with bacterial regeneration*

Aqueous ferric iron processes have been applied for H_2S treatment from natural gas and coke oven gas emissions (Bedell et al., 1988). In these processes H_2S is absorbed in an aqueous slurry of chelated iron ($Fe_2(SO_4)_3$). H_2S is converted into elemental sulfur while $Fe_2(SO_4)_3$ is reduced to Fe_2SO_4 in the absorber (Equation 23.1). Sweet gas exits at the top of the absorber, and elemental sulfur is separated from the liquor slurry by physical methods:

$$H_2S + Fe_2(SO_4)_3 \leftrightarrow S\downarrow + 2FeSO_4 + H_2SO_4 \tag{23.1}$$

In the traditional process, the $FeSO_4$ solution is chemically reoxidized with air (Equation 23.2) and recycled to the H_2S absorber. Imaizumi (1986) proposed a biological regeneration of Fe_2SO_4 with *Thiobacillus ferrooxidans* under acidic conditions (pH 2–3) as given in equation:

$$2FeSO_4 + H_2SO_4 + \tfrac{1}{2}O_2 \rightarrow Fe_2(SO_4)_3 + H_2O \tag{23.2}$$

Under optimal conditions, the overall reaction is:

$$H_2S + \tfrac{1}{2}O_2 \rightarrow S\downarrow + H_2O \tag{23.3}$$

In the process described above, the biological oxidation of the ferrous iron is the rate-limiting step (Kim et al., 2002) and several experimental systems and reactor configurations have been assessed to overcome this limitation. Higher densities of iron-oxidizing bacteria have been obtained by cell immobilization (Nemati & Webb, 1996; Pagella & De Faveri, 2000). Nemati and Webb (1996) reported a ferrous oxidation rate of $35\,g\,L^{-1}h^{-1}$ using a fixed bed with a polyurethane foam packing. Sonta and Shiratori (1990) reported 98.8% desulfurization with the system shown in Figure 23.1 treating a gas containing 0.06 vol.% H_2S. On a large scale, the Bio-SR™ process has been applied for the treatment of H_2S emissions from sewage treatment plant (OECD, 1994) where the inlet concentrations were reduced from 400–2000 ppm_v to less than 10 ppm_v.

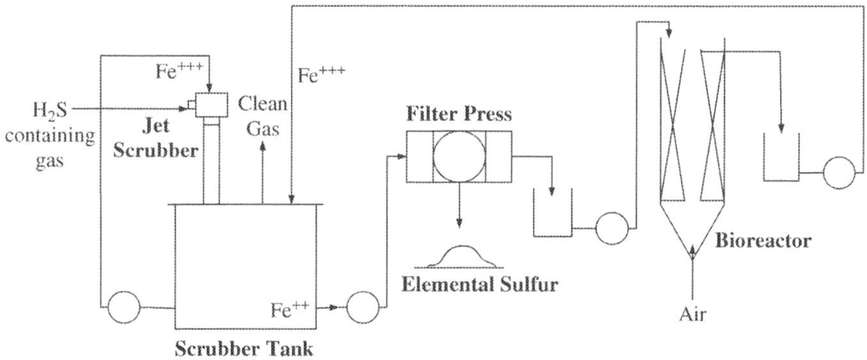

Figure 23.1 Aqueous-iron processes with bacterial regeneration. (Sonta and Shiratori, 1990).

23.2.1.2 Phototrophic microbial conversion of H_2S to elemental sulfur

Purple sulfur bacteria (*Chromatiaceae*) and green/brown sulfur bacteria (*Chlorobiaceae*) can grow with light, carbon dioxide as carbon source and H_2S as electron donor (van Gemerden, 1986). In the purple sulfur bacteria, globules of elemental sulfur are stored inside the cells, whereas it is excreted in the green sulfur bacteria. The phototrophic system is independent of oxygen, as sulfide is anaerobically converted to elemental sulfur with CO_2. The green sulfur bacterium *Chlorobium limicola* has been used for the removal of acid gases (H_2S and CO_2) generated by fossil fuels processing (Cork et al., 1983). In this anaerobic process elemental sulfur accumulates as an intermediate (Equation 23.4), but sulfate is the ultimate oxidation product in both groups (Equation 23.5):

$$2H_2S + CO_2 + \text{light} \longrightarrow CH_2O + H_2O + 2S^0 \qquad (23.4)$$

$$2H_2S + 2CO_2 + 2H_2O + \text{light} \longrightarrow 2CH_2O + H_2SO_4 \qquad (23.5)$$

Maka and Cork (1990) reported for *Chlorobium limicola forma thiosulfatophilum* that simultaneous control of molar flow rates of input gases and photon flux was important to optimize sulfur production. The major disadvantages of using photosynthetic bacteria on a large scale are related with their anaerobic nature and their requirement for radiant energy, and hence the need for an extremely large transparent area (Kim & Chang, 1991).

23.2.1.3 Chemotrophic microbial conversion of H_2S to elemental sulfur

The use of chemotrophic sulfide-oxidizing bacteria in fuel gas desulfurization has shown to be a very promising process, with some successful industrial applications. Heterotrophs and autotrophs can be used, but heterotrophs require organic substrates to maintain a stable population for desulfurization, thus limiting their application (Jensen & Webb, 1995).

Thiobacilli are the most widely studied chemoautotrophic bacteria and have frequently been applied for H_2S removal processes. Elemental sulfur production from the partial oxidation of reduced sulfur compounds is the objective (Buisman et al., 1989; Janssen et al., 1997; Visser et al., 1997b). Sulfur production (Equation 23.6) from the partial oxidation of sulfide instead of a complete oxidation to sulfate (Equation 23.7) has important environmental implications as elemental sulfur can be removed by sedimentation and no further treatment of sulfate is required. Additionally, lower energy consumption for aeration is required because the oxidation to sulfur requires four-fold less oxygen. Furthermore, hydroxide ions that are consumed during the absorption of H_2S (see Equation 23.9 later) are regenerated when sulfur is formed (Equation 23.6), reducing the costs of neutralizing solutions for pH control:

$$HS^- + \tfrac{1}{2}O_2 \rightarrow S^0\downarrow + OH^- \qquad \Delta G^0 = -129.50\,\text{kJ mol}^{-1} \tag{23.6}$$

$$HS^- + 2O_2 \rightarrow SO_4^{2-} + H^+ \qquad \Delta G^0 = -732.58\,\text{kJ mol}^{-1} \tag{23.7}$$

According to Equations (23.6) and (23.7), oxygen is the key parameter controlling the extent of oxidation. Janssen et al. (1997) reported that the optimal oxygen to sulfide molar ratio to improve the sulfur production was about 0.7. Stefess et al. (1996) confirmed that *Thiobacillus* and *Thiobacillus neapolitanus* produced elemental sulfur from the partial oxidation of H_2S or thiosulfate. Sulfide and thiosulfate proved to be interchangeable substrates for chemolithotrophic bacteria producing sulfur and sulfate. They also found that elemental sulfur was formed under oxygen limitation (0.1% saturation) or at substrate loading rates higher than 18 mmol $L^{-1}h^{-1}$. Visser et al. (1997a) reported that elemental sulfur formation in *Thiobacillus* sp. W5 was directly related to its maximum electron transporting capacity or oxygen consumption rate. Consequently, elemental sulfur formation occurred after the maximum oxidative capacity of the culture was attained.

At pH > 9 a different group of chemolithotrophic bacteria grow, which are generally grouped as alkaliphilic sulfide-oxidizing bacteria (Table 23.2). Under alkaline conditions, H_2S absorption (Equation 23.8), is enhanced as its pKa which is around 7 (Equation 23.9), favours the chemical equilibrium towards the more soluble hydrosulfide (HS^-) or sulfide (S^{2-}). These bacteria have shown promising capacities to oxidize H_2S under extreme alkalinity and salinity (Banciu et al., 2004):

$$H_2S_{(g)} \leftrightarrow H_2S_{(liq)} \tag{23.8}$$

$$H_2S_{(liq)} + OH^- \leftrightarrow HS^- + H_2O \tag{23.9}$$

Reactor configurations There are several reactor configuration systems for the chemotrophic removal of H_2S from a gas stream such as bioscrubbers, biotrickling filters, biofilters and membrane bioreactors (De Bo et al., 2002; Degorce-Dumas et al., 1997; Gabriel & Deshusses, 2003; Janssen et al., 1997).

Janssen et al. (1997) described the performance of an expanded-bed reactor designed to control the oxygen/sulfide molar ratio and improve elemental sulfur formation. In this reactor, the aeration of the liquid phase and the oxidation of sulfide were spatially separated and the oxygen was controlled by the recirculation flow and its

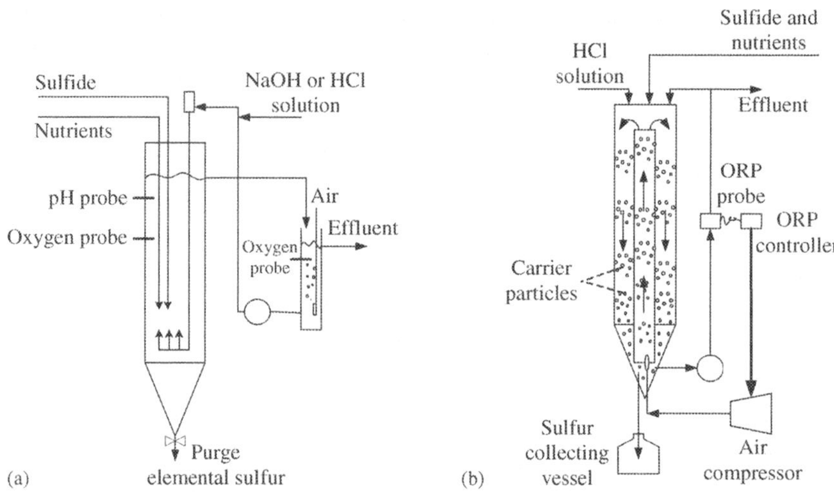

Figure 23.2 Treatment of sulfide-containing wastewater with elemental sulfur recovery. (a) Bioreactor–settler system and (b) reverse fluidized loop reactor.

concentration in the aerator. The system allowed sulfur formation at oxygen/sulfide molar ratio from 0.4 to 0.7 with a maximum (80% yield) at a molar ratio of 0.7.

Alcántara et al. (2004) investigated the formation and the recovery of elemental sulfur in a bioreactor–settler system shown in Figure 23.2a. The reactor has a dynamic flow zone and a static zone, which allowed sulfur sedimentation as the liquid phase aeration and the sulfide oxidation were spatially separated. Steady-state sulfide-oxidizing cultures were established at different O_2/S^{2-} molar ratios. At a O_2/S^{2-} ratio > 1.0, sulfate was the main product while with an O_2/S^{2-} ratio of 0.5, elemental sulfur formed attaining 85% of the theoretical value shown in Equation (23.6). With this system most of the sulfate formed was in the aeration vessel and a dual oxygen–sulfide limitation was observed in the settler (Gonzalez-Sanchez et al., 2005; Velasco et al., 2004).

Krishnakumar et al. (2005) used an aerobic bioreactor called reverse fluidised loop reactor shown in Figure 23.2b, which was developed to recover elemental sulfur from aqueous sulfide. The reactor contained buoyant carrier particles on which sulfide-oxidizing bacteria formed a biofilm. The process allowed the elemental sulfur to be efficiently separated from the aqueous phase.

Figure 23.3 shows the bioscrubbing system employed by Thiopaq (Janssen et al., 2000b), consists of a gas absorber and a bioreactor. In the gas absorber, gaseous H_2S is scrubbed (Equations (23.8) and (23.9)) using an alkaline solution (pH 8.0–8.5). The dissolved sulfide is then oxidized by microorganisms in the bioreactor to elemental sulfur and sulfate according to Equations (23.6) and (23.7). The Thiopaq desulfurization process has been applied to the removal of H_2S from high-pressure natural gas (Janssen et al., 2000b). In a demonstration plant, natural gas with a H_2S content of up to 8 vol.% was desulfurized at high gas pressures (between 5 and 53 bar) without any negative effect on the bacteria. The H_2S content in the treated gas was below 1 ppm_v, while all absorbed sulfide was oxidized into elemental sulfur. A drawback of the

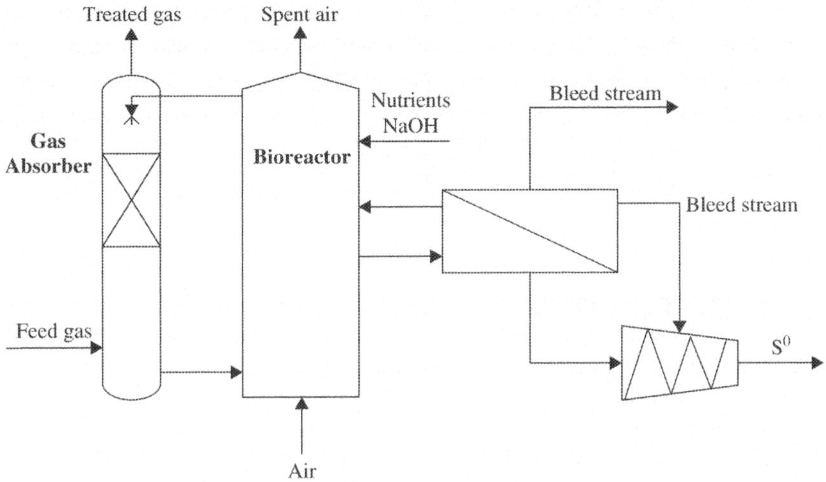

Figure 23.3 Bioscrubber system diagram for the treatment of H_2S-containing gas streams (Thiopaq process).

system is that sodium bicarbonate can precipitate at high CO_2 partial pressures. To avoid this, low CO_2 partial pressures should be present. Alternately, lower pH can be used in the absorber but higher liquid flows or larger scrubbers would likely be required to get similar absorption capacity. At pH below 5.8, the formed sulfur might lose its hydrophilic character, which helps to reduce clogging (McIntush et al., 2001).

23.2.2 Post-combustion process

All fuels used such as coal, oil, natural gas, peat, wood and other organic materials contain sulfur, which is released as SO_2 upon combustion. One approach consists in absorbing SO_2 either in water or in aqueous slurries of limestone in a chemical scrubber and converting it to sulfide and sulfate (Janssen et al., 1995). These compounds can then be reduced to sulfide by sulfate-reducing bacteria under anaerobic conditions by adding an electron donor. The formed sulfide can be partially oxidized, usually by *Thiobacillus* spp., to elemental sulfur under the environmental conditions described above. Philip and Deshusses (2003) reported a complete treatment of sulfur dioxide (SO_2) from flue gases in a two-stage process consisting of a biotrickling filter, followed by a biological post-treatment unit.

23.2.3 Elemental sulfur applications

Possible applications of the biologically produced elemental sulfur can be found in mining and agriculture, where the specific properties of biologically produced sulfur (small particle size, hydrophilic surface) have clear advantages over the use of inorganic sulfur (Janssen et al., 1996). Bioleaching is used in mining to dissolve metals from sulfidic ores by reducing the pH and in some cases, additional elemental sulfur is added as the substrate (Tichý, 2000). Tichý et al. (1994) found that the hydrophilic properties of the biologically produced sulfur increased the rate of sulfuric acid production compared to the use of chemical sulfur flower. In agriculture, the properties of the biologically recovered sulfur can be used for the controlled release of sulfate to the soil. Kijlstra et al. (2001) reported that the grain yield was higher when biologically produced sulfur was used as compared to other commercially available sulfur sources.

23.3 CONCLUSIONS

There are a number of processes that can transform environmentally hazardous sulfides into elemental sulfur. This is the most desirable product as it can be contained and sequestered from the sulfur geochemical cycle. Biological processes can be effective under some conditions and several industrial scale systems have been implemented. Improvements in reactor configurations and control and the use of more diversified microorganisms can bring about exciting new applications.

REFERENCES

Alcántara, S., Velasco, A., Muñoz, A., Cid, J., Revah, S. and Razo-Flores, E. (2004) Hydrogen sulfide oxidation by a microbial consortium in a recirculation reactor system: sulfur formation under oxygen limitation and removal of phenols. *Environ. Sci. Technol.* **38**, 918–923.

Amelin, A.G., Liberzon, L.M. and Rodov, A.B. (1966) Optimization of the process of sulfuric acid production by the contact method. *Khim. Prom.* **42**(4), 290–292.

Banciu, H., Kleerebezem, R., Muyzer, G., Kuenen, J. and Sorokin, D. (2004) Application of haloalkaliphilic sulfur-oxidizing bacteria for the removal of H_2S from gas streams. *European Symposium on Environmental Biotechnology*, pp. 345–348.

Bedell, S.A., Kirby, L.H., Buenger, C.W. and McGaugh, M.C. (1988) Chelates role in gas treating. *Hydrocarb. Process.* **76**, 63–66.

Buisman, C., Post, R., Ijspeert, P., Geraats, G. and Lettinga, G. (1989) Biotechnological process for sulfide removal with sulfur reclamation. *Acta Biotechnol.* **9**, 255–267.

Cohen, Y., Padan, E. and Shilo, M. (1975) Facultative anoxygenic photosynthesis in the cyanobacterium *Oscillatoria limnetica*. *J. Bacteriol.* **123**, 855–861.

Cork, D.J., Garunas, R. and Sajjad, A. (1983) *Chlorobium limicola forma thiosulfatophilum* biocatalyst in the production of sulfur and organic carbon from a gas stream containing H_2S and CO_2. *Appl. Env. Microbiol.* **45**, 913–918.

De Bo, I., Van Langenhove, H. and Heyman, J. (2002) Removal of dimethyl sulfide from waste air in a membrane bioreactor. *Desalination* **148**, 281–287.

Degorce-Dumas, J., Kowai, S. and Le Clorec, P. (1997) Microbiological oxidation of hydrogen sulfide in a biofilter. *Can. J. Microbiol.* **43**, 264–271.

Fenouil, L.A., Towler, G.P and Lynn, S. (1994) Removal of H_2S from coal gas using limestone: kinetic considerations. *Ind. Eng. Chem. Res.* **33**, 265–272.

Gabriel, D. and Deshusses, M. (2003) Retrofitting existing chemical scrubbers to biotrickling filters for H_2S emission control. *Proc. Natl. Acad. Sci. USA.* **100**(11), 6308–6312.

Gonzalez-Sanchez, A., Alcántara, S., Razo-Flores, E. and Revah, S. (2005) Oxygen transfer and consumption in a thiosulfate oxidizing bioreactor with sulfur production. *Lett. Appl. Microbiol.* **41**, 141–146.

Hurse, T.J. and Keller, J. (2004) Reconsidering the use of photosynthetic bacteria for removal of sulfide from wastewater. *Biotechnol. Bioeng.* **85**, 48–55.

Imaizumi, T. (1986) Some applications of inorganic microbial oxidation in Japan. *Biotech. Bioeng. Symp.* **16**, 363.

Janssen, A., Sleylter, R., van der Kaa, C., Jochemsen, J., Bontsema, J., Ma, S. and Lettinga, G. (1995) Biological sulfide oxidation in a fed-batch reactor. *Biotechnol. Bioeng.* **47**, 327–333.

Janssen, A., de Keizer, A., van Aelst, A., Fokkink, R., Yangling, H. and Lettinga, G. (1996) Surface characteristics and aggregation of microbiologically produced sulfur particles in relation to the process conditions. *Colloid Surf. B Biointerf.* **6**, 115–129.

Janssen, J.A., Ma, S., Lens, P. and Lettinga, G. (1997) Performance of a sulfide-oxidizing expanded-bed reactor supplied with dissolved oxygen. *Biotechnol. Bioeng.* **53**, 32–40.

Janssen, A.J.H., Dijkman, H. and Janssen, G. (2000a) Novel biological processes for the removal of H_2S and SO_2 from gas streams. In *Environmental Technologies to Treat Sulfur Pollution* (eds. N.L. Lens and H.L. Pol), pp. 265–279. *Principles and Engineering*, IWA Publishing, London.

Janssen, A.J.H., Arena, B. and Kijlstra, W.S. (2000b) New developments of THIOPAQ technology for the removal of H_2S from gaseous steams. *Proceedings of Sulphur 2000, San Francisco.*

Jensen, A.B. and Webb, C. (1995) Treatment of H_2S containing gases: a review of microbiological alternatives. *Enzyme Microb. Technol.* **17**, 2–10.

Kijlstra, S.W.S., Janssen, A.J.H. and Arena, B. (2001) Biological process for H_2S removal from (high pressure) gas: the Shell-Thiopaq gas desulfurization process. In *Proceedings of the Laurance Reid Gas Conditioning Conference*, University of Oklahoma, Norman, OK.

Kim, B and Chang, H. (1991) Removal of hydrogen sulfide by *Chlorobium thiosulfatophilum* in immobilized-cell and sulfur-settling free-cell recycle reactors. *Blotechnol. Prog.* **7**, 495–500.

Kim, T.W., Kim, C.J., Chang, Y.K., Ryu, H.W. and Cho, K.-S. (2002) Development of an optimal medium for continuous ferrous iron oxidation by immobilized *Acidothiobacillus ferrooxidans* cells. *Biotechnol. Prog.* **18**, 752–759.

Krishnakumar, B., Majumdar, S., Manilal, V.B. and Haridas, A. (2005) Treatment of sulfide containing wastewater with sulfur recovery in a novel reverse fluidized loop reactor (RFLR). *Water Res.* **39**, 639–647.

Ma, X., Kaneko, T., Tashimo, T., Yoshida, T. and Kato, K. (2000) Use of limestone for SO_2 removal from flue gas in the semidry FGD process with a powder-particle spouted bed. *Chem. Eng. Sci.* **55**, 4643–4652.

Maka, A. and Cork, D. (1990) Quantum efficiency requirements for an anaerobic photobioreactor. *J. Ind. Microbiol.* **5**, 337–354.

McIntush, K., Seeger, D., Rueter, C. and DeBerry, K. (2001) Comparison of technologies for removing sulfur from high pressure sour natural gas with sulfur throughputs between 0.1 and 30 long Tons/Day. *Proceedings of 80th Annual GPA Convention.*

Nemati, M. and Webb, C. (1996) Effect of ferrous iron concentration on the catalytic activity of immobilized cells of *T. ferrooxidans*. *Appl. Microbiol. Biotechnol.* **46**, 250–255.

OECD (1994) Biotechnology for a clean environment – prevention, detection, remediation. OECD Reports. p. 39.

Pagella, C. and De Faveri, D. (2000) H_2S gas treatment by iron bioprocess. *Chem. Eng. Sci.* **55**, 2185–2194.

Pandey, R.A. and Malhotra, S. (1999) Desulfurization of gaseous fuels with recovery of elemental sulfur: an overview. *Crit. Rev. Environ. Sci. Technol.* **29**, 229–268.

Philip, L. and Deshusses, M. (2003) Sulfur dioxide treatment from flue gases using a biotrickling filter–bioreactor system. *Environ. Sci. Technol.* **37**,1978–1982.

Schlegel, H.G. (1995) *General Microbiology*, 7th edn. Cambridge University Press, Cambridge.

Sonta, H. and Shiratori, T. (1990) Method of treating H_2S containing gases. US Patent 4931262.

Sorokin, D.Y., Kuenen, J.G. and Jetten, M.S. (2001) Denitrification at extremely high pH values by the alkaliphilic, obligately chemolithoautotrophic, sulfur oxidizing bacterium *Thioalkalivibrio denitrificans* strain ALJD. *Arch. Microbiol.* **175**, 94–101.

Sorokin, D.Y., Gorlenko, V.M., Tourova, T.P., Tsapin, A.I., Nealson, K.H. and Kuenen, G.J. (2002) *Thioalkalimicrobium cyclicum* sp.nov. and *Thioalkalivibrio jannaschii* sp. nov., novel species of haloalkaliphilic, obligately chemolithoautotrophic sulfur-oxidizing bacteria from hypersaline alkaline Mono Lake (California). *Int. J. Syst. Evol. Microbiol.* **52**, 913–920.

Stefess, G., Torremans, R., de Schrijver, R., Robertson, L. and Kuenen, J. (1996) Quantitative measurement of sulfur formation by steady state and transient state continuous cultures of autotrophic *Thiobacillus* species. *Appl. Microbiol. Biotechnol.* **45**, 169–175.

Syed, M.A. and Henshaw, P.F. (2003) Effect of tube size on performance of a fixed-film tubular bioreactor for conversion of hydrogen sulfide to elemental sulfur. *Water Res.* **37**, 1932–1938.

Tichý, R., Janssen, A., Grotenhuis, J.T.C., Lettinga, G. and Rulkens W.H. (1994) Possibilities for using biologically-produced sulfur for cultivation of *Thiobacilli* with respect to bioleaching processes. *Bioresour. Technol.* **48**, 221–227.

Tichý, R. (2000) Treatment of solid materials containing inorganic sulfur compounds. In: *Environmental Technologies to Treat Sulfur Pollution* (eds. N.L. Lens and H.L. Pol), pp. 329–354. *Principles and Engineering*, IWA Publishing, London.

van Gemerden, H. (1986) Production of elemental sulfur by green and purple sulfur bacteria. *Arch. Microbiol.* **146**, 52–56.

Velasco, A., Alcántara, S., Razo-Flores, E. and Revah, S. (2004) Partial thiosulfate oxidation by steady-state continuous culture in a bioreactor–settler system. *J. Chem. Technol. Biotechnol.* **79**, 132–139.

Visser, J.M., Stefess, G.C., Robertson, L.A. and Kuenen, J.G. (1997a) *Thiobacillus* sp. W5, the dominant autotroph oxidizing sulfide to sulfur in a reactor for aerobic treatment of sulfidic wastes. *Antonie van Leeuwenhoek* **72**, 127–134.

Visser, J., Robertson, L., Verseveld, H. and Kuenen, J.G. (1997b) Sulfur production by obligately chemolithoautotrophic *Thiobacillus* species. *J. Appl. Environ. Microbiol.* **63**, 2300–2305.

Wen, G., Schoenau, J.J., Mooleki, S.P., Inanaga, S., Yamamoto, T., Hamamura, K., Inoue, M. and An, P. (2003) Effectiveness of an elemental sulfur fertilizer in an oilseed-cereal-legume rotation on the Canadian prairies. *J. Plant Nutr. Soil Sci.* **166**, 54–60.

PART V: Implementation of resource recovery from gas streams

24

CO_2 recovery from flue gas by membrane gas absorption and use in the horticultural industry

A.E. Jansen and P.H.M. Feron

24.1 INTRODUCTION

CO_2 membrane gas absorption (MGA) has been under development at the Netherlands Organization for Applied Scientific Research (TNO) since 1991. The potential of this technology has been identified in a number of assessment studies (Feron, 1992, 1994; IEA Report, 1992). The ensuing customer-driven development has been focusing on the production of carbon dioxide from flue gas with the aim of supplying high purity carbon dioxide to the horticultural industry to increase crop yield (Feron & Jansen, 1995, 1997). Other commercial applications are in the field of life support, spacecrafts, submarines, recovery of anaesthetics, consumer products like beverage production in house, biogas treatment and natural gas treatment at elevated pressures. Figure 24.1 shows an overview of the atmospheric pressure applications currently being investigated.

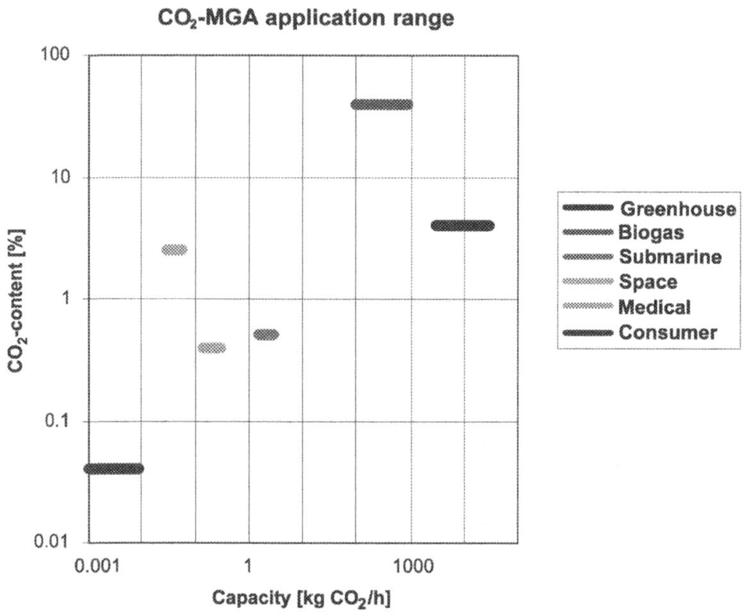

Figure 24.1 CO_2 MGA applications (atmospheric pressure) currently under development.

This chapter discusses the technical aspects of CO_2 production from flue gas for the horticultural industry using MGA. The process is based on a novel membrane absorption method patented by TNO (Jansen & Feron, 1995), which employs dedicated aqueous absorption liquids, called CORAL (an acronym for CO_2 removal absorption liquid), based on amino acids and alkaline salts in combination with membranes in a contactor function. Benefits of these liquids over conventional amines are:

- Very low vapour pressure, hence no loss of absorption liquids through evaporation.
- Stable operation with cheapest membranes available viz. polyolefin membranes.
- Oxygen stability.
- Excellent corrosion resistance.

24.1.1 CO_2 in the horticultural industry

In the Netherlands, carbon dioxide is used to promote plant growth in greenhouses. Depending on the crop, the production can be 25% higher by increasing the CO_2 concentration to 500 ppm. Figure 24.2 shows the influence of the CO_2 content on plants (Proefstation, 1999).

24.1.2 Integrated supply of heat and CO_2

In order to be economically viable, greenhouses need to be heated in temperate climates. This makes the horticultural industry in the Netherlands a major consumer of natural gas. The required temperature levels in greenhouses are only slightly above ambient. This means that low-grade heat suffices to maintain the required temperature

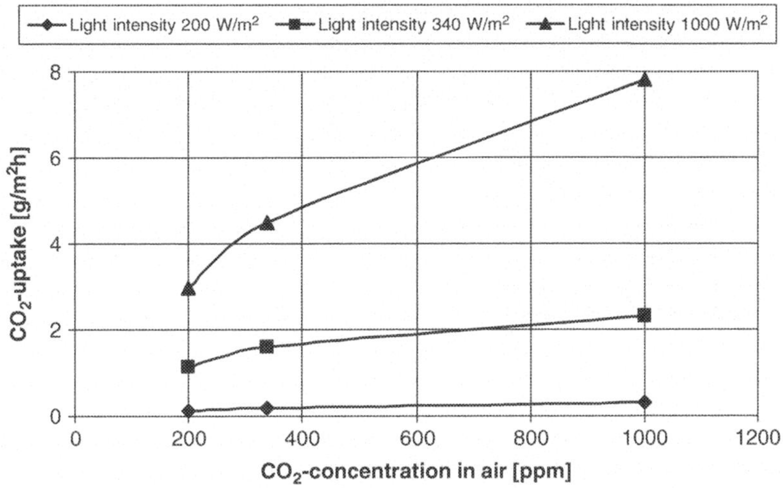

Figure 24.2 Influence of CO_2 content in air on plants growth.

levels. Future heat supply systems will therefore be based on cogeneration plants producing both electricity and heat. The heat can be used to maintain temperature levels in the greenhouse. Significant energy savings can be achieved compared to the situation where natural gas fired boilers are used for greenhouse heating.

Energy savings in the horticultural industry can be increased further if the heat supply is integrated with the carbon dioxide supply. This will incite growers to buy heat from the cogeneration plant rather than to use their own gas fired boilers. These boilers are needed to cover peak demands and emergencies. Also the amount of available carbon dioxide per unit heat will be larger in case of combined production of heat and electricity, thus allowing for an improved production. Unfortunately, the heat and CO_2 demand respectively are generally anti-cyclic. During periods of high heat demand (during winter and at night) little or no CO_2 is needed to maintain the desired CO_2 levels in the greenhouse. During periods of high CO_2 demand (during summer and during the day) little or no heat is required. The thermal energy of the cogeneration plant can either be used for greenhouse heating or for the production of carbon dioxide by using the heat for the controlled release of dissolved CO_2 from the (partially) loaded absorption liquid. This will also extend the operating time of the cogeneration plant to daytime hours, which is an additional benefit in terms of capacity credits and efficiency. A schematic drawing of a cogeneration plant supplying heat and carbon dioxide to greenhouses is shown in Figure 24.3.

24.2 MEMBRANE GAS ABSORPTION

24.2.1 Principle and process

The principle of MGA is shown in Figure 24.4. The essential element in a membrane absorber is a porous, water repellent or hydrophobic, polymeric membrane. The gas

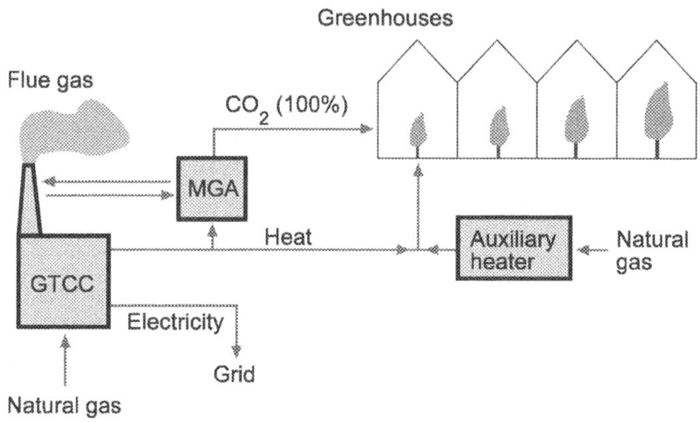

Figure 24.3 Cogeneration plant supplying heat and carbon dioxide to greenhouses (GTCC: gas turbine combined cycle; MGA = membrane gas absorption).

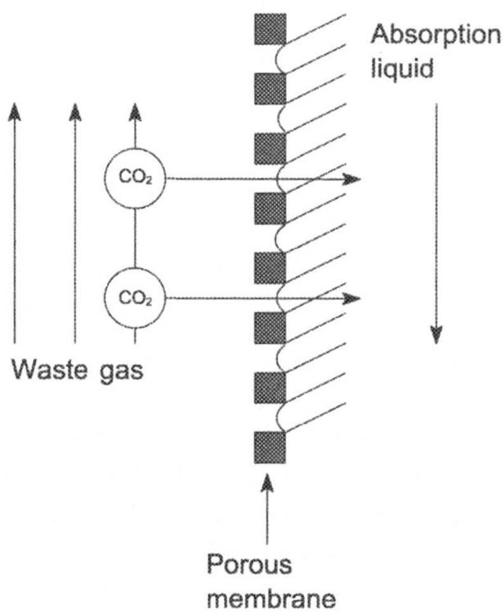

Figure 24.4 Principle of membrane gas absorption.

phase remains separated from the liquid absorbent as a result of the hydrophobicity of the membrane. A CO_2 containing flue gas stream is fed along one side of the membrane. The components to be removed from the gas stream diffuse through the gas-filled pores of the membrane. On the other side of the membrane, they are absorbed into the absorption liquid.

The use of a membrane absorber leads to the following advantages:

- Gas and liquid flow are independent resulting in avoidance of problems encountered in packed/tray columns such as flooding, foaming, channelling and entrainment.
- No need to have a wash section after the absorber to recover absorption liquid which is entrained.

- Operation at thermodynamically optimal conditions, not dictated by the hydrodynamic conditions of the contacting equipment.
- Operation not influenced by the orientation or movement of the absorber which is important in offshore applications and zero-gravity applications.
- Equipment will be compact because small hollow fibre or dense-packed flat sheet membranes have high interfacial areas.

The choice of the combination of absorption liquid and membrane is extremely important. The hydrophobicity and pore characteristics of the membrane material should be such that the pores remain gas filled even at an excess pressure of, preferably, more than 1 bar on the liquid side. Commercially available porous polyolefin membranes appear not to be suitable for operation with the usual absorption liquids, that is aqueous solutions of alkanolamines (Feron, 1994; Matsumoto et al., 1994; Nishikawa et al., 1995; Rangwala, 1996). In the course of time such liquids will penetrate through the pores of these membranes as a result of the low surface tension of the organic liquid/water mixtures. This can be resolved by the utilization of more hydrophobic expensive PTFE membranes (Falk-Pedersen & Dannström, 1997). An alternative solution to this is the use of an absorption liquid with a surface tension similar to water. TNO has developed an MGA process for acid gas removal using dedicated absorption liquids based on mixtures of salts and amino acids (CORAL). This substantially facilitates the use of inexpensive polyolefin membranes, developed for, for example, medical applications (artificial lungs), which is believed to be essential for the economic success of MGA. These membranes are available on the market and are known as microfiltration (MF) membranes. The use of salt solutions also has the advantage that the vapour pressure of the active components is basically zero, owing to the inorganic character of the reactive species, and no solvent loses occur because of evaporation.

24.2.2 Membrane flow sheet

The overall flow sheet of an MGA process, shown in Figure 24.5 is similar to a conventional absorption process flow sheet. The feed gas containing carbon dioxide is counter-currently brought into contact with the absorption liquid in the membrane absorber. The loaded, rich absorption liquid is subsequently fed via a heat exchanger to the regenerator where concentrated, pure carbon dioxide is released at an elevated temperature (118°C).

The design of the membrane absorber cannot be based on the conventional rules developed for the more traditional membrane module applications. Historically, these designs have been based on the use of hollow fibre membranes for filtration duties with not well-defined flow conditions on the shell side of the membrane. When membranes are used for contacting duties, as is the case for MGA, it is necessary that flow conditions are well defined and known on either side of the membrane in order to achieve optimal mass transfer rates. Furthermore, it is essential that, given the large volume flows of flue gas streams, the module design is easy to scale up. TNO has patented a transversal-flow membrane module design (Figure 24.6), which has good mass transfer rates and well-defined scale-up rules (Ter Meulen, 1991). Simulations have been done to compare the performance of packed columns and membrane

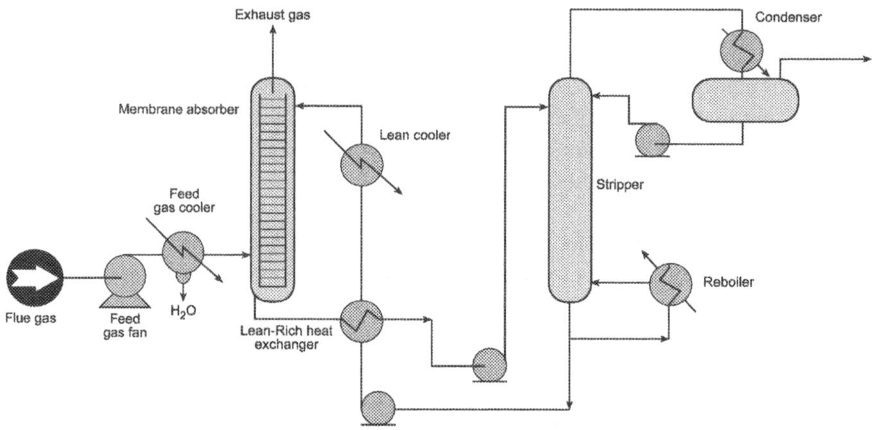

Figure 24.5 Flow sheet of the MGA process.

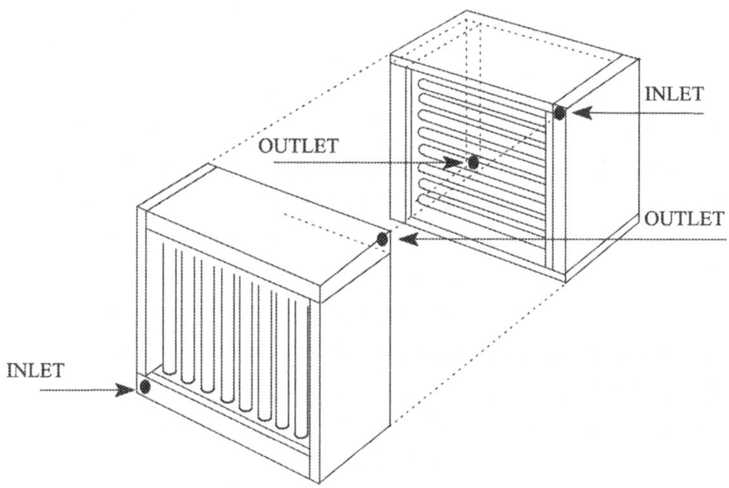

Figure 24.6 Transversal-flow membrane module design, patented by TNO (Feron & Jansen, 1995).

absorbers. It has been shown that a membrane absorber can lead to an up to 10-fold reduction in absorber size for CO_2 removal from flue gas (Feron & Jansen, 1995), illustrating the potential of gas absorption carried out in membrane contactors.

24.2.3 Mass transfer analysis

The most important performance parameter is the CO_2 flux through the membranes, because it determines the total membrane area requirement for a given application and specification, respectively. The total required membrane area is expected to represent the major investment component in a commercial plant. The CO_2 flux is influenced by:

- The chemical reaction kinetics and the reaction mechanism, which are dependent on the nature and characteristics of the active components and their concentration levels.
- The absorption temperature. Higher temperature leads to faster reaction rates but this will be counteracted by thermodynamic equilibrium which is at elevated temperatures.
- Fluid dynamics. The fluid dynamic conditions (gas and liquid) will influence the mass transfer characteristics.
- Membrane type. The membrane type should be sufficiently porous not to add substantial mass transfer resistance.
- CO_2 concentrations in both gas and liquid. If the liquid is already substantially loaded, this will reduce the CO_2 flux due to the reduced concentration differences, or driving force, of active components. An optimized regeneration process is therefore also required, which will include a trade-off with the energy consumption of the regeneration process. At higher CO_2 concentrations in the gas the CO_2 flux will be higher as a result of the increased driving force.

The CO_2 flux, φ_{CO_2}, through the membranes can be expressed by:

$$\varphi_{CO_2} = K(c_g - c_l^*) = P_m(p_g - p_l^*) \qquad (24.1)$$

where:

φ_{CO_2}: CO_2 flux (kmol/m^{2}s)
K: overall mass transfer coefficient (m/s)
c_g: concentration in gas phase (kmol/m^3)
c_l^*: gas phase concentration in equilibrium with bulk liquid phase (kmol/m^3)
P_m: permeance (kmol/m^{2}s Pa)
p_g: partial pressure in gas phase (Pa)
p_l^*: partial pressure in equilibrium with bulk liquid phase (Pa)

Mass transfer is determined by the following, consecutive processes occurring in the three phases:

(1) Diffusion of CO_2 from the bulk in the gas phase towards the outer membrane surface
(2) Diffusion of CO_2 through membrane pores
(3) Dissolution of CO_2 into the absorption liquid followed by liquid phase diffusion/chemical reaction. These transport processes proceed in parallel.

The overall mass transfer coefficient with respect to the inner surface of the hollow fibre membrane (K) can then be expressed by a resistance, in series model:

$$1/K = 1/(k_g d_o/d_i) + 1/(k_m) + 1/(mEk_l) \qquad (24.2)$$

where:

k_g: gas side mass transfer coefficient (m/s)
k_m: membrane mass transfer coefficient (m/s)

k_l: liquid phase mass transfer coefficient (m/s)
d_o: outer diameter of hollow fibre membrane (m)
d_i: inner diameter of hollow fibre membrane (m)
m: distribution coefficient between gas and liquid phase (–)
E: enhancement factor due to chemical reaction (–)

The mass transfer coefficient k_g in the gas phase can be expressed by:

$$k_g = ShD_{g,b}/d_o \tag{24.3}$$

where:
Sh: Sherwood number (–)
$D_{g,b}$: CO_2 diffusion coefficient in gas phase (m^2/s)

TNO has evaluated the mass transfer performance of transversal-flow membrane modules for several applications. An engineering correlation for Sh has been developed for these modules, which is adequate for mass transfer assessments:

$$Sh = 0.9Re^{0.5}Sc^{0.33} \tag{24.4}$$

where:
Re: Reynolds number
Sc: Schmidt number

The mass transfer k_m in the membrane is dependent on the porosity, thickness and tortuosity.

$$k_m = D_{g,e}\, \varepsilon/\tau d_m \tag{24.5}$$

where:
$D_{g,e}$: effective CO_2 diffusion coefficient in gas-filled pores (m^2/s)
d_m: membrane thickness (m)
ε: membrane porosity (–)
τ: membrane tortuosity (–)

The effective diffusion coefficient $D_{g,e}$ is determined by the combined processes of bulk diffusion and Knudsen diffusion:

$$1/D_{g,e} = 1/D_{g,b} + 1/D_k \tag{24.6}$$

where:
D_k: Knudsen diffusion coefficient of CO_2 (m^2/s)

which can be written as:

$$D_k = 1/3\, d_p(8RT/\pi M)^{0.5} \tag{24.7}$$

d_p: pore diameter (m)
R: gas constant (J/(kmol K))
T: temperature (K)
M: molecular mass (kg/kmol)

Physical mass transfer in the liquid phase can be described by:

$$k_l = Sh D_l / d_i \tag{24.8}$$

where:
Sh: Sherwood number (–)
D_l: bulk diffusion coefficient in liquid phase (m^2/s)

An engineering correlation for mass transfer in tubes is:

$$Sh = (4^3 + 1.8^3 Re\, Sc\, d_i/l)^{0.33} \tag{24.9}$$

where:
Re: Reynolds number
Sc: Schmidt number
l: Length of hollow fibre membrane (m)

The constants 4^3 and 1.8^3 in Equation (24.9) follows from the Greatz–Leveque equation described by Dindore (2005).

The enhancement factor E in Equations (24.10) and (24.11) depends on the chemical reaction kinetics. For CO_2 MGA using alkanolamines as the absorption liquid the reaction data and equilibrium data are well documented and mass transfer modelling is straightforward. Results of such modelling work can be found in Nishikawa et al. (1995), Matsumoto et al. (1994) and Nii and Takeuchi (1994) showing a good correlation with experimental results. A more detailed and general numerical model incorporating the effect of the laminar flow profile is presented in Kreulen et al. (1993). As mentioned before alkanolamines cannot be used in conjunction with polyolefin membranes, which sparked off the development of novel absorption liquids. For these absorption liquids details on the reaction mechanism and rates are as yet unknown, but studied at Twente University in the Netherlands (Kumar et al., 2003). At this stage an assessment of the expected contributions towards the mass transfer resistances can be made, assuming that the chemically enhanced mass transfer can be described by an irreversible pseudo first-order reaction. For monoethanolamine (MEA) this assumption is valid at lower temperatures. Standard theory for this situation can be found in, for example, Westerterp et al. (1984). Two extremities can be defined, one where the reaction is considered to be fast and one where the reaction is instantaneous compared to the mass transfer rate. In the former case, mass transfer is governed by the reaction kinetics, in the latter case the mass transfer is governed by the diffusion of the reaction species to the reaction plane. In case of a fast reaction the enhancement factor E is given by Equation (24.10):

$$E = (k_r C_b D_{l,a})^{1/2} / k_l \tag{24.10}$$

where:
k_r: second-order reaction rate constant ($m^3/(kmol\,s)$)
C_b: bulk concentration of active component ($kmol/m^3$)
$D_{l,a}$: CO_2 diffusion coefficient in liquid phase (m^2/s)

In case of an instantaneous reaction, the enhancement factor E can be described by Equation (24.11):

$$E = (1 + (D_{l,b}C_b)/(nD_{l,a}C_{a,i}))\,(D_{l,a}/D_{l,b})^{1/3} \tag{24.11}$$

where:
$D_{l,b}$: diffusion coefficient of active component in liquid phase (m^2/s)
n: number of molecules active in reaction with CO_2 as determined by stoichiometry
$C_{a,i}$: CO_2 concentration at the gas–liquid interface ($kmol/m^3$)

24.3 CASE STUDY: CO_2 REUSE IN HORTICULTURE

24.3.1 Bench scale studies

24.3.1.1 Experimental set-up

A bench scale experimental set-up has been designed and built for testing the complete system during a pre-feasibility study. This means that the set-up is a small-scale version of a real plant, including the CORAL regeneration process. The basic specifications of the bench scale set-up are shown in Table 24.1. The bench scale flow sheet is shown in Figure 24.7.

During the experiments CO_2 and air are mixed via mass flow controllers and fed to a prototype transversal-flow membrane absorber (called DAM-module in flow sheet).

Table 24.1 Basic specifications of CO_2 removal bench scale unit.

CO_2 concentration	0.05–10%
Gas flow rate	0.05–4 m^3/h
Liquid flow rate	0.5–20 l/h

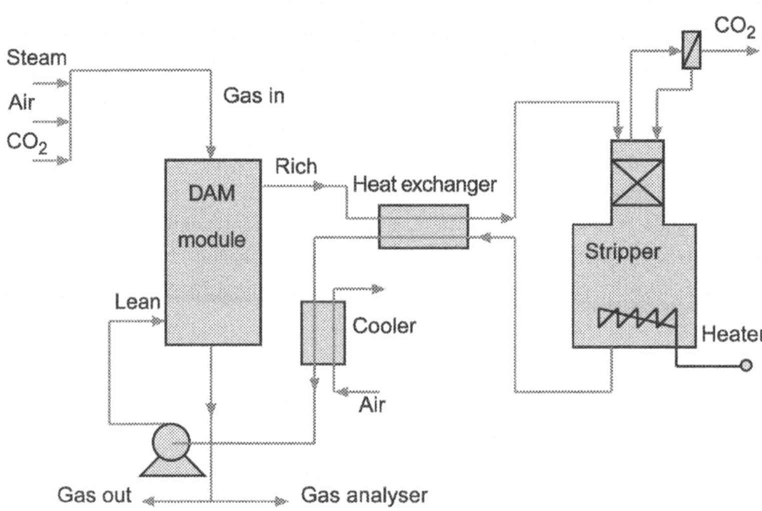

Figure 24.7 Flow diagram of bench scale set-up.

Table 24.2 Specifications of the prototype transversal-flow membrane module.

Module membrane area (internal)	0.27 m^2
Fibre material	Polypropene
Fibre type	Accurel Q3/2
Average pore size	0.265 μm
Porosity fibre material	0.79
Tortuosity	4
Fibre diameter (inside/outside)	0.6/1.0 mm
Fibre length/channel dimension	0.05 m
Number of fibres	700
Number of elements	4

In this absorber, the CO_2 present in the gas stream is transferred to a liquid stream in a counter-current fashion. Liquid is flowing through the membrane fibres or lumen, whereas the gas flow outside the fibres is perpendicular to membranes. Within a single membrane element, the flow pattern is cross-flow but overall the flow pattern is counter-current because a membrane module consists of a number of the elements. The rich absorption liquid is fed to the top of a stripper via a plate heat exchanger. In the present project, the stripper temperature is maintained at 105°C by an electric heating element. As a result of the temperature increase of the absorption liquid, CO_2 is desorbed from the solution. Water is recovered from the CO_2 flow in the condenser and is fed back into the stripper. The lean liquid exits the stripper at the bottom and is fed back to the membrane absorber via the heat exchanger. Temperatures, pressures, dew points, flow rates and carbon dioxide concentrations at inlet and outlet of the membrane absorber are measured continuously and are used for the performance assessment.

During all experiments, air was used as the main feed gas component, which did not result in a decrease in performance during the 4-months testing period using the same absorption liquid. This illustrates that the CORAL liquid is superior in terms of chemical stability over, for example monoethanolamine, which degrades substantially in the presence of oxygen.

The specifications of the prototype transversal-flow membrane (DAM) module designed, according to the TNO patent (Ter Meulen, 1991), are shown in Table 24.2.

A large number of experiments were carried out to obtain realistic data for the mass transfer under a variety of conditions using CORAL20 as the absorption liquid, which is a 2 M aqueous solution containing a proprietary component. The density of this solution is 1140 kg/m^3 and its kinematical viscosity is $1.57 \cdot 10^{-6} m^2/s$ and $0.84 \cdot 10^{-6} m^2/s$ at 20°C and 40°C, respectively.

24.3.1.2 Bench scale experimental results

The mass transfer performance, expressed as a CO_2 flux through the membrane, is shown in Figure 24.8 for the CORAL20 liquid (2 M of active components) at room temperature as a function of the logarithmic CO_2 partial pressure (i.e. assuming plug flow). The data have been gathered from a variety of experiments at different gas and liquid flow rates, resulting in gas flow velocities between 0.05 and 0.45 m/s and liquid

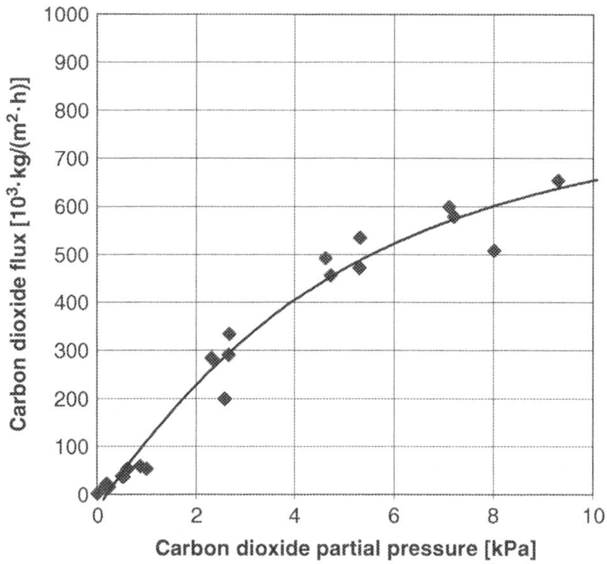

Figure 24.8 CO_2 membrane mass flux for CORAL20 at room temperature ($T = 20°C$) and low cyclic loading (<0.11 mol/mol) as a function of the logarithmic mean CO_2 partial pressure.

Table 24.3 Mass transfer analysis of the DAM.

Overall mass transfer coefficient (K)	Gas phase mass transfer coefficient ($k_g d_o/d_i$)	Membrane mass transfer coefficient (k_m)	Liquid phase mass transfer coefficient (mEk_l)
1.6×10^{-3} m/s	13–70 $\times 10^{-3}$ m/s	14×10^{-3} m/s	1.8–2.1 $\times 10^{-3}$ m/s

flow velocities between 0.7 and 30 mm/s. CO_2 partial pressures are typical for flue gas and life support applications. Cyclic loadings were up to 0.11 mol CO_2/mol of active component, which is much lower than the theoretically maximum loading of 0.50 mol/mol. The lean liquid loading at the absorber inlet was, typically, around 0.055 mol CO_2/mol of active component.

The experimentally observed CO_2 fluxes show a near linear increase at low CO_2 partial pressure, with a tendency to level off towards an asymptotic limit at higher CO_2 partial pressures (Figure 24.8). At low CO_2 partial pressure, the behaviour is described by the fast reaction regime with an overall mass transfer coefficient of 1.6×10^{-3} m/s, which is derived from the tangent line at the axis. This can be understood by the fact that at low carbon dioxide partial pressure, the concentration of the active components in the liquid is much larger than the CO_2 concentration in the liquid at the interface.

Table 24.3 shows the mass transfer analysis carried out using the mass transfer theory given in Section 24.2.3. Estimates for the mass transfer coefficient in the gas phase and the membrane were based on Equations (24.3)–(24.7). The liquid side mass transfer coefficient was derived from Equation (24.2) using the other mass transfer coefficient estimates of the DAM unit. The mass transfer analysis shows that the gas phase

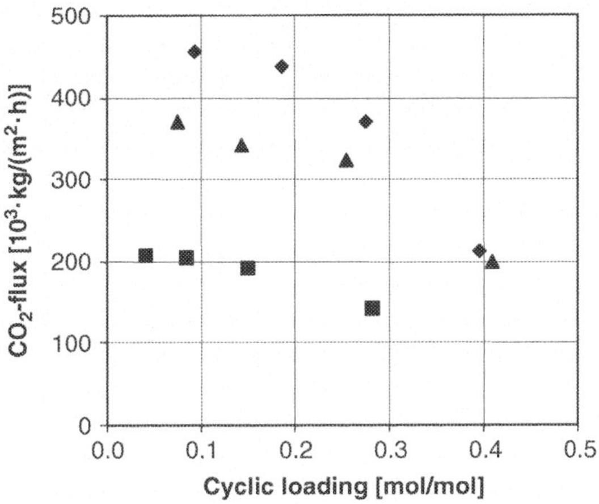

Figure 24.9 CO_2 membrane mass flux at room temperature (T = 25°C) for CORAL20 at three levels of the CO_2 concentration at the module inlet as a function of cyclic liquid loading. CO_2 concentration of 8.4 kPa (◆), 5.6 kPa (▲) and 2.8 kPa (■) at module entrance.

mass transfer resistance and the membrane mass transfer are generally much higher than the liquid phase mass transfer resistance. Nevertheless, they cannot be neglected totally and for further development it is necessary that the gas side hydrodynamics are optimized (increase in gas velocity) and the membrane mass transfer resistance is further reduced, for example, by using membranes with a smaller wall thickness.

The effect of the cyclic liquid loading on the CO_2 flux was investigated in a series of experiments at constant gas flow (4 (STP) m^3/h) with three levels of the CO_2 concentration at the module inlet during which the liquid flow rate (1.5–15 l/h) was varied (Figure 24.9).

The CO_2 flux decreases as the cyclic loading increased, which is expected because the active component will be consumed more and therefore less of it is available for the reaction with carbon dioxide at high liquid loadings. This will lead to a lower mass transfer coefficient, as the enhancement factor will be lowered (see e.g. Equation (24.11)). High cyclic loading is achieved by reducing the liquid flow rate and this will also lead to a lowering of the physical mass transfer coefficient (Equations (24.8) and (24.9)) as a result of the lower liquid velocity. This effect is, however, small as the effect of the *Re*-number on mass transfer is governed by a 1/3 power, at the most. Also at high loadings the CO_2 driving force will be affected and lowered. Cyclic loadings up to 0.4 mol/mol can be achieved, but at the expense of an up to 50% reduction in the CO_2 membrane mass flux. In an optimization of the overall process, the most favourable liquid flow rate (and hence liquid loading) will need to be defined. This will include the effect of the liquid flow rate on the desorption process (equipment size and energy consumption) and the associated heat transfer equipment.

Next, the effect of absorption temperature on the mass transfer performance was investigated. Higher reaction rates and diffusion coefficients are expected to result in higher mass transfer rates. This will be offset by the increased CO_2 back pressure at higher temperatures, which leads to a lower driving force. The measured CO_2 mass flux

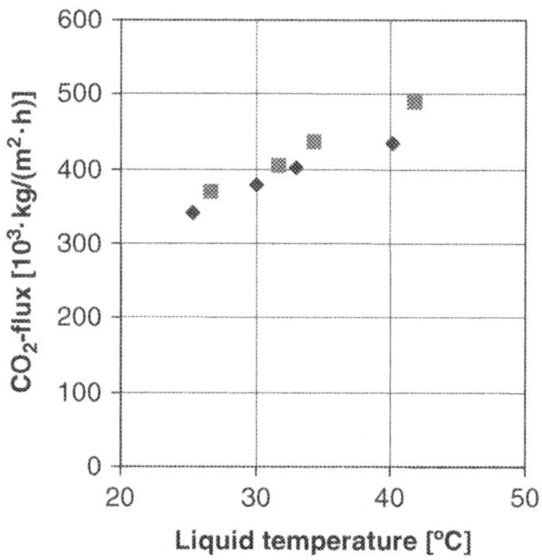

Figure 24.10 CO_2 membrane mass flux for CORAL20 at two flow rates at a CO_2 – partial pressure of 4.4 kPa at the module inlet as a function of temperature. Liquid flow rate of 15 l/h (◆) and 7.5 l/h (▩).

is shown in Figure 24.10 as a function of liquid temperature for two different flow rates, representing a liquid velocity of 0.01 and 0.02 m/s through the fibre lumen. The CO_2 partial pressure at the inlet was 4.4 kPa during these experiments. The cyclic liquid loading was 0.14–0.18 mol/mol, 0.07–0.10 mol/mol for the low and high flow rate, respectively. Figure 24.10 shows the benefits of operation at elevated temperature. Apparently, the combination of a higher reaction rate and faster diffusion rates is more important than the increased back pressure and the resulting reduction in driving force.

Figure 24.11 shows the benefits of operation at elevated temperature. Apparently, the combination of a higher reaction rate and faster diffusion rate are more important than the increased CO_2 back pressure and the resulting reduction in driving force. Using the experimental results, a comparison can be made with the information achieved by other researchers using various MGA systems. Experimentally determined overall mass transfer coefficients, K (Equation 24.2) are presented in Table 24.4.

Mass transfer coefficients are influenced by many variables, such as temperature, CO_2 partial pressure and liquid loading. This makes a comparison of experimental results of a MGA system from various literature sources very complicated. Nevertheless, Table 24.4 illustrates that the CORAL MGA system seems to give the highest mass transfer coefficient, despite the lower concentration of active components (2 M versus 5 M). Furthermore, the literature data were obtained from experiments in which the liquid was fed through the membrane modules and then discarded. Hence, the solvent did not contain CO_2 at the absorber inlet. In contrast with this, the results of this study were obtained in a system in which the liquid was continuously regenerated and reused. Hence, it gives a good representation of the performance in an actual plant. As the performance for the CORAL MGA system was determined for 2 M concentrations of the active components, it can be expected that the absorption rates still can be improved by using more concentrated CORAL and/or more reactive additives.

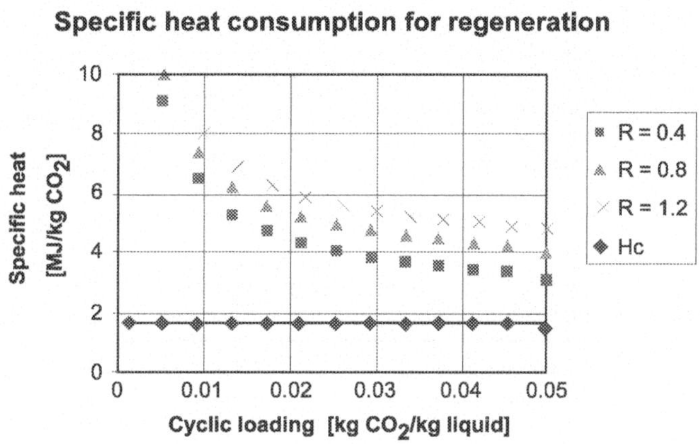

Figure 24.11 Specific heat requirement for CORAL20.

Table 24.4 Comparison of overall mass transfer coefficient at different operating conditions.

K (10^{-3} m/s)	MGA system	CO_2 at inlet (kPa)	Liquid loading (mol/mol)	Source
0.8 at 50°C	MEA (5 M) – PTFE Hollow fibres (d_i = 1.0 mm)	15	Ca. 0.02	Matsumoto et al. (1994)
1.3 at 50°C	MEA (5 M) – PTFE Hollow fibres (d_i = 1.0 mm)	15	Ca. 0.03	Nishikawa et al. (1995)
0.8–1.3 at 20°C	MEA (5 M) – PTFE Hollow fibres (d_i = 1.0 mm)	5.5–15.5	0.05–0.09	Nii and Takeuchi (1994)
0.3 at 25°C	MEA (5 M) – PTFE	0–40	Not known	Ledjeff (1999)
0.03 at 20°C	AMP (4%) – PTFE Hollow fibres (d_i = 0.9 mm)	40	Not known	Kim and Yang (2000)
0.15–1.0 at 25°C	K_2CO_3 (15%) – PP Hollow fibres 23 (d_i = 0.4 mm)	23	Not known	Chun et al. (1997)
1.6 at 20–25°C	CORAL (2 M) – PP Hollow fibres (d_i = 0.6 mm)	2–6	<0.11	This work

24.3.2 Pilot plant study

24.3.2.1 Experimental set-up

During the feasibility study, a pilot plant has been designed and built for testing in the lab as well as for testing on location with a real flue gas. The basic specifications of the pilot plant are shown in Table 24.5. Table 24.6 gives the specifications of the prototype cross-flow membrane module designed according to a TNO patent. The pilot plant

Table 24.5 Basic specifications of CO_2 removal pilot plant.

CO_2 concentration	3.5%
Gas flow rate	25 m^3/h
CO_2 removal	80%
Liquid flow rate	20–100 l/h

Table 24.6 Specifications prototype cross-flow membrane module.

Membrane area	7.5 m^2
Fibre material	Polypropene
Pore size	0.2 μm

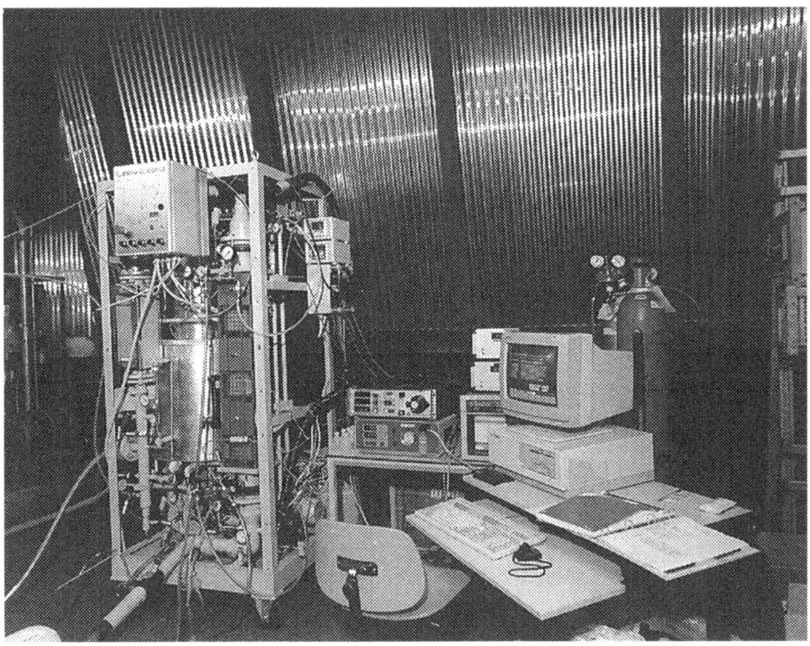

Figure 24.12 Photograph of pilot plant on location.

flow sheet is identical to the bench scale flow sheet. Figure 24.12 shows a picture of the pilot in operation using a side stream from a power plant exhaust.

24.3.2.2 *CO_2 product quality*

Contaminants and boundary conditions Applying the produced CO_2 as a fertilizer in the greenhouses means that the CO_2 product has to meet certain purity requirements. Concentrations of NO, NO_2, C_2H_4, NH_3 and O_3 in the greenhouse are not allowed to rise above defined boundary values. Nitrous oxides (NO, NO_2) and ethene

Table 24.7 Boundary values (vppm) for harmful components in fertilizer gas (1000 ppm CO_2 in greenhouse).

Component	100% CO_2	11.7% CO_2[a]	10% CO_2[b]	7.3% CO_2[c]	3.5% CO_2[d]
NO	290.9	34.0	29.1	21.2	10.2
NO_2	156.2	18.3	15.6	11.4	5.5
C_2H_4	7.4	0.9	0.7	0.5	0.3
NH_3	252.6	25.3	25.3	18.4	8.8

(a) Flue gas from a gas engine (stoichiometrical)
(b) Flue gas composition to pilot plant
(c) Flue gas from a gas engine (with overdose of air; $\lambda = 1.55$)
(d) Flue gas from a gas turbine combined cycle

(C_2H_4) are present in the flue gases while ammonia (NH_3) can be present as a product of the degradation of the absorption liquids used. Ozone (O_3) is kept out of this evaluation, because there is no reason to expect this molecule which is very instable.

The boundary conditions for the allowed concentrations of these components in the fertilizer gas in general depend on the CO_2 concentration in the fertilizer gas and the desired CO_2 concentration in the greenhouse. Therefore, a procedure has been defined in which two processes for CO_2 transport are described (Proefstation, 1999): air exchange outside the greenhouse and CO_2 intake by the plants (photosynthesis). In winter time with closed windows, ventilation of 0.5 h^{-1} and little photosynthesis there is a 0.5 g/(m^2h) CO_2 intake. The methodology of Proefstation is applied to calculate the allowable concentrations of harmful substances in a fertilizer gas consisting of flue gas or mainly out of CO_2 existing gas. In Table 24.7 the results of these calculations are presented with a desired value of 1000 ppm CO_2 in the greenhouse. The calculated values are only indicative. Interaction of a cocktail of harmful components is not taken into account. Table 24.6 shows that when CO_2 has a higher concentration in the fertilizer gas, the allowable concentration of contaminants may also be higher. This is the result of the fact that the ratio between CO_2 and contaminants determines when the limits are reached. At lower CO_2 concentrations in the flue gas, the limits will also be lower. The contaminants in almost 100% CO_2 may have considerable higher values.

The measured concentrations of the contaminants were, as well for the averages as for the maximum concentrations, for all components lower than the limits calculated for pure CO_2 (Table 24.7). The ammonia content corresponds to a solvent consumption of 0.1 kg/ton CO_2. This is deduced from the 40 ppm in the CO_2 product, assuming that all the ammonia is derived from the amino acid. This chemical consumption is significantly lower compared to the conventional MEA-installation of 1.5 kg solvent per ton CO_2. For ethene it was expected that the concentration in the CO_2 product also would also be lower than the concentration in the flue gas, because ethene can only be transported by a physical solution. However, ethene occurs in a higher concentration in the CO_2 product. This suggests that the ethene is formed in the liquid or equipment. No limit is determined for CO with the above described method, because CO_2 dosing from central heating units are mostly secured with a flue gas fan which will stop operating if the CO content rises above 30 ppm.

Table 24.8 Experimental conditions and feed gas composition during CO_2-quality experiments.

Gas flow rate	10–15 (STP)m^3/h
Liquid flow rate	60 l/h
Absorber temperature	40°C
Regeneration temperature	120°C
Inlet composition	
N_2	75.0 mol%
CO_2	15.0 mol%
H_2O	6.40 mol%
O_2	3.50 mol%
SO_2	490 ppm
NO	94 ppm
CO	80 ppm

Table 24.9 Purity of CO_2 product.

Absorption liquid	Number of measurements	Average purity	Standard deviation	Minimum purity	Theoretical minimum purity
CORAL20	13	99.7%	0.1%	99.55%	98.8%

Contaminant concentration in simulated flue gases from solid fuel firing In a separate experiment the pilot plant was operated in the laboratory with a special feed gas mixture to which impurities were added. These impurities were chosen to mimic a certain flue gas composition from solid fuel firing. In Table 24.8 the experimental conditions and feed gas composition are presented. These experiments were aimed at the determination of the influence of flue gas impurities on the CO_2 purity, as well as O_2 and SO_2 behaviour in the processor.

It is expected that the absorption liquids for CO_2 removal are generally inert towards most gases present in the flue gas except SO_2 which is an acid gas like CO_2. As such it is expected to react with the alkaline absorption liquids where it can be accumulated or alternatively it might be released into the CO_2 product during the regeneration process. Gas components other than CO_2 and SO_2 are expected to be physically absorbed into the liquid and will be released into the CO_2 product during regeneration of the liquid. The physical absorption of these gases into the liquids is however very limited compared to the chemical sorption.

The overall concentration of CO_2 in the CO_2 product was determined using a CO_2 purity tester (Table 24.9). The theoretically expected minimum purity is estimated under the following assumptions: (1) liquid leaving the absorber is completely saturated with the inert gases O_2 and N_2 present in flue gas, (2) solubility coefficients for water are applicable and (3) liquid regeneration leads to complete desorption of inert gases. The measured CO_2 purity is higher than the theoretically estimated purity. This might be caused by the salting-out effect of the CORAL liquid leading to lower solubility or the liquid is not completely saturated with inert gases in the membrane absorber due to a short residence time. It is apparent from Tables 24.9 and 24.10 that there is less transfer of inert gases from the feed gas stream to the CO_2 product than expected from theory.

Table 24.10 Measured O_2-concentration in CO_2 product.

Absorption liquid	Number of measurements	Average O_2 concentration	Standard deviation	Theoretical O_2 concentration
CORAL20	3	460 ppm	10 ppm	1020 ppm

Table 24.11 Measured SO_2 concentration levels.

SO_2 at membrane absorber inlet	490 ppm
SO_2 at membrane absorber outlet	130 ppm
SO_2 in CO_2 product	1.5 ppm

Table 24.12 Average concentrations (vppm) of contaminants in flue gas and CO_2 product.

Component	Flue gas	CO_2 product
CO	25.1	0.4
NO	5.3	0.35
NO_x	7.4	0.50
NH_3	0.21	27
C_2H_4	0.2	2.0

SO_2 present in the flue gas is substantially removed (Table 24.11) and accumulates in the absorption liquid. It is expected that it will react to sulfite (SO_3^{2-}) under the alkaline conditions. The presence of oxygen might also lead to the formation of sulfate (SO_4^{2-}). It is not known whether the SO_2 will also directly react with the amino acids present in the solution. In both cases a decrease of absorption capacity will be the end result. This means that the solutions will have to be treated in a reclaimer or they will have to be renewed more frequently. The expected SO_2-contamination of the CO_2 product is rounded off and taken to be 2 ppm in a real plant.

Contaminant concentration in flue gases from gas firing The CO_2 product quality from the MGA process was determined experimentally with the pilot plant operated on site with flue gas from a gas fired boiler (Figure 24.12). Table 24.12 shows that for CO and NO_x the concentrations in the CO_2 product are lower than in the feed gas, whereas for NH_3 and ethane the concentrations in the CO_2 product are higher than in the feed gas. However, for all contaminants the concentrations in the CO_2 product are lower than the boundary values mentioned in Table 24.7.

24.3.3 Heat requirement

The energy consumption of the process is dominantly determined by the heat required for the regeneration of the absorption liquids. Heat is required for:

- Desorption of carbon dioxide. The type of absorption liquid will determine the desorption heat.

- Evaporation of water to carry out the desorption or stripping process. The type of absorption liquid and the design of the stripping process will determine the required amount of water to be evaporated. This is expressed by the reflux ratio, Rf, representing the amount of water vapour per amount of carbon dioxide (kg H_2O/kg CO_2).
- Heating of liquid. Liquid loading and heat exchanger size will determine the required net energy for the liquid heating.

In Figure 24.11 the specific heat consumption is calculated for a number of reflux ratios as a function of liquid loading. The desorption energy (H_c = 1.36 MJ/kg) is determined from equilibrium measurements. Furthermore a temperature difference of 10 K is assumed in the heat exchanger. Figure 24.12 shows that decreasing the reflux ratio is an equally effective method for reducing the heat consumption as increasing the liquid loading. For the economic evaluation conservative figures are used, that is a specific heat consumption of 5 MJ/kg based on a liquid loading of 0.020 kg/kg and a reflux ratio equal to 1.

24.4 CONCLUSIONS

- The mass transfer resistance in the CORAL MGA process is dominated by the transport rates in the liquid phase. The measured CO_2 membrane absorption rates increase with the CO_2 partial pressure, but start to level off towards an asymptotic value which is in accordance with standard theory for mass transfer enhanced by a chemical reaction. The CO_2 membrane mass flux decreases as the cyclic liquid loading is increased. Loadings of up to 0.4 mol CO_2/mol of active component have been achieved in the pilot plant experiments. The CO_2 membrane mass flux increases with increasing temperature from 25°C to 40°C.
- The CORAL liquids also show a higher mass transfer rate when compared with other MGA systems, currently under investigation. Given the scope for further improvement, the CORAL MGA process is an interesting candidate for CO_2 removal from a large number of CO_2 containing gases.
- The measured concentrations of the contaminants both the averages and the maxima, were for all components lower than the limits calculated for pure CO_2.
- At long operating times (>4000 h) the MGA technology in combination with storage will be competitive with the usual practice of feeding the products of natural gas combustion directly in the greenhouse.

NOMENCLATURE

$C_{a,i}$ — CO_2 concentration at gas–liquid interface (kmol/m^3)

C_b — bulk concentration of active component (kmol/m^3)

c_g — concentration in gas phase (kmol/ m^3)

c_l^* — gas phase concentration in equilibrium with bulk liquid phase (kmol/m^3)

$D_{g,b}$ bulk CO_2 diffusion coefficient in gas phase (m^2/s)
$D_{g,c}$ effective CO_2 diffusion coefficient in gas-filled pores (m^2/s)
D_l bulk diffusion coefficient in liquid phase (m^2/s)
$D_{l,a}$ CO_2 diffusion coefficient in liquid phase (m^2/s)
$D_{l,b}$ diffusion coefficient of active component in liquid phase (m^2/s)
D_k Knudsen diffusion coefficient of CO_2 (m^2/s)
d_i inner diameter of hollow fibre membrane (m)
d_m membrane thickness (m)
d_o outer diameter of hollow fibre membrane (m)
d_p pore diameter (m)
E enhancement factor due to chemical reaction (–)
Hc desorption energy (MJ/kg)
K overall mass transfer coefficient (m/s)
k_g gas side mass transfer coefficient (m/s)
k_m membrane mass transfer coefficient (m/s)
k_l liquid phase mass transfer coefficient (m/s)
k_r second-order reaction rate constant (m^3/(kmol s))
l length of hollow fibre membrane (m)
M molecular mass (kg/kmol)
m distribution coefficient between gas and liquid phase (–)
n number of molecules active in reaction with CO_2 (–)
P_m permeance (kmol/(m^2 s Pa))
p_g partial pressure in gas phase (Pa)
p_l^* partial pressure in equilibrium with bulk liquid phase (Pa)
R gas constant (J/(kmol K))
Rf reflux ratio (–)
Re Reynolds number (–)
Sh Sherwood number (–)
Sc Schmidt number (–)
T temperature (K)

Greek

ε membrane porosity (–)
φ_{CO_2} CO_2 flux (kmol/(m^2s))
τ membrane tortuosity (–)

REFERENCES

Chun, M.S., Lee, K.H. and Lee. K.H. (1997) Analysis on a hydrophobic hollow-fiber membrane and experimental observations of CO_2 removal by enhanced absorption. *Separ. Sci. Technol.* **32**(15), 2445–2466.

Dindore, V.Y., Brilman, D.W. and Versteeg, G.F. (2005) Modelling of cross flow membrane contactors: physical mass transfer processes. *J. Membr. Sci.* **251**, 209–222.

Eckhard, F., Feron, P.H.M. and Witt, J. (1996) Evaluation of a membrane based carbon dioxide absorber for spacecraft ECLS applications. Paper no. 961369. *Proceedings of the 26th International Conference on Environmental Systems*, California.

Eckhard, F., Feron, P.H.M., Jacobs, P., Savage, Ch. and Witt, J. (1999) Integrated CO_2, humidity and thermal control by membrane gas absorption. *Proceedings of Workshop on Advanced Life Support*. ESTEC, Noordwijk, The Netherlands.

Falk-Pedersen, O. and Dannström, H. (1997) Separation of carbon dioxide from offshore gas turbine exhaust. *Energ. Convers. Mgmt.* **38**(Suppl.), 81–86.

Feron, P.H.M., Jansen, A.E. and Klaassen, R. (1992) Membrane technology in carbon dioxide removal. *Energ. Convers. Mgmt.* **33**(5–8), 421–428.

Feron, P.H.M. (1994) Membranes for carbon dioxide recovery from power plants. In *Carbon Dioxide Chemistry "Environmental Issues"* (eds. J. Paul and C.M. Pradier), Vol. 24, pp. 236–249. The Royal Society of Chemistry, Cambridge.

Feron, P.H.M. and Jansen, A.E. (1995) Capture of carbon dioxide using membrane gas absorption and reuse in the horticultural industry. *Energ. Convers. Mgmt.* **36**(6–9), 411–414.

Feron, P.H.M. and Jansen, A.E. (1997) The production of carbon dioxide from flue gas by membrane gas absorption. *Energ. Convers. Mgmt.* **38**(Suppl.), 93–98.

Feron, P.H.M. and Jansen, A.E. (1999) *Techno-economic Assessment of Membrane Gas Absorption for the Production of Carbon Dioxide from Flue Gas*, pp. 53–58. In: (ed. B. Eliasson et al.) Greenhouse Gas Control Technologies, Pergamon, Amsterdam.

IEA Report (1992) Carbon dioxide capture from power station. Obtainable from *IEA Greenhouse Gas R&D Programme*, IEA Publication http://www.ieagreen.org.uk/ (study OE8).

Jacobs, P., Feron, P.H.M., Savage, C.J. and Witt, J. (1997) Integrated CO_2 and humidity control by membrane gas absorption. *Proceedings of the 6th European Symposium on Space Environmental Control Systems*.

Jansen, A.E. and Feron, P.H.M. (1995) Method for gas absorption across a membrane. EP 0751 815 and USP 5,749,941.

Kim, Y.S. and Yang, S.M. (2000) Absorption of carbon dioxide through hollow fibre membranes using various aqueous absorbents. *Separ. Purif. Technol.* **21**, 101–109.

Kreulen, H., Smolders, C.A., Versteeg, G.F. and Van Swaaij, W.P.M. (1993) Microporous hollow fibre membrane modules as gas–liquid contactors, Part 2: Mass transfer with chemical reaction. *J. Membr. Sci.* **78**, 217–238.

Kumar, P.S., Hoogendoorn, J.A., Feron, P.H.M. and Versteeg, G.F. (2003) Reaction of carbon dioxide with aqueous potassium salt solutions of taurine and glycine. *AIChE. J.* **49**(1), 203–213.

Ledjeff-Hey, K., Roes, J. and Wolters, R. (1999) Gasreinigungsverfahren für de PEFC-Brenstoffzelle – Membrankontaktoren für die Absorption van Kohlendioxid 385–388 in Preprints Aachener Membran Kolloquium, 9–11 März 1999, VDI GVC, Düsseldorf.

Matsumoto, H., Kamata, T., Kitamura, H., Ishibashi, M., Ohta, H. and Nishikawa, N. (1994) Fundamental study of CO_2 removal from the flue gas of thermal power plant by hollow-fiber gas–liquid contactor. In *Carbon Dioxide Chemistry: Environmental Issues* (eds. J. Paul, and C.M. Pradier), Vol. 27, pp. 270–281. The Royal Society of Chemistry, Cambridge.

Nii, S. and Takeuchi, H. (1994) Removal of CO_2 and/or SO_2 from gas streams by a membrane absorption method. *Gas Separ. Purif.* **8**(2), 107–114.

Nishikawa, N., Ishibashi, M., Ohta, H., Akutsu, N., Matsumoto, H., Kamata, T. and Kitamura, H. (1995) CO_2 removal by hollow-fiber gas–liquid contactor. *Energ. Convers. Mgmt.* **36**(6–9), 415–418.

Proefstation voor Bloemisterij en Glasgroente (1999) CO_2 in de glastuinbouw, 3e druk. Aalsmeer/Naaldwijk, The Netherlands.

Rangwala, H.A., (1996) Absorption of carbon dioxide into aqueous solutions using hollow fibre membrane contactors. *J. Membr. Sci.* **112**, 229–240.

Ter Meulen, B.Ph., (1991) Transfer device for the transfer of matter and/or heat from one medium flow to another medium flow, WO 91/09668.

Westerterp, K.R., Van Swaaij, W.P.M. and Beenackers, A.A.C.M. (1984) *Chemical Reactor Design and Operation*. Wiley & Sons.

25

Treatment of carbon disulfide from industrial waste gas emissions: pollutant converted into reusable sulfuric acid

N.J.R. Kraakman and K.J.A. de Waal

25.1 INTRODUCTION

25.1.1 Quantities of carbon disulfide emissions

Carbon disulfide (CS_2) is released into the atmosphere from a variety of natural and anthropogenic sources. Industry is the main emission source of CS_2 and contributes to 58% of the total release to the atmosphere (Smet et al., 1998). CS_2 serves as an excellent solvent for most organic chemicals and is therefore used in different production processes (Plas et al., 1993). Industries emitting CS_2 include production of cellulose-based products such as rayon fibers and filaments, cellophane, casings, sponges, and other manufacturing processes such as the production of fungicides or vulcanizing. However, the main contributions are from the processing of cellulose-based products. It is estimated that worldwide more than 100 industrial rayon plants are emitting a total of more than 250,000 tons of CS_2 per year (Bioway, 1999) and that the total anthropogenic emissions could be as high as 1 million tons per year (Kelly & Smith, 1990).

25.1.2 Adverse effects of carbon disulfide

Treatment of carbon disulfide emissions is required for health, safety, and environmental reasons. Emissions of CS_2 can create health problems since it is a potent neurotoxic for humans (Beauchamp et al., 1993; WHO, 1996). Acute poisoning from CS_2 is rare and most reported toxicity has occurred from repeated exposure to low concentrations.

CS_2 emissions contribute to the greenhouse effect. CS_2 is unstable in the atmosphere and has a lifetime of 7–12 days. It forms carbonylsulfide (COS) which is the most abundant atmospheric sulfur gas with a lifetime of at least 1 year (Kelly & Smith, 1990). COS may be transported to the stratosphere where its oxidation is thought to be a major source for stratospheric sulfur aerosol. This aerosol influences the earth's radiation balance and hence the climate (Chin & Davis, 1993).

There are also nuisance problems related to the odor of CS_2. It has a low odor threshold (9.6 ppb_v) similar to hydrogen sulfide (1 ppb_v), which is commonly present in CS_2-contaminated waste gas streams.

25.1.3 Carbon disulfide treatment

Regulations of CS_2 emissions have been recently updated in the USA (EPA, 2002) and in Europe (EU, 2005). The basis for regulating CS_2 emissions is its toxicity, its environmental impact and explosion, and fire hazards. The focus is in the first place on occupational hazards. In the EU and the USA the Occupational Exposure Limit, defined as the exposure during an 8-h period, is set at a maximum of 10 ppmv for CS_2. In the near future the Committee for Occupational Exposure Limit strongly recommends to reduce the maximum to 1–5 ppmv CS_2. In the EU, emissions to the atmosphere from industries are set at maximum concentration of 20 mg CS_2/m^3 or about 6.5 ppmv.

Traditionally, industrial emissions of CS_2 have been treated by adsorption or incineration. These conventional treatment systems have relatively high operational costs and use substantial amounts of energy. They also require many safety measures for explosion-proof operation and generate waste streams (e.g. SO_2 emissions or spent activated carbon). Biological processes for waste gas treatment represent an interesting alternative. They have been applied more frequently for industrial applications, because they eliminate many of the drawbacks of classical physical–chemical techniques (Kennes & Veiga, 2001; Shareefdeen & Singh, 2005).

This chapter describes a newly developed biological technology for the treatment of carbon disulfide vapors. Microbiological conversion of carbon disulfide is discussed, biological and traditional carbon disulfide waste gas treatment is illustrated, examples of full-scale bioreactor applications are shown and opportunities to reuse the biologically produced dilute sulfuric acid are discussed.

25.2 BIOLOGICAL CARBON DISULFIDE DEGRADATION

25.2.1 Microbial carbon disulfide degradation

The biological formation and degradation of hydrogen sulfide (H_2S) is well understood, but not much knowledge is available on the biotransformations of CS_2. The conversions of organic and inorganic sulfur compounds are performed primarily by archaea and bacteria (Lomans, 2001). Microbial growth on CS_2 as sole source of carbon and energy appears to be very limited. CS_2 has both bacteriostatic and fungicidal properties, and has been shown to be a strong inhibitor of nitrification (Hartel & Haines, 1992). One of the first articles describing CS_2 degrading microorganisms was by Smith and Kelly (1988). The autotrophic *Thiobacillus thioparus* strain TK-m degraded CS_2 to CO_2 and H_2S. This relatively small rod-shaped bacterium showed a very slow growth rate on CS_2. For optimal growth of this bacterium, a pH between 7.0 and 7.4 is required. At a pH of 6.1 and 7.7, the growth rate was halved and below a pH of 5.6, no growth occurred. The subsequent oxidation of H_2S is the only energy source in the process. CS_2 and COS must therefore in that case be regarded exclusively as chemolithophic substrates. Elemental sulfur, which is chemically stable in the presence of oxygen in most pH neutral environments, is an intermediate in the degradation of H_2S and CS_2, but it is readily oxidized by many sulfur-oxidizing bacteria. Carbon for cell growth comes from CO_2, of which about 90% is from the air and 10% from the CS_2 after oxidation to CO_2. The affinity for CS_2 is relatively low ($K_s \sim 16.5\,\mu M$), the maximum activity observed (V_{max}) was 524 nmol O_2 per min per mg protein and the growth yield (Y) appeared to be 7.9 ± 0.9 g cell carbon per mol CS_2.

Since then, several researchers have shown that the microbiological degradation of CS_2 is possible by obligate chemolithoautotroph as well as facultative heterotrophic bacteria (Alcantara et al., 1999; Hartikainen et al., 2000; Jordan et al., 1995; Plas et al., 1991). Most studies indicate that a neutral pH is required for the oxidation of CS_2. The optimum temperature for these mesophilic bacteria is in general between 25°C and 35°C. The self-inhibition of the process by CS_2 differs greatly: Plas et al. (1993) reported total inhibition at CS_2 concentrations higher than 2 mM (corresponding to 95 g/m^3 in a gas phase at 25°C), whereas Jordan et al. (1995) reported CS_2 inhibition above 47 μM (corresponding to 2.2 g/m^3 in a gas-phase at 25°C).

25.2.2 Carbon disulfide degrading microorganisms isolated from full-scale bioreactors

25.2.2.1 Physiological characteristics

An adapted bacterium called strain G8 (Figure 25.1) was isolated from one of the full-scale bioreactors featured in Section 25.5.2. Isolation was difficult and special methods were required to obtain pure cultures. The Gram-negative rod-shaped motile bacterium degraded CS_2 via COS and H_2S as illustrated in Figure 25.2. Different fungi were also found both in the full-scale and in laboratory scale bioreactors, but were not capable of degrading CS_2.

Strain G8 can treat CS_2 concentrations up to at least 3000 ppmv in the gas-phase (data not shown). The guideline for the ventilation of areas without human activities calls for CS_2 concentration in the area to remain below 10% of the lower explosive limit of CS_2 (1500 ppmv). Thus, the bacterium can treat CS_2 concentrations greater than those expected in the off-gases from an industrial plant emitting CS_2. The bacterium was able to degrade CS_2 at a very low pH (<0.5). Therefore, it can be considered as an extreme acidotolerant.

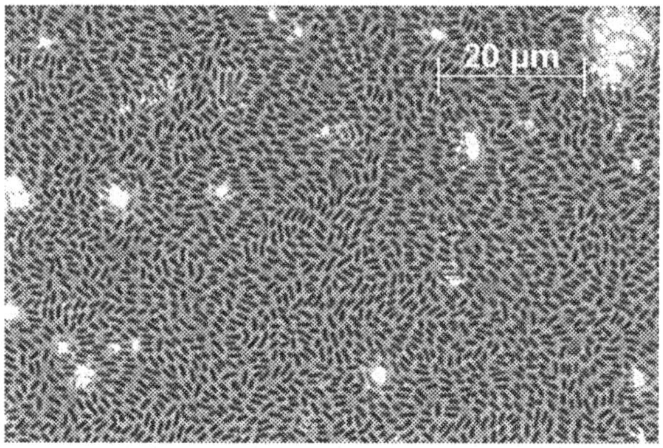

Figure 25.1 Optical microscope picture of the extreme acidophile bacterium (strain G8) isolated from full-scale bioreactors treating CS_2 emissions.

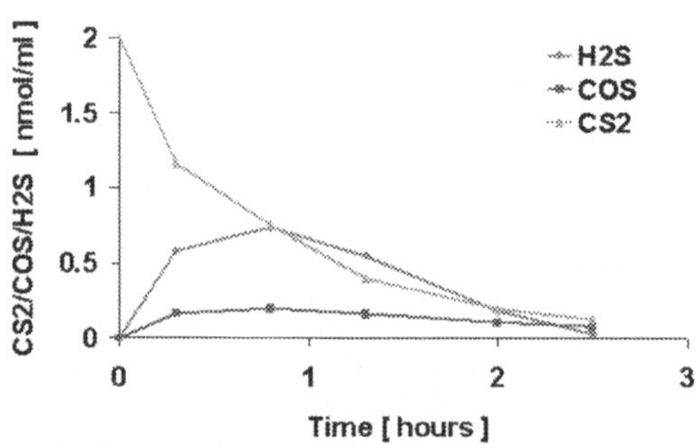

Figure 25.2 Degradation of CS_2 by strain G8 in shake flask. Breakdown of CS_2 proceeds via COS and H_2S as intermediates.

25.2.2.2 Kinetic parameters

Normally pollutants in waste gases need to be reduced below a certain concentration, in general determined by regulatory agencies. This maximum concentration defines the treatment objective for an effective waste gas treatment. The required maximum emission concentration in the gas phase can be converted to a required maximum liquid-phase concentration using the gas–liquid partition coefficient. For an optimal microbial degradation rate, the substrate affinity constant of a microorganism should be lower or of the same order of magnitude than the above required maximum liquid-phase concentration. In other words, a biological waste gas treatment process can only be effective if the substrate affinity constant of a microorganism multiplied by the gas–liquid partition coefficient results in a low enough bioreactor effluent concentration.

The affinity for CS_2 of the isolated bacterium was relatively high when compared to other bacteria described in the literature. Even so, the affinity constant (K_s) was lower than 1 μM (data not shown), which corresponds to a gas-phase concentration of 10 ppmv at 20°C. Consequently, treatment leading to effluent concentrations lower than 10 ppmv can be accomplished if required.

25.3 BIOREACTORS TREATING CARBON DISULFIDE WASTE GAS EMISSIONS

25.3.1 Introduction

The microbiological degradation of CS_2 has only recently been investigated. Consequently, bioreactors treating CS_2 emissions are not yet very common. Only a few pilot- or full-scale biological air treatment applications have been described (Hugler et al., 1999; Windsperger, 1990; Yang & Alibeckoff, 1995). In many cases, problems such as slow start-up, unstable operations, pressure drop increase over time, and the requirement to frequently change the bioreactor medium were reported. In addition, most of these pilot- or full-scale biological treatment systems included caustic addition to operate at a neutral pH, making it difficult to reuse the dilute sulfuric acid produced from the degradation of CS_2.

25.3.2 New developed bioreactors for full-scale applications

The full-scale bioreactors treating CS_2 emissions illustrated in Section 25.5.3 are counter-current bioreactors with three layers of packing (Kraakman, 2003). Each layer can be irrigated separately with water and nutrients. The process water in the bioreactor can be recirculated. Each layer consists of an inert structured packed media to provide surface for biofilm growth and to provide an optimal contact between the gas phase and the biofilm. The pH, freshwater supply and nutrient dosing are automated. All process parameters such as temperature, pH, salt content, water, and nutrient flow as well as air and water pressure are measured continuously and recorded by means of a programmable logic controller and a SCADA operator interface system. The process air is normally humidified prior to the bioreactor to prevent drying of the biofilm in the lower part of the bioreactor. The humidification also serves for dust and particulate matter control. A standard bioreactor has a footprint of $11\,m^2$. The design of the bioreactor typically presumes a removal efficiency of 80% for CS_2 and 90% for H_2S, but higher removal efficiencies are possible.

25.3.2.1 Start-up and removal efficiencies

The start-up period depends on the inlet air conditions, such as temperature and CS_2 concentrations. A typical start-up of a bioreactor takes about 6–8 weeks (Figure 25.3). The biodegradation of CS_2 is much more difficult than that of H_2S, as can be seen by the removal efficiencies of both compounds during the start-up phase.

The removal capacity depends to a great extent on the pollutant load of the bioreactor. Removal capacity data collected over 2 years of operation are plotted versus the loading in Figure 25.4. A removal capacity of more than 10 kg per bioreactor per hour can be reached. These data were obtained from a bioreactor system treating CS_2 concentrations between 100 and 500 ppmv and H_2S concentrations between 125 and 550 ppmv. The temperature of the inlet air fluctuated between 22°C and 41°C and the pH of the drain water was around 0.5. Figure 25.4 shows that the overall sulfur removal exceeded 80%.

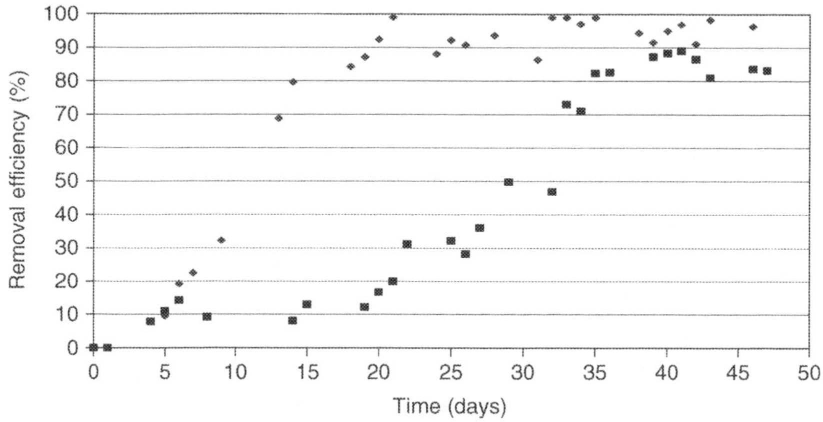

Figure 25.3 A typical start-up of the bioreactor treating CS_2 (square symbols) and H_2S (diamonds) emissions at a viscose plant.

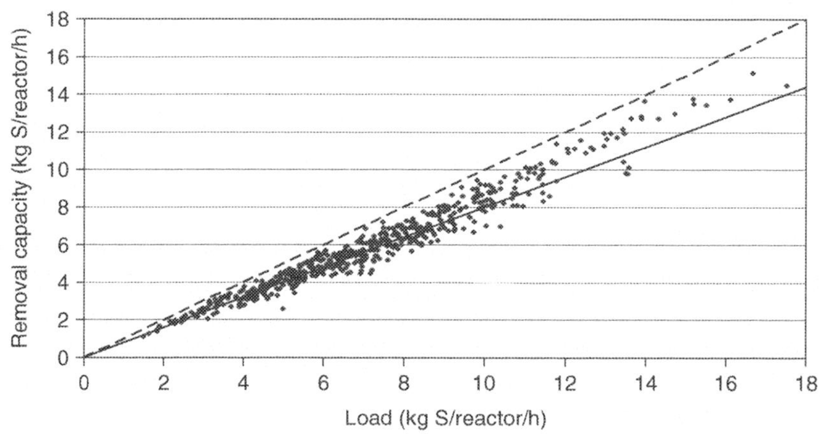

Figure 25.4 Removal capacity of sulfur versus loading in a full-scale bioreactor treating CS_2 and H_2S emissions at a viscose plant. The lines represent 80% (—) and 100% (- - -) removal.

25.3.2.2 Robustness of a biological system

Robustness is important and any biological system for air pollution control must be reliable under the variable conditions experienced at industrial sites. Robustness is defined here to reflect the ability of the biological system to deal with fluctuations and the ability to recover after operational failures. Biological systems are influenced by parameters such as pH, water content, temperature and the concentrations of the carbon sources, energy sources, and nutrients. The robustness of biological systems can be demonstrated with well-designed experiments in the laboratory or with pilot- and full-scale experiments in an industrial environment.

Robustness can be quantified by determining the probability of a negative effect on the biological system for each possible upset, multiplied by the frequency of the particular upset, and summing over all possible upsets. The risk of negative effects on a biological system (R) can be defined as:

$$R = \Sigma(p \times E)$$

P = the probability of occurrence of a fluctuation or upset
E = the negative effect of the fluctuation or upset

The probability of the occurrence of a fluctuation (P) can be expressed as the expected number of occurrences per year (number/year) or as the percentage of operating time during which it is likely to occur. The negative effect of the fluctuation (E) can be expressed as the loss of the removal efficiency (%), the loss of the total removal (kg/day or kg/year), or the impact on the people living near the installation (e.g. the number of occurrences during which the concentration of the emitted air stream exceeds the odor threshold in the neighborhood).

The microbiological community in the bioreactor system is subjected to fluctuations related to the production process (changes of pollutant inlet concentrations and operating conditions such as flow and temperature). These fluctuations result from continuous or discontinuous processes, unplanned shut-downs (production process failures), planned shut-downs (holiday/maintenance), and diurnal fluctuations. There may also be fluctuations related to the operation of the waste gas purification system, for example, associated with failure of the nutrient supply, water recirculation flow or with changes in the pH of the support medium.

A bench-scale experiment illustrated that after a shutdown period of 8 days, the removal efficiency was about 20% shortly after the restart and the full recovery took about 3–4 days (Figure 25.5). This lab experiment was in good agreement with the data obtained in a full-scale bioreactor system at an industrial site for which the recovery of full removal efficiency after 9 days of pollutant starvation took about 4 days (Kraakman, 2003).

Many examples of the responses of bioreactors to fluctuations or upsets are shown elsewhere (Kraakman, 2003). Based on field experience, the overall risk for operating this type of bioreactor system at a viscose plant is estimated at less than 4% loss of the total pollutant removal per year.

When the robustness is quantified and the risks are understood, extra measures can be taken to reduce the risks. The following measures are recommended in full-scale bioreactor systems:

- *Automate the process control by introducing two or more levels of alarm for the critical process parameters:* the first level of alarm could warn the operator (e.g. "the inlet air temperature is higher than normal") and the second step could be an automatic action (e.g. automatic opening of a valve to reroute the inlet air in order to prevent hot air from coming in contact with the biomass in the air purification system).
- *Install critical backup equipment.* Pumps and fans are critical for maintaining the required optimal environmental conditions for the biomass. Installation of a second fan ensures that the microorganisms in the system will have enough oxygen and that damaging anaerobic conditions will be avoided. If one fan breaks down, the

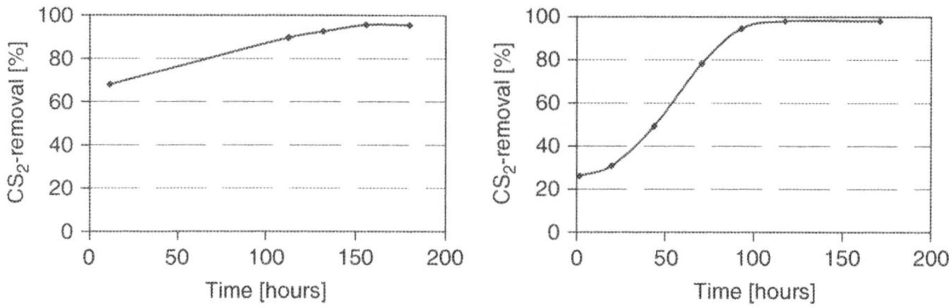

Figure 25.5 Recovery of a bench-scale bioreactor after 84 h (3.5 days; left) and 192 h (8 days; right) of absence of approximately 200 ppmv CS_2 in the inlet air stream.

backup fan should be turned on automatically. Backup equipment is operated in tandem to ensure that the backup equipment is maintained properly.

- *Design the bioreactor system for situations in which failures can occur.* A valve is installed in such a way that in the event of an electrical power failure, the valve will automatically turn to the position that is least damaging for the microorganisms (e.g. some water inlet valves are opened when the electrical power is off to prevent drying of the biofilm).
- *Specify specific operational procedures (SOPs) for operators.* SOPs are provided to operators of industrial installations with appropriate directions for actions to be taken during process interruptions or fluctuations. These procedures are developed in close cooperation with specific plant personnel.

25.3.2.3 Formation of sulfuric acid and possibilities for reuse

The biological degradation of CS_2 and H_2S in the bioreactor produces sulfuric acid (see Equations 25.1–25.4). During the first hydrolysis of CS_2, COS and H_2S are formed:

$$CS_2 + H_2O \rightarrow COS + H_2S \tag{25.1}$$

COS is further hydrolyzed to CO_2 and H_2S:

$$COS + H_2O \rightarrow CO_2 + H_2S \tag{25.2}$$

Finally, the H_2S is oxidized to sulfuric acid:

$$H_2S + 2O_2 \rightarrow SO_4^{2-} + 2H^+ \tag{25.3}$$

The overall process reaction is thus:

$$CS_2 + 2H_2O + 4O_2 \rightarrow CO_2 + 2H_2SO_4 \tag{25.4}$$

The chemolitho-autotrophic microorganisms are acidophiles that can degrade CS_2 at pH around 0.5. The bioreactors can be operated at low pH in order to concentrate the produced sulfuric acid. 1 kg CS_2 forms 2.6 kg H_2SO_4 and 1 kg H_2S forms 2.9 kg H_2SO_4.

The produced sulfuric acid is a clear liquid with a concentration of approximately 3–5% (w/w). This dilution limits the applicability for acid reuse. As such, the dilute acid can be used as rinsing liquid in a production process. Another possible application is the use as pickling acid in the metal cleaning industry. Neutralization of alkaline waste is also a possibility.

The sulfuric acid that is produced in a bioreactor can be neutralized by caustic soda in the bioreactor to form Na_2SO_4. The recovery of sulfuric acid and caustic from this salt using electrodialysis was investigated (Melse & Kraakman, 1998). Electrodialysis is an electrochemical separation process in which ions are removed from a less concentrated feed solution into a more concentrated solution as a result of the driving force of an applied electrical potential gradient. Tests at a rayon fiber production plant of this electrodialysis process in combination with a pilot bioreactor showed that the production of acid and bases from the corresponding salt is technically possible. Economical evaluations show, however, that the production of acid and base from the corresponding salt via electrodialysis is expensive and probably not very interesting to apply in these types of applications.

To increase acid reuse options, a further concentration step is required. This can be done by evaporation. Concentrating to 12–15% is a possible application for acid coagulation baths in some of the viscose industries. Evaporation is most economical in cases where excess steam is available. As standard equipment, multiple effect falling film evaporators are used to concentrate solutions. Energy and capital investments are required in this case, but will result in cost savings on the intake of sulfuric acid, as sulfuric acid is required for viscose production processes.

Below is a sample calculation of the energy required to concentrate the sulfuric acid produced biologically so that it can be used in acid coagulation baths. To concentrate the biologically produced sulfuric acid from 40 g/L (=4% w/w) to 14% (w/w), 674 L of water needs to be evaporated. To achieve this, it is assumed that the acid solution needs to be heated from 20°C to 120°C. Due to use of multiple effect evaporation, an efficiency factor of 2 for steam requirement is assumed. The energy requirement is calculated from the vaporization enthalpy (2260 kJ/kg) and the heat capacity (4.18 kJ/(kg K)) of water. This means that to concentrate 1000 L of a 4% acid solution to 286 L of a 14% acid solution $(2260 + 4.18 \times 100) \times 674 \times 0.5 = 910$ MJ energy is required. This translates into the use of 400 kg of steam. In other words, the recovery of 1 kg CS_2 per hour results in 64 L of 4% H_2SO_4 solution per hour. This requires 26 kg steam per hour to bring the 4% to a 14% acid solution. Assuming 8500 h production per year, this amounts to approximately 220 tons of steam per year for concentrating the biologically produced acid from 4% to 14% via multiple effect evaporation. The cost of 1 ton of steam ranges from €7.10 to €10.50 (DACE, 2005). Summarizing, the recovery of 1 kg CS_2 per hour results in 64 L of a 4% sulfuric acid solution via the biological process. To increase reuse applications, a further concentration to 14% solution (coagulation bath concentration) is required which will cost about € 2000 per year.

If reuse of the biologically produced sulfuric acid is not preferred or not possible, the production of gypsum is probably the most economical option to process the produced acid. Limestone ($CaCO_3$) reacts with the sulfuric acid solution and calcium sulfate dihydrate ($CaSO_4 \cdot 2H_2O$) is formed. The gypsum contains no impurities such as heavy metals unlike gypsum derived from SO_2-flue gas treatment. Thus, it can be used for various applications that require high purity gypsum such as plaster, building material, and soil conditioner. In this case, approximately 2.6 kg limestone is required per kg CS_2 emitted.

25.4 TRADITIONAL TREATMENT OF CARBON DISULFIDE WASTE GAS EMISSIONS

25.4.1 Introduction

Several abatement technologies can be used for the treatment of carbon disulfide waste gas emissions. In most production processes in which CS_2 is emitted, H_2S is also formed and therefore both CS_2 and H_2S have to be controlled. In the following sections, the pros and cons of adsorption and incineration are discussed.

25.4.2 Adsorption

Carbon disulfide can be adsorbed on activated carbon and when the carbon bed is saturated, CS_2 is stripped with steam or nitrogen. After the regeneration step, the activated carbon can be reused and the CS_2 is recovered and recycled.

When H_2S is also present in the emission, it interferes with the CS_2 adsorption process. Consequently, H_2S has to be removed first. This is done by absorption in a caustic solution. The resulting sulfide solution has to be treated further, for example, with sodium hypochlorite to oxidize the sulfide to sulfate. With the current high energy prices, the cost of caustic and sodium hypochlorite has increased significantly and the chemical scrubbing of H_2S has become more expensive.

CS_2, apart from being toxic, is also flammable and explosive. Therefore, the traditional regeneration of the carbon bed with steam is not allowed anymore, because this is too hazardous. Indeed, several installations had fires and explosions. Nowadays, the regeneration has to be carried out with hot nitrogen gas. This means high energy costs as after the gas stripping process, the N_2/CS_2 gas mixture has to be cooled to low temperatures in order to condensate sufficient CS_2. In addition, the capital investment of carbon adsorption systems has increased because of these safety measures.

A main advantage of adsorption is the recovery of CS_2 that can be recycled. However, the disadvantages might outweigh this advantage as scrubbing of H_2S is costly because of increased prices of chemicals, the recovery cycle with nitrogen (hot flushing and then deep cooling) results in high energy costs and the investment cost of a carbon adsorption system has increased because of the complexity of the installation for safe operation.

25.4.3 Incineration

Several types of incineration systems are being used. The process is carried out in two steps: in the first step, CS_2 and H_2S are oxidized to SO_2 and in the second step SO_2 is treated.

The simplest type is thermal incineration at approximately 900°C and the resulting SO_2 is chemically adsorbed with caustic to sodium sulfite, and further oxidized to sodium sulfate. In general, at low concentrations of CS_2 and H_2S, the heat of combustion of the CS_2 and H_2S mixture is insufficient to reach a temperature of 900°C and additional fuel (usually natural gas) must be supplied.

An improvement is the application of a catalyst system that lowers the conversion temperature to approximately 400°C. Energy use can be further improved by heat recovery via a heat exchange system in which the incoming CS_2 and H_2S gas mixture is preheated by the exhaust. In this way, a waste air containing approximately 3–4 g CS_2 and H_2S mixture per m^3 waste gas could be self-sustaining. Below this concentration, additional fuel is still required.

The final improvement is the catalytic conversion of the SO_2 containing combustion gases to SO_3 and subsequent absorption in water to form a pure (>90%) concentrated H_2SO_4 solution that can be reused in the production process.

The integrated catalytic conversion system that produces concentrated H_2SO_4 is in use today. Several manufacturers produce and sell these modern installations. A main advantage is that production of pure and concentrated H_2SO_4 from gas mixtures containing more than 3 to 4 g/m^3 CS_2 and H_2S do not require additional fuel.

A possible drawback is the requirement of additional fuel for lean gas mixtures; the lower the concentrations of CS_2 and H_2S, the more fuel will be required. Fully integrated systems are not modular and therefore not flexible. An increase in production capacity will require an additional installation. Safe operation requires measures and procedures to minimize risk for explosions.

25.5 EXAMPLES OF FULL-SCALE BIOREACTORS APPLIED AT INDUSTRIES EMITTING CS_2

25.5.1 Introduction

As mentioned earlier, the major use of CS_2 is in the manufacturing of cellulose-based products. Cellulose is reacted with caustic and CS_2 to form the water-soluble sodium cellulose xanthogenate (viscose). Viscose is coagulated subsequently in a dilute sulfuric acid bath. During coagulation, excess CS_2 and by-product H_2S are emitted. The coagulated products can be of various shapes such as fibers, sheet, etc. The following viscose-based products are economically important: rayon (viscose filament yarn and viscose staple fiber), cellophane (wrapping applications), cellulose casings, cellulosic sponges, and household products. Other processes where CS_2 is used include the synthesis of specialty chemicals like fungicides, pesticides based on thiocarbamates, vulcanizing agents in the rubber industry, and the production of mercaptans.

The producers of CS_2 are increasingly recognizing the problems of CS_2 emissions. For example, Akzo Nobel (2001), a major producer of CS_2 worldwide, adopted the concept "from cradle to grave" for CS_2. This means that the range of services they provide to their customers such as safety, handling, etc. now also includes advice on how to reduce CS_2 and H_2S emissions.

Several full-scale bioreactors treating CS_2 emissions have been operating since the late 1990s. Below, examples of applications at different production plants emitting CS_2 are shown.

Table 25.1 Case study of a bioreactor system operated at a cellulose sponge manufacturer.

Type of production plant	Cellulose sponge manufacturer	
Type of air stream	Waste gas from different sources in the production hall	
	Contaminants treated Year Air flow rate Average operational temperature	Carbon disulfide ($\approx$250 ppmv) and hydrogen sulfide ($\approx$250 ppmv) 1999 51,000 m^3/h 25–40°C
Reactor system	9 bioreactors (V-spring™), upstream wet scrubber, control room	
Design and acceptance criterion	Carbon disulfide removal >80% and hydrogen sulfide removal >90%	

25.5.1 Treatment of CS_2 emissions from a cellulose sponge manufacturer

The bioreactor system shown in Table 25.1 treats 51,000 m^3/h off-gases containing CS_2 and H_2S, and was installed after pilot-scale studies for over a year. Like all bioreactors shown in this section, the system operates with the recirculation water at a pH around 0.5. The CS_2 concentrations of the inlet air fluctuate from 100 to 500 ppmv and the temperature of the inlet air stream varies from 25°C to 40°C. The bioreactor installation has been running successfully and results have been reported earlier (Kraakman, 2003). At this sponge production plant the diluted sulfuric acid produced in the bioreactors is reused at one stage in the manufacturing process. In total, 50% of the dilute acid is reused in the production process. The bioreactor installation has been expanded recently as production was increased.

25.5.2 Treatment of CS_2 emissions at fungicide manufacturers

The waste gas treated in the installation shown in Table 25.2 comes directly from two production reactors operated batch-wise. These produce daily about 10–12 emission peaks. Therefore, the waste gas contains extremely fluctuating concentrations of CS_2 and H_2S.

The bioreactor system is capable of removing on average about 60–65% of the daily load of CS_2 and H_2S (Figure 25.6). In 2005 the installation was upgraded to increase the treatment capacity to about 70–90% of the daily loads of CS_2 and H_2S. Part of the upgrade was the installation of a caustic scrubber to remove the residual

Table 25.2 Case study of a bioreactor system operated at a fungicide production plant.

Type of production plant	Fungicide manufacturer
Type of air stream	Waste gas from the two parallel batch production processes
	Contaminants treated: Carbon disulfide (0–1000 ppmv) and hydrogen sulfide (0–2500 ppmv) Year: 1998 and upgraded in 2005 Air flow rate: 9000 m^3/h Average operational temperature: 25–30°C
Reactor system	4 bioreactors (V-spring™) with chemical scrubber for H_2S peak polishing
Design and acceptance criterion	Carbon disulfide removal >60% and hydrogen sulfide removal >90%

H_2S emitted during the peak emissions. The recirculation liquid from the chemical scrubber containing dissolved sulfide is continuously mixed with the acidic blowdowns from the bioreactors to release the H_2S from the liquid to the gas-phase. The H_2S released is directed to the air inlet of the bioreactors for treatment. In doing so, the extreme peak emissions of H_2S are treated.

The waste gas from the fungicide production plant shown in Table 25.3 contains very small amounts of dust particles of fungicide and also compounds other than CS_2. Therefore, several laboratory studies and bench-scale studies were performed first to determine whether biological treatment of this waste gas was possible. The results showed that dust (fungicide powder) had to be removed from the air stream and that pre-treatment of the waste gas was necessary to remove some of the traces of the volatile organic compound (VOC) pollutants present in the waste gas.

The initial removal of CS_2 in the bioreactor system was low and problems were encountered for obtaining a stable operating system. Additional field research was combined with laboratory studies to improve the system stability. It was found that the stability problems were related to waste gas pooled from an unwanted source, a sub-optimal pre-treatment upstream of the scrubber and dust accumulation in the scrubber. Unwanted compounds in the waste gas (most likely chlorinated compounds released from the fungicide powder) probably affected the CS_2 degrading bacteria.

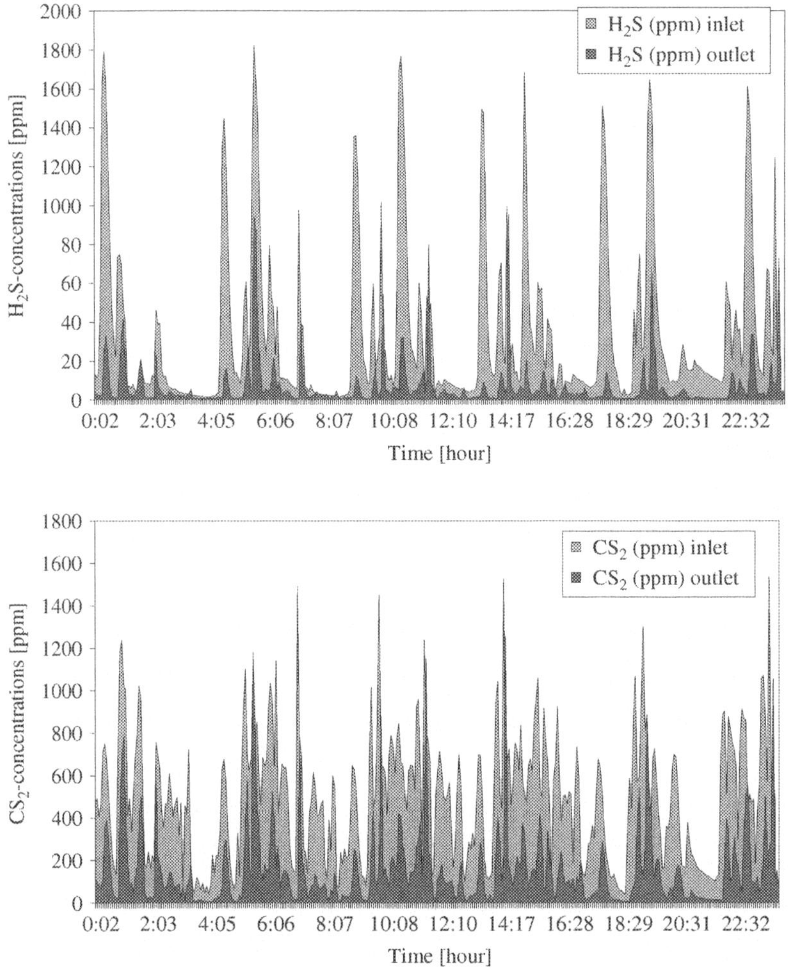

Figures 25.6 CS_2 (top) and H_2S (bottom) concentrations at the inlet and outlet of the bioreactor system.

After optimization, the system was able to operate according to specification (see Table 25.4). The total CS_2 emission was below 10 ppmv at the design inlet CS_2 loadings.

A CS_2-dosing system was part of the treatment system to maintain effective biodegradation of CS_2 with irregular production schedules. During production shutdowns, CS_2 can be dosed in small amounts to keep the microorganisms active and to shorten start-up times after resuming production. It has been determined that the CS_2-dosing system should be started about 12 h prior to resuming production in order for the CS_2 emissions from the process to be removed effectively directly after restart of the production. The bioreactor system showed also to be robust for situation like

Table 25.3 Case study of a bioreactor system operated at a fungicide production plant.

Type of production plant	Fungicide manufacturer	
Type of air stream	Waste gas from discontinuously operated process reactors, drying screens and vacuum pumps	
	Contaminants treated	Carbon disulfide (200–300 ppmv) and traces of other VOCs (<5 ppmv)
	Year	2002
	Air flow rate	8500 m^3/h
	Average operational temperature	20–25°C
Reactor system	2 bioreactors (V-spring™) with upstream scrubber, control room	
Design and acceptance criterion:	97% removal of carbon disulfide with an average outlet concentration of less than 10 ppmv	

Table 25.4 CS_2-removal in the bioreactors measured at different days.

Measurement	Inlet CS_2 concentration (ppmv)	Outlet CS_2 concentration (ppmv)	Removal (%)
1	273	2.6	99.1
2	315	2.0	99.4
3	404	9.4	97.7
4	457	7.2	98.4
5	456	13.3	97.1

the extensive power outage of 8 h in the North America on August 14, 2002 (data not shown here).

25.5.3 Treatment of CS_2 emissions from manufacturers of cellulose casings

The CS_2 concentrations emitted from a cellulose casing producing plant are in general relatively constant. Carbon disulfide is emitted in combination with hydrogen sulfide. Table 25.5 shows a bioreactor system operated at a cellulose casing production plant that was recently tested by the legislator for the required CS_2 removal. The bioreactors were fitted with the possibility to use steam for additional heating of the inlet air if needed during the winter in order to meet the minimum temperature of about 25°C.

Table 25.5 Case study of a bioreactor system operated at a cellulose casing production plant.

Type of production plant	Manufacturer of cellulose casings	
Type of air stream	Different sources in the production process	
	Contaminants treated	Carbon disulfide (≈50–200 ppmv) and hydrogen sulfide (≈100–200 ppmv)
	Year	2005
	Air flow rate	60,000 m^3/h
	Average operational temperature	25–35°C
Reactor system:	6 bioreactors (V-spring™) with upfront humidifier and control room	
Design and acceptance criterion	Carbon disulfide removal >70% and hydrogen sulfide removal >90%	

25.6 CONCLUSION

Treatment of carbon disulfide emissions in a bioreactor system operated under extremely acidic conditions (pH 0.5) was presented. The different applications demonstrate the technical and economical feasibility to reduce carbon disulfide emissions from industrial waste gases. Opportunities to reuse the final degradation product of carbon disulfide (i.e. sulfuric acid) was exploited as sulfuric acid is one of the raw materials in many production processes emitting carbon disulfide. Future development will focus on achieving effective treatment under even more acidic conditions for increased water savings and to extend the applicability of the produced acid.

REFERENCES

AkzoNobel (2001) Internal memo.

Alcantara, S., Estrada, I., Soleda Vasquez, M. and Revah, S. (1999) Carbon disulfide oxidation by a microbial consortium from a trickling filter. *Biotechnol. Lett.* **21**, 815–819.

Beauchamp, R.O., Bus, J.S., Propp, J.A. and Boreiko, C.J. (1993) A critical review on carbon disulphide toxicity. *CRC Rev. Toxic.* **11**, 169–278.

Bioway, b.v. (1999) Market research: CS_2 emissions (internal report).

Chin, M. and Davis, D.D. (1993) Global sources and sinks of COS and CS_2 and their distributions. *Global Biogeochem. Cycles* **7**, 321–337.

DACE (2005) Dutch Association of Cost Engineers/International Cost Engineering Council. *Pricepocket*, 24th edn, May 2005, Klein Gennewiek, Doetinchem, The Netherlands.

EPA (2002) http://www.epa.gov/ttn/atw/cellulose/cellupg.html

EU (2005) http://www.europa.eu.int/comm/environment/ippc/index.htm

Hartel, P.G. and Haines, B.L. (1992) Effects of potential plant CS_2 emissions on bacterial growth in the rhizosphere. *Soil Biol. Biochem.* **24**, 219–224.

Hartikainen, T., Ruuskanen, J., Räty, K., von Wright, A. and Martikainen (2000) Physiology and taxonomy of *Thiobacillus* strain TJ330, which oxidizes carbon disulphide (CS_2). *J. Appl. Microbiol.* **89**, 580–586.

Hugler, W., Acosta, C. and Revah, S. (1999) Biological removal of carbon disulfide from waste air streams. *Environ. Progress* **18**(3), 173–177.

Jordan, S.L., Kraczkiewicz-Dowjat, A.J., Kelly, D.P. and Wood, A.P. (1995) Novel eubacteria able to growth on carbon disulphide. *Arch. Microbiol.* **163**, 131–137.

Kelly, D.P. and Smith, N.A. (1990) Organic sulfur compounds in the environment. *Adv. Microbiol. Ecol.* **11**, 345–385.

Kennes, C. and Veiga, M.C. (eds.) (2001) *Bioreactors for Waste Gas Treatment*, Kluwer Acadamic Publishers, Dordrecht, The Netherlands.

Kraakman, N.J.R. (2001) New bioreactor system for treating suphur- or nitrogen-compounds. In: *Bioreactors for Waste Gas Treatment* (eds. C. Kennes and M.C. Veiga) pp.270–284, Kluwer Acadamic Publishers, Dordrecht, The Netherlands.

Kraakman, N.J.R. (2003) Full-scale biological treatment of industrial CS_2-emissions at extreme conditions. The robustness of a biological system and its risks to the waste gas purification. *Environ. Eng.* **22**(2), 79–86.

Kraakman, N.J.R., Melse, R.W., Koers, B.B. and van Dijk, J. (1998) Biological treatment of waste gases containing H_2S in combination with either odor or CS_2. In: *Proceedings of USC-TRG Conference on Biofiltration*, Los Angeles, October 1998. 91–98.

Lomans, B.P. (2001) Microbial cycling of volatile organic sulfur compounds in anaerobic fresh-water sediments, PhD Thesis, Department of Microbiology, Radboudt University Nijmegen, The Netherlands.

Melse, R.W. and Kraakman, N.J.R. (1998) Biological treatment of waste gases containing H_2S and CS_2 combined with the production of concentrated NaOH and H_2SO_4. In: *Proceedings of FAB 12th Forum of Applied Biotechnology*, 24 and 25 September 1998, Brugge, Belgium.

Plas, C.H., Wimmer, K., Harant, H., Holubar, P. and Braun, R. (1991) Biological Oxidation von schwefelwasserstoff durch ein *Thiobacillus*-Isolat. *GWF Wasser Abwasser* **132**(7), 419–421.

Plas, C.H., Wimmer, K., Holubar, P., Mattanovicj, D., Danner, H., Jelinek, E., Harant, H. and Braun, R. (1993) Degradation of carbon disulfide by a *Thiobaccillus* isolate. *Appl. Microbiol. Biotechnol.* **38**, 820–823.

Shareefdeen, Z. and Singh, A. (eds.) (2005) *Biotechnology for Odour and Air Pollution Control*, Springer-Verlag, Heidelberg, Germany.

Smet, E., Lens, P. and Van Langenhove, H. (1998) Treatment of waste gases contaminated with odorous sulfur compounds. *Crit. Rev. Environ. Sci. Technol.* **28**, 89–117.

Smith, N.A. and Kelly, D.P. (1988) Oxidation of carbon disulphide as the sole source of energy for the autotrophic growth of *Thiobacillus thioparus strain TK-m*. *J. Gen. Microbiol.* **134**, 3041–3048.

WHO (1996) *Biological Monitoring of Chemical Exposure in the Workplace*, Guidelines Vol. 1, pp. 156–167. Geneva: WHO.

Windsperger, A. (1990) Anwendung eines biologischen tropfkorpereactors zur abluftreinigung eines viscosebetreibes. *Chem. Ing. Tech.* **62**(12), 1033–1034.

Yang, Y. and Alibeckoff, D. (1995) Biofiltration for carbon disulfide and hydrogen sulfide vapors. *Proceedings 1995 Conference on Biofiltration*, USC, Los Angeles CA. 165–172.

26

Volatile organic compound treatment by adsorption onto activated carbon fiber cloth and electrothermal regeneration

Albert Subrenat and Pierre Le Cloirec

26.1 INTRODUCTION

Due to increased atmospheric pollution, Directives (1999) in Europe or the Clean Air Act Amendments (1990) in USA, gradually regulate the emissions of hazardous air pollutants. Several industrial activities are directly involved, such as painting, chemistry, petrochemistry, pharmacy, degreasing and dry cleaning manufactures as well as storage or utilizations of organic solvents in general (Le Cloirec, 1998; Liu & Liptàk, 1997). In 2006, in some cases a volatile organic compound (VOC) mass flow of 2 $kg\,h^{-1}$ or 100 $g\,h^{-1}$ and a concentration of 150, 20 or 2 $mg\,m^{-3}$, depending on the chlorinated organic compound emissions, are required. In this context, it is necessary to propose original treatment processes, efficient and well adapted to the technical and economic constraints of industry.

 Waste Gas Treatment for Resource Recovery edited by Piet N.L. Lens, Christian Kennes, Pierre Le Cloirec and Marc Deshusses. ISBN: 1843391279. Published by IWA Publishing, London, UK.

Activated carbon fiber cloths (ACFC) were recently developed, which seem very promising for air or water treatment (Le Cloirec et al., 1997). The physicochemical and thermal properties of ACFC reveal the possibility to develop new and original air treatment processes (Brasquet et al., 2000; Cal et al., 1997). Laboratory studies have shown that these adsorbents have good adsorption capacities for a large number of organics used as solvents (Brasquet & Le Cloirec, 1997). Their mechanical characteristics make it possible to produce compact filters with low pressure drops (Brasquet & Le Cloirec, 2000; Goodings, 1964; Subrenat et al., 2000). Moreover, their electrical properties (Petkovska & Tondeur, 1991; Rood et al., 2003; Subrenat et al., 2001) allow fast regenerations of the saturated filter in a dry way, by direct electrical heating, thus opening the way of their use in industrial treatment processes (Subrenat & Le Cloirec, 2000).

The objective of this chapter is to describe the properties of ACFC and to show the performance of a temperature swing adsorption (TSA) process using ACFC and electrothermal regeneration. The system is applied at an industrial site emitting chlorinated compounds in the atmosphere.

26.2 ACTIVATED CARBON FIBER CLOTH

26.2.1 Macroscopic and microscopic characteristics

The studied materials are manufactured starting from woven rayon (Bacon, 1973). First, the precursor is carbonized. This stage makes it possible to obtain a material whose carbon level is higher than 95%. This operation is carried out in an inert atmosphere (nitrogen) at temperatures ranging between 400°C and 1200°C. Hence, the thermal stresses support the inter-chain interactions of the macromolecules and make it possible to extract the volatile matter from the precursor. Then, it is physically activated (activation with CO_2 or steam) to increase pore surface and volume by extracting amorphous carbon. In the presence of an oxidizing gas (steam or carbon dioxide), oxidation occurs at typical temperatures ranging between 700°C and 1500°C.

The operational parameters of the carbonization and activation process of the studied material (Manufactured by Actitex, Saint Maurice, France) are presented in Table 26.1. These parameters play an essential role in the structural organization of the material, its surface quality, its macroscopic properties (weight, permeability) and its pore structure (Brasquet et al., 2000).

From a macroscopic point of view, ACFC are presented in the form of traditional textiles (Figure 26.1) composed of multi-fiber threads (Brasquet et al., 2000a). Thus,

Table 26.1 Main characteristics of the ACFC.

Name	Texture	Pore volume ($cm^3 g^{-1}$)	Micro-pore volume (%)	BET (Brunauer–Emmett–Teller) surface area ($m^2 g^{-1}$)	Specific weight ($g m^{-2}$)
S-15	8-ply satin	0.543	96	1600	150
WWP-3	Woven	0.417	80	870	140

there are as many textures (e.g. twill, satin, fabric, woven or knitting) as the textile industry can produce. Typically, the thread diameter is between 0.4 and 1 mm, composed of many fibers with a diameter between 4 and 15 μm.

26.2.2 Electrical properties

The material is composed of more than 99% carbon. Its electrical behavior is similar to that of a classical ohmic conductor. Thus, ACFC heating can be performed using the Joule effect or electromagnetic induction (Mocho et al., 1996). In terms of material processing design, the electrical resistance of fabric filters can be deduced from the media resistivity and its dimensions (Figure 26.2). This property, corresponding to the definition of electrical resistivity, is known for traditional conducting materials in the algebraic form:

$$R_{L/l} = \frac{\rho}{e}\frac{L}{l} \tag{26.1}$$

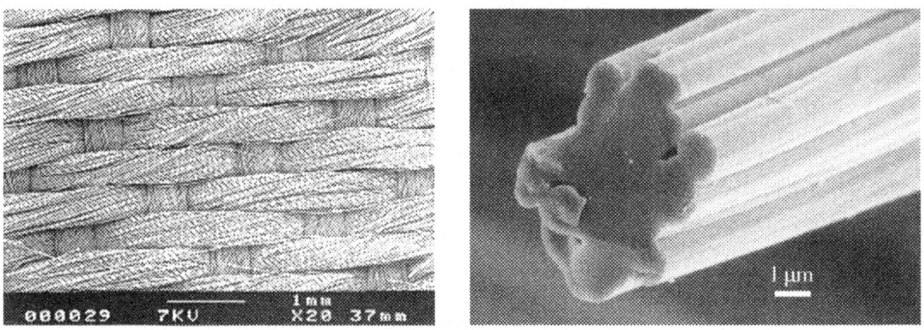

(a) (b)

Figure 26.1 Microscopic electronic scanning photographs of ACFC: (a) WWP-3 (Woven) and (b) details of a WWP-3 fiber.

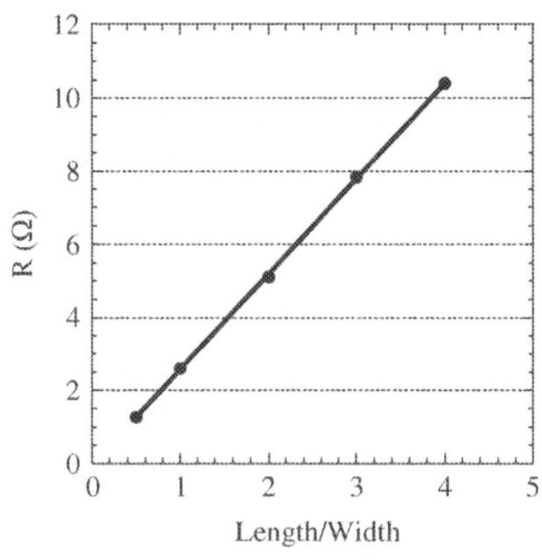

Figure 26.2 ACFC electrical resistance versus the ratio (length/width).

with

ρ : electrical resistivity of the material (Ω m)
$R_{L/l}$: electrical resistance (Ω)
e : thickness of the material (m)
L : length of the sample (m)
l : width of the sample (m)

The electrical resistivity values of the activated carbon fabrics have been deduced from measurements of electrical resistance of different ACFC samples. In the studied temperature range (290–450 K), a decreasing linear law for the resistivity, usually found with traditional conductors (Nasar, 1989; Subrenat et al., 2001), is used in order to fit the experimental data. It is given by the relation:

$$R_{L/l} = \frac{\rho}{\rho_0} = 1 + \alpha_0(\bar{T} - T_0) \quad (26.2)$$

with

ρ : Electrical resistivity at temperature T (Ω m)
ρ_0 : Electrical resistivity at temperature T_0 (Ω m)
α_0 : Thermal coefficient (K^{-1})
T : Activated carbon cloth (ACC) average temperature (K)
T_0 : 273.15 K

The measured values of the activated carbon media presented in this chapter are given in Table 26.2.

The model parameters (ρ_0 and α_0) enable the comparison of the electrical conductivity of the various carbon cloths studied. The electrical resistivity (ρ_0) and the thermal coefficient (α_0) increase with the specific surface area (Subrenat et al., 2001). Moreover, the electrical resistivity of these media depends significantly on the fiber arrangement.

26.2.3 Thermal properties

In an electric circuit composed of a resistance and a power supply as a source of voltage, the current resulting in the circuit increases the temperature in the resistance. This phenomenon is known as the Joule effect. In this case, the ACFC is the resistance which is heated by applying a voltage drop. The heating of these ACFC materials is performed by the Joule effect (between 20°C and 250°C).

The heating is generally highly homogeneous and the spatial distribution of the temperatures (observed with an Infra Red camera, Figure 26.3) is directly connected

Table 26.2 Electrical resistivity of two different cloths.

Adsorbent	VS-1501	WWP-3
$R_{L/l}$ (Ω m)	2.83	70.75
$\alpha_0 \times 10^4$ (K^{-1})	−14.39	−55.08

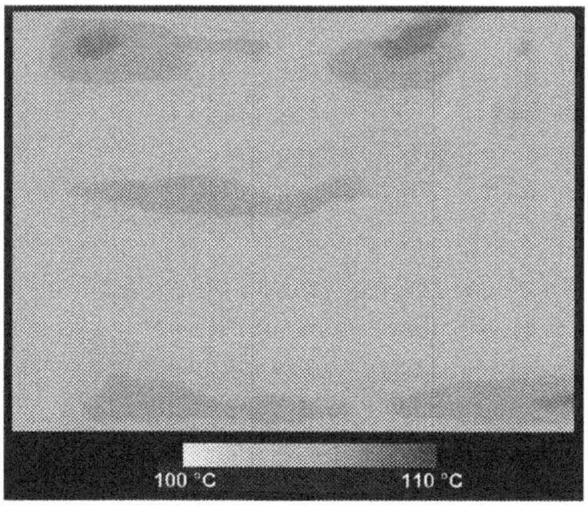

Figure 26.3 Homogeneous temperature distribution in an ACFC (infrared camera picture).

to the activation level distribution of the material. In the experimental conditions used, the evaluation of the thermal losses of activated carbon fabric according to its equilibrium temperature is obtained through the evaluation of the thermal heat transfer surface area using the classical transfer model:

$$P = UI = A_{\text{th}}(\sigma_s\, \varepsilon(T^4 - T_{\text{r}}^4) + h(T - T_r)) \tag{26.3}$$

with:

P : electrical power (W)
U : voltage (V)
I : current-intensity (A)
A_{th} : heat transfer surface area (m^2)
T : ACC average temperature (K)
T_{r} : room temperature (293.15 K)
ε : emissivity (=1 for ACC equivalent to a black body)
σ_{s} : Stefan coefficient ($5.77 \cdot 10^{-8}\ \text{W m}^{-2}\text{K}^{-4}$)
h : convective transfer coefficient ($\text{W m}^{-2}\text{K}^{-1}$) $h = 2.16\,(T - T_{\text{r}})^{0.25}$

The interest of this empirical model is not to describe precisely the heat transfer phenomena but to give a rough idea of the electrical power required to reach a desorption temperature. It is used to fit the experimental data ($P = f(T)$) and then estimate the heat transfer surface area. A good agreement between the experimental data and the heat transfer model is observed (Figure 26.4). According to the results obtained, the heat transfer surface area can be estimated from the apparent fabric surface area with an average correction of +10%.

Knowledge of the electrical behavior and the electrical power required for the heating of the whole quantity of ACC used in the process facilitates the sizing and the design of the filters and the electrical generator. Hence, a pilot unit was set up to perform and validate adsorption–desorption cycles using the Joule effect.

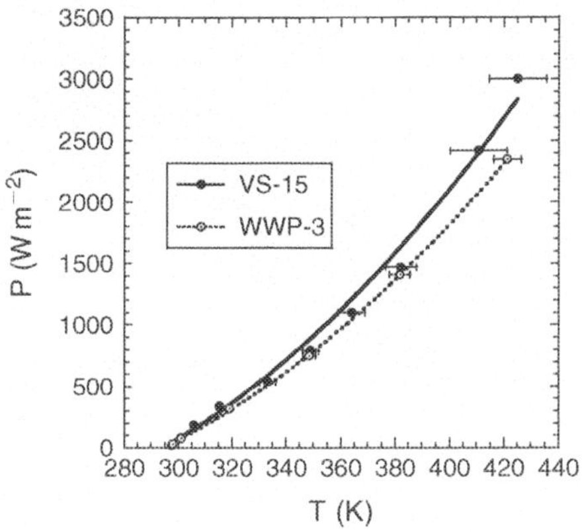

Figure 26.4 Electrical power dissipated by convection and radiation versus the mean temperature (Equation (26.3)).

26.3 ADSORPTION ONTO ACFC–ELECTROTHERMAL DESORPTION PROCESS

The experiments of adsorption and desorption by electrical heating in a unit ($100\,m^3h^{-1}$) have been previously published (Subrenat & Le Cloirec, 2006). However, an example of adsorption–desorption in a laboratory system ($10\,m^3h^{-1}$) is presented in this chapter.

Joule effect regeneration is a dry gas desorption process particularly adapted to the treatment of chlorinated compounds (especially used in industry). This solvent recovery is quite difficult, due to the formation of hydrochloric acid by hydrolysis of molecule. A laboratory pilot unit is used to perform adsorption–desorption stages for the treatment of dichloromethane and perchloroethylene vapors. The objective here is to concentrate the industrial effluent in the gas phase to recover the solvents using a condensation process.

A cylindrical rolled filter (Figure 26.5) using ACFC (reference WWP-3 for its good adsorption capacities) is used to perform the adsorption and Joule effect regeneration stages. The characteristics of this filter and the operating conditions of the process are given in Table 26.3.

The adsorption parameters are given by the industrial application. The desorption process consists of two steps: first, the filter is heated by the Joule effect, then a high flow rate pulse of carrier gas is used to evacuate the desorbate molecules (without heating the filter) in a very short time to obtain a high concentration peak of solvent and then a good concentration ratio.

Figure 26.6 gives results of the adsorption–desorption cycles. The adsorption capacity in a dynamic system could be evaluated with the first adsorption step. The maximal adsorption capacity is found to be about $205\,mg\,g^{-1}$. But during each regeneration, the amount of desorbed solvent is only around $150\,mg\,g^{-1}$. After each desorption step, the solvent quantities adsorbed are also close to $150\,mg\,g^{-1}$, corresponding to the working capacity of the process. In these conditions, the adsorption time is 20 min and the

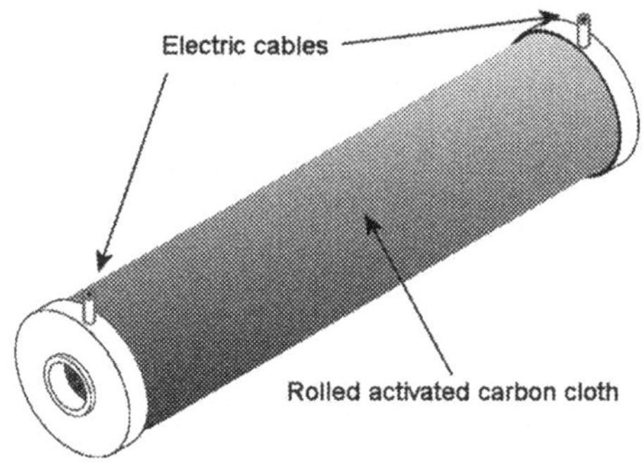

Figure 26.5 Technical drawing of a cylindrical rolled adsorber.

Table 26.3 Operating conditions of the chlorinated compound removal process.

Filter device	
Length (mm)	250
Diameter (mm)	55
External surface area (m^2)	0.043
Number of rolled layers	10
Cloth reference	WWP-3
ACC mass (g)	66
Adsorption	
Inlet concentration ($g\,m^{-3}$)	6
Superficial velocity ($m\,s^{-1}$)	0.03
Flow rate (m^3h^{-1})	4.65
Outlet concentration ($g\,m^{-3}$)	<0.01
Desorption	
Heating temperature (°C)	100
Heating duration (min)	2
Carrier gas flow rate (m^3h^{-1})	154

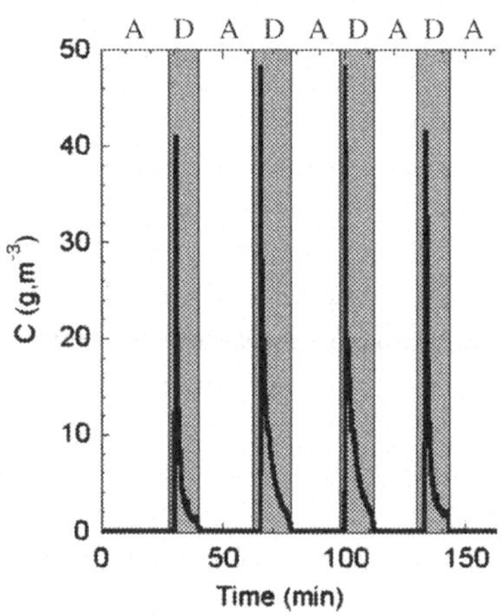

Figure 26.6 Adsorption–desorption stages for the treatment of dichloromethane present in air. A: adsorption and D: desorption.

desorption time is about 10 min. Thus, adsorption–desorption cycles may perform alternatively with two reactors for a continuous treatment of the industrial effluent. The concentration peak in the desorption gas is $50\,g\,m^{-3}$. Hence, a condensation process (liquid nitrogen cryocondenser or water cooler) may be used to recover and recycle the solvent.

26.4 DEVELOPMENT OF AN INDUSTRIAL TREATMENT PROCESS

Previous studies have shown a good adsorption of VOC present in air onto ACFCs (Brasquet, 1997). In a chemical plant, a treatment process was carried out to control a gaseous emission loaded with methylene chloride. The air treatment process consists of two modules of ACFC alternatively adsorbing pollutant present in the gaseous phase and being regenerated by an electrothermal swing desorption. In 2006, three industrial treatment processes are working to remove chlorinated solvents, alcohols and aromatic compounds alone or in complex mixture.

26.4.1 Characterization of the industrial polluted emission

A gaseous emission (flow close to $100\,m^3\,h^{-1}$) loaded with methylene chloride at concentrations ranging between 100 and $700\,g\,m^{-3}$ (overall influent), that is mass flow rates from 10 to $70\,kg\,h^{-1}$, had to be treated to get a final mass flow lower than $100\,g\,h^{-1}$ (less than $1\,g\,m^{-3}$). This value is required by the regulations in force. The flow and the VOC concentrations are variable as a function of the chemical production. The system is working 24 hours a day and 7 days a week during 320 days per year.

26.4.2 VOC removal system

The air cleaning system (Figure 26.7) consists of a first step to treat the inlet air loaded by methylene chloride with a nitrogen cryogenic condenser and in a second step with an adsorber to remove the solvent present in the vent of the condenser (several g/Nm^3) by VOC transfer onto an ACFC module. Only the last process will be described in this chapter. This ACFC system is composed of two compact reactors as shown in Figure 26.7.

26.4.3 ACFC adsorber module

From the laboratory studies, a VOC removal process using an adsorption and desorption system was carried out and studied in this industrial application. A mass of 1.6 kg of a microporous ACFC was wound around two identical cylindrical modules of 17 cm diameter and 50 cm length (Figure 26.8). Some characteristics of the ACFC are given in Table 26.4. This microporous material has been studied previously (Brasquet & Le Cloirec, 1997; Brasquet et al., 2000a; Fournel et al., 2005). Due to its specific surface area and porous volume, it is well known to adsorb methylene chloride with a maximum adsorption capacity close to $300\,mg\,g^{-1}$ (Brasquet et al., 1999). Its electrical and

Figure 26.7 Industrial ACFC with electrothermal swing filters.

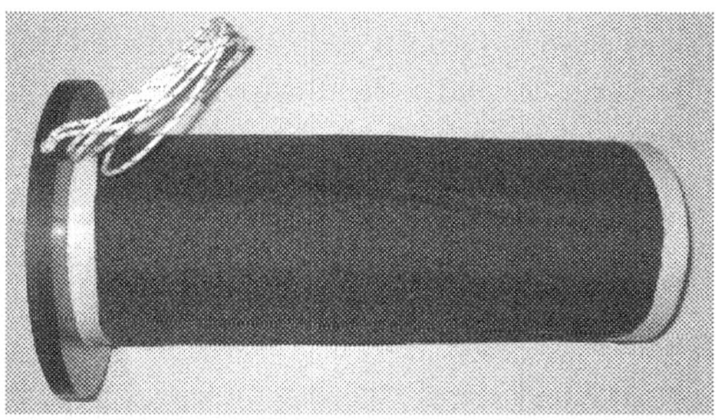

Figure 26.8 Adsorption onto ACFC – electrothermal desorption module.

Table 26.4 Characteristics of the ACFC and module used in the adsorber.

Fiber cloth	
Origin	Rayon
Texture	Woven
Specific weight ($g\,m^{-2}$)	140
Thickness (mm)	0.4
Electrical resistivity (Ω mm)	28.4
Specific surface area ($m^2 g^{-1}$)	873
Porous volume ($cm^3 g^{-1}$)	0.435
Microporous volume ($cm^3 g^{-1}$)	0.417
Module	
Module diameter (cm)	17
Module length (m)	0.50

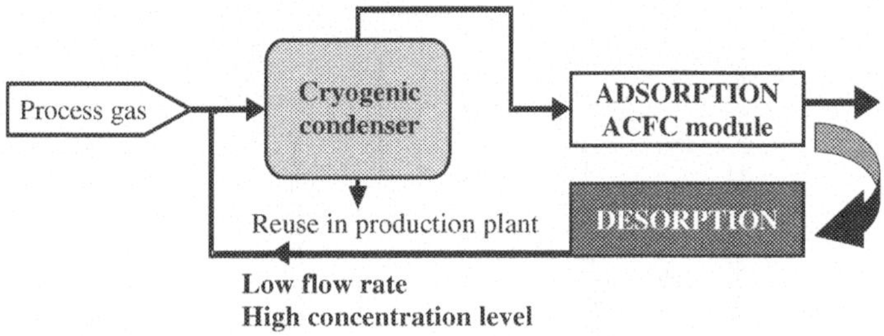

Figure 26.9 Simplified diagram of the VOC removing process.

thermal properties allow homogeneous electrical heating (Subrenat et al., 2001). The electrical resistance of the filter is 1 Ω at ambient temperature.

26.4.4 Adsorption: electrothermal desorption procedures

The modules worked alternatively in an adsorption–desorption cycle of about 1 h (Figure 26.9). Data (inlet and outlet pollutant concentrations analyzed by a flame ionization detector (FID), adsorption and desorption temperatures and flow rate) were collected during several days to determine the best operating conditions and the performances of the system. The polluted air velocity through the module was 400 m h^{-1}.

While the first filter treats the effluent of the condenser, the second filter is regenerated by direct electrical heating. After saturation of the adsorbent, an electric current was applied to the module to get an ACC surface temperature close to 60°C. The electric power applied to the filter was about 200 W. A flow rate of gaseous nitrogen (5 $m^3 h^{-1}$, i.e. 20 m h^{-1}) was used to put the desorbate out of the system and to pass it through a liquid nitrogen trap. The adjustments of the regeneration operating conditions allow to generate a desorbate effluent strongly loaded with the solvent at a nitrogen flow rate 20 times lower than the inlet effluent treated upstream the condenser.

The use of two filters, working in parallel, alternating adsorption phases and short duration of the regeneration allows an uninterrupted treatment of the condenser vent. Moreover, the filter cycles are completely independent of variations of the gaseous emission in terms of flow rate and concentrations. The short cycle time (about 1 h) allows to design a small filter, to save ACFC and to get a small reactor volume (lower than 100 L). The reduction of the regeneration time is due to the Joule's effect, heating desorption. The electric power injected directly in the cloth controls the desorption temperature.

The desorbed concentrated gas is treated by cryogenic condensation with the process gas in the first step (Figure 26.9). Liquid methylene chloride was recovered and re-used in the chemical processes of the plant.

26.4.5 Process validation

After 7 months, concentration and flow rate measurements were performed during a 4-day period. The cryogenic condenser removed about 99% of the initial solvent

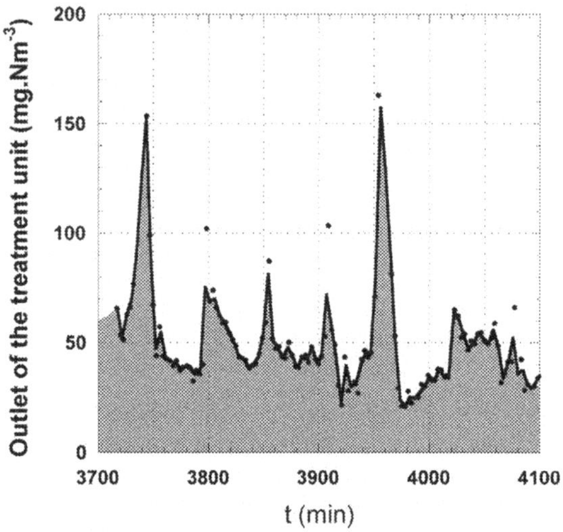

Figure 26.10 Outlet concentration of methylene chloride.

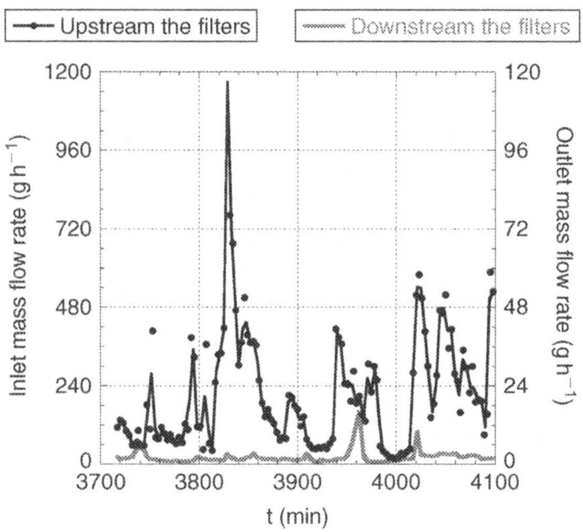

Figure 26.11 Mass flow rate upstream and downstream the filter.

concentration in the process gas. The measurements, presented in Figures 26.10 and 26.11, are performed upstream and downstream the filters for a 6-h period during which the two filters worked alternatively in adsorption and regeneration phases. The low outlet concentrations of solvent show the good adsorption capacities of the filters after the regeneration steps.

The instantaneous concentrations at the outlet emission, after treatment onto ACFC filters, are lower than 150 mg Nm^{-3}. The mean value (calculated during 1 day) downstream the filters is around 60 mg Nm^{-3}, whereas the inlet concentration upstream the filters is close to 6 g Nm^{-3}. In spite of flow and concentration variations of the process gas, the adsorption onto the ACFC filters shows a great stability and low levels of the outlet solvent emissions. The instantaneous mass flow rate measurements at the treatment unit outlet (Figure 26.11) give data less than $15\,g\,h^{-1}$ with a mean value of about $2.5\,g\,h^{-1}$ for about $200\,g\,h^{-1}$ in the inlet of the filter. The treatment unit is in conformity with the French regulations, which impose by regulations a maximal mass flow rate of $100\,g\,h^{-1}$.

26.5 CONCLUSION

Previous studies on ACC have shown that it is an interesting porous material to control VOC present in air. Heating of this material by electricity allows to get a good regeneration. Applied to industrial emissions loaded with a chlorinated solvent, the system of electrothermal-swing adsorption gives good results in terms of outlet concentration and outlet mass flow rate.

The compactness of the adsorbers and the simplicity of the direct electrical regeneration facilitate the implementation of the filters in terms of engineering and process operation. The desorption procedure offers several advantages: a great flexibility in the regeneration operating conditions, a reduced maintenance and low operating costs.

The industrial demonstration is an important step in the diffusion of this technology and opens the way for many other applications:

- The pre-treatment of effluents with high flow rates, but low concentrations upstream of a recovery or destructive process.
- The realization of very compact unit for the treatment of local emissions at specific industrial units.

ACKNOWLEDGMENTS

Some parts of this chapter have been previously published in *Journal of Environmental Engineering* (2004) and *Chemical Engineering Communication* (2006). The authors gratefully appreciated financial support of the laboratory studies by the French Environmental and Energetic Agency (Ademe). Many thanks to Air Liquide Italy, Actitex (Saint Maurice, France) and Sofrance (Nexon, France) for fruitful collaboration.

REFERENCES

Bacon, R. (1973) Carbon fibres from rayon precursors. *Chem. Phys. Carbon* **9**, 1.

Brasquet, C. and Le Cloirec, P. (1997) Adsorption onto activated carbon fibers: application to water and air treatments. *Carbon* **9**, 1307–1313.

Brasquet, C. and Le Cloirec, P. (2000) Pressure drop through textile fabrics – experimental data modelling using classical models and neural networks. *Chem. Eng. Sci.* **55**, 2767–2778.

Brasquet, C., Bourges, B. and Le Cloirec, P. (1999) Quantitative structure properties relationship (QSPR) for the adsorption of organic compounds onto activated carbon cloth: comparison between multilinear regression and neural network. *Environ. Sci. Technol.* **33**, 4226–4231.

Brasquet, C., Rousseau, B., Estrade-Szwarckopf, H. and Le Cloirec, P. (2000) Observations of activated carbon fibers with SEM and AFM. *Carbon* **38**(3), 407–422.

Cal, M.P., Rood, M.J. and Larson, S.M. (1997) Gas phase adsorption of volatile organic compounds and water vapor on activated carbon cloth. *Energ. Fuel.* **11**, 311–315.

Clean Air Act Amendement (1990).

Directive 1999/13/CE du Conseil de l'Union Européenne du 11 mars 1999 relative à la réduction des émissions de composés organiques volatils dues à l'utilisation de solvants organiques dans certaines activités et installations, *Journal Officiel des Communautés Européennes*, 29 mars 1999.

Fournel, L., Mocho, P., Fanlo, J.L. and Le Cloirec, P. (2005) External capillary condensation and adsorption of VOCs onto activated carbon fiber cloth and felt. *Environ. Technol.* **26**, 1277–1287.

Goodings, A.C. (1964) Air flow through textile fabrics. *Text. Res. J.* **34**, 713–724.

Le Cloirec, P. (1998) *Les composés organiques volatils dans l'environnement*. Tec & Doc, Lavoisier, Paris, France.

Le Cloirec, P., Brasquet, C. and Subrenat, E. (1997) Adsorption onto fibrous activated carbon: applications to water treatment. *Energ. Fuel* **11**(2), 331–336.

Liu, D.H.F. and Liptàk, B.G. (1997) *Environmental Engineers' Handbook*, 2nd edn. Lewis Publishers, Boca Raton, FL, USA.

Mocho, P., Bourhis, J.Ch. and Le Cloirec, P. (1996) Heating activated carbon by electromagnetic induction. *Carbon* **34**, 851–856.

Nasar, A.S. (1989), *Electrical Circuits.* McGraw Hill, Boston, MA, USA.

Petkovska, M.D. and Tondeur, D. (1991) Temperature swing gas separation with electrothermal desorption step. *Separ. Sci. Technol.* **26**, 425–444.

Rood, M.J., Ramirez, D., Vidal, E.X., Hashisho, Z., Emanmipour, H., Sullivan, P.D. and Hay, K.J. (2003) Organic vapor recovery using activating carbon fiber cloth electrothermal-swing adsorption. *Macro Rev.* **16**(1), 504–508.

Subrenat, A. and Le Cloirec, P. (2000) Removal of VOC by adsorption–desorption cycles using activated carbon cloth filter: regeneration by Joule effect. In *Adsorption Science and Technology* (ed. D.D. Do), pp. 361–365. World Scientific, Singapore. ISBN 981-02-4263-8.

Subrenat, A. and Le Cloirec, P. (2006) Volatile organic compound (VOC) removal by adsorption onto activated carbon fiber cloth and electrothermal desorption: an industrial application. *Chem. Eng. Comm.* **193**, 1–9.

Subrenat, A., Baléo, J.N. and Le Cloirec, P. (2000) Analysis of pressure drops in pleated activated carbon cloth filters. *J. Environ. Eng.* **156**(6), 562–568.

Subrenat, A., Baléo, J.N. and Le Cloirec, P. (2001) Electrical behaviour of activated carbon cloth heated by the Joule effect. *Carbon* **39**, 707–716.

27

Treatment of waste gases from formaldehyde- and formaldehyde-resin-producing industries

Óscar Prado, María C. Veiga and Christian Kennes

27.1 INTRODUCTION

27.1.1 Production of formaldehyde and formaldehyde-based resins

27.1.1.1 Formaldehyde usage

The formaldehyde- and formaldehyde-resin-producing industries have significantly expanded over the last decades. This is a consequence of the high demand of these compounds which, in turn, derives from their wide range of uses. Formaldehyde is mainly employed as a chemical intermediate for the production of different organic compounds and also to control microbial growth, through its microbicidal effect. The

 Waste Gas Treatment for Resource Recovery edited by Piet N.L. Lens, Christian Kennes, Pierre Le Cloirec and Marc Deshusses. ISBN: 1843391279. Published by IWA Publishing, London, UK.

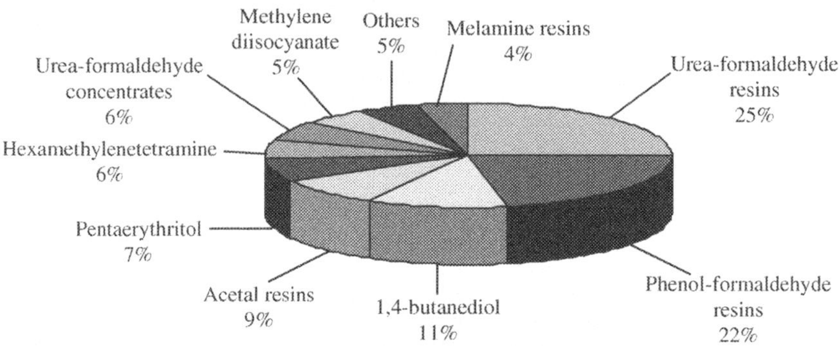

Figure 27.1 Uses of formaldehyde in the United States, 1989 (*Kirk-Othmer Encyclopedia of Chemical Technology*, 1998).

uses of formaldehyde-based resins include the stabilization of fuels, the production of solvents, plastics and adhesives as well as the impregnation of materials in order to make them fire, moisture or corrosion-resistant. Figure 27.1 shows the main uses of formaldehyde produced in the United States during 1989 (Kirk-Othmer Encyclopedia of Chemical Technology, 1998). According to data presented by the Agency for Toxic Substances and Disease Registry (1999), formaldehyde production in the United States has increased from around 0.8 million tons, in the early sixties up to around 3.0 million tons in the early nineties. This trend kept growing on during the first years of the new century, reaching a global annual production around 15 million tons in 2005.

There is a wide variety of formaldehyde-based resins, as urea–formaldehyde, melamine–formaldehyde and phenol–formaldehyde resins (Figure 27.1). In most cases, the formaldehyde-resin-producing industries generate themselves the formaldehyde that will serve as prime matter for the production of synthetic resins. Although a few other methods exist, most of the formaldehyde produced worldwide is obtained from a chemical reaction or a series of reactions involving gas-phase methanol, air and a catalyst. The overall reaction, which has a conversion efficiency of around 90%, can be performed by using metals (mainly silver or copper) or metal oxides as catalysts. These processes are described in more detail below.

27.1.1.2 Production of formaldehyde

Formaldehyde production using a metal as catalyst Although copper has been the most commonly used catalyst for the oxidation of methanol to formaldehyde in the past, nowadays silver is largely preferred. In both cases, a rich mixture of methanol with air is needed. Besides formaldehyde, by-products are formed during the reaction due to catalyst inefficiencies. The process takes place in a fixed bed catalytic reactor, at ambient pressure and a temperature between 450°C and 650°C (Figure 27.2). The chemical process that takes place in the catalytic reactor consists of two simultaneous reactions (27.1) and (27.2). The oxidation reaction is exothermic and produces between 50% and 60% of the formaldehyde, while the rest is generated by the endothermic reaction of dehydrogenation (Kirk-Othmer Encyclopedia of Chemical Technology, 1998). The net balance of the global process is exothermic.

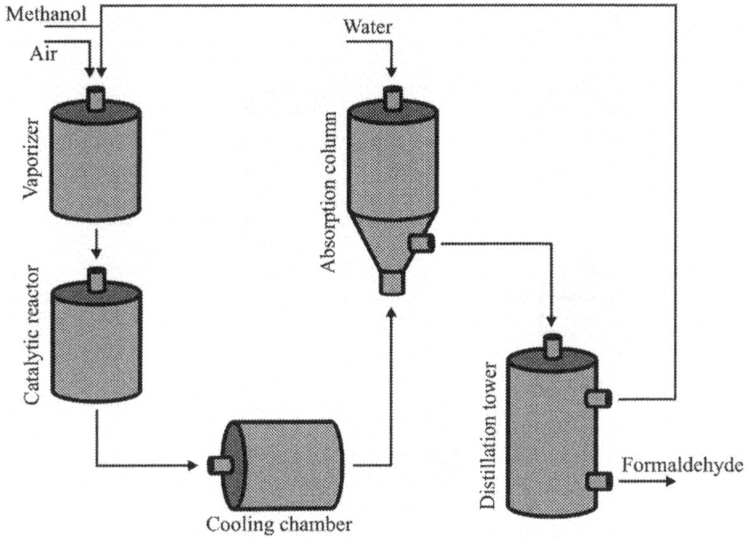

Figure 27.2 Metal-catalyzed method for formaldehyde production.

$$CH_3OH + \tfrac{1}{2}O_2 \rightarrow CH_2O + H_2O \quad (DH = -37.3\,\text{kcal}) \tag{27.1}$$

$$CH_3OH \rightarrow CH_2O + H_2\ (\Delta H = 20.3\,\text{kCal}) \tag{27.2}$$

Afterwards, the off-gas crosses a cooling chamber, reaching a temperature around 150°C. Finally, an absorption column is used to transfer this gas-phase formaldehyde to liquid water, that leads to the production of an aqueous formaldehyde solution with typical concentrations around 55%. The hopper of the absorber is connected to a distillation tower, in which methanol is recovered and recirculated to the first production stages. Commonly, the final formaldehyde solution contains less than 1.5% methanol.

Formaldehyde production using a metal oxide as catalyst When a metal oxide is employed as catalyst, the methanol concentration in the mixture is lower, and the off-gas does usually not contain unreacted methanol in significant amounts. The most commonly used catalysts of this type are vanadium pentoxide and iron-molybdate oxides. The method of production is similar as in the previous case, although the absence of methanol in the resulting solution avoids the need of a distillation tower (Figure 27.3). In this case, formaldehyde is entirely produced by means of reaction (27.1), at ambient pressure and temperatures between 300°C and 400°C, with almost quantitative methanol conversion at selectivities above 90%.

27.1.1.3 Production of formaldehyde derivates

Under adequate conditions, this formaldehyde can react with a wide variety of compounds to yield different resinous polymers. As an example, phenol–formaldehyde resins are produced by means of a polymerization of formaldehyde and phenol. Two main types of these resins can be distinguished: resols, formed in alkaline media with an

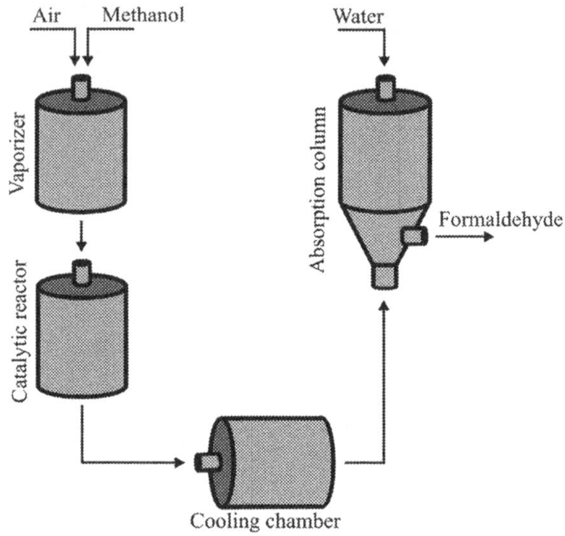

Figure 27.3 Metal oxide-catalyzed method for formaldehyde production.

excess of formaldehyde, and novolaks, formed in acid media with a low formaldehyde content. Urea and melamine–formaldehyde resins are produced by mixing formaldehyde with urea or melamine, respectively, to yield methylol monomer units. These units are condensed afterwards under acidic conditions. Polyacetal resins are produced by the polymerization of anhydrous formaldehyde or its trimer, trioxane.

27.1.2 Emission of volatile pollutants in formaldehyde-resin-producing facilities

As a consequence of their activity, the formaldehyde-resin-producing industries generate and emit undesirable gaseous compounds to the atmosphere. The major source of waste gas is the formaldehyde-absorption column. The temperature of that waste is typically in the "mesophilic" range (<40°C). It contains formaldehyde, methanol, dimethylether and carbon monoxide as pollutants. Emissions from the storage tanks, containing mainly formaldehyde and methanol, need also to be taken into account. Both the amounts emitted of each substance and their relative concentrations in the mixture are variable, but the concentration of each pollutant does generally not exceed 200 mg/m^3 in the worst case. The gas flow rate is typically around 20000 ± 5000 m^3/h. The waste gas does not carry any particulate matter, meaning that no preliminary treatment is required to eliminate particles. Table 27.1 shows some of the main physico-chemical characteristics of those pollutants.

Considering the huge production levels of formaldehyde and formaldehyde-based resins worldwide, as well as the emission rates of undesirable and potentially toxic compounds in waste gases from these facilities, it is clear that a big effort is needed in order to diminish their environmental impact. Two main groups of technologies for the removal of gas-phase pollutants may be distinguished: traditional technologies, based on physico-chemical principles, and biological technologies, using microorganisms as

Table 27.1 Physico-chemical properties of the main pollutants emitted from formaldehyde-resin-producing industries.

	Formaldehyde	Methanol	Dimethylether	Carbon monoxide
Formula	CH_2O	CH_3OH	$(CH_3)_2O$	CO
MW (g/mol)	30.03	32.04	46.07	28.01
BP (°C)	−19.5	64.7	−24.8	−191.5
FP (°C)	−92.0	−97.8	−141.5	−205.0
IT (°C)	300	470	235	700
H_{25} (–)	$8.1 \cdot 10^{-6}$	$1.86 \cdot 10^{-4}$	0.31	42.8
RD (–) (air = 1)	1.067	1.110	1.617	0.968
Toxicity	High	Low	Low	Very high

biocatalysts to convert pollutants into simple molecules, as carbon dioxide and water. In the next sections, the major air pollution control alternatives will briefly be described first, before presenting novel results employing bioreactors for the treatment of the waste gases. More detailed information on the fundamentals of the different technologies can be found elsewhere (Kennes & Veiga, 2001).

27.2 NON-BIOLOGICAL TECHNOLOGIES FOR THE ABATEMENT OF OFF-GASES FROM THE FORMALDEHYDE INDUSTRY

Traditional technologies, as incineration, absorption and adsorption, have been used extensively in the past, and are, sometimes, still preferred nowadays. Their advantages include the possibility of treating wide ranges of pollutant concentrations and their high efficiencies, unaffected by the biotoxicity of the compounds to be treated. Also, under certain conditions, the recovery of some valuable compounds from the off-gas is possible. However, the costs associated to these technologies are usually high and, in some cases, they merely transfer the pollutant to a different phase instead of eliminating it, generating a liquid (absorption) or solid (adsorption) waste that needs further treatment.

Incineration has proven to be a reliable alternative, although usually expensive both in terms of investments and operation costs. Nevertheless, incineration is a highly used technology. The most conventional system is based on the oxidation of the pollutants mixture in a catalytic oxidizer using a precious metal catalyst attached on a monolithic honeycomb structure, that ensures a minimal pressure drop. Although the oxidation reaction is exothermic, a heating device may be necessary mainly during periods of low pollutant loads, start-up or after a shut-down period. Removal efficiencies exceeding 95% are generally reached.

Regarding absorption, Lundquist (1996) obtained good results with a system which combined scrubbing and cryogenic condensation for the abatement of mixtures of formaldehyde and methanol from a resin manufacturing facility. In a subsequent work (Lundquist, 1998), the same author found that scrubbing with a solution containing sodium metabisulfite could efficiently remove more than 95% of the formaldehyde present in actual emissions.

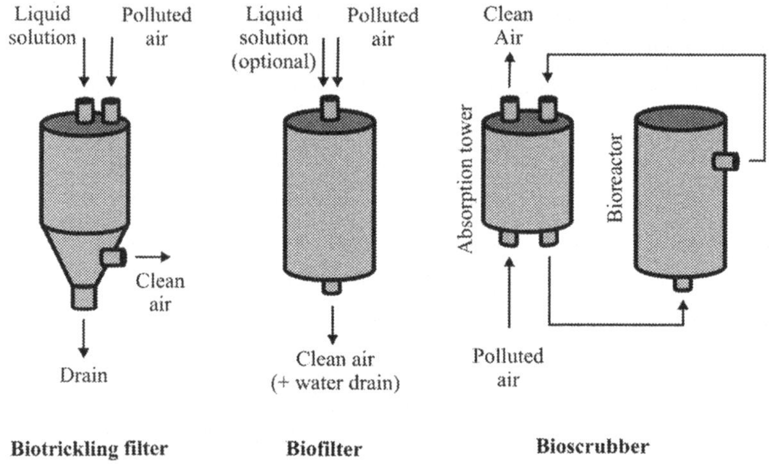

Figure 27.4 Schemes of the main types of bioreactors.

Also, some works have been published on the elimination of formaldehyde and methanol by adsorption, employing mainly activated carbon, ceramic material or chemicals as potassium permanganate or manganese oxides as adsorbents (Sekine & Nishimura, 2001). Another example is the use of ash obtained from burning wood residue to treat gaseous emissions containing mixtures of formaldehyde and methanol (Cowan et al., 2005). The authors showed that the formaldehyde concentration can be efficiently reduced in the waste gas from about 20 to less than 1 ppmv, while methanol is retired to a significantly lower extent. In that work, adsorption occurred at temperatures of up to 165°C, while oxidation predominated at higher values.

27.3 BIOLOGICAL TECHNOLOGIES FOR THE ABATEMENT OF OFF-GASES FROM THE FORMALDEHYDE INDUSTRY

27.3.1 Bioreactors for waste gas treatment

Biological technologies are based on the capacity of certain microorganisms to use volatile pollutants as carbon and/or energy source, converting them into simple inorganic products. The microbes can be used in bioreactors, in which the biocatalyst is kept under adequate conditions to carry out the elimination of the harmful compounds. Three main types of bioreactors are being used in practice, biotrickling filters, biofilters and bioscrubbers (Figure 27.4), although new reactor configurations as membrane bioreactors and rotating biological contactors are also being tested at laboratory scale (Kennes & Veiga, 2001). The three most common reactor types are characterized by the presence of a mobile liquid phase in biotrickling filters and bioscrubbers, and a solid filter bed in biofilters and biotrickling filters.

27.3.2 Previous studies

Formaldehyde, methanol, dimethylether and carbon monoxide have all proven to be, at least to some extent, biodegradable by different microbial strains. A high number of studies have been published in the literature on the treatment of formaldehyde and methanol as single pollutants in bioreactors, while studies with mixtures of them are much scarcer. Very few data have been published on the biofiltration of carbon monoxide (Munz et al., 2002; Robra et al., 1997). We have no knowledge of previous research on the treatment of dimethylether or dimethylether-containing mixtures in bioreactors. An in-depth review on formaldehyde treatment in bioreactors, alone and in mixtures with other organic compounds, can be found in Prado et al. (2003). Also, previous studies on biofiltration of methanol are numerous (Cornabé et al., 2002; Krailas et al., 2000; Mohseni & Allen, 2000; Prado et al., 2005a; Shareefdeen et al., 1993).

27.3.3 Optimization of the biological treatment

The biological treatment of gaseous mixtures of formaldehyde, methanol, dimethylether and carbon monoxide in biological reactors has been studied extensively over the past years. The general purpose is to optimize the operational parameters to achieve a high treatment performance, in order to eventually implement the biological technologies in formaldehyde and resin-producing industries.

27.3.3.1 Batch studies

Prado et al. (2005b) evaluated the capacity of a sludge obtained from the wastewater treatment plant of a formaldehyde-resin-producing industry to perform the treatment of the four compounds individually (Figure 27.5). The main characteristics of the sludge have been described elsewhere (Prado et al., 2002). A series of batch assays at 30°C and at pH 7.5 were performed in closed flasks, containing 3.5 g/L volatile suspended solids (VSS) in 50 mL of a nutrient solution described elsewhere (Prado et al., 2005a, b). All the vials were prepared in duplicate, including "blanks" and "controls". The "blanks" had been sterilized, and the "control" vials contained distilled water instead of sludge. The experiment was carried out using the four above-mentioned compounds as carbon source. In the case of carbon monoxide, a pure culture of a carboxidotrophic bacterium, *Oligotropha carboxidovorans* OM5, was also tested. In this latter study the culture concentration was of 0.35 g VSS/L. Reactions (27.3)–(27.6) show the stoichiometries of biodegradation for formaldehyde, methanol, dimethylether and carbon monoxide, respectively.

$$CHOH + O_2 \rightarrow CO_2 + H_2O \quad (27.3)$$

$$CH_3OH + \frac{3}{2}O_2 \rightarrow CO_2 + 2H_2O \quad (27.4)$$

$$(CH_3)_2O + 3O_2 \rightarrow 2CO_2 + 3H_2O \quad (27.5)$$

$$CO + \frac{1}{2}O_2 \rightarrow CO_2 \quad (27.6)$$

Figure 27.5a–d shows that the sludge employed as biocatalyst in the four first sets of assays can efficiently carry on the degradation of the four pollutants. Formaldehyde and methanol were readily eliminated, whereas the biodegradation of dimethylether

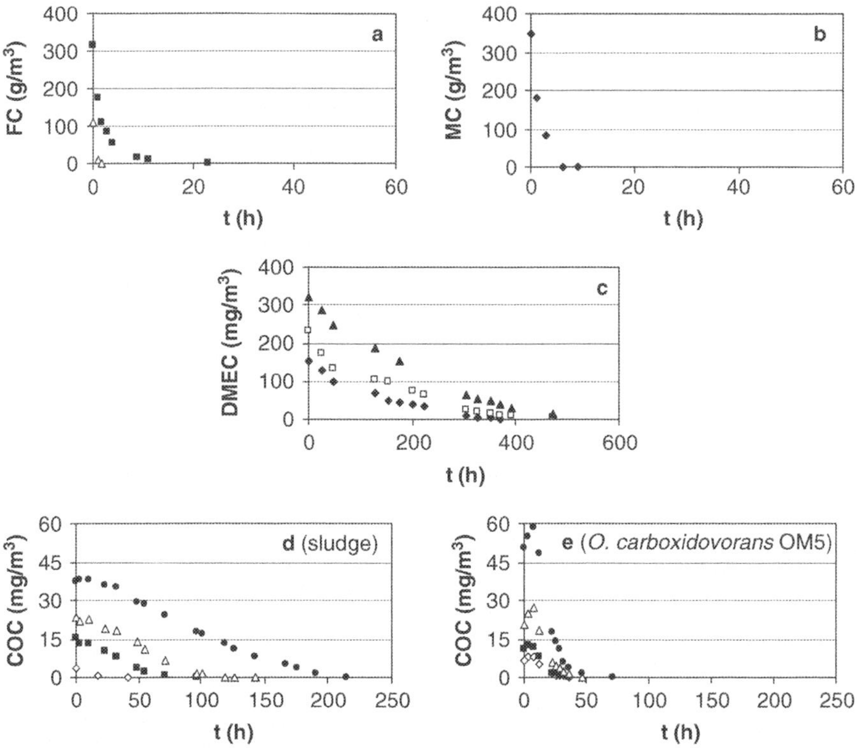

Figure 27.5 Elimination of (a) formaldehyde, (b) methanol, (c) dimethylether and (d, e) carbon monoxide in batch assays.

and carbon monoxide needed a significantly longer time. One reason for this is that formaldehyde and methanol have a lower air/water partition coefficient and are thus more soluble in the liquid phase than dimethylether and carbon monoxide. Their biodegradability by the biomass in the sludge was presumably also higher compared to dimethylether and carbon monoxide. Therefore, a more significant growth of formaldehyde and methanol degraders is expected than for the other two pollutants. Also, comparing Figure 27.5d, e shows that carbon monoxide removal was significantly faster when using the *O. carboxidovorans* OM5 culture as inoculum instead of the sludge, even though the VSS concentration of the pure culture was 10 times lower than for the sludge in these experiments. No elimination was observed in the "blank" and "control" vials (data not shown).

27.3.3.2 Treatment of formaldehyde and methanol in single-stage bioreactors

Effect of the carrier material Subsequently, a series of experiments was performed in bioreactors, with the purposes of evaluating their performance and of optimizing the biological treatment (Prado et al., 2004). Initially, mixtures of formaldehyde and methanol were selected as sole carbon sources, because of their somewhat higher concentration in the off-gases from formaldehyde-resin-producing industries. These compounds are also released from other sources, as the storage tanks. The first experiment undertaken in bioreactors was aimed at comparing the performance of three conventional biofilters packed with different inert filter bed materials, namely lava rock, perlite and activated carbon. The same sludge as used in the batch experiments (Figure 27.5a–d) was employed as inoculum for the reactors. An empty bed residence time (EBRT) of 80 s was set in all systems. No nutrient, water or sludge was added after inoculation. During the first week of operation, formaldehyde and methanol loads, respectively, of around 15 g/m^{3}h and from 0.5 g to 26.0 g/m^{3}h were fed to each reactor. Formaldehyde removal efficiencies of $63.2 \pm 10.9\%$, $55.8 \pm 15.3\%$ and $57.0 \pm 13.9\%$ were achieved for the lava rock, perlite and activated carbon reactors, respectively, while the methanol removal efficiency exceeded 80% in all cases. Global efficiencies were very similar in each system, showing that the three packing materials employed during the study are equally suitable. After the first week of operation the methanol loads were increased in all the biofilters up to values around 640, 370 and 580 g/m^{3}h for the lava rock, perlite and activated carbon biofilters, respectively. Formaldehyde loads were not varied. This increase did not affect the elimination efficiency of methanol to a great extent, which remained above 90% in all biofilters, but led to a dramatic decrease in the removal of formaldehyde. Such an effect was not observed in subsequent similar studies performed in mature bioreactors, which could handle important fluctuations of the toxic load. It can be concluded that during start-up bioreactors are more sensitive to shock loads than once steady state is reached.

Effect of moisture, pH and nutrient supply In a subsequent experiment, the performance of four different bioreactor configurations was compared (Prado et al., 2004). System (a) was an upflow conventional biofilter packed with 3.4 L lava rock which was not fed any liquid solution. Systems (b)–(d) were downflow biotrickling filters packed with, respectively, 3.4, 3.4 and 2.0 L lava rock. System (b) was continuously fed distilled water at a rate of 3 L/h, while Systems (c) and (d) were fed a nutrient solution described in Prado et al. (2005a) at rates of 9 and 3 L/h, respectively. This solution was only regularly renewed in System (d). In this case the renewal was done weekly, and the pH of the solution was adjusted to 7.5 each time. All bioreactors were inoculated and operated under similar conditions for more than 1 month.

Satisfactory results were obtained with System (a), reaching average removal efficiencies around 65% for formaldehyde and around 96% for methanol, at formaldehyde and methanol loads of, respectively, around 47 and 5 g/m^{3}h. Nevertheless, it was considered that its performance could be limited as a consequence of a deficient amount and distribution of moisture in the filter bed. System (b) showed a slightly better performance than System (a), reaching average elimination efficiencies for formaldehyde and methanol, respectively, of 79% and 96%, at average loads of each compound of 42 and 8 g/m^{3}h. System (c) was designed and operated, with the objective of checking if the lack of nutrients was affecting the performance of the former bioreactors. Contrary to what was expected, the removal efficiencies of both formaldehyde and methanol were lower than in the previous case, with average values of, respectively, 60% and 6% for similar pollutant loads. Analysis of the trickling liquid medium showed that its pH was acidic (generally around 4.2), as a consequence of the supply of

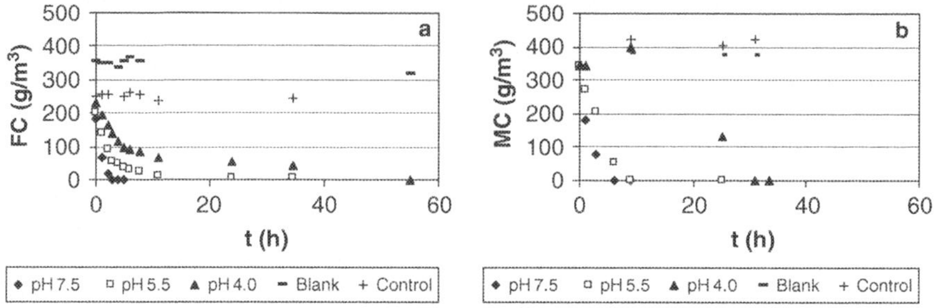

Figure 27.6 Effect of pH on the biodegradation of (a) formaldehyde and (b) methanol.

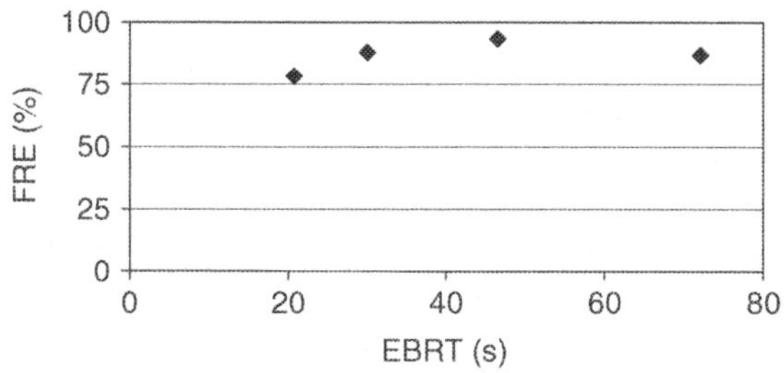

Figure 27.7 Effect of the EBRT on the formaldehyde removal efficiency in a biotrickling filter.

a non-buffered nutrient solution rich in ammonium. It was considered that this low pH could affect the treatment efficiency. Hence, a series of batch experiments was performed in order to confirm this hypothesis. Figure 27.6 confirms that formaldehyde and methanol removal was slower at such low pH values compared to a near neutral pH. Also, experiments performed in a formaldehyde-degrading biotrickling filter showed that the elimination efficiency was negatively affected at pH values below 4.2 (data not shown).

The operation of System (c) suggested that the high trickling rate of 9 L/h could induce mass transfer limitations of the pollutants from the gas phase to the biofilm as well as removal of some active biomass. In order to overcome these problems, System (d) was developed. In this biotrickling filter, the same liquid solution as in System (c) was used, but being recirculated at a lower trickling rate of 3 L/h and renewed weekly, with its pH set at 7.5. This bioreactor proved to be the most efficient of the four systems studied.

Effect of the EBRT A subsequent study was aimed at determining the effect of the EBRT on the treatment of formaldehyde in System (d). The operation of the biofilter was the same as described in the previous section. A constant formaldehyde concentration of $0.9 \pm 0.1\,g/m^3$ was fed to the reactor, and the EBRT varied between 21 and 72 s, corresponding to average formaldehyde loads between 45 and $145\,g/m^3h$. As Figure 27.7 shows, the biofilter efficiency was high in all cases, decreasing slightly only when the EBRT dropped to its smaller value of 21 s.

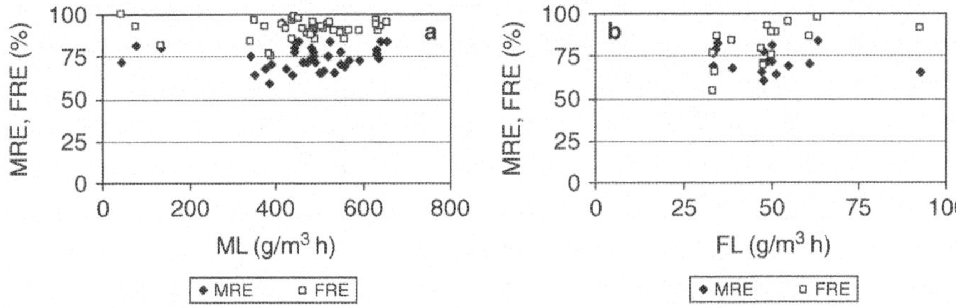

Figure 27.8 (a) Effect of the inlet methanol load on the removal of the mixture, when feeding a constant inlet formaldehyde load of 50 g/m^{3}h. (b) Effect of the inlet formaldehyde load on the removal of the mixture, when feeding a constant inlet methanol load of 700 g/m^{3}h.

Effect of the substrate composition It is well known that the global efficiency of the treatment of mixtures of gaseous compounds in biological reactors can sometimes dramatically be affected by the concentration of any of them. The aim of the next experiments was to study if, during the treatment of mixtures of formaldehyde and methanol, some interaction could be found between these pollutants. In a first study, a biotrickling filter similar to the one previously described as System (d) was operated for more than 4 months at a constant formaldehyde load of 50 g/m^{3}h and variable methanol loads, ranging from 40 to 600 g/m^{3}h. The aim was to check the effect of different methanol loads on the formaldehyde and methanol elimination. Figure 27.8a shows that no significant interaction occurred, even at methanol loads around 600 g/m^{3}h. A similar experiment, in which a constant methanol load of 700 g/m^{3}h and variable formaldehyde loads ranging from 30 to 90 g/m^{3}h were fed simultaneously, proved that the elimination of the mixture seemed not to be affected by the pollutant load under these conditions.

27.3.3.3 Treatment of formaldehyde and methanol in two-stage bioreactor configuration

A last experiment was aimed at optimizing the elimination of the mixture of both compounds. Previous results showed that biotrickling filters can achieve higher maximum elimination capacities than biofilters. However, in biotrickling filters, removal efficiencies rarely reach 100%, probably due to pollutant transfer limitation from the gas phase to the biofilm. On the other hand, very high removal efficiencies can be achieved in biofilters.

In order to achieve high elimination capacities associated to removal efficiencies close to 100%, a two-stage hybrid system was developed. It was composed of a biotrickling filter connected in series to a biofilter. According to previous studies, it was expected that most of the removal would take place in the first stage, the biotrickling filter, while the rest of the waste gas would be fully degraded in the second stage, a biofilter. The use of such multistage systems has proved to be useful for the treatment of mixtures of organic vapors and hydrogen sulfide (Chitwood et al., 2001), odorous sulfur emissions (Li et al., 2003) and styrene vapors (Choi et al., 2004), among others. Each bioreactor

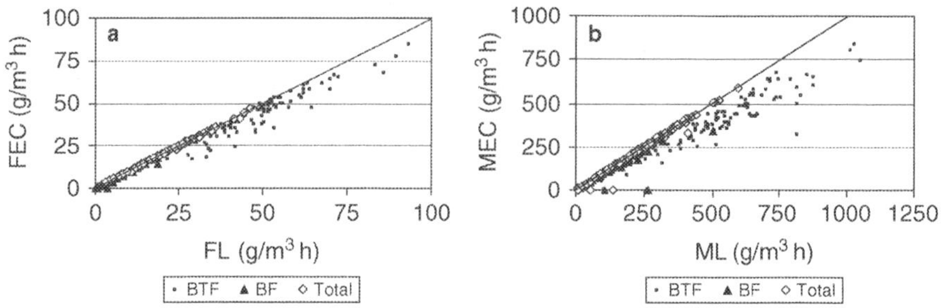

Figure 27.9 Relationship between the pollutant loads and their elimination capacities. (a) formaldehyde and (b) methanol. The lines indicate 100% removal.

was packed with 1 L lava rock, and operated at room temperature with a downward air flow during 11 months. The air flow rate supplied to the system was 100 L/h during the first 160 days of operation, and 67 L/h during the rest of the experiment. The total EBRT of the system was 72 s during the first 160 days, and 108 s between days 160 and 330. The biotrickling filter was continuously fed 2 L of a nutrient solution, recirculated at a rate of 3 L/h and renewed monthly, with its pH adjusted to 7.5. The biofilter was weekly fed 2 L of that same solution, again, with its pH adjusted to a value of 7.5. Both reactors had previously been employed for the biodegradation of mixtures of formaldehyde and methanol, showing high efficiencies. Hence, a short start-up phase was expected. Average pollutant loads fed to the global system were 22.5 (formaldehyde) and 268.9 (methanol) $g/m^3 h$.

For the combined system, removal efficiencies around 100% were commonly found both for formaldehyde and for methanol (Figure 27.9). As expected, most of the degradation occurred in the first stage, in which removal efficiencies of 87.7 ± 9.7% and 72.3 ± 15.0% were obtained for formaldehyde and methanol, respectively. In the second stage, formaldehyde and methanol removal efficiencies of 95.8 ± 14.0% and 97.6 ± 5.6%, respectively, were obtained.

27.3.3.4 Treatment of mixtures of formaldehyde, methanol, dimethylether and carbon monoxide

In a last study, a biofilter was fed a mixture of formaldehyde, methanol, dimethylether and carbon monoxide at concentrations similar to those commonly found in actual formaldehyde-resin-producing industries (50–100 mg/m^3), and its performance was studied. The reactor was packed with 1 L lava rock and operated with a downward flow, at a temperature around 25°C, during 1 month. It was weekly fed a nutrient solution (Prado et al., 2005a). Three different EBRT values were assayed: 58 s (between days 0 and 14), 41 s (days 14 and 21) and 30 s (days 21 and 28). Figure 27.10 shows the removal of all four compounds during the experiment.

Both formaldehyde and methanol were degraded with efficiencies close to 100% during most of the experiment (Figure 27.10). This is consistent with the results obtained in Section 27.3.3.3. Also, the elimination of low concentrations of carbon

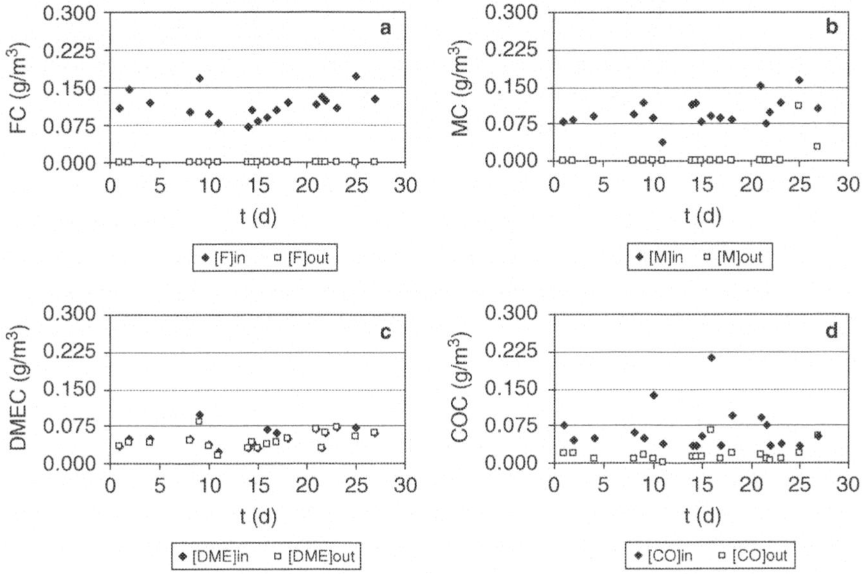

Figure 27.10 Concentrations of each pollutant at the inlet and at the outlet of the bioreactor. (a) formaldehyde, (b) methanol, (c) dimethylether and (d) carbon monoxide.

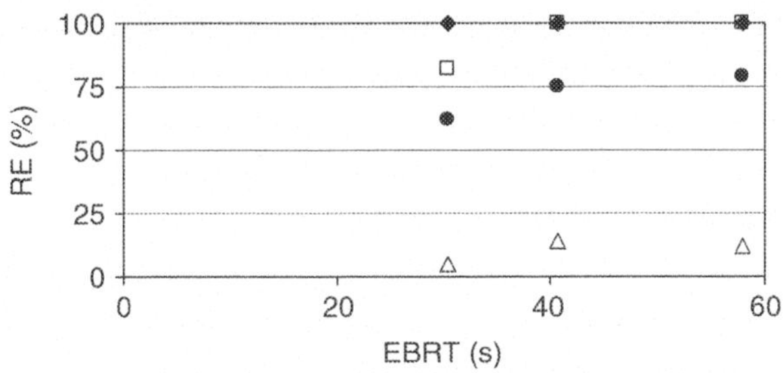

Figure 27.11 Effect of the EBRT on the removal efficiency of formaldehyde (diamonds), methanol (squares), dimethylether (triangles) and carbon monoxide (circles).

monoxide was possible with quite good efficiencies, proving that, at low inlet loads, its removal is feasible in biofilters, even mixed with organic compounds. On the other hand, the elimination of dimethylether was quite low during the study, due to its poor biodegradability, and probably also due to its relatively low solubility in water. However, as its environmental impact is much weaker than that of the other compounds of the mixture, these results were considered satisfactory. Figure 27.11 shows how the decrease in the EBRT, from its original value of 58–30 s, led to a slight decay in the removal efficiencies of methanol, dimethylether and carbon monoxide. The removal of formaldehyde, however, was not clearly affected, which is consistent with the results presented above.

27.4 CONCLUSIONS

Biological technologies are useful alternatives to physico-chemical technologies for the abatement of harmful compounds commonly present if off-gases from formaldehyde- and formaldehyde-resin-producing industries. Mixtures of formaldehyde, methanol, dimethylether and carbon monoxide, at concentrations similar to those commonly found in actual emissions from formaldehyde-resin-producing facilities, can be efficiently treated by biological reactors. Complete elimination can be achieved for formaldehyde and methanol, whereas dimethylether was only moderately eliminated. Relatively good values (around 75%) were also found for carbon monoxide, although its removal was only studied for a short period of time. An aerobic sludge obtained from the wastewater treatment plant of a formaldehyde-resin-producing industry proved to be highly adequate as inoculum for the system.

Commonly, higher formaldehyde and methanol maximum elimination capacities can be achieved when employing biotrickling filters than when using biofilters. However, when treating synthetic formaldehyde waste gas, removal efficiencies close to 100% were hardly obtained in the biotrickling filter configuration. On the other hand, biofilters usually reached 100% removal efficiencies, but only when being fed low or moderate pollutant loads (<100–200 g/m^{3}h). The application of a hybrid two-stage bioreactor system (BTF-BF) could overcome the limitation of each reactor configuration individually, combining high maximum elimination capacities and removal efficiencies around 100% in the two-stage configuration. The efficiency of these systems can be improved by feeding periodically a nutrient solution with a neutral or slightly alkaline pH. Other operation parameters which affected the performance of the bioreactors to a high extent were the EBRT and the moisture content.

ACKNOWLEDGMENTS

The studies presented herein were financed by the Spanish Ministry of Science and Technology (project PPQ 2001-0557). ÓJP would like to express his gratitude to the Xunta de Galicia for the financial support of his Ph.D. research. The authors are also indebted to Dr. Ortwin Meyer and his group for supplying the original *Oligotropha carboxidovorans* OM5 culture.

NOMENCLATURE

BF	Biofilter
BP	Boiling point [T]
BTF	Biotrickling filter
COC	Carbon monoxide concentration [M/L^3]
DMEC	Dimethylether concentration [M/L^3]
EBRT	Empty bed residence time [t]
FC	Formaldehyde concentration [M/L^3]
FEC	Formaldehyde elimination capacity [M/L^{3}t]
FL	Formaldehyde load [M/L^{3}t]
FP	Flaming point [T]
FRE	Formaldehyde removal efficiency [–]

H_{25}	Henry's constant at 25°C [–]	MW	Molecular weight [M]
IT	Ignition temperature [T]	MRE	Methanol removal efficiency [–]
MC	Methanol concentration [M/L^3]	RD	Relative density [–] (air RD = 1)
MEC	Methanol elimination capacity [M/L^3t]	RE	Removal efficiency [–]
ML	Methanol load [M/L^3t]	*t*	Time [t]

REFERENCES

Agency for Toxic Substances and Disease Registry (1999) Toxicological profile for formaldehyde. Available at www.atsdr.cdc.gov

Chitwood, D.E. and Devinny, J.S. (2001) Treatment of mixed hydrogen sulfide and organic vapors in a rock medium biofilter. *Water. Environ. Res*. **73**(10), 426–435.

Choi, D.S., Devinny, J.S. and Deshusses, M.A. (2004) Behavior of field-scale biotrickling filter under nonsteady state conditions. *J. Environ. Eng*. **130**(3), 322–328.

Cornabé, T., Auria, R., Christen, P., Ferrer, H., Revah, S. and Moukha, S. (2002) Biofiltration of methanol by a recombinant *Pichia pastoris* strain. *Proceedings of the 2002 USC-TRG Conference on Biofiltration*, Newport Beach, CA, USA, 31 October to 1 November.

Cowan, J., Abu-Daabes, M. and Banerjee, S. (2005) Controlling formaldehyde emissions with boiler ash. *Environ. Sci. Technol*. **39**, 5101–5104.

Kennes, C. and Veiga, M.C. (2001) *Bioreactors for Waste Gas Treatment*. Kluwer Academic Publishers, Dordrecht, The Netherlands.

Kirk-Othmer Encyclopedia of Chemical Technology, 4th edn. (1998) Eds. J.I. Kroschwitz and M. Howe-Grant. John Wiley & Sons Inc., New York, USA.

Krailas, S., Tuan Pham, Q., Amal, R., Jiang, J.K. and Heitz, M. (2000) Effect of inlet mass loading, water and total bacteria count on methanol elimination using upward flow and downward flow biofilters. *J. Chem. Technol. Biotechnol*. **75**(4), 299–305.

Li, H., Mihelcic, J.R., Crittenden, J.C. and Anderson, K.A. (2003) Field measurements and modeling of two-stage biofilter that treats odorous sulfur air emissions. *J. Environ. Eng*. **129**(8), 684–692.

Lundquist, P.R. (1996) Controlling VOC and HAP emissions with scrubbing and condensation recycles solvents directly into plant chemical feedstocks. *Proceedings of the 89th Annual Meeting of the Air and Waste Management Association*, Nashville, TN, USA, 23–28 June.

Lundquist, P.R. (1998) Controlling formaldehyde emissions with MBS scrubbing. *Proceedings of the 91st Annual Meeting of the Air and Waste Management Association*, San Diego, CA, USA, 14–18 June.

Mohseni, M. and Allen, D.G. (2000) Biofiltration of mixtures of hydrophilic and hydrophobic volatile organic compounds. *Chem. Eng. Sci*. **55**(9), 1545–1558.

Munz, G., Dixon, M.A. and Darlington, A.B. (2002) Indoor carbon monoxide removal using plant-based biofilters. *Proceedings of the 2002 USC-TRG Conference on Biofiltration*, Newport Beach, CA, USA, 31 October to 1 November.

Prado, Ó.J., Mendoza, J.A., Veiga, M.C. and Kennes, C. (2002) Optimization of nutrient supply in a downflow gas-phase biofilter packed with an inert carrier. *Appl. Microbiol. Biotechnol*. **59**(4–5), 567–573.

Prado, Ó.J., Eiroa, M., Veiga, M.C. and Kennes, C. (2003) Bioreactors for the treatment of industrial waste gases containing formaldehyde and other aliphatic compounds. In *Focus on Biotechnology, Volume 3C. Biotechnology for the Environment: Wastewater Treatment and Modeling, Waste Gas Handling* (eds. S.N. Agathos and W. Reineke), pp. 259–273, Kluwer Academic Publishers, Dordrecht, The Netherlands.

Prado, Ó.J., Veiga, M.C. and Kennes, C. (2004) Biofiltration of waste gases containing a mixture of formaldehyde and methanol. *Appl. Microbiol. Biotechnol*. **65**(2), 235–242.

Prado, Ó.J., Veiga, M.C. and Kennes, C. (2005a) Treatment of gas-phase methanol in conventional biofilters packed with lava rock. *Water Res*. **39**(11), 2385–2393.

Prado, Ó.J., Veiga, M.C. and Kennes, C. (2005b) Biological treatment of mixtures of toxic compounds emitted from formaldehyde resin-producing industries. In *Biotechniques for Air Pollution Control* (eds. C. Kennes and M.C. Veiga), pp. 391–396, University of La Coruña, La Coruña, Spain.

Robra, K.H., Wellacher, M., Kirchmeir, F., Leistentritt, R. and Pucher, K. (1997) Abluftreiningung für Straßentunnel und Tiefgaragen durch carboxidotrophe Mischpopulationen. *Proceedings of the International Symposium on Biological Waste Gas Cleaning*, Maastricht, The Netherlands.

Sekine, Y.S. and Nishimura, A. (2001) Removal of formaldehyde from indoor air by passive type air-cleaning materials. *Atmos. Environ.* **35**, 2001–2007.

Shareefdeen, Z., Baltzis, B.C., Oh, Y.S. and Bartha, R. (1993) Biofiltration of methanol vapor. *Biotechnol. Bioeng.* **41**, 512–524.

28

Upgrading of biogas for its usage in high-temperature fuel cells and gas grid injection

W. Ahrer, F. Accettola, S. Trogisch and R. Schöftner

28.1 INTRODUCTION

The upgrading of biogas from anaerobic digestion enables its usage in highly innovative applications such as fuel cells and natural gas grid injection. Both applications require similar treatment of the biogas especially concerning the removal of hydrogen sulfide and silicon-containing compounds (siloxanes).

Carbon dioxide as the main component in biogas does not need to be removed as long as the biogas is fed into high-temperature fuel cells. However, when injecting biogas into the natural gas grid, CO_2 removal is required. Therefore, the carbon dioxide content of biogas is of particular interest in terms of technological and economic figures, since the removal of CO_2 is in general an expensive technology.

While there are various technologies available on the market for purification of biogas from CO_2 (pressure/temperature swing adsorption, scrubbing, selective absorption, ...)

technologies are still under development for the removal of H_2S and siloxanes. Microbiological means for the oxidation of H_2S to produce elemental sulfur or sulfate are applied in combination with air injection into the digester and also into separated biogas cleaning units, but they cannot be applied in combination with gas grid injection. Because of diluting the biogas with nitrogen from the air, the heating value of natural gas according to national regulations (DVGW, ÖVGW 31) cannot be guaranteed anymore. Therefore research efforts have been dedicated toward the development of advanced biogas upgrading units allowing the removal of sulfide from biogas without introduction of nitrogen from air.

28.2 BIOGAS UPGRADING TECHNOLOGIES

28.2.1 Advanced biotrickling filter for sulfide removal

Hydrogen sulfide, which is normally present in biogas at concentrations between 200 and 4000 ppm_v, is a toxic, corrosive and smelly compound, poisonous for fuel cells, damaging for gas motors and a detrimental compound in the natural gas grid. That means that it has to be removed from the biogas so that it can be used as fuel in engines or fuel cells or introduced into the natural gas grid. Due to its unpleasant characteristics a large variety of methods have been developed to remove H_2S from different gas streams. These methods can be classified according to their principle of action in two big groups: physical–chemical and biotechnological methods. Among the physical–chemical methods, which are the traditional and nowadays still the most commonly applied, the adsorption with activated carbon and iron oxide, and the iron salt and alkaline absorption are the most common methods used to remove H_2S from biogas (Langnickel, 2000; Prabucki et al., 2001; Stoddart et al., 2001).

Recently, biotechnological methods have experienced a great development because of their high efficiency (similar or higher than the physical–chemical), their lower investment cost and because they lead to savings on energy, avoid catalysts and the formation of secondary contaminants. The most important of these biotechnologies are the biofilters, bioscrubbers and biotrickling filters. The latter are the best in terms of efficiency, costs and sustainability.

Profactor's system (Figure 28.1) has been designed to remove H_2S from biogas which requires aerobic bacteria and hence oxygen has to be provided. This is achieved by forcing the recirculated liquid through a bubble column where it is saturated with oxygen instead of injecting air directly in the stream gas, which is the normal method. This procedure has several advantages: Firstly, if the end application is a fuel cell, it avoids the risk of explosions due to a dangerous biogas and O_2 mixture. Secondly, damage to the anode of the fuel cells, due to the presence of oxygen, is also avoided. Finally, the quality of the biogas is better because it is not diluted with the N_2 from air. In order to achieve the aim of reducing the concentration of H_2S from about 2000 ppm to below 3 ppm_v, the biotrickling filter is divided into several packed beds or sections. In this way, a higher elimination capacity is achieved and oxygen transfer problems are avoided. Profactor has worked during the past 4 years on the biotrickling filter and has set up several laboratory units and one laboratory scaled field trial system with a capacity of approximately 200 L/h biogas flow. This has been done in the EU RTD

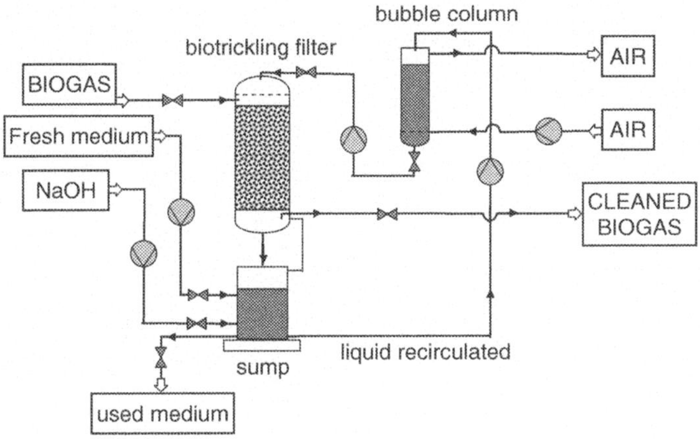

Figure 28.1 Schematic drawing of the biotrickling filter for removal of hydrogen sulfide.

project EFFECTIVE (ERK5-CT-1999-0007). The system for the field trials has run continuously for over 2 years.

The main innovations are: reduction of the H_2S content down to under 5 mg H_2S/m^3 of biogas, a new design in the column for improved microbial efficiency, no wastes are produced with the operating system, no air input into the gas stream (dilution of biogas, explosions danger) and a simple construction. This enables biotrickling filters to compete with other products on the market and to use the prototype as technical basis for market penetration of the system (Trogisch & Baaske, 2004a).

28.2.2 Microbial removal of siloxanes from biogas

28.2.2.1 Measurement of siloxanes

From the analytical point of view, biogas is a rather complex matrix because of the large number of components with a broad range of volatility, concentration and differing polarity. In addition, moisture and reactivity of some components may interfere with the analytical process. The most widely used sample preparation method in the analysis of landfill gas is the enrichment of analytes on solid sorbent materials.

To extract siloxanes from gas samples like biogas, materials like activated carbon, Tenax, Chromosorb, Carbosieve, silica gel, XAD resins, polyurethane foam and tetradecane have been tested by Huppmann and Lohoff (1996).

Schweigkofler and Niessner (1999) presented another technique for sampling, identification and quantification of siloxanes and volatile organic carbon (VOC) in landfill gas and sewage gas. After sample collection using evacuated stainless steel canisters, biogas was analyzed by gas chromatography-mass spectrometry/atomic emission spectroscopy (GC-MS/AES). Using gas canisters, the sampling process was simplified (no vacuum pump needed) and multiple analysis was possible. The simultaneous application of a mass selective detector (MSD) and an atomic emission detector (AED) allowed a rapid screening of silicon compounds in complex biogas

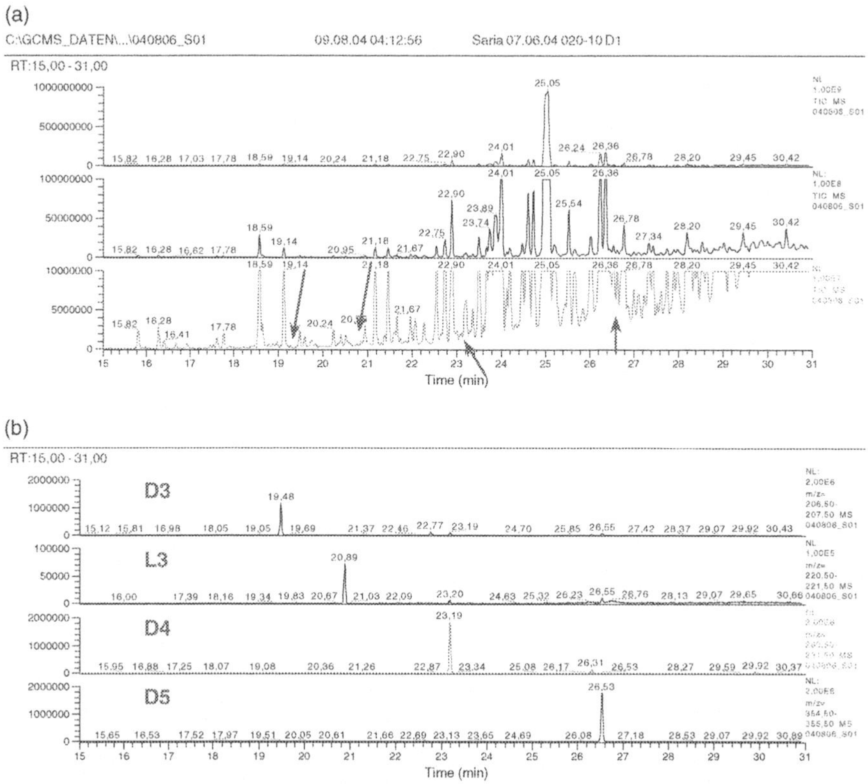

Figure 28.2 Typical chromatogram (TD-GC-MS) TIC (a) and extracted ion (b) of biogas and siloxanes.

samples. Precision, linearity and detection limits have been studied. In real samples, concentrations of silicon-containing compounds in the mg/m^3 range have been observed (Schweigkofler & Niessner, 1999).

Analysis of linear and cyclic siloxanes in plasma and blood employing GC-MS has been reported by Flassbeck et al. (2001) in urine by Varaprath et al. (1999) and tissue by Lykissa et al. (1997). Dagger and Semlyen (1998) investigated small cyclic and linear dimethylsiloxanes using analytical techniques like GC, Fourier transform infrared (FT-IR) spectroscopy and nuclear magnetic resonance spectroscopy (^{1}H, ^{13}C and ^{29}Si NMR).

Linear and cyclic siloxanes have been determined in the biogas of plants in Austria, Denmark, Germany and Slovakia using thermal desorption (TD) coupled to GC-MS (Hoffmann, 2005; Trogisch & Baaske, 2004b). The sampling was carried out on site by the plant personnel using thermal desorption tubes filled with Tenax® and the determination was performed by TD-GC-MS, using external calibration. Depending on the type of substrate, different amounts of each siloxane were observed. A typical chromatogram of a biogas sample is shown in Figure 28.2. The arrows in the total ion current (TIC) mode indicate the position of the siloxanes D3, L3, D4 and D5. Due to the presence of numerous compounds in biogas the quantification is facilitated after extracting the respective ion-tracks of the analytes (represented in Figure 28.2 at the bottom).

28.2.2.2 Microbial degradation of siloxanes

So far, there has been little research concerning biodegradation of organosiloxanes. Moreover, the difficulty in measuring them due to their low solubility and high vaporization rate makes it hard to perform degradation tests. Nevertheless, degradation of polydimethylsiloxane (PDMS) in soil has been proved in the past few years (Lehmann and Griessbach, 1999; Lehmann et al., 1995). It seems that degradation takes place into two steps. In the first step, rearrangement and hydrolysis of PDMS occurs with the production of lower molecular weight compounds: primarily octamethylcyclotetrasiloxane (D4), decamethyl-cyclopentasiloxane (D5), linear volatile methylsiloxanes (VMSs) and silanols (mostly dimethylsilanediol, DMSD). Upon this partial hydrolysis of PDMS to DMSD the second step of degradation may proceed, giving inorganic water-soluble compounds as the final product. Higher rates have been detected for the first step when decreasing the water content, leading to the general idea that this step is purely chemical and catalyzed by clays minerals. In contrast, the second step, degradation of DMSD should be biological (Lehmann et al., 1998). Sabourin et al. (1996) identified a bacterium, *Arthrobacter* sp., and a fungus, *Fusarium oxysporum* using DMSD as co-substrate (degradation rates: 2.4% per month for *F. oxysporum* when adding 2-propanol; 4% per month for *Arthrobacter* with dimethylsulfone). Wasserbauer and Zadák (1990) demonstrated that bacteria from the genus *Pseudomonas* (*P. putida* and *P. fluorescens*) are capable of growing on oligomeric and polymeric dimethylsiloxanes, the growth being a function of the increasing viscosity (hence of the chain length) of the siloxanes. Biodegradation occurs in aerobic conditions. Nevertheless, degradation products in this study were not identified. Grümping et al. (1999) found DMSD in sewage sludge spiked with D4 and incubated in anaerobic conditions. D4 degradation after 100 days was found to be 3% of the initial quantity added. PDMS degradation was not detected.

Biodegradation studies have been carried out in order to isolate bacteria able to grow on siloxane as the only carbon source from the sludge. Bacteria growth was evaluated by measuring the optical density at 270 and 600 nm, respectively (Figure 28.3), and the protein concentration. A community of bacteria isolated from activated sludge can grow on D4 which can be seen from the figure as a steady increase of the optical density for both wavelengths. An accurate identification, performed by 16S rRNA- sequencing, revealed that *Pseudomonas* is the dominating species in the cultures, while other bacteria are: *Rhodanobacter*, *Zooglea*, *Mesorhizobium* and *Xanthomonadacea*.

28.3 HIGH-TECH APPLICATIONS OF BIOGAS

28.3.1 Fuel cells

Fuel cells are efficient energy convertors, which all work on the hydrogen principle. They are currently seen as a key element within a sustainable energy scenario (Wallmark & Alvfors, 2002). However, there is a large difference in their requirements when using biogas.

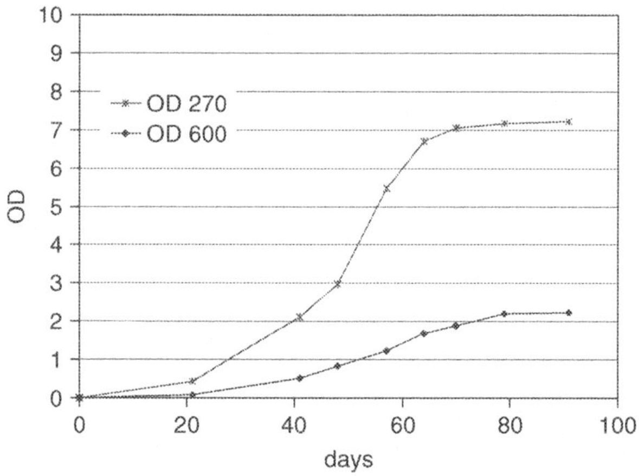

Figure 28.3 Optical density (OD) at 270 and 600 nm as a function of time of bacteria growing on siloxanes isolated from a wastewater treatment plant.

Low-temperature fuel cells, mainly polymer electrolyte membrane (PEM) fuel cells, are most suitable for mobile and portable applications, and are mainly powered with pure hydrogen. Low-temperature fuel cells cannot stand CO_2 and/or CO and will not be able to use CH_4 directly as a fuel. Phosphorous acid fuel cells (PAFCs) are already on the market with relative low efficiencies of around 40% due to their high (external) fuel processing requirements.

High-temperature fuel cells have been developed using natural gas as a fuel and are mainly used for stationary energy production. Molten carbonate fuel cells (MCFCs) are very well suited for biogas because they can use CH_4 as fuel and stand the approximately 35% CO_2 and CO that accompanies CH_4 in biogas (Bischoff & Huppmann, 2002). Further, several tests have been made where it has been demonstrated, that CO_2 acts as a reactant in the process, increasing the electrical efficiency of the system by approximately 2% (Trogisch & Baaske, 2004c). MCFC are currently well developed and in a phase of market penetration (Figure 28.4). The MCFC is seen by many experts as an ideal source for power generation scaled between 250 kW and 3 MW.

The alternative to MCFCs concerning the direct utilisation of methane are solid oxide fuel cells (SOFCs) which operate on a very high temperature level (≈1000°C) and, therefore, face material- and sealing challenges. MCFCs operate at a temperature of about 650°C, producing heat at a useful level.

The fuel of an MCFC or SOFC must be reformed to a hydrogen-rich gas to feed to the stack (*Fuel Cell Handbook*, 2000). The high-temperature fuel cells (MCFC and SOFC) produce heat and water vapor at the anode, which can be used for the steam reformation of CH_4. This means that it is inherently more efficient than fuel cells requiring external fuel processing (PEMFC and PAFC).

In summary, all fuel cells can use biogas as fuel. The difference between low- and high-temperature fuel cells lies in the requirements of the gas quality. This has a strong effect on the gas upgrading technology and costs and the possibility for high-temperature fuel cells to efficiently internally reform biogas with the waste heat

Figure 28.4 The HotModule of MTU CFC Solutions, operated with natural gas, at the Rhön-Klinikum Hospital in Bad Neustadt (Germany).

produced during the process. Thus, these high-temperature fuel cells reach electrical efficiencies between 50% and 60%.

28.3.2 Natural gas grid injection

The cleaning and injection of biogas into the existing natural gas grid is an interesting alternative for biogas utilization. This application leads to multiple advantages, like flexibility of use (electric power, generation of heat, process energy, fuel for transportation). An improved overall efficiency can be expected by producing energy from biogas in locations where exhaust heat can also be used. The possibility of employing electrical conversion technologies with higher efficiencies than gas engines (e.g. gas turbines) enables a further increase in the efficiency of the biogas usage.

In the first realized pilot plant in Austria for the injection of cleaned biogas into the natural gas grid, real data of biogas cleaning and injection into the grid are acquired. The plant consists of a biotrickling filter to remove the hydrogen sulfide and a pressure swing adsorption for removal of carbon dioxide and for drying. Finally the upgraded biogas is odorized and injected into the gas grid (Figure 28.5).

The plant is designed to deliver 6 m^3/h upgraded biogas, which is sufficient to supply 40 households in a selected area with energy during the summer. In winter, the higher demand is covered by natural gas. In Figure 28.6 a photo of the hydrogen sulfide removal system (left) and the CO_2 removal system (right) is shown.

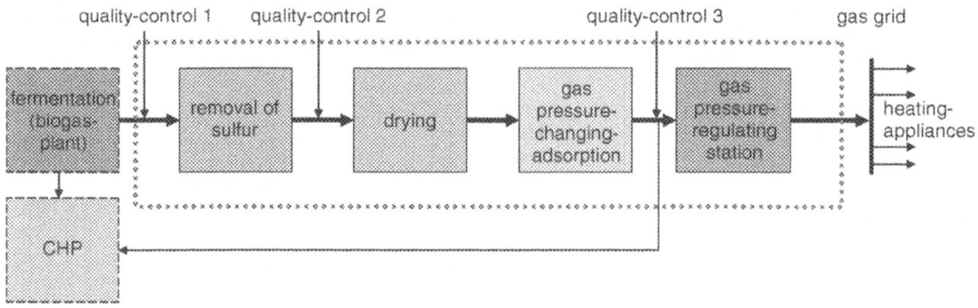

Figure 28.5 Schematic drawing of a gas grid injection system for upgraded biogas.

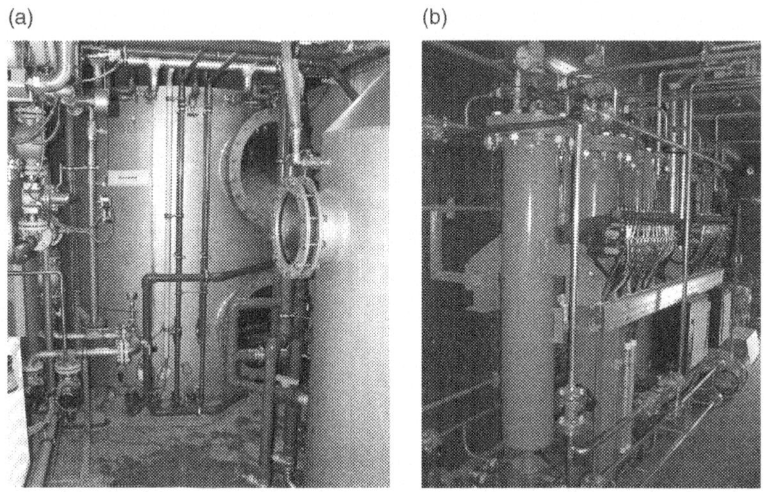

Figure 28.6 H_2S-removal system using a biotrickling filter (left) and CO_2-removal system using pressure swing adsorption (right).

28.4 CONCLUSIONS

Biogas can be applied to high-tech applications similar to natural gas. A pre-requirement is an extensive gas cleaning, which has to be performed in the case of fuel cells and natural gas grid injection. The development of inexpensive and sustainable technologies for biogas upgrading is an important task where biological methods have been successfully applied for the removal of hydrogen sulfide. The removal of siloxanes, which occur mainly in sewage and landfill gas, is more complicated due to the low biodegradation rates. However, proof of the principle has been shown, on which a viable plant concept can be developed.

REFERENCES

Bischoff, M. and Huppmann, G. (2002) Operating experience with a 250 kWel molten carbonate fuel cell (MCFC) power plant. *J. Power Sources* **105**, 216–221.

Dagger, A.C. and Semlyen, J.A. (1998) Studies of cyclic and linear poly(dimethylsiloxanes): 33. Preparation and characterization of per-deuterated linears and cyclics. *Polymer* **39**(12), 2621–2627.

Flassbeck, D., Pfeiderer, B., Grümping, R. and Hirner, A.V. (2001) Determination of low molecular weight silicones in plasma and blood of women after exposure to silicone breast implants by GC/MS. *Anal. Chem.* **73**, 606–611.

Fuel Cell Handbook (2000) Fifth Edition by EG&G Services, National Technical Information Service, US. Department of Commerce, 6–29, 6–32.

Grümping, R., Michalke, K., Hirner, A.V. and Hensel, R. (1999) Microbial degradation of octamethylcyclotetrasiloxane. *Appl. Environ. Microb.* **65**, 2276–2278.

Hoffmann, J. (2005) *Biofuels for Fuel Cells* (eds. P. Lens, P. Westermann, M. Haberbauer and A. Moreno), pp. 478–494. ISBN: 1843390922.

Huppmann, R. and Lohoff, H.W. (1996) Cyclic siloxanes in the biological waste water treatment process determination, quantification and possibilities of elimination. *Fresen. J. Anal. Chem.* **354**, 66–71.

Langnickel, U. (2000) First European fuel cell application using digester gas. *Fuel Cell Bull.* **3**(22), 10–12.

Lehmann, R.G., Varaprath, S., Annelin, R.B. and Arndt, J.L. (1995) Degradation of silicone polymer in a variety of soils. *Environ. Toxicol. Chem.* **14**, 1299–1305.

Lehmann, R.G., Miller, J.R. and Collins, H.P. (1998) Microbial degradation of dymethylsilanediol in soil. *Water Air Soil Pollut.* **106**(1–2), 111–122.

Lehmann, R.G. and Griessbach, F.C. (1999) Degradation of polydimethylsiloxane fluids in the environment – a review. *Chemosphere* **38**(6), 1461–1468.

Lykissa, E.D., Kala, S.V, Hurley, J.B. and Lebovitz, R.M. (1997) Release of low molecular weight silicones and platinum from silicone breast implants. *Anal. Chem.* **69**, 4912–4916.

Prabucki, M.J., Doczyck, W. and Asmus, D. (2001) Removal of organic silicon compounds from landfill and sewer gas. *Proceedings Sardinia 2001, Eighth International Waste Management and Landfill Symposium*, Vol. 2, pp. 631–639.

Sabourin, C.L., Carpenter, J.C., Leib, T.K. and Spivack, J.L. (1996) Biodegradation of dymethylsilanediol in soils. *Appl. Environ. Microb.* **62**(12), 4352–4360.

Schweigkofler, M. and Niessner, R. (1999) Determination of siloxanes and VOC in landfill and sewage gas by canister sampling and GC-MS/AES analysis. *Environ. Sci. Technol.* **33**, 3680–3685.

Stoddart, J., Zhu, M., Staines, J., Rothery, E. and Lewicki, R. (1999) Experience with halogenated hydrocarbons removal from landfill gas. *Proceedings Sardinia 99, Seventh International Waste Management and Landfill Symposium*, Vol. 2, pp. 489–498.

Trogisch, S. and Baaske, W.E. (2004a) *Biogas Powered Fuel Cells, Case Studies*, pp. 36–41, ISBN: 3854876262.

Trogisch, S. and Baaske, W.E. (2004b) *Biogas Powered Fuel Cells, Case Studies*, pp. 43–46, ISBN: 3854876262.

Trogisch, S. and Baaske, W.E. (2004c) *Biogas Powered Fuel Cells, Case Studies*, pp. 71–74, ISBN: 3854876262.

Varaprath, S., Salyers, K.L., Plotzke, K.P. and Nanavati, S. (1999) Identification of metabolites of octamethylcyclotetrasiloxane (D4) in rat urine. *Drug Metab. Dispos.* **27**(11), 1267–1273.

Wallmark, C. and Alvfors, P. (2002) Design of stationary PEFC system configurations to meet heat and power demands. *J. Power Sources* **106**, 83–92.

Wasserbauer, R. and Zadák, Z. (1990) Growth of *Pseudomonas putida* and *P. fluorescens* on silicone oils. *Folia Microbiol.* **35**, 384–393.

Index

IWA Publishing's authorised EU representative for General Product Safety Regulations is Diane D'Arras, 15 rue Duret, 75116 Paris, France, e-mail: safety@iwap.co.uk.

Printed and bound by CPI Group (UK) Ltd, Croydon, CR0 4YY
16/06/2026
02140401-0002